TIMBER DESIGNERS' MANUAL

Third Edition

Also of Interest

Structural Timber Design to Eurocode 5
Jack Porteous & Adby Kermani
978 14051 4638 8

Structural Masonry Designers' Manual
Third Edition
W.G. Curtin, G. Shaw, J.K. Beck & W.A. Bray
Revised by David Easterbrook
978 06320 5612 5

Structural Foundation Designers' Manual
Second Edition
W.G. Curtin, G. Shaw, G.I. Parkinson & J.M. Golding
Revised by N.J. Seward
978 14051 3044 8

Steel Designers' Manual
Sixth Edition
The Steel Construction Institute
978 14051 3412 5

TIMBER DESIGNERS' MANUAL

E. C. Ozelton & J. A. Baird

Third Edition
revised by

E. C. Ozelton

Blackwell
Science

Blackwell Science Ltd, a Blackwell Publishing company
Editorial offices:
Blackwell Science Ltd, 9600 Garsington Road, Oxford OX4 2DQ, UK
Tel: +44 (0) 1865 776868
Blackwell Publishing Inc., 350 Main Street, Malden, MA 02148-5020, USA
Tel: +1 781 388 8250
Blackwell Science Asia Pty, 550 Swanston Street, Carlton, Victoria 3053, Australia
Tel: +61 (0)3 8359 1011

First edition published in Great Britain by Crosby Lockwood Staples 1976
Reprinted by Granada Publishing Ltd 1981, 1982
Second edition published 1984
Reprinted by Collins Professional and Technical Books 1987
Reprinted with updates by BSP Professional Books 1989
Reprinted 1990
Reprinted by Blackwell Science 1995
Third edition published 2002
Reprinted 2004
Reissued in paperback 2006

3 2008

ISBN: 978-1-405-14671-5

Library of Congress Cataloging-in-Publication Data is available

A catalogue record for this title is available from the British Library

Set in 10 on 12 pt Times
by SNP Best-set Typesetter Ltd., Hong Kong

The publisher's policy is to use permanent paper from mills that operate a sustainable forestry policy, and which has been manufactured from pulp processed using acid-free and elementary chlorine-free practices. Furthermore, the publisher ensures that the text paper and cover board used have met acceptable environmental accreditation standards.

For further information on Blackwell Publishing, visit our website:
www.blackwellpublishing.com

Contents

Preface

It is 18 years since the publication of the previous edition of this manual, during which time the timber engineering industry has undergone many changes. Most notable is the harmonizing of British Standards with European practice and the recent release of BS 5268-2:2002 'Code of Practice for permissible stress design, materials and workmanship' which moves design concepts closer to those given in Eurocode 5 'Design of Timber Structures'. This third edition is updated to reflect the changes that are introduced in BS 5268-2:2002.

Recent years have seen the introduction of a range of composite solid timber sections and I-Beams from the USA and Europe. There are now a number of suppliers offering a range of products supported by technical literature in the form of safe load tables and section performance properties with in-house staff providing a comprehensive design service to the construction industry. This greatly reduces the input required from the timber designer and is reflected in Chapter 8, Thin Web Beams, and Chapter 10, Structural Composite Lumber.

The load capacities for nails, screws, bolt and dowel joints are now taken from Eurocode 5 and are discussed in detail in Chapter 18, Mechanical Joints, and Chapter 27, Design to Eurocode 5. Code bolt capacities are based on grade 4.6 bolts and Chapter 18 reviews and simplifies the formulae given in Annex G for the derivation of bolt capacities and considers the improved performance that may be achieved using grade 8.8 bolts.

This manual should be read in conjunction with BS 5268-2:2002. Reproduction of Code text and tables is kept to a minimum. As with earlier editions, tables and coefficients are provided to save the practising engineer many design hours and should prove indispensable time-savers.

The timber engineering industry is constantly changing and it is hoped that this latest edition will give the reader an overview of current practice.

E. C. Ozelton

Acknowledgements

This manual could not have been published without considerable help from companies and individuals active within the timber engineering industry.

I am particularly grateful to a number of organizations and people for their assistance in editing this third edition. Taking the chapters progressively I offer my thanks to:

Peter Steer BSc, CEng, MIStructE, MIMgt, Consulting Engineer and Chairman of the code drafting committee to BS 5258-2 for his invaluable contribution to Chapters 1–5.

Truss-Joist MacMillan, Boisse Cascade, James Jones & Sons Ltd, Fillcrete Masonite and Finnforest Corporation for their contributions to Chapters 8 and 10.

Janet Brown, Andrew Hughes and Richard Adams at Arch Timber Protection, Knottingley, for their contribution on Preservation in Chapter 24.

Abdy Kermani BSc, MSc, PhD, FIWSc, Napier University, Edinburgh, and author of *Structural Timber Design* (Blackwell Science) for his comprehensive review of Eurocode 5 in Chapter 27.

The cover photographs were kindly provided by Constructional Timber (Manufacturers) Ltd, Barnsley.

Finally, special thanks to my wife Joan for her support during the period of revising the manual.

About the Authors

E. Carl Ozelton is a consulting engineer specialising in the design and detailing of all forms of timber engineering and timber frame construction. Prior to setting up his own practice in 1977 he was Technical Director of Walter Holme & Sons Ltd, Timber Engineers, Liverpool and Technical Director of Prestoplan Homes Ltd, Timber Frame Manufacturers, Preston.

He is a Chartered Structural Engineer, a Fellow of the Institution of Structural Engineers and an Associate of the Institute of Wood Science. He was awarded first prize in the Plywood Design Award 1966/7 sponsored by the Timber Trade Federation.

Jack A. Baird, a Chartered Structural Engineer, specialised initially in structural steel work before becoming Technical Manager of Newsum Timber Engineers, following which he worked on BSI documents such as design code BS 5268. In 1970 he started the Swedish Timber Council, subsequently to become the Swedish Finnish Timber Council, in which role he produced factual information on many aspects of timber such as structural timber, and he helped to persuade Nordic sawmillers to machine stress grade at source to BS 4978 under the Kitemark scheme. He co-authored the first edition of *Timber Designers' Manual* with Carl Ozelton and was responsible seeing the second edition through the press.

Chapter 1
The Materials Used in Timber Engineering

1.1 INTRODUCTION

The decision by the European Commission in the early 1980s to have common material and design standards for the various construction materials has led to the withdrawal of many long-established British Standard Specifications and their replacement with standards produced under the auspices of the Comité Européen de Normalisation (CEN).

Whereas it was common practice for a British Standard Specification to describe a product or group of products, and indicate how to control manufacture, test the output, mark the product or products and even how to use it, the CEN rules for drafting standards mean that each one of these procedures becomes a 'stand alone' document. As a consequence there has been a proliferation of European standards relating to timber and its associated products. A European standard is identified by a number preceded by the letters 'EN' (Europäische Norm) in a similar way to the prefix 'BS' to British standards.

Under European legislation any conflicting part of a national standard has to be withdrawn within a specified time period from the publication date of the European standard. Because the use of a series of interrelated standards may well rely upon the completion of one particular document, the standards for timber are being collated and released in batches. The first of these batches relating to solid timber brought about the major revision of BS 5268-2 in 1996. The release of the second batch relating to panel products (plywood, particleboard, etc.) will give rise to a further revision.

European standards are published in the UK by the British Standards Institution with the prefix 'BS EN'. To facilitate use in the UK these BS EN documents can have a UK National Foreword explaining how the standard fits into the existing UK legislation and methods of working. There can also be UK National Annexes giving 'custom and practice' applications of the standard (e.g. BS EN 336: 1995 'Structural timber – coniferous and polar – Sizes – Permissible deviations' gives the cross-sectional dimensions of timber usually held by UK merchants). Neither the National Foreword nor a National Annex can alter the content or intent of the original European standard.

The European standards, particularly for the panel products, have broadened the range of materials available to the designer. Unfortunately reliable strength properties are not presently available for many of the newer panel products. Nevertheless these materials have been described in this chapter and it is left to the reader to assimilate the appropriate design values when they become available.

In addition to the extension of the ranges of the existing materials, new materials that may be generally described as 'engineered wood products' have become available, viz. Laminated Veneered Lumber (LVL, see section 1.5.2), Parallel Strand Lumber (PSL, see section 1.5.3) and Laminated Strand Lumber (LSL, see section 1.5.4). These materials are not currently covered by either British or European standards but most products have an Agrement certificate that allows their use in the UK.

The European standards are intended to support the limit state timber design code Eurocode 5 (EC5), the initial draft of which is available in the UK as DD ENV 1995-1 (published by the British Standards Institution). The values expressed in these supporting standards are 'characteristic values' set at the fifth percentile level, i.e. in statistical terms, 1 in 20 of the test values could fall below the characteristic value. For the purposes of the permissible stress design code (BS 5268), these characteristic values are further reduced by including safety factors to arrive at grade stress values or strength values for use in design.

The materials covered by the European standards have new European designations, e.g. softwood timber strength classes are described as C16, C24, etc., and these are the only ones now available in the market place. BS 5268 therefore uses the new European designations and descriptions albeit with permissible stress values rather than the characteristic values.

1.2 TIMBER

1.2.1 General

The many species of timber used in timber engineering can be divided into two categories: softwoods and hardwoods. Softwood is the timber of a conifer whereas hardwood is that of a deciduous tree. Some softwoods can be quite hard (e.g. Douglas fir), and some hardwoods can be quite soft (e.g. balsa).

This manual deals almost entirely with design in softwood, because nearly all timber engineering in the UK is carried out with softwood. Hardwoods are used, however, for certain applications (e.g. harbour works, restoration works, farm buildings, etc.), and Chapter 25 deals with aspects that must be considered when using hardwood.

The UK is an importer of timber even though the proportion of home-grown softwood for structural applications has risen steadily in recent years from 10% to about 25% of the total requirement. About 80% of the imported softwood comes from Norway, Sweden, Finland, Russia, Poland and the Czech Republic with the balance mainly from Canada and the USA, although imports from other parts of Europe, and from New Zealand, Southern Africa and Chile, do occur.

The European imports are usually the single species European whitewood (*Picea abies*) or European redwood (*Pinus sylvestris*).

Canada and the USA supply timber in groups of species having similar properties, e.g.

- Spruce–pine–fir consisting of
  Engelmann spruce (*Picea engelmannii*)
  lodgepole pine (*Pinus contorta*)

alpine fir (*Abies lasiocarpa*)
red spruce (*Picea rubens*)
black spruce (*Picea mariana*)
jack pine (*Pinus banksiana*)
balsam fir (*Abies balsamea*)

- Hem–fir consisting of
  western hemlock (*Tsuga heterophylla*)
  amabilis fir (*Abies amabilis*)
  grand fir (*Abies grandis*)
  and additionally from the USA
  California red fir (*Abies magnifica*)
  noble fir (*Abies procera*)
  white fir (*Abies concolor*)
- Douglas fir–larch
  Douglas fir (*Pseudotsuga menziesii*)
  western larch (*Larix occidentalis*)

The USA provides the following groupings

- Southern pine consisting of
  balsam fir (*Abies balsamea*)
  longleaf pine (*Pinus palustris*)
  slash pine (*Pinus Elliottii*)
  shortleaf pine (*Pinus echinata*)
  loblolly pine (*Pinus taeda*)
- Western whitewoods consisting of
  Engelmann spruce (*Picea engelmannii*)
  western white pine (*Pinus monticola*)
  lodgepole pine (*Pinus contorta*)
  ponderosa pine (*Pinus ponderosa*)
  sugar pine (*Pinus lambertiana*)
  alpine fir (*Abies lasiocarpa*)
  balsam fir (*Abies balsamea*)
  mountain hemlock (*Tsuga mertensiana*)

The UK and Ireland provide

- British spruce consisting of
  Sitka spruce (*Picea sitchensis*)
  Norway spruce (*Picea abies*)
- British pine consisting of
  Scots pine (*Pinus sylvestris*)
  Corsican pine (*Pinus nigra var. maritima*)

The UK provides

- Single species
  Douglas fir (*Pseudotsuga menziesii*)

- Larch consisting of
  hybrid larch (*Larix eurolepsis*)
  larch (*Larix decidua*)
  larch (*Larix kaempferi*)

From the publication of the first UK timber code, CP 112: 1952, through the 1960s and 1970s the usual practice for obtaining timber for structural use was to purchase a 'commercial grade' (see section 1.2.3) of a particular species, often from a specific source and then by visual assessment to assign the timber to an appropriate structural grade. Today timber is strength graded either visually or by machine with the specification aimed essentially towards the strength of the timber and then towards the species only if there are requirements with regard to specific attributes such as appearance, workability, gluability, natural durability, ability to receive preservative treatments, etc.

The most commonly used softwood species in the UK are European whitewood and redwood. These species have similar strength properties and by virtue of this and their common usage they form the 'reference point' for European strength-grading practices. The designer can consider the two species to be structurally interchangeable, with a bias towards whitewood for normal structural uses as redwood can be more expensive. Redwood, on the other hand, would be chosen where a 'warmer' appearance is required or if higher levels of preservative retention are needed.

1.2.2 Strength grading of timber

1.2.2.1 General

There is a need to have grading procedures for timber to meet the requirements for either visual appearance or strength or both. Appearance is usually covered by the commercial timber grades described in section 1.2.3. Strength properties are the key to structural design although other attributes may well come into consideration when assessing the overall performance of a component or structure.

Although readers in the UK may well be familiar with the term 'stress grading', 'strength grading' is the European equivalent that is now used. Strength grading may be described as a set of procedures for assessing the strength properties of a particular piece of timber. The strength grade is arrived at by either *visual grading* or *machine grading*.

It is convenient to have incremental steps in these strength grades and these are referred to as 'Strength Classes'. The European strength class system is defined in BS EN 338: 1995 'Structural timber. Strength classes' and this has been adopted for use in BS 5268. There is a set of classes for softwoods – the 'C' classes ('C' for conifer) – and a set for hardwoods – the 'D' classes ('D' for deciduous).

Through referenced codes of practice and standards, the various Building Regulations in the UK require timber used for structural purposes to be strength graded and marked accordingly. In addition to the BS EN 338 requirements, certain grading rules from Canada and the USA may be used. The acceptable grading rules are listed in BS 5268-2.

Strength grading of solid timber can be achieved in one of two ways:

- Visual means, using the principles set out in BS EN 518: 1995 'Structural timber. Grading. Requirements for visual strength grading standards' with the requirements for timber to be used in the UK given in detail in BS 4978 'Specification for softwood grades for structural use'.
- Machine methods, in accordance with the requirements of BS EN 519: 1995 'Structural timber. Grading. Requirements for machine strength graded timber and grading machines'.

Visual grading will give as an output the strength related to the visual characteristics and species while machine grading will grade directly to a strength class. From tables in BS 5268-2 it is possible to arrive at an equivalent grade or strength class, whichever method of grading is used. Thus from Table 2 of BS 5268-2, SS grade redwood is equivalent to strength class C24.

The grading of timber to be used for structural purposes in the UK is controlled by the United Kingdom Timber Grading Committee (UKTGC). They operate Quality Assurance schemes through authorized Certification Bodies. These Certification Bodies may be UK or overseas based but each overseas Certification Body must have nominated representation in the UK, which may be through personnel resident in the UK or through another Certification Body that is UK based. The Certification Bodies are responsible for licensing persons as visual graders and for the approval and continuing inspection of approved types of grading machine. As a large part of the UK softwood requirement is imported, it follows that these approval schemes operate world wide.

Each piece of graded timber is marked to give the grade and the species or species combination of the timber, whether it was graded 'dry' (at or below 20% moisture content) or 'wet', the standard to which the timber was graded (BS 4978 for visual or BS EN 519 for machine grading) and with sufficient information to identify the source of the timber, i.e. the Certification Body and reference of the grader. In certain circumstances marking may be omitted, e.g. aesthetic reasons, in which case each parcel of a single grade has to be issued with a dated certificate covering the above information plus the customer's order reference, timber dimensions and quantities together with the date of grading.

Only timber that has been graded and marked in accordance with the procedures described above should be used for designed timber structures in the UK and in particular for structures and components purporting to be in accordance with BS 5268. The requirement for marking includes timber sized in accordance with the span tables given in the various Building Regulations.

1.2.2.2 *Visual strength grading to European standards*

As each country in the European Union has its own long-established visual grading rules it is not surprising that a common European visual grading standard could not be agreed. Instead BS EN 518 gives the principles for visual grading that national standards should achieve. In the UK the national standard is BS 4978: 1996 'Specification for visual strength grading of softwood'. Before the introduction of the European grading standards, BS 4978 also covered the machine grading of timber to be used in the UK.

BS 4978 describes two grades for visual strength grading: General Structural and Special Structural, which are abbreviated to GS and SS respectively. For visual strength grades, bending strength is influenced mainly by the presence of knots and their effective reduction of the first moment of area of the timber section, so the knot area ratio (KAR) and the disposition of the knots are important.

While knots may be the most critical aspect, the rules in BS 4978 also include limitations for the slope of grain relative to the longitudinal axis of the piece of timber, the rate of growth (as given by average width of the annual rings), fissures, wane, distortion (bow, spring, twist and cup), resin and bark pockets and insect damage.

The knot area ratio is defined in BS 4978 as 'the ratio of the sum of the projected cross-sectional areas of the knots to the cross-sectional area of the piece'. In making the assessment, knots of less than 5 mm may be disregarded and no distinction need be made between knot holes, dead knots and live knots. Figure 1.1 illustrates some typical knot arrangements and their KAR values.

As a knot near an edge has more effect on the bending strength than a knot near the centre of the piece, the concept of a margin and a margin condition is introduced. For the purposes of BS 4978 a margin is an outer quarter of the cross-sectional area, and the margin knot area ratio (MKAR) is the ratio of the sum of the projected cross-sectional areas of all knots or portions of knots in a margin to the cross-sectional area of that margin. Likewise the total knot area ratio (TKAR)

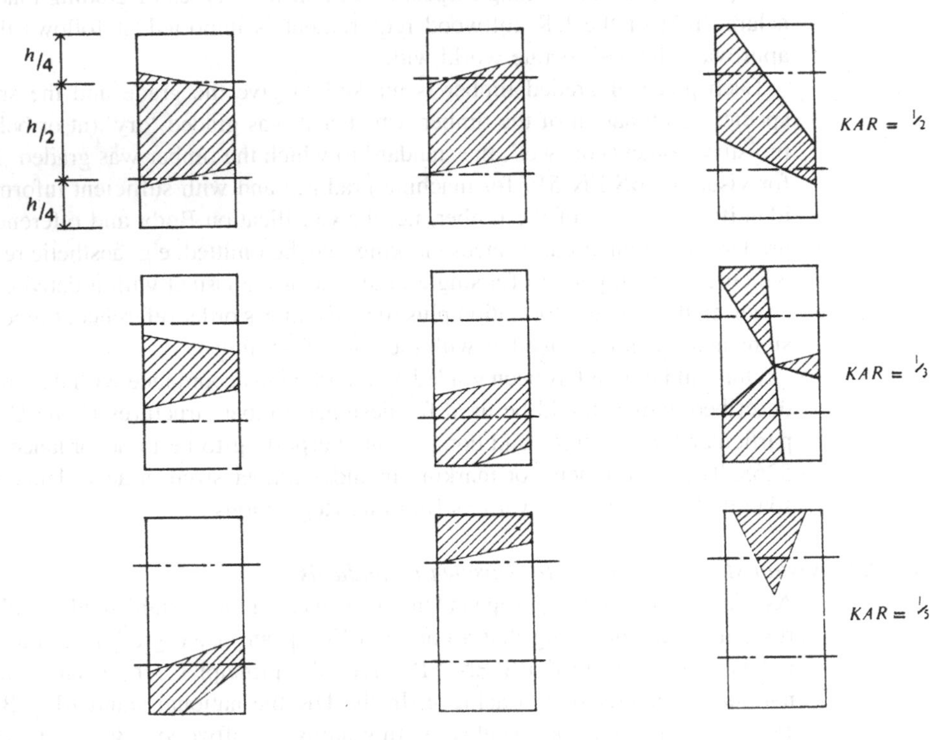

Fig. 1.1

is the ratio of the projected cross-sectional area of all knots to the cross-sectional area of the piece.

To qualify as SS grade, the MKAR must not exceed $\frac{1}{2}$, and the TKAR must not exceed $\frac{1}{3}$; where the MKAR exceeds $\frac{1}{2}$, the TKAR must not exceed $\frac{1}{5}$. For GS grade, the MKAR must not exceed $\frac{1}{2}$ and the TKAR must not exceed $\frac{1}{2}$; where the MKAR is greater than $\frac{1}{2}$, the TKAR must not exceed $\frac{1}{3}$.

The most onerous (theoretical) arrangement of knots corresponding to these limits is illustrated in Fig. 1.2. The ratio Z_{net}/Z_{gross} is shown alongside each sketch. From these ratios it can be deduced that the ratio of bending stresses between SS and GS grades would be in the order of

$$\frac{0.44}{0.64} = 0.69 \quad \text{or} \quad \frac{0.384}{0.464} = 0.83$$

depending on the extent to which a margin condition is relevant. The ratio of the bending strengths in BS 5268-2 is 0.7 for all softwood species.

Providing that any processing does not remove more than 3 mm from an initial dimension of 100 mm or less, and 5 mm from larger dimensions, then according to BS 4978 the grade is deemed not to have been changed. If a graded piece is re-sawn or surfaced beyond these limits then it must be regraded and re-marked if it is to be used structurally. If a graded piece is cut in length then the grade of each piece is not reduced. The grade could well be increased if a critical defect is removed by this means! The strength grading can be carried out in the country of origin or in the UK.

As visual grading gives simply the projected area of knots, to establish the grade strength of timber from the GS or SS rating, the species of the timber has also to be given. Thus SS grade redwood/whitewood lies in strength class C24 while SS

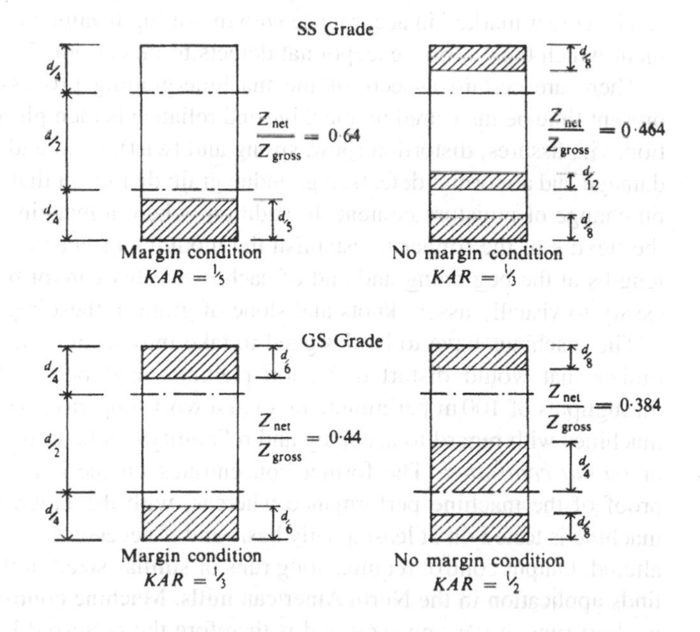

Fig. 1.2

grade British Sitka spruce is in a lower class, C18. Correlation of the European and North American visually graded species to a strength class is given in Tables 2, 3, 4 and 5 of BS 5268-2. The correlation between various European national grades and the BS EN 338 strength classes is given in BS EN 1912 'Structural timber – Strength classes – Assignment of visual grades and species'. It should be noted that certain timbers, e.g. Radiata pine from New Zealand, can only be graded for use in the UK by machine strength grading.

1.2.2.3 *Machine strength grading to European standards*

Grading machines were introduced commercially in the 1960s so there is now some 30 years' background experience in their operation.

Machine strength grading relies essentially on the relationship between the modulus of elasticity, *E*, and the modulus of rupture of a particular species of timber from a particular geographical location. The modulus of elasticity may be determined in a number of ways such as: applying a known force and measuring the corresponding deflection; applying a known displacement and measuring the force to achieve this; and by establishing the modulus of elasticity from dynamic measurement. A statistical population, i.e. many hundreds of pieces, has to be tested in the laboratory to establish the relationship between *E* and the modulus of rupture. From background research these initial findings can be interpolated and extrapolated to different cross-sectional sizes of timber of the same species. Where the relationship between these properties is determined by deflection measurement, it is normal practice to bend the timber about its minor axis, i.e. as a plank rather than as a joist.

For machines using flexure to determine *E*, each piece of timber is graded in increments of, say, 150 mm of its length as it passes through the grading machine and the minimum value obtained is given to that piece. Knowing this grade value allows automatic sorting to stock piles of similar grade. Alternatively, the piece can be colour marked in accordance with the strength value measured at each increment which then allows exceptional defects to be cut out.

There are certain aspects of the machine-grading process that cannot at the present time be measured by machine and reliance is then placed on visual inspection, viz. fissures, distortion (bow, spring and twist), resin and bark pockets, insect damage and abnormal defects, e.g. undue grain distortion that could cause damage on change of moisture content. In addition, where a machine process relies upon the flexure of the timber to establish the modulus of elasticity there will be short lengths at the beginning and end of each piece that cannot be tested so it is necessary to visually assess knots and slope of grain in these regions.

The machines have to be designed to take into account any natural bow in the timber that would distort deflection readings, and be capable of operating at throughputs of 100 m per minute or so in a workshop environment. The control of machines with regard to accuracy and reliability can be either *machine controlled* or *output controlled.* The former concentrates on the continual assessment and proof of the machine performance whereas, with the latter, the output from the machine is tested on at least a daily basis and if necessary the machine settings are altered. Output control requires long runs of similar-sized timber to be effective so finds application in the North American mills. Machine control is more applicable to short runs of varying sizes and is therefore the preferred European method.

The requirements for marking the timber with the source, grading machine reference as well as whether graded dry or wet, the grade, species and the standard to which the grading has been made (BS EN 519) are similar to those for visual grading.

1.2.2.4 Strength classes

Strength classes were originally introduced in CP 112: 1967 as a means of simplifying the specification of structural timber where species, appearance and similar attributes would not be critical. A strength class is the grouping of timbers that possess similar strength characteristics irrespective of species. By specifying a strength class the designer knows that the timber selected will be a reasonably economic solution. Left to his own devices there is the possibility that the designer may specify an exotic species such as pitch pine where the more readily available and cheaper redwood or whitewood would be acceptable.

For many years, until the introduction of BS EN 338 in 1996, grading machines were set to produce the equivalent of a visual grade, e.g. GS visual grade and MGS machine grade redwood/whitewood, as well as specific grades only achievable by machine grading. As machine grading became more widespread the logical move was towards strength classes rather than species/strength grade combinations.

The European standard BS EN 338 'Structural timber. Strength classes' gives nine softwood grades (C14 to C40) and six hardwood grades (D30 to D70). The number after the initial letter is the characteristic bending strength of the timber. This is a value intended for use with the limit state Eurocode 5 and has a duration of load equivalent to the time of testing, i.e. a few minutes. It can, however, be roughly translated to a long-term strength value for use in a permissible stress code such as BS 5268 by dividing by 3.20. In practice the procedures for arriving at the tabulated grade bending strength values in BS 5268-2 are rather more sophisticated.

To meet the requirements for a particular strength class the bending strength, the characteristic density and the measured mean modulus of elasticity of the timber had to be equal or greater than the values given for the strength class in BS EN 338. This had a number of problems in that many of the strength and stiffness values were set on the basis of visual grading. It is of relatively little consequence if the modulus of elasticity in a visual grading system is optimistic, i.e. high in its value, but in machine grading this is the essential parameter and too high a value will reduce the yield of a particular strength class from a parcel of timber. This occurred with timber used in trussed rafter manufacture where the industry looked to the new strength class C27 to replace the former redwood/whitewood M75 grade. Both have the same bending strength of 10.0 N/mm^2 but the mean E values are 12 300 N/mm^2 and 11 000 N/mm^2 respectively. A particular timber sample when machine graded would have a lower yield of the stiffer C27 material than the M75 grade. BS 5268-2 therefore lists an additional strength class TR26 that has similar properties to redwood/whitewood M75.

When strength classes were determined using the machine-grading procedures set out in BS 4978, the former edition of BS 5268-2 allowed timber to be admitted to a strength class if (a) the strength in compression parallel to the grain, (b) the strength in shear parallel to the grain and (c) the mean modulus of elasticity were not less than 95% of the values required for the strength class. No two species of timber are identical in the relationships between their various mechanical properties, so although two species may have identical modulii of rupture, one may have a higher E value and the other a higher compression strength parallel to the grain. The pragmatic UK approach smoothed out these minor anomalies. Unfortunately the European standards do not at the present time contain this flexibility but it is likely that some progress towards the former UK approach will be made.

Designers should note that in the UK it is still possible to design for a specific species/grade combination using the associated strength values. Where a particular species/grade lies just above the boundary between strength classes, then there will be benefits in using these strength values.

1.2.2.5 *Strength graded North American timber*

Timber graded to North American standards (National Lumber Grades Association, NLGA, in Canada and National Grading Rules for Dimension Lumber, NGRDL, in the USA) falls into three visual grading groups and five machine grades. The three visual grading groups are

- Joist and plank (J&P) graded as Select Structural, No. 1, No. 2 and No. 3.
- Structural light framing (SLF) graded as Select Structural, No. 1, No. 2 and No. 3.
- Light framing (LF) and stud grades graded as Construction, Standard, Utility and Stud.

These grading groups relate to end usage and the visual grading rules reflect this. In J&P, where bending of the extreme fibres at the top and bottom of the section is critical, then limitations on knot sizes at these positions are defined, whereas stud grade is used for compression members so the distribution of the critical size of knot is uniform across the section. Within each group there are separate grades (e.g. No. 1, No. 2) and the strength values are then related to the species groupings described in section 1.2.1 as well, for example the complete specification could be 'Hem-fir Structural Light Framing No. 2'. To further complicate matters, before 1996 the groupings SLF, LF and Stud related tabulated strength values to a single size, 38 × 89, and for other sizes, e.g. 89 × 89, reference had to be made to an additional table for size modification factors. BS 5268-2: 1996 rationalized the presentation of the strength values so they are now tabulated and used in the same manner as the values for timber from the rest of the world, taking into account the size modification factors K_7 and K_{14}. In the original North American strength values for the various grades, tension was not allowed in SLF No. 3, all LF grades and Stud grade. BS 5268-2 does now give tension values for these grades but their use in tension members should be avoided where possible.

It is common practice to purchase mixed parcels of No. 1 and No. 2 of J&P or SLF (the individual pieces are graded and marked accordingly so they may be easily identified and used for specific situations if required). In design the strength values used generally would be No. 2 but the higher grade pieces would be available if necessary and could also be sorted on site for members likely to be more heavily loaded, e.g. trimmers.

The North American machine grading can be either Machine Stress Rated (MSR) or Machine Evaluated Lumber (MEL). Only MSR lumber is listed in BS

5268-2. The grades are given as a combination of bending strength and E value for the various species groupings, e.g. Douglas fir–larch 1200f–1.2E has a flexural strength of 1200 lbf/in^2 and E of 1 200 000 lbf/in^2 (converted to 7.9 N/mm^2 and 8000 N/mm^2 respectively in BS 5268). The North American grading machines are output controlled.

Canadian timber is also available from the East coast visually graded to BS 4978 and sawn to European sizes.

1.2.3 Commercial grades

1.2.3.1 General

Commercial grades are of less importance for today's structural work than in the past when only commercial grades were available and the assessment of the value of the timber as a structural material rested with the user. Commercial grades may still need to be purchased in order to obtain the size of timber required or perhaps the volume needed, particularly with species normally associated with appearance rather than structural properties, e.g. western red cedar. Strength-grading rules described in section 1.2.2 seldom go beyond a cross-sectional dimension of 250 mm, whereas 400 mm or so may be required for repair or renovation purposes. Commercial timbers are available in large balks or flitches that can then be converted to the appropriate dimension for structural use.

1.2.3.2 Swedish and Finnish commercial grades

For the whitewood and redwood from Sweden and Finland there are six grades numbered I, II, III, IV, V and VI (known as 'firsts', 'seconds', etc.). There are agreed descriptions for these grades, but these are only guiding principles and most well-known mills will practise a stricter sorting regime.

The basic qualities I–IV are usually grouped together for export. Because they are not sorted into separate grades, this grouping is sold with the title 'unsorted'. This is traditionally a joinery grade. The V quality is sold separately (or, if a mill has little unsorted it may sell 'fifths and better') and is traditionally a grade for building and construction. The VI quality is sold separately and is traditionally a grade for lower quality uses in building and for packaging.

The mill grading for whitewood and redwood from Poland and Norway, and the whitewood from the Czech Republic, may be considered to be based on similar rules to those of Sweden and Finland.

1.2.3.3 Russian commercial grades

The Russian commercial grades are somewhat similar to those of Sweden and Finland except that they are divided into five basic grades. The basic qualities I to III form the unsorted, and IV and V are similar to the Swedish/Finnish V and VI respectively.

1.2.3.4 Canadian commercial grades

Some of the timber end products imported from Canada of non-strength graded timber are given below. These timbers are usually sold and shipped in the 'green' condition, i.e. at a moisture content of 30% or more.

- Spruce–pine–fir and hem–fir in sawn sizes up to a maximum green size of 4″ × 12″ in a mix of 'No. 1 Merchantable' and 'No. 2 Merchantable' and occasionally in 'No. 3 Common' grade to the 'Export R List Grading and Dressing Rules'.
- Douglas fir–larch, hemlock or western red cedar to the 'Export R List' in 'No. 2 Clear and Better', 'No. 3 Clear' and 'No. 4 Clear', imported in sawn sizes up to a maximum green size of 4″ × 12″.
- Western red cedar 'Shop Lumber' in green sawn sizes up to a maximum green size of 2″ × 12″.
- 'Beams and Stringers' to the 'Export R List' up to a maximum green size of 6″ × 16″ in hemlock, Douglas fir–larch and western red cedar.
- 'Posts and Timbers' to the 'Export R List' up to a maximum quoted size of 16″ × 16″ in hemlock, Douglas fir–larch, western red cedar and 8″ × 8″ in spruce–pine–fir.

Canadian timber is often costed at a price per 'Standard'. This refers to a 'Petrograd Standard' which is 165 ft^3 or 4.671 m^3.

1.2.3.5 *Joinery grades in the UK*

There will be cases where a designer will be asked to design using a joinery grade timber. The grading rules are given in BS EN 942: 1996 'Timber in joinery – General classification of timber quality'. The rules are applicable to softwoods and hardwoods and cover knots, shakes, resin and bark pockets, discoloured sapwood, exposed pith and beetle damage.

The grading reference is based on the maximum dimension of a knot or knot cluster on a particular surface. This size is expressed as a percentage of the face width together with a maximum dimension. The highest grade is J2 which allows knots not exceeding 2 mm measured in accordance with BS EN 1310, then J10 maximum 30% of face or 10 mm maximum, J30 – 30% and 30 mm, J40 – 40% and 40 mm and J50 – 50% and 50 mm.

Note that there is no restriction on the slope of grain for joinery grades. It is therefore possible to get 'clear' or effectively knot-free timber. To achieve this it is likely that the slope of grain will be as steep as 1 in 4 whereas for SS grade to BS 4978 the limit is 1 in 10 and for GS 1 in 6. For this reason the designer must not be tempted to use 'clear' timber in a structural application.

1.2.4 Sizes and processing of timber

1.2.4.1 *General*

This is probably the most confusing aspect of the timber trade for persons ranging from the uninitiated through to the lawyers attempting to unravel the niceties of a shipping contract. The alternative descriptions of the sizing of cross-sections as 'sawn', 'regularized' and 'planed all round' or 'surfaced four sides' may be hard to grasp but when the complication is added of the actual size of a supposed 2″ × 4″ cross-section being $1\frac{5}{8}$″ × $3\frac{5}{8}$″ the novice is understandably confused.

1.2.4.2 *Processing of timber*

When a log is delivered to a mill from the forest it is sawn to either large balks or flitches for shipment and subsequent re-sawing by the purchaser or it may be sawn to smaller, directly usable sections. Timber can be used in this sawn condition where dimensional tolerances are of no significance, e.g. for rafters to receive roofing tiles where out of line of the top edge of the rafter by 3 mm or so is no problem. Timber can be either rough or fine sawn. The former is the condition that is normally seen, for fine sawn timber can be difficult to tell from planed timber.

Where there is a dimensional requirement with regard to the depth or width of a member – say, for a floor joist that has to provide surfaces on both the top and bottom edges to receive a boarded walking surface and a ceiling – then it is obviously advantageous to have all the joists the same depth. Planing two parallel edges to a specified dimension is termed 'regularizing' and may be coded S2S or surfaced on two faces.

Where a member is exposed or there is a need for dimensional accuracy on all four faces, or if it makes handling easier (as with CLS and ALS timber), all four faces can be planed. This process is known as 'planed all round' or 'surfaced four sides' and may be coded S4S.

1.2.4.3 *The sizing and tolerances for European timber*

The move to European standards has paradoxically simplified the problem of sizing outlined in section 1.2.4.1. BS EN 336 'Structural timber – Coniferous timber – Sizes – Permissible deviations' together with its National Annexes gives straightforward rules for defining and arriving at the dimensions of timber. It is necessary to first set out the relevant definitions:

- *Work size*: the size at 20% moisture content of the section before any machining that may be required to achieve the *target size*.
- *Target size*: the size at 20% moisture content the designer requires and it is the size to be used in any calculations; the deviations or tolerances specified below are not included in the value of the target size.
- *Cross-sectional deviations (tolerances)*: the amounts by which the actual dimensions may vary from the target size:
  - tolerance class 1 (T1): for dimensions up to 100 mm (+3/–1) mm; dimensions over 100 mm (+4/–2) mm
  - tolerance class 2 (T2): for dimensions up to 100 mm (+1/–1) mm; dimensions over 100 mm (+1.5/–1.5) mm.

Tolerance class 1 is applicable to sawn sections while tolerance class 2 would be used with the surfaced, processed or planed dimensions. It is normal UK practice to assume that 3 mm is removed when surfacing or processing timber up to 150 mm initial dimension and 5 mm for timber over 150 mm initial dimension. This does not mean that greater amounts cannot be removed, but beware that the grading of the piece is not compromised (see section 1.2.2.2).

Note that these tolerances are exactly what they say: they are *tolerances*. It is incorrect to take a piece with a work size of, say, 38 mm surfaced to 35 mm and say that all pieces are intended to be 35 – 1 = 34 mm; the average of all pieces is intended to be 35 mm and some may be as low as 34 but some can also be as high as 36 mm.

Taking as an example a timber section obtained as a sawn section (work size) of 47 × 150 mm at 20% moisture content, then

- used without any further processing or machining, the *target size* would be specified as 47 (T1) × 150 (T1) with the minimum size as 46 × 148, and the maximum size 50 × 154; the designer would use the dimensions 47 × 150 in calculations;
- used after machining on the width (commonly known as regularizing) the *target size* would be specified as 47 (T1) × 145 (T2) with the minimum size as 46 × 143.5, and the maximum size 50 × 146.5; the designer would use 47 × 145 in calculations;
- used after machining on all four faces (commonly known as planed all round or surfaced four sides) the *target size* would be specified as 44 (T2) × 145 (T2) with the minimum size as 43 × 143.5, and the maximum size 45 × 146.5; the designer would use 44 × 145 in calculations.

1.2.4.4 *Sizes available from European mills*

The dimensions of timber commonly available in the UK based on the National Annex in BS EN 336 are given in Table 1.1.

Cross-sectional dimensions above 200 mm are less readily available so could attract a cost premium. Scandinavian mills also supply timber to North American CLS or ALS sections (see section 1.2.4.5).

Smaller sizes can be obtained by re-sawing any of the tabulated sizes and in these cases a reduction of at least 3 mm should be allowed for the saw cut thickness. A 100 mm size, if sawn equally into two, would normally be considered to yield two 48 mm pieces. Remember too that the re-sawn pieces will need to be re-graded if they are to be used structurally.

The moisture content of the timber (see section 1.2.5) is assumed to be at 20% for the cross-sectional dimensions and tolerances given above. For other moisture contents the change in dimension from the base value at 20% can be taken as 0.25% change in the face dimension for 1% change in moisture content. It can be assumed that no increase in dimension will take place for moisture content changes above 30%. These movements in cross-sectional dimensions can be assumed applicable to all softwoods. For more detailed information relating to moisture movements radially and tangentially to the growth rings, see section 1.2.5.3.

Table 1.1 Customary target sizes of European timber available in the UK

	Dimension (mm)														
Sawn dimension	22	25	38	47	63	75	100	125	150	175	200	225	250	275	300
Surfaced dimension	19	22	35	44	60	72	97	122	147	170	195	220	245	270	295

The lengths of timber are available in increments of 300 mm with 3.9 m being the limit of commonly available timber. Greater lengths will likely attract a premium cost. In practical terms there is no change in length with change in moisture content.

1.2.4.5 Sizes available from North America

Many sections are marketed in the 'green' condition in imperial (inch) dimensions so the designer has to be careful when assessing whether the timber should be dried by kilning or open sticking before use, and what the final dimensions are likely to be when actually used. Fortunately most of the timber in the J&P, SLF, LF and Stud grades (see section 1.2.2.5) are available kiln dried to a moisture content of 19%.

North American timber in these particular grades is surfaced to Canadian Lumber Standards, CLS, or American Lumber Standards, ALS. The thickness of the timber is usually 38 mm but depending on the grade can also be 63 or 89 mm and the widths are 63, 89, 114, 140, 184, 235 or 285 mm. The timber is planed on four surfaces and the arrises are slightly rounded, which makes the handling of the timber easier. These timber sizes are the target sizes and are assumed to be in tolerance class 2. The sizes are available in European redwood/whitewood from Scandinavian mills. Likewise Canadian mills on the east coast do saw European sizes.

'Occasional' (i.e. 10%) pieces of consignments of sawn, green timber graded to the NLGA rules are permitted to have the full sawing tolerances, which are given in section 747 of the NLGA rules as:

under 2″	$\frac{1}{16}$″ under or $\frac{1}{8}$″ over
2″ and larger not including 5″	$\frac{1}{8}$″ under or $\frac{1}{4}$″ over
5″ and larger not including 8″	$\frac{3}{16}$″ under or $\frac{3}{8}$″ over
8″ and larger	$\frac{1}{4}$″ under or $\frac{1}{2}$″ over

These tolerances are in addition to any reductions on drying.

1.2.5 Moisture content and movement of timber

1.2.5.1 Measurement of moisture content

The moisture content of timber for precise measurement or instrument calibration purposes is obtained by weighing a sample of the timber, oven drying and then re-weighing thus establishing the weight of water in the sample. From these measurements

$$\text{moisture content} = \frac{\text{weight of water}}{\text{weight of oven dry sample}} \times 100\%$$

The method is destructive, therefore it is usually used only in the case of calibration or dispute. The result is an average value for the particular sample, so achieving a reasonable assessment for a large member can require a large number of samples.

Alternatively, the moisture content can be determined by a moisture meter. These meters measure the electrical resistance of the timber between two probes inserted into the timber. The wetter the timber the lower the electrical resistance. The performance of this type of meter is therefore dependent upon a number of factors that the user should be aware of:

- the various species of timber have different resistances; meter manufacturers overcome this by having different scales on the meter or using different calibration 'keys' for insertion into the meter;
- the depth of penetration of the probes will affect the readings, as the surface layers of the timber may be at a different moisture content to those 20 mm below the surface depending upon the environment or the exposure of the timber to the weather;
- organic solvent preservatives are said not to influence readings to any significant extent, but readings are affected by water-borne preservatives and fire retardants; the manufacturer of the meter may be able to give a correction factor.

Provided that a moisture meter is calibrated correctly and used correctly, it is usually accurate enough for most checking required in a factory or on site. If insulated probes are used it is possible to plot the moisture gradient through a section.

Measurement of the moisture content of board materials with a moisture meter designed for use on solid timber can at best be described as 'indicative' due to the relatively small thickness of the boards and the effects of any adhesives that may have been used.

1.2.5.2 *Service classes*

BS 5268-2 now follows Eurocode 5 in defining three service classes:

- *service class 1* is characterized by a moisture content in the materials corresponding to a temperature of 20 °C and the relative humidity of the surrounding air only exceeding 65% for a few weeks per year;
- *service class 2* is characterized by a moisture content in the materials corresponding to a temperature of 20 °C and the relative humidity of the surrounding air only exceeding 85% for a few weeks per year;
- *service class 3* indicates climatic conditions leading to higher moisture contents than in service class 2.

For most softwoods the conditions for service class 1 will result in an average moisture content that does not exceed 12% and the corresponding moisture content for service class 2 will not exceed 20%. Note that these values are for softwoods such as redwood and whitewood. In the environmental conditions for the service classes the moisture contents of hardwoods and board materials will differ from those for softwoods, e.g. plywood at the top end of service class 2 would be around 16%.

At the present time all materials for design purposes are assumed to be either 'dry' (service classes 1 or 2) or 'wet' (service class 3). The standardization at 20% moisture content for softwoods for sizing purposes as well as design is an improvement over the previous UK practice whereby dimensions were measured at 20% and designs to BS 5268-2 were assumed to be at 18%. The North American 'dry' condition at 19% still applies but it is at least on the conservative side of 20% rather than midway between the previous UK limits!

1.2.5.3 Moisture movement of timber

When a tree is felled the moisture content could be well over 100%. There is effectively no dimensional movement of the timber as it dries until it reaches 'fibre saturation'. This is the point when further drying removes water from within the cell walls of the timber fibres, all previous drying having been of 'free' water in the cells themselves. The fibre saturation point for most softwoods is taken as 25 to 30% moisture content.

With change in moisture content below the fibre saturation point timber changes in dimension. This movement may be assumed to be linearly related to the change in moisture content for moisture contents between 8% and the fibre saturation point. In timber-framed buildings due account has to be taken of such movements, particularly in the vertical direction as timber installed at a moisture content of 20% could dry down to an equilibrium moisture content of 12% in a heated building.

The effects of timber shrinkage with such a change of moisture content can be made more onerous in buildings where, for example, clay brickwork is used as an external cladding which could expand between 0.25 mm and 1.00 mm per metre height of masonry. BS 5268-6.1 recommends an allowance of 6 mm per storey height (say 2.6 m) for the differential movements at external door and window

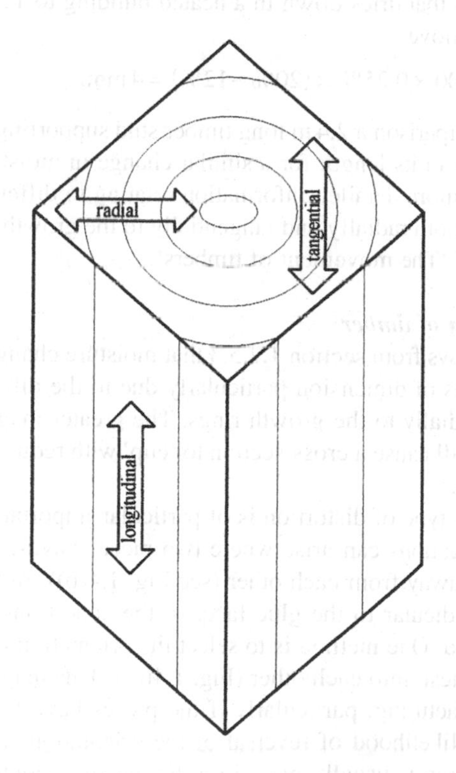

Fig. 1.3

openings, at lift installations, for vertical services and for the wall ties between masonry cladding and the timber frame.

The magnitude of moisture movement depends upon the orientation of the timber. In cross-section, i.e. across the grain, the movement in the tangential direction is about twice that in the radial direction for a given change in moisture content.

For European redwood the tangential movement is 0.28% change in dimension for 1% change in moisture content while in the radial direction it is 0.12% in dimension for 1% moisture change. The corresponding change along the grain, i.e. along the length of the timber, is much smaller being around 0.004% in dimension for 1% moisture content change, which is so small that it is usually neglected.

Today the conversion of timber from the log to structural size means that in any cross-section the moisture movement parallel to a particular surface will be a combination of tangential and radial effects. One method of assessing this condition is to take the average of the two movements, e.g. for European redwood:

$$0.5(0.28\% + 0.12\%) = 0.20\% \text{ change in dimension for } 1\% \text{ moisture content}$$

In a timber-framed platform construction (the most common form of timber-framed house construction) the largest single contribution to cross grain movement is the floor joist. The conversion of the timber means there will be a tendency for more tangential movement than radial (see Fig. 1.3 above) so the combined cross-grain shrinkage of the common softwoods is often quoted as 0.25% movement for 1% moisture content change. A 200 mm floor joist installed at 20% moisture content that dries down in a heated building to 12% equilibrium moisture content could move

$$200 \times 0.25\% \times (20\% - 12\%) = 4\,\text{mm}.$$

By comparison a 2.4 m long timber stud supporting the floor joist would move only 0.8 mm in its length for a similar change in moisture content.

For more detailed information relating to different species and moisture movements both radially and tangentially to the growth rings, see FPRL Technical Note No. 38 'The movement of timbers'.

1.2.5.4 Distortion of timber

It follows from section 1.2.5.3 that moisture change in a piece of timber will cause changes of dimension particularly due to the different characteristics tangentially and radially to the growth rings. The greater magnitude of the tangential component will cause a cross-section to 'cup' with reduction in moisture content (see Fig. 1.4(a)).

This type of distortion is of particular importance when gluing timber together, for situations can arise where two pieces having a common glueline will tend to move away from each other (see Fig. 1.4(b)). Adhesive joints are relatively weak perpendicular to the glue face, so the conditions shown in Fig. 1.4(b) should be avoided. One method is to select the orientation of the pieces so that, on cupping, they 'nest' into each other (Fig. 1.4(c)). This may not be practicable in large-scale manufacturing, particularly if the pieces have been end jointed for there is then every likelihood of reversal of the orientation within the length of one piece. A limitation is usually placed on the moisture content so that the maximum differ-

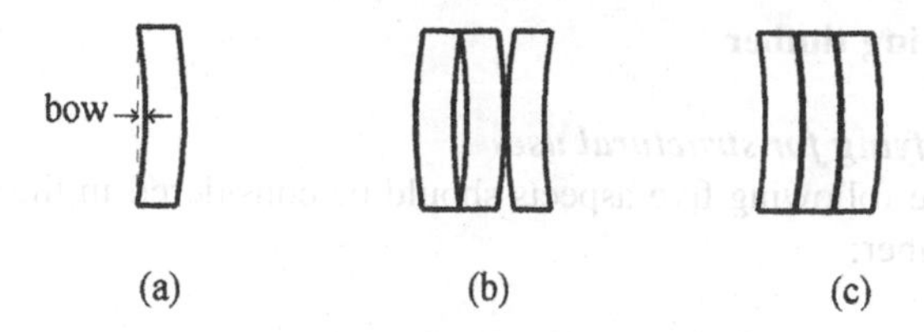

Fig. 1.4

ence between any piece at the time of gluing and in its equilibrium moisture content in service does not exceed 5% and that the maximum difference between adjacent pieces at the time of gluing does not exceed 3%.

It can be seen that if the grain direction in a piece of timber does not coincide exactly with the geometrical axes of the timber, then any moisture changes will be accompanied by corresponding asymmetrical movements in cross-section and along the length giving rise to bow, spring and twist of the piece. These movements can also be made worse locally due to grain swirl around knots. The rules for machine grading have visual overriding limits on bow, spring, twist and fissures (splits and shakes) in the timber at the time of grading. These limits are similar to those specified for visual grading. More often than not in machine grading the timber is run through the machine in the wet condition and only the timber passing this test is then dried. As the magnitudes of deformations and distortions depend on the moisture content of the timber it is essential that when machine grading for the 'dry' requirement, i.e. 20% moisture content or less, that the visual overrides are assessed with the timber at the required 20% moisture content or less.

1.2.5.5 *Thermal movement of timber*

The coefficient of linear thermal expansion is approximately $4 \times 10^{-6}/°C$, which is about 33% of the value for steel and concrete. Under normal climatic conditions the movement of timber will be small enough to be neglected. It is likely that only very long span beams will require special consideration.

The low thermal movement of timber is a significant asset when a timber member is subject to fire. Much of the damage in a fire is due to the thermal expansion of the construction materials causing disruption at junctions with other materials and at the supports of members long before the effects of temperature rise initiates failure of the members themselves. In a fire situation the 'core' of the timber section remains effectively at ambient temperature due to the good thermal insulation of the charred outer surfaces. This, coupled with the low coefficient of thermal expansion, means that there is little if any disruption due to thermal movement.

1.2.5.6 *Thermal conductivity of timber*

Timber has low thermal conductivity, so it is not usually necessary to consider 'thermal bridging' through wall studs or roof joists where these components form part of a thermally insulated wall or roof construction. Values are typically 0.14 W/m K for timber, plywood and tempered hardboard. Other fibreboards are of lower thermal conductivity and can make a significant contribution to the thermal insulation of a building element.

1.2.6 Specifying timber

1.2.6.1 Specifying for structural uses

The following five aspects should be considered in the specification for structural timber:

1. Strength class, with a limit on the species permitted if this is relevant (section 1.2.2.4) *or* the strength grade and species (abbreviations for various species and species groups are given in BS 4978 for softwoods and in BS 5756 for hardwoods).
2. Target sizes and tolerances (section 1.2.4.3). This defines the surface finishes, e.g. if T1 then it is a sawn finish and, if T2, a fine sawn or planed surface with the target size derived from commercially available work sizes (sections 1.2.4.4 and 1.2.4.5).
3. Length in increments of 300 mm for loose timbers sent to site for cutting and fitting *or* the actual lengths required where these are given on relevant drawings or schedules.
4. Whether the timber is to be graded for use in service class 2 *or* whether it can remain unclassified, i.e. assumed to be 'wet' *or* whether a specific moisture content is required, say 14% ± 3% (for example, for avoiding undue distortion when gluing, see section 1.2.5.4).
5. Preservation – where necessary (see Chapter 24 for guidance).

Examples

Requirement: Regularized timber grade C24, work size 47 × 200, for floor joists in a domestic property, cut on site to bear into masonry walls with a nominal length of 3.140 m:

C24 – 47 (T1) × 195 (T2) × 3300 long, DRY

Requirement: Exposed floor joists of European redwood SS grade for centrally heated building, work size 47 × 200, cut to exact length 3.140 m:

European redwood (ER), SS grade, 44 (T2) × 195 (T2) × 3140 mm, DRY

Requirement: Greenheart timber piles for use in sea water, work size 300 × 300, grade D70 cut to length 5050 mm:

Greenheart D70, 300 (T1) × 300 (T1) × 5050 mm

Requirement: European whitewood, C24 timber for glued laminated timber construction in a covered environment with an anticipated equilibrium moisture content of 16%, work size 47 × 100, component length 4.145 m trimmed to length on completion of the laminating:

European whitewood (EW) C24, 44 (T2) × 100 (T1) × 4200 mm, moisture content 16% ± 3%

Note that the 100 mm breadth would be planed after laminating to remove hardened adhesive, 'squeeze out' and misalignment of the laminates to give a finished beam breadth of 90 mm.

Requirement: Canadian Douglas fir floor joists, J&P No. 2, CLS size 38 × 235 × 3.255 m (exact length):

Douglas fir Canada (NA/DFL) J&P No. 2, 38 × 235 CLS × 3255 mm, DRY

Note the CLS dimensions are assumed to have a tolerance of T2.

Requirement: TR26 timber, target size 35 × 120 for use in trussed rafter manufacture, maximum length of piece 4.200 m:

TR26 – 35 (T2) × 120 (T2) × 4200 mm, DRY

1.2.6.2 Specifying for uses other than structural

Depending on circumstance it may be necessary to specify one or more of the following:

- Species of timber, if this is important for appearance, surface finish, workability, etc.
- Grade. If an appearance grade is important it may be relevant to refer to one of the classes of BS EN 942 (see section 1.2.3.5). If grade is not particularly important it may be adequate to refer to a commercial grade (see section 1.2.3), remembering that re-sawing invalidates the original grading, or indeed simply to state what is important (e.g. straight, no wane, no decay).
- Moisture content, where important.
- Finished size, surface and tolerance. If no surface finish is specified a sawn surface will normally be assumed by the supplier.
- Length, either as a nominal dimension for cutting to fit or a specific dimension as given on drawings or schedules.
- Preservation, where necessary. (See Chapter 24 for guidance.)

The specifications for non-structural timber will tend to be more descriptive than those for structural timber given in section 1.2.6.1.

Examples

Preamble: Timber to be European whitewood of V quality, tolerances to BS EN 336, dried to 22% moisture content or less with the following dimensions:

32 × 200 × 4.20 m
32 × 195 regularized × 4.20 m
43 × 191 planed all round × 1.80 m

Preamble: Timber to be European redwood to class J30 of BS EN 942; exposed and concealed surfaces as indicated on drawing nos. XYZ/1 and 2;

preservation to be by organic-solvent double-vacuum process to Table 2 of BS 5589, desired service life 60 years.

1.3 PLYWOOD

1.3.1 General

Several countries produce plywood either from softwood or hardwood logs or a mixture of both; however, very few plywoods are considered suitable for structural use. As a general rule only plywood bonded with an exterior quality resin adhesive should be specified for structural use. As far as the UK is concerned, the principal structural plywoods are:

Canadian Douglas fir-faced plywood commonly referred to as Canadian Douglas fir plywood.
Canadian softwood plywood
Finnish birch plywood (birch throughout)
Finnish birch-faced plywood
Finnish conifer plywood
American construction and industrial plywood
Swedish softwood plywood.

Sections 1.3.2–1.3.5 give outline details of the sizes and qualities of these plywoods and the name of the organizations from which further details can be obtained. Before specifying a particular plywood it is as well to check on its availability. The strength values for various plywoods are tabulated in BS 5268-2. Those values are at a moisture content of 15%.

When designing with plywood, it is necessary to distinguish between the 'nominal' thickness and the actual thickness. Designs to BS 5268-2 must use the 'minimum' thickness which is based on the assumption that all veneers have their minus tolerance deducted. The designer should also check to see whether sanded or unsanded sheets will be supplied as this can make a difference of several percent to the thickness and strength values.

There are a number of ways of expressing plywood strength values. One is to assume that the whole cross-sectional area of the plywood has equal strength (i.e. the 'full area' method). Another is to assume that the veneers which run perpendicular to the direction of stress have no strength at all (i.e. the 'parallel ply' method). Yet another method is the 'layered' approach in which the perpendicular veneers are assumed to have some strength but much less than the parallel veneers. The designer must be careful to use the correct geometrical properties (either full area or parallel ply) with the corresponding strength values. If used correctly in calculations these three methods should give similar results. BS 5268-2 uses the full area method which is also the method used in this manual. The strength values in the full area method differ with thickness and quality of the plywood.

If appearance is important it is prudent to check whether the description of the face veneer relates to both outer veneers or only to one. In the latter case the better veneer is described as the 'front' face with the other outer veneer being described as the 'back' veneer.

When a plywood is described as an exterior grade of plywood, this reference is to the durability of the adhesive and not to the durability of the species used in the plywood construction.

It is common practice when giving the dimensions of a plywood sheet (length and breadth) to state firstly the dimension of the side parallel to the grain of the face veneer. Thus a sheet 2440 × 1220 will have the face grain of the outer veneer parallel to the 2440 dimension. This is an important description to recognize as structural plywoods are stronger in the direction of the face grain.

If plywood is to be glued and used as part of an engineered component (e.g. a ply web beam or stress skin panel) the designer should satisfy himself as to the suitability of the face quality and integrity of the chosen plywood for gluing. Some plywoods are intended mainly for wall or floor sheathing rather than glued components.

1.3.2 Available sizes and quality of Canadian plywood

The size of plywood sheets usually available is 2440 × 1220 mm (8 ft × 4 ft) but metric size sheets 2400 × 1200 mm are available. The face veneer runs parallel to the longer side (Fig. 1.5).

Two basic types of Canadian exterior grade plywood are available in the UK. These are a plywood with one or both outer veneers of Douglas fir and inner veneers of other species which is known commercially as Douglas fir plywood, and an exterior Canadian softwood plywood which is made from much the same species as Douglas fir plywood but without Douglas fir face veneers. Appendix D of BS 5268-2 lists the species that are used.

The face quality grades available for both Douglas fir and softwood plywoods are:

- 'Sheathing Grade' with grade C veneers throughout, unsanded.
- 'Select Grade' with grade B face veneer and grade C inner and back veneers, unsanded.
- 'Select Tight Face Grade' with grade B 'filled' face veneer and grade C inner and back veneers, unsanded.
- 'Good One Side Grade' (G1S) with grade A face veneer and grade C inner and back veneers, sanded (strength values for this grade are not given in BS 5268-2).

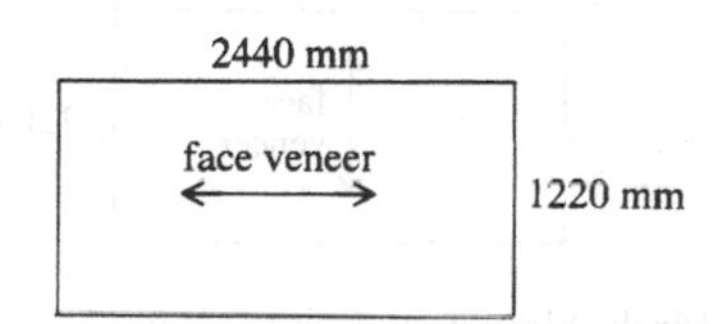

Fig. 1.5 Canadian fir-faced plywood.

- 'Good Two Sides Grade' (G2S) with grade A face and back veneers and grade C inner veneers, sanded (strength values for this grade are not given in BS 5268-2).

The Douglas fir grades G1S and G2S are required where a good appearance of the plywood surface is necessary but because the minimum thickness requirements for the face veneers could not be guaranteed, strength values are not given in BS 5268-2.

For further information refer to BS 5268-2.

1.3.3 Available sizes and quality of Finnish construction plywood

The most commonly available size is 1220 × 2440 mm. The face veneer runs parallel to the shorter side (Fig. 1.6). Other sizes which are quoted as being generally available are:

1200 × 1200/2400/2500/3000/3600
1220 × 1220/2440/2500/3050/3660
1250 × 1250/2400/2500/3000/3600
1500 × 1500/2400/2500/3000/3600
1525 × 1525/2440/2500/3050/3660

The face veneer runs parallel to the first dimension.

Three basic types of Finnish exterior grade construction plywood are available in the UK as well as a flooring plywood. The three construction plywoods are:

- Birch plywood made entirely of birch veneers (commonly described as birch 'thro and thro').
- Birch-faced plywood which has the outer veneers on both faces made from birch and the inner veneers of conifer and birch. There are three variations of this type:
  - Combi, having two birch veneers on each face with the inner veneers alternately conifer and birch;
  - Combi Mirror with one birch veneer on each face and the inner veneers alternately conifer and birch;
  - Twin, with one birch veneer on each face and conifer veneers internally.
- Conifer plywood with outer veneers of spruce or pine and the inner veneers from conifers.

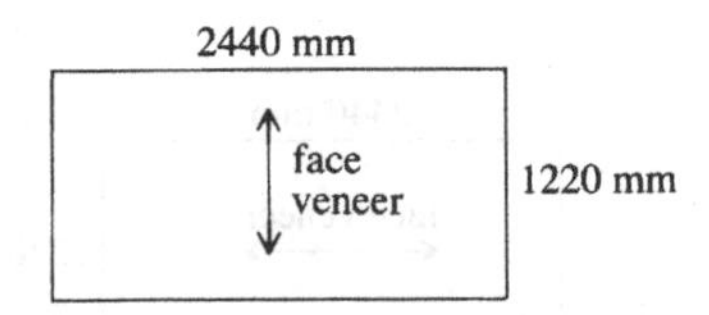

Fig. 1.6 Finnish birch, birch-faced or conifer plywood.

The thickness of birch veneers is usually 1.4 mm and the conifer veneers range from 1.4 to 2.8 mm.

The face qualities available for birch face veneers, based on practice nearly a century old, are A (practically without defect), through B, S, BB to WG (sound knots up to 65 mm diameter, open splits and checks up to 4 mm wide, etc.). Today the grading follows ISO standards for the qualities of both birch and conifer face veneers being E (practically without defect), through I, II, III to IV in a similar manner to the older birch grading (WG equivalent to IV). The reader may well find reference to either grading system for birch face veneers in technical literature and the market place. BS 5268-2 refers only to the number grading system.

It is normal practice to give the grades of the face veneer and then the back veneer, e.g. I/III. Because the veneers are thin the quality of the face veneer does not have any significant effect on the strength properties of the plywood. For this reason Conifer plywoods I/I and IV/IV are given the same values in BS 5268-2.

For further information refer to BS 5268-2 or contact Finnish Plywood International.

1.3.4 Available sizes and quality of American construction plywood

The volume of American Construction and Industrial plywood exported to the UK has increased in recent years, mainly in the C–D grade for use as sheathing. The most commonly available sheet size is 2440 × 1220 mm. The face veneer runs parallel to the longer side (as shown in Fig. 1.5).

The type of American exterior grade construction plywood covered in BS 5268-2 is a plywood with both outer veneers of Group 1 species (usually Douglas fir for export to the UK) and inner veneers of either Group 1 or Group 2 species. Annex D of BS 5268-2 lists the species that are used in Groups 1 and 2. The acceptable species listed in Annex D are more restrictive than those given in the American standard PS 1-95.

The grading of the face veneers is similar to the Finnish approach with the best quality A through to D. The quality of the plywood sheet is described by the grading of the two outer veneers. The following are the grades listed in BS 5268-2:

- 'C–D grade' unsanded, with the face veneer grade C of species Group 1, back veneer grade D of species Group 1 and the inner veneers grade D of species Groups 1 or 2.
- 'C–C grade' unsanded, which has veneers throughout of grade C of species Group 1.
- 'B–C grade' sanded, which has a face veneer of grade B and back and inner veneers of grade C, all veneers being of species Group 1.
- 'A–C grade' sanded, which has a face veneer of grade A and back and inner veneers of grade C, all veneers being of species Group 1.
- 'Underlayment C–D plugged', with the face veneer grade C or grade C plugged of species Group 1, back veneer grade D of species Group 1 and the inner veneers of grade D and species Group 1 or 2.

- 'Underlayment C–C plugged', with the face veneer grade C or grade C plugged of species Group 1, back veneer grade C or C plugged of species Group 1 and the inner veneers of grade D and species Group 1 or 2.

BS 5268-2 requires that these plywoods must be marked to show compliance not only with the American standard but with the specific restrictions of the UK code, viz. the limitation on the timber species acceptable for Groups 1 and 2.

BS 5268 states that the C–D grade is not suitable for use either in service class 3 conditions or as gussets for trussed rafters. For all the plywood grades listed above there is a reduction in bending, tension and compression strengths for panel dimensions less than 600 mm varying linearly from zero reduction at 600 mm to 50% at 200 mm and less.

For further information refer to BS 5268-2 or contact the American Plywood Association.

1.3.5 Available sizes and quality of Swedish construction plywood

Sheet sizes of 2400 × 1200 mm and 2440 × 1220 mm are available. The face veneer is parallel to the longer side (Fig. 1.7). Spruce (whitewood) is mainly used in manufacture, but fir (redwood) is permitted.

There are several face qualities, the most common for structural uses being C/C. However, the important reference for structural purposes in the UK is 'P30 grade'. The figure 30 is the bending strength in N/mm^2 at the lower 5% exclusion limit. P30 can be obtained either unsanded or sanded but only unsanded values are given in BS 5268-2.

For further information refer to BS 5268-2 or the Swedish Forest Products Laboratory.

1.4 PARTICLEBOARD, ORIENTED STRAND BOARD, CEMENT-BONDED PARTICLEBOARD AND WOOD FIBREBOARDS

1.4.1 General

All the familiar British standards relating to these materials have been replaced by European standards. In each of the following material sections the closest possible BS EN material designation to the former BS material is given. This allows future work to be specified with the correct material grade and allows an

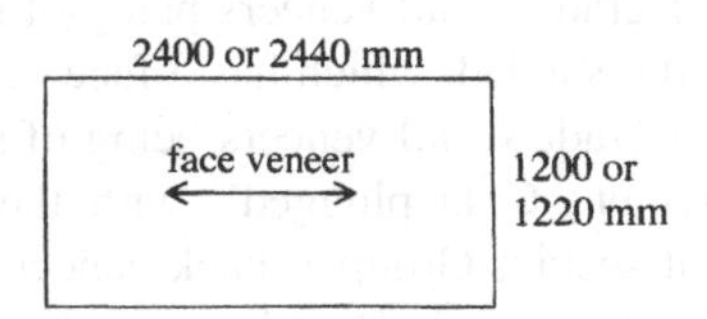

Fig. 1.7 Swedish softwood plywood.

historical comparison between, say, the new materials and those that were described some years previously.

These European standards give little if any advice on the use or application of the new materials. As a consequence, the user friendly advice given previously in BS 5669 'Particleboards' and BS 1142 'Specification for fibre building boards' has been redrafted and published in BS 7916 'Code of practice for the selection and application of particleboard, oriented strand board (OSB), cement-bonded particleboard and wood fibreboards for specific purposes'. Annex A of BS 7916 also gives the 'nearest fit' between the former BS board designations and the European standards designations. Further information on the effect of these changes is given in *Wood-Based Panels – Transition from British Standards to European Standards* published by the Wood Panels Industry Federation.

1.4.2 Environmental conditions

The environmental conditions in which the various types and grades of boards can be used are similar to solid timber but with a different terminology, i.e.

timber service class 1	is equivalent to *dry conditions*
service class 2	is equivalent to *humid conditions*
service class 3	is equivalent to *exterior conditions* (Note that none of the products covered in section 1.4 except cement-bonded particleboard meet the requirements for this condition.)

There are additionally ratings for biological hazard (risk of decay), viz.

class 1	undercover, fully protected and not exposed to wetting
class 2	undercover, fully protected where high environmental humidity can lead to occasional but not persistent wetting
class 3	not covered and not in contact with the ground; either continuously exposed to the weather or protected but subject to frequent wetting
class 4	in contact with the ground or fresh water and exposed to permanent wetting.

Formaldehyde-based bonding agents are frequently employed in the manufacture of these products. There are stringent rules in the European standards limiting the rate and quantity of formaldehyde emission from the products.

1.4.3 Structural usage

BS 5268-2: 1996 refers to one particleboard grade, C5, and one tempered hardboard grade, TE. The C5 grade particleboard has never been produced in commercial quantities and the hardboard found only limited use as wall sheathing and the webs of I beams.

Eurocode 5 admits certain particleboards, oriented strand boards, high-density medium fibreboards, tempered hardboards and medium-density fibreboards.

Reliable strength data values are not currently available for many of these products and likewise the present information on creep movements has to be treated with caution.

As the use of many of these products in the UK is for floor or roof decking and wall sheathing it is possible to use them quite satisfactorily with limited design data based on experience in use and proof by testing. For example, for domestic flooring a P5 or P7 grade particleboard should be 18/19 mm thickness where the supporting joists are spaced at 450 mm centres.

1.4.4 Particleboard

Particleboard is defined as a panel material manufactured under pressure and heat from particles of wood (wood flakes, chips, shavings, saw dust, wafers, strands and similar) and/or other lignocellulosic material in particle form (flax shives, hemp shives, bagasse fragments and similar) with the addition of an adhesive. Wood chipboard is a particular derivative of this family, being made from small particles and a binder.

The history of the use of particleboards in the UK for structural purposes and floor/roof decking in particular has not been good. Unless a degree of inherent moisture resistance is provided by the board there are potential problems in use from water spillage or condensation. Boards without moisture resistance properties will suffer irrecoverable swelling and loss of strength when wetted – a situation that may be described as 'the soggy Weetabix syndrome'. This condition is unknown in Scandinavia where fully moisture-resistant grades are available capable of total immersion for months without significant change of dimension or strength. This problem was overcome to some extent with the introduction in BS 5669 of moisture-resistant boards such as C3(M) and C4(M). The C5 grade board was intended to have moisture-resistant characteristics but in commercial terms the closest a designer could get to the BS 5268-2 requirements was C4(M).

The correlation between BS grades and the European grades for structural applications is given in Table 1.2.

Chipboard finds many applications as decking ranging from domestic applications to heavy duty decking in mezzanine floor construction in thicknesses from 18 mm to 38 mm. In all cases it is advisable to use a board suitable for humid conditions, i.e. the P5 or P7 type.

Table 1.2 Correlation of types of particleboard

British standards		European standards	
Use	BS type	BS EN type	Use
Load bearing (moisture resistant)	C4(M)	P5	Load bearing (humid conditions)
No equivalent	–	P6	Heavy duty load bearing (dry conditions)
Heavy duty load bearing (moisture resistant)	C5	P7	Heavy duty load bearing (humid conditions)

1.4.5 Oriented strand board

Oriented strand board (OSB) is a multilayer board made from strands of wood of a predetermined shape together with a binder. The strands in the external layers are generally aligned with the long edge of the board although alignment with the board width is also allowed. The strands in the centre layer or layers may be randomly oriented, generally at right angles to the external layers. The layup has a similarity to plywoods and the use for decking follows similar rules, i.e. the 'face grain' should lie perpendicular to the deck supports.

The equivalent grades of OSB are given in Table 1.3. Very little, if any, differentiation of the type of OSB was made in the application of BS 5669. The European classifications require more care in the specification of the material for a particular application.

OSB finds many applications as a substitute for plywood for decking or wall sheathing. As the number of logs available for peeling veneers for plywood falls with the passage of time, so OSB will become more widely used particularly as it is possible to obtain it in very large sheets, e.g. 4.8 × 2.4 m for wall sheathing.

Table 1.3 Correlation of types of oriented strand board

British standards		European standards	
Use	BS type	BS EN type	Use
General purpose (unconditioned)	F1	OSB/2	Load bearing (dry conditions)
Load bearing (humid conditions) (conditioned)	F2	OSB/3	Load bearing (humid conditions)
No equivalent	–	OSB/4	Heavy duty load bearing (humid conditions)

1.4.6 Cement-bonded particleboard

Cement-bonded particleboard is defined as a sheet material manufactured under pressure, based on wood or other vegetable particles bound with hydraulic cement and possibly containing additives.

Cement-bonded particleboard is specified in a single grade in the European standards. It is suitable as a decking and cladding in a humid environment and is finding application in external situations where the board is not immersed in water but subject to intermittent wetting. The drawback to the use of this type of board is its self-weight: at 1000 kg/m^3 it is much heavier than plywood, particleboard and OSB (commonly 700 kg/m^3).

The comparable grades are given in Table 1.4.

Table 1.4 Correlation of types of cement-bonded particleboard

British standards		European standards	
Use	BS type	BS EN type	Use
Internal use (dry conditions)	T1	–	No equivalent
Internal and external use	T2	OPC bonded	Cement-bonded particleboard (ordinary) Portland cement (dry, humid and exterior conditions)

1.4.7 Wood fibreboards

Wood fibreboards are a generic family of products manufactured from lignocellusosic fibres with the application of heat and/or pressure. Synthetic resin adhesives and other additives may be included in the process. The classification of the fibreboards is into two groups:

- *wet process* fibreboards, which have a moisture content of at least 20% at the time of manufacture;
- *dry process* fibreboards, where the moisture content is below 20% at manufacture and the density is greater than 450 kg/m^3.

Wet process fibreboards are further classified by density:

- *hardboards*, coded HB and having density greater than 900 kg/m^3, can be given additional properties such as fire retardancy, moisture resistance, resistance against biological attack, etc., by specific treatments such as 'tempering' or 'oil tempering' or the addition of adhesives or additives;
- *high-density medium boards*, coded MBH and density >560 kg/m^3 but ≤900 kg/m^3;
- *low-density medium boards*, coded MBL and density ≥400 kg/m^3 but ≤560 kg/m^3.

The medium boards have found application as a breathable wall sheathing material having a degree of structural strength that gives racking resistance and also allows use as the web member of a composite beam. As with the hardboards their properties can be improved during manufacture.

- *softboards*, coded SB and density ≥230 kg/m^3 but <400 kg/m^3, find application as thermal and sound insulation materials and, in addition to the improvements previously described for wet process boards, impregnation with a petrochemical yields the well-known bitumen impregnated fibreboard.

Dry process fibreboards, generally known as medium-density fibreboards (MDF), are also classified by density:

- *HDF*, an MDF board with density ≥800 kg/m^3
- *light MDF*, with density ≤650 kg/m^3
- *ultra-light MDF*, with density ≤550 kg/m^3

The properties of these boards can be modified during manufacture in a similar manner to the other fibreboards.

There is further classification and marking:

- for conditions of use (see section 1.4.2): dry no symbol
  humid H
  exterior E
- for application: general no symbol
  load bearing L
  load bearing for all durations of load A
  load bearing for instantaneous or short-term durations of load S

The load-bearing capacity is further classified as 1 for normal loading and 2 for heavy duty loading. The various symbols described are then combined to give a particular specification as shown below:

Fibreboard type: condition of use + application purpose + load duration + load-bearing category

(the last two are optional codes) so, for example, a heavy duty load-bearing hardboard for use in humid conditions for all types of loading is HB.HLA2.

1.4.7.1 Tempered hardboard

Sometimes known as 'oil tempered hardboard' the strength and durability characteristics are excellent. There are three drawbacks to its use. The movement with change in moisture content is very high so boards coming directly from manufacture at a low moisture content will pick up atmospheric moisture until the equilibrium moisture content is reached with a consequent expansion. The magnitude of this movement can cause buckling of the hardboard in wall panels of about 50 mm when fastened to studs spaced at 600 mm. It is often a user recommendation that boards should be conditioned by applying water and allowing the boards to reach equilibrium. The boards then tend to shrink 'drum tight' onto the supporting frame. The second problem is in hand nailing the boards. The board surfaces are so hard that starting a nail is difficult. Tempered hardboard has a smooth surface on one face and the other is 'rough' – technically described as the mesh face. It is somewhat easier to nail into the mesh face but mesh surface appearance is usually unacceptable. The third limitation is in board thickness. The maximum thickness is around 8 mm, thus limiting its application to wall sheathing or the webs of built-up beams.

The BS 1142 grade THE has been replaced by the BS EN 316 grade HB.HLA2, the code that was described in the previous paragraph. For further information refer to the Wood Panel Industries Federation.

1.5 ENGINEERED WOOD PRODUCTS

1.5.1 General

The various panel products described in the previous section are all made by reconstituting wood fibre. It is not surprising therefore that thoughts have turned to the manufacture of larger sections that can themselves be used as compression, tension and flexural members without any further manufacture or assembly such as required by a plywood box beam. The first of these products, Laminated Veneer Lumber (LVL), was simply the extension of plywood technology, whereas the two later products, Parallel Strand Lumber (PSL) and Laminated Strand Lumber (LSL), required the development of new techniques and processes. The procedures of reconstituting wood fibre in these different ways gives products that are so uniform and consistent in their mechanical properties that the variations normally associated with sawn timber can be ignored. For example, there is a single E value equivalent to E_{mean}, for the statistically derived E_{min} is practically no different from the mean value.

1.5.2 Laminated Veneer Lumber

The simplest description of Laminated Veneer Lumber (LVL) is a unidirectional plywood, i.e. the grain direction of all the veneers is the same. There are LVL products for specific applications where a small proportion of the veneers are set at right angles to the main body to give cross bonding for stability.

The standard dimensions available are widths up to 900 mm, thicknesses from 19 to 89 mm and lengths up to 20 m, although transport and handling becomes a problem over 12 m. The boards can be used on edge as a beam (similar to a vertically laminated beam with the laminates 3 to 4 mm thick) or flat as a plank particularly where concentrated wheel loads are applied. One of the problems with the beam application is that the potential lateral buckling instability for many applications requires a section perhaps 75 mm wide × 600 mm deep, i.e. a depth/breadth ratio of 8 to 1. The BS 5268-2 rule of thumb limit for this ratio is 7 to 1 so care has to be exercised in detailing the lateral restraints. As a plank, deflection can be a problem as the ratio of *E*/bending strength is typically 650 for LVL whereas softwood is 1100, so LVL used as a plank is more likely to be controlled by deflection than by bending strength.

The strength of LVL makes it particularly useful for I beams and built-up lattice trusses and girders. There is a single modulus of elasticity value due to its consistency (statistically the 5th percentile value is so close to the mean that differentiation would be unreasonable) and likewise the load-sharing factor, K_8, has the value 1.04. The end connections can be a problem since, with split rings, nail plates and such items that do not have full penetration through the section, there is a potential failure plane at each glueline. The cross-banded LVL finds particular application in these circumstances. Joints should be designed assuming the C27 strength class values of BS 5268-2 with restriction on the diameter of nail or screw when driven parallel to the glue line.

EN standards are being drafted for the production and use of LVL. All the present products in the market place, e.g. Kerto and Microllam, have Agrement certificates describing the product and giving design information.

1.5.3 Parallel Strand Lumber

Parallel Strand Lumber (PSL) is made by a patented process and marketed by Trus Joist MacMillan as Parallam. The process starts by peeling veneers in a similar way to plywood from Douglas fir, Southern pine or Western hemlock, drying and then grading to remove excessive defects. The veneers are clipped into strands 13 mm or so in width × 3 mm thickness up to 2.5 m long. These strands are then coated with a phenol formaldehyde adhesive, the strands are aligned and pressed and the adhesive cured at high temperature to form a billet. The resulting product is not solid as there are voids between the compressed, bonded strands.

The standard size of billets are 68, 89, 133 and 178 mm in thickness and 241, 302, 356, 457 and 476 mm depth. The maximum size of billet that can be produced is 285 × 488 mm × 20 m length.

The moisture content on completion of the production of PSL is approximately 10% and as the dimensional changes arising from exposure to high atmospheric relative humidity are small the material may be considered dimensionally stable in most environments. On the other hand, prolonged exposure to liquid water – for example, from rain or a leaking pipe – can cause dimensional changes of up to 12% perpendicular to the axis of the strands and 5% in the direction of the strands due to water penetration into the voids in the material. Recovery on drying from this wet condition will leave a residual, permanent swelling of 50% of the maximum value. It follows that PSL used in an external environment must be protected from the direct action of the weather.

In all other ways PSL can be considered as a stronger more consistent form of the parent timber. Like LVL there is a single modulus of elasticity value and because of its consistency the load-sharing factor, K_8, has the value 1.04. Otherwise the design of PSL follows the procedures of BS 5268-2. Joints should be designed assuming strength class C27.

Because PSL is the product of a single manufacturer located outside the European Union, it is unlikely that EN standards will be drafted for the product. Reliance currently has to be placed on an Agrement certificate for the properties of the material.

1.5.4 Laminated Strand Lumber

Laminated Strand Lumber (LSL) is made by a patented process and marketed by Trus Joist MacMillan as TimberStrand. Logs are soaked, debarked and flaked into strands about 220 mm long and 1 mm thick. The strands are dried, coated with an isocyanate (polyurethane) adhesive, aligned and deposited as a mat 2.5 m wide × 10.7 m long. This is heated under pressure to cure the adhesive and creates a billet which is then sanded and trimmed to the required size.

The billet sizes created are 2.4 m × 10.7 m in thicknesses for structural use ranging from 30 to 140 mm. The timber species used is Aspen, which is a fast-growing timber of relatively poor structural quality. As solid timber, LSL has more consistent, higher properties than the parent material. Aspen is one of the least durable timbers, so exposure to environments that could promote insect or fungal attack should be avoided.

The material is more akin to OSB in that it does not have voids as in PSL. It is dimensionally stable and starts life at approximately 8% moisture content. The result of prolonged exposure in a high humidity environment is to take up moisture through the edges rather than the top and bottom surfaces. As a consequence, dimensional change is more noticeable at the corners of a sheet where moisture take up is through two edges.

LSL, like PSL, has a single modulus of elasticity value and the consistency is reflected in a K_8 value of 1.04. Joint design is somewhat more complicated than with the other two engineered wood products as limitations are imposed by the Agrement certificate on the fastener diameter, type and orientation to the top and bottom surfaces of the board. For example, bolts and dowels up to 12 mm diameter should be used with C24 values but not into the edges of a board, and for diameters greater than 12 mm C27 values should be used but not into the edges of a board.

For the same reasons as PSL, the only source of technical information for designs in the UK and Europe are an Agrement certificate.

1.6 MECHANICAL FASTENERS

1.6.1 General

There are several mechanical fasteners which can be used in timber constructions. Some are multipurpose fasteners, and some are specially produced for timber engineering. These are described briefly here and covered in more detail in Chapter 18.

1.6.2 Nails

Nails are satisfactory for lightly loaded connections where the nails are in shear. Ordinary nails used structurally in the UK are usually circular as they are cut from wire coil. Nails can be unprotected or treated against corrosion, and, increasingly, use is being made of stainless steel for nail manufacture. Nails will slightly indent the timber when loaded in shear. This is not usually serious but must be appreciated because, for example, in a stress skin panel, a fully rigid joint cannot be claimed between the plywood skin and the timber joists if the joint is made only by nails. Nails can be used to give close contact during curing of a glued–nailed joint. Withdrawal loads are allowed for nails driven at right angles to the side of timber but no allowance can be made when driven into end grain. Nails can be driven mechanically from 'nailing guns'.

1.6.3 Improved nails

Improved nails are used to a limited extent. Square nails (twisted or untwisted) are permitted to take higher lateral loads than round wire nails as the 'working' to form the nail improves the strength of the steel. Annular-ringed shank or helical-threaded shank nails are permitted to take higher withdrawal loads, and can be useful in situations where tension or vibration can occur during construction, such as will occur during the construction of the membrane of a shell roof. Improved nails can be unprotected or treated against corrosion and are available in stainless steel. Like ordinary round wire nails they do not give a fully rigid joint.

1.6.4 Staples

Staples can be used for lightly loaded connections between a sheet material and solid timber, and to give close contact during curing of a glued joint. It is important that they are not over-fired by a stapling gun. Their use in environments where corrosion of the steel can occur is not recommended as there is a high stress concentration due to the forming of the staple at the shoulders. The combined effects of stress corrosion can lead to fracture at the shoulder resulting in the fastening becoming two slender pins.

1.6.5 Screws

Conventional wood screws manufactured to BS 1210 require preboring and are therefore very much slower to insert than nails or improved nails. Newer types of screw with a modified thread pattern may be described as 'self-tapping' and do not require preboring. The use of screws is therefore increasing in general applications as well as special situations, for example, where security against vibration is required or demountability is a feature. Screws are available in various finishes or materials to cover different environmental conditions.

1.6.6 Bolts and dowels

The bolts and dowels usually used in timber engineering are of ordinary mild steel, either 'black' or galvanized. Part of the strength of a bolted joint is achieved by the bearing of the bolt shank onto the timber, therefore there is no advantage in using friction-grip bolts except where a special joint can be designed as timber shrinkage could negate the clamping force completely. The washers used within bolted connections should have a diameter of three times the bolt and a thickness of 0.25 × bolt diameter (Form G washers to BS 4320) under the head and nut. If a bolt is galvanized the designer should ensure that the nut will still fit. It may be necessary to 'tap' out the nut or 'run down' the thread before galvanizing. Stainless steel bolts may also be used in corrosive environments.

1.6.7 Toothed plate connector units

The relevant British standard is BS 1579. Toothed plate connectors (see Fig. 1.8) can be either single or double sided, square or round, and can be placed on one or both sides of the timber. This type of connector is always used in conjunction with a bolt, complete with washers or steel plate under head and nut. The flat part of the connector sits proud of the timber.

The hole in the connector is a tolerance hole, therefore with single-sided connectors there is always bolt slip. Using double-sided connectors it may be possible with careful detailing to eliminate slip in the connection. The British standard

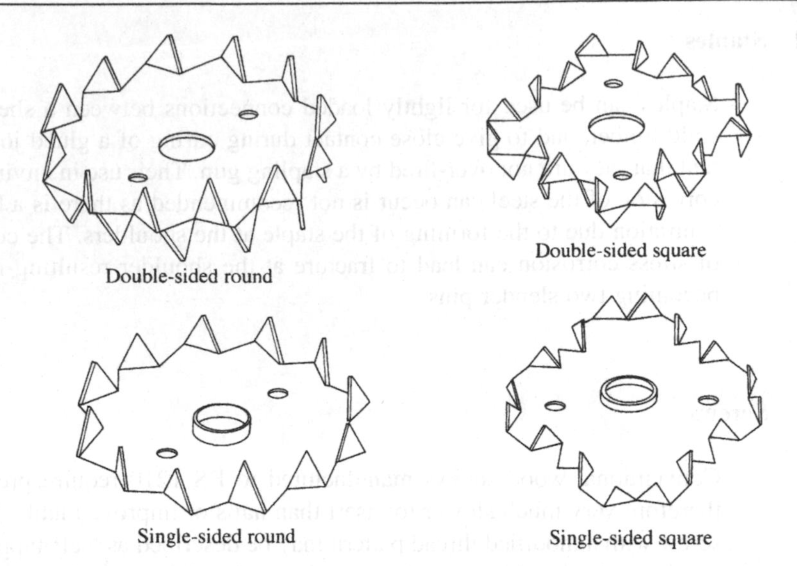

Fig. 1.8 Toothed plate connectors.

calls for an anti-corrosion treatment, but does not specify which one. This is left to the individual manufacturers.

BS 1579: 1960 lists connector reference numbers for each connector detailed in the standard, but this classification system is not used widely enough to enable a specifier to use it and rely on its being understood by all concerned. The standard has not been metricated but BS 5268-2 uses metric equivalents in referring to connectors.

1.6.8 Split ring connector units

The relevant British standard is BS 1579. A split ring may have bevelled sides (Fig. 1.9) or parallel sides. It takes shear and necessitates the timbers being held together by another connector – usually a bolt (Fig. 1.10). A small amount of joint slip is likely to take place. The British standard requires the connectors to have an anti-corrosion treatment. A special tool is required to groove out the timber for the shape of the split ring and a special 'drawing' tool is required to assemble the unit under pressure. Units can carry a relatively high lateral load.

1.6.9 Shear plate connector units

The relevant British standard is BS 1579. Shear plate connectors are of pressed steel (Fig. 1.11) or malleable cast iron (Fig. 1.12). A special cutter is required to cut out the timber to take each connector but no tool is required to draw the timber together. A connector surface is flush with the surface of the timber. The timbers must be held together with a bolt. A small amount of joint slip is likely to take

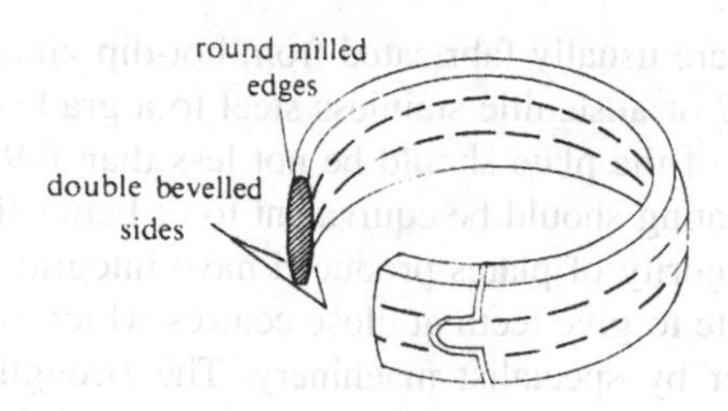

Fig. 1.9 A split ring connector.

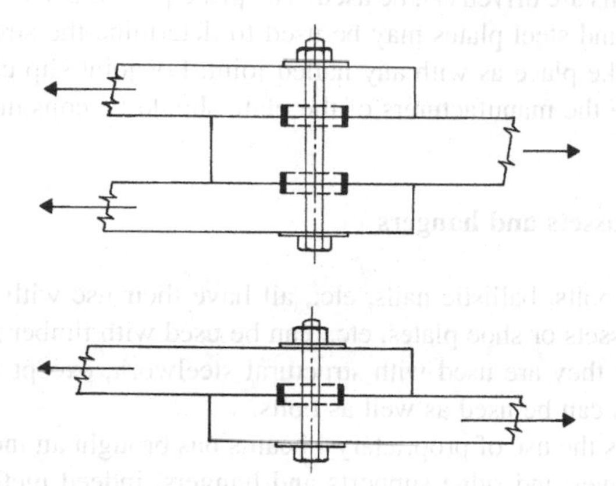

Fig. 1.10 A bolt connector.

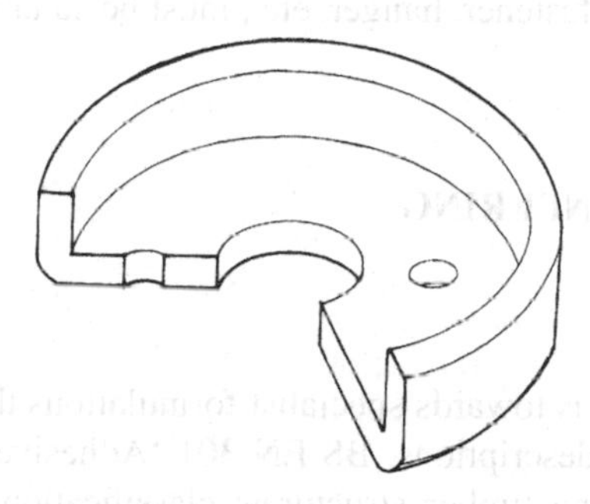

Fig. 1.11 Pressed steel shear plate 67 mm outside diameter.

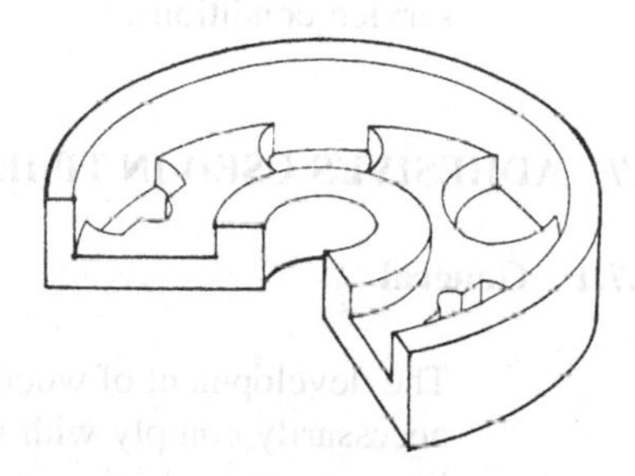

Fig. 1.12 Malleable cast iron shear plate 102 mm outside diameter.

place. Shear plate connectors are located in place for transit with two locating nails. The connectors must be given an anti-corrosion treatment.

1.6.10 Punched metal plate fasteners

Most applications of punched metal plate fasteners are for the manufacture of trussed rafters. The requirements and specifications given in the following are based on BS 5268-3 'Code of practice for trussed rafter roofs'. Punched metal plate fasteners are usually fabricated from hot-dip zinc-coated steel sheet or coil to BS EN 10147 or austenitic stainless steel to a grade defined in BS EN 10088-2. The thickness of the plate should be not less than 0.9 mm and not more than 2.5 mm. A zinc coating should be equivalent to or better than Z275 of BS EN 10147.

The majority of plates produced have integral teeth pressed out at right angles to the plate to give teeth at close centres which will subsequently be pressed into the timber by specialist machinery. The strength capacities of the various nail configurations are determined by test and published in the plate manufacturer's Agrement certificate. As an alternative, plates with pre-punched holes through which treated nails are driven can be used. The principles of BS 5268-2 with regard to nailed joints and steel plates may be used to determine the strength of a joint. Joint slip will take place as with any nailed joint. For joint slip characteristics of a particular plate the manufacturers of the plate should be consulted.

1.6.11 Other fasteners, gussets and hangers

Rag bolts, rawl bolts, ballistic nails, etc., all have their use with connections for timber. Steel gussets or shoe plates, etc., can be used with timber in a similar way to that in which they are used with structural steelwork, except that in this case nails and screws can be used as well as bolts.

In recent years the use of proprietary I beams has brought an increased range of joist hangers, eaves and ridge supports and hangers, indeed methods of jointing timber members for almost any imaginable situation (for example, seismic conditions). These products are the result of extensive testing, confirmed by use in practice. Obviously the finish of any fastener, hanger, etc., must be suitable for the service conditions.

1.7 ADHESIVES USED IN TIMBER ENGINEERING

1.7.1 General

The development of wood adhesives is towards specialist formulations that do not necessarily comply with traditional descriptions. BS EN 301 'Adhesives, phenolic and aminoplastic, for load-bearing timber structures; classification and performance standards', as its title suggests, is moving towards performance rather than prescriptive specifications. Adhesive manufacturers produce products to their own formulations but from the point of view of a structural designer these fall into three main categories, as described in sections 1.7.2–1.7.4. See Chapter 19 for detailed information on glue joints.

1.7.2 Weatherproof and boil proof glues

When a component or joint is likely to be directly exposed to weather, either during its life or for lengthy periods during erection, it is necessary to use a BS EN 301 type I adhesive which can match the 'weatherproof and boil proof' (WBP) requirements of BS EN 391. These glues are normally gap-filling thermosetting resorcinol–formaldehyde (RF), phenol–formaldehyde (PF) or phenol/resorcinol–formaldehyde (PF/RF) resin adhesives. They are required where there is high hazard from weather, temperatures approaching 50 °C or higher, high humidity, or a chemically polluted atmosphere. They are often used (sometimes with a filler) for structural finger jointing. These glue types require careful quality control during storage, mixing, application and curing, and are therefore normally intended for factory use.

1.7.3 Boil-resistant or moisture-resistant glues

When a component is unlikely to be subjected to any serious atmospheric conditions once in place, and is only likely to receive slight wettings during transit and erection, it may be possible to use an adhesive which matches the 'boil resistant' (BR) type formerly defined in BS 1204: Part 1: 1979. These are normally gap-filling melamine/urea–formaldehyde (MF/UF) or certain other modified UF adhesives which can match the BR or MR requirements. They are usually cheaper than WBP adhesives. For certain low hazard situations it might be possible to use an adhesive which matches only the 'moisture-resistant and moderately weather-resistant' (MR) requirements of BS 1204: Part 1. These are normally urea–formaldehyde (UF) adhesives, perhaps modified, to match the MR requirements. They require careful quality control during storage, mixing, application and curing, and are therefore normally intended for factory use.

1.7.4 Interior glues

A structural component is not normally bonded with an adhesive which matches only the 'interior' (INT) requirements of BS 1204: Part 1. There are exceptions, perhaps where a site joint is required or where control of the gluing cannot be guaranteed to a sufficient degree to enable a WBP, BR or MR glue to be used. An 'interior' glue should be used only when the designer can be certain that the glue joint will not be subjected to moisture or high temperature in place.

A typical non-structural adhesive commonly available in the market place is polyvinyl acetate (PVA). This adhesive is subject to high creep and deterioration in a moist environment. Developments of the internal bonding characteristics of PVA have led to the 'cross-linked PVA' that has good weathering characteristics but is slightly more flexible ('rubbery') than the traditional wood adhesives. It therefore has application for joinery rather than structural members but it is an indication of future adhesive development and the reasons for moving towards performance specifications.

1.7.5 Epoxy resins

Epoxy resins are finding increasing application in timber engineering, particularly in the bonding of steel and timber. They can be extremely useful in localized situations such as ensuring true bearing in a compression joint between timber and a steel bearing plate or, for example, in sealing the possible gaps between a timber post and a steel shoe. To reduce the cost in such compression joints the resin can be mixed with an inert filler such as sand.

1.7.6 Gluing

Gluing, including gluing of finger joints, is described in more detail in Chapter 19. Basically the quality control requirements to obtain a sound glue joint are correct storage, mixing and application of the adhesive, correct surface conditions, moisture content and temperature of the timber, and correct temperature of the manufacturing environment during application of the adhesive and during curing. The instructions of the adhesive manufacturer must be followed with regard to adhesive mixing, pot life, open assembly time, closed assembly time, setting and curing periods. The joints must be held together during gluing, either by externally applied pressure (usually necessary where solid timber is being glued in order to achieve close contact – the pressure for softwoods is usually quoted as $0.7\,N/mm^2$) or by nails or staples (glued–nailed construction for bonding sheet material to timber with the fasteners left in, although their only duty is to hold the surfaces in close contact during curing, the design shear force being taken only by the adhesive).

Adhesives and the use of adhesives must be compatible with any preservative used. Water-borne preservative will increase the moisture content of the timber quite considerably and may raise grain. If the timber is to be glued after treatment (and re-drying) it is likely that the surfaces will have to be processed, even if lightly, which will remove some of the preservation. If a fire retardant containing ammonia or inorganic salts is to be used, then gluing by resorcinol types (and perhaps others) should not take place after treatment, nor should treatment take place until at least seven days after gluing. Certain organic solvent preservatives contain water-repellant additives which may affect the bonding characteristics.

For certain components or joints, particularly in a mass production situation, accelerated curing by radio frequency heating of the gluelines may be economical. Alternatively, a heated environment perhaps at 60 °C can be created to reduce the setting and curing times.

Chapter 2
Stress Levels for Solid Timber

2.1 INTRODUCTION

To achieve the most economical design of timber components it is necessary to understand the way in which the strengths and E values for timber are derived. This is explained in this chapter in relation to the grade stresses quoted in BS 5268-2:2002. For further information the reader is referred to the publications of the Building Research Establishment.

Comment is often made on the variable nature of timber, but it must be emphasized that the properties and characteristics of most commercial timbers have been obtained from extensive testing, so the variability can be established by statistical methods. The methods of arriving at strength values used in the design of structural components give much higher factors of safety than the factors normally required for structures as a whole.

There is no comparable margin of safety in the modulus of elasticity or shear modulus of timber and other wood-based products. When coupled with the possible variations in loading and time-dependent movements such as creep and shrinkage, deformations and deflections calculated on this information are at best an estimate or prediction of what may happen in reality. The various Building Regulations in use in the UK are concerned with the safety of people, so adequate margins of safety with regard to the strength of a component or construction are essential, but the same criteria only rarely apply to deflection. The most likely situation where deflection could affect the strength of a structure would be, say, a statically indeterminate structure where undue deformation or deflection of a timber member could affect a support condition. Calculations that give the deflection of a timber member to perhaps 0.01 mm can at best be described as 'optimistic'.

All materials are variable and an understanding of the method of deriving grade stresses and E values for timber can often help in an appreciation of the performance in practice of other materials. A close statistical examination of the properties of other materials will often reveal them to be more variable than timber even though the presumption is they have little if any variability!

The statistical analysis of the results from tests of commercial sizes of timber with the corresponding measurements of actual defects has led to the setting of strength grades which can be selected by practical visual methods. Such strength grades are described in BS 4978 'Specification for visual strength grading of softwood' and are given the titles 'General Structural' (GS) and 'Special Structural' (SS). Strength grading by machine is carried out to the requirements of BS EN 519

'Structural timber – Grading – Requirements for machine strength graded timber and grading machines'. Visual grading relates the sizes of defects in the timber to limiting values for each grade. The grade, when associated with a particular timber species, gives the grade stress values in BS 5268-2. For example, a particular knot size in Western red cedar will give the same visual grade as the same sized knot in Douglas fir timber, but a lower grade stress because of the inherent difference in the basic strengths of the two timbers. Machine grading on the other hand effectively assigns strength values directly to the timber in the process of grading.

2.2 DERIVATION OF BASIC STRESS AND CHARACTERISTIC STRENGTH VALUES

2.2.1 Background

Originally the mean and basic stress values were obtained from the statistical analysis of test results from small, clear, green (moisture content above the fibre saturation point) specimens. This procedure was replaced with the testing of full-sized, graded pieces of timber, a practice which is continued today. Since the introduction of full size 'in grade' testing the only changes have been to the statistical methods of analysis employed.

2.2.2 The UK methods up to 1984

Values were based on the 'basic stress', i.e. 'the stress which could safely be permanently sustained by timber containing no strength-reducing characteristics'. The basic stress was then modified to allow for the particular grade characteristics and moisture content. The basic stresses were derived from small (20 mm × 20 mm cross-section), clear (no strength-reducing defects), green (moisture content above the fibre saturation point) specimens. Many hundreds of samples of a particular species would be tested. Because of the 'uniformity' of the sample, statistical analysis was possible assuming a normal, Gaussian distribution.

A well-quoted example of tests on small clear specimens which illustrates this method for deriving basic stresses was carried out at the Forest Products Research Laboratory, Princes Risborough (now incorporated in the Building Research Establishment) to establish the modulus of rupture of 'green' Baltic redwood. Two thousand, seven hundred and eight specimens, each 20 mm square × 300 mm long, were tested in bending under a central point load. (BS 373 detailed the standard test method.) The failure stresses for all specimens were plotted to form a histogram as shown in Fig. 2.1. This demonstrates the natural variation of the modulus of rupture about the mean value for essentially similar pieces of timber. Superimposed upon the histogram is the normal (Gaussian) distribution curve and this is seen to give a sufficiently accurate fit to the histogram to justify the use of statistical methods related to this distribution to derive basic stresses. In the example shown, the mean modulus of rupture for the green timber is 44.4 N/mm^2 and the standard deviation is 7.86 N/mm^2.

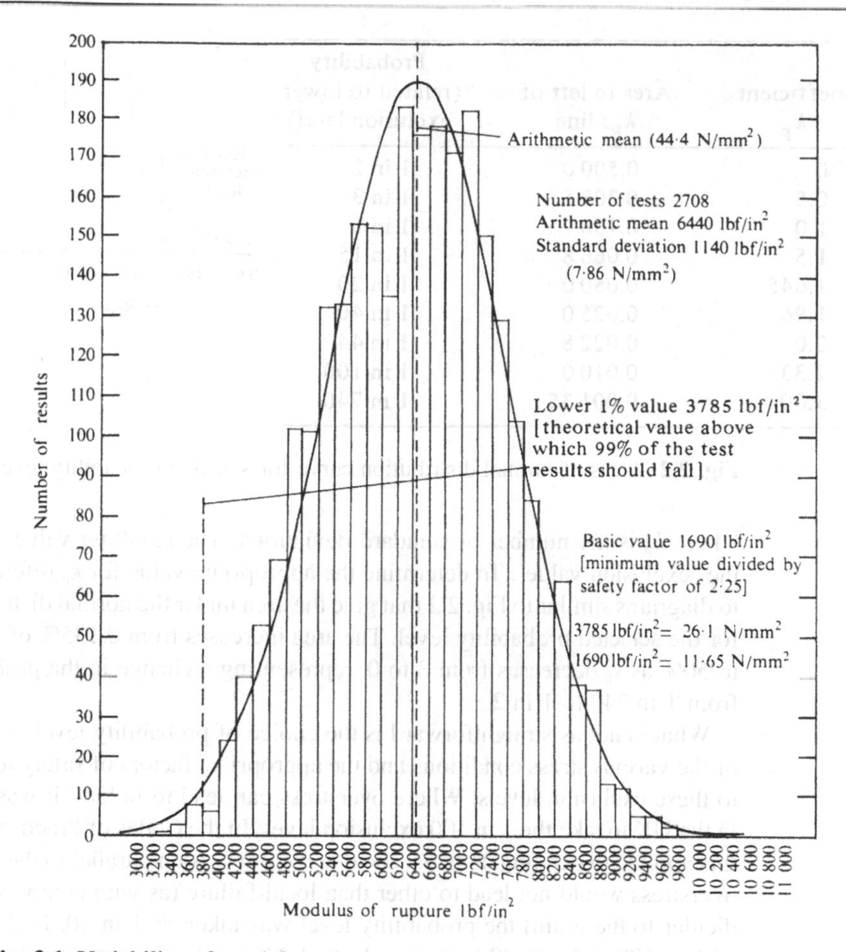

Fig. 2.1 Variability of modulus of rupture of wet Baltic redwood.

In mathematical terms the standard deviation of a normal distribution is given by

$$s = \sqrt{\frac{\sum (x - \bar{x})^2}{N - 1}}$$

where s = standard deviation
x = individual test value
$\bar{x}$ = mean of the test values
N = number of tests.

The normal distribution curve extends to infinity in either direction but for practical purposes it may be regarded as terminating at three times the standard deviation on each side of the arithmetic mean. Likewise it is possible to set probability levels below which a certain proportion of the overall sample would be expected to lie. For example, the exclusion value is given by

$$\sigma = \bar{x} - k_p\, s$$

Coefficient k_p	Area to left of $k_p s$ line	Probability (related to lower exclusion level)
0	0.500 0	1 in 2
0.5	0.308 5	1 in 3
1.0	0.158 7	1 in 6
1.5	0.066 8	1 in 15
1.645	0.050 0	1 in 20
1.96	0.025 0	1 in 40
2.0	0.022 8	1 in 44
2.33	0.010 0	1 in 100
3.00	0.001 35	1 in 740

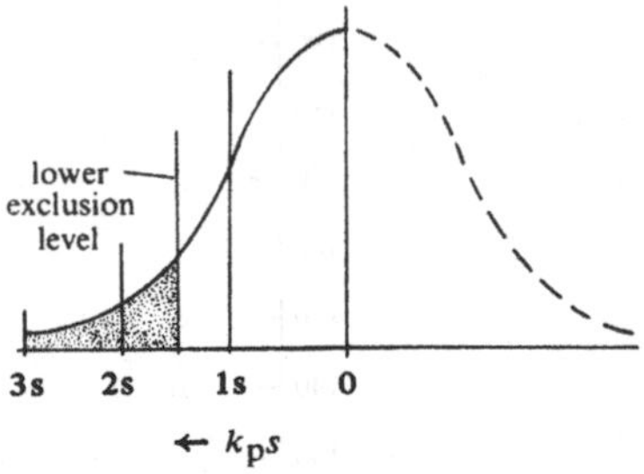

Fig. 2.2 Areas of normal distribution curve for selected probability levels.

where k_p is the number of standard deviations. The resulting value, σ, is termed the 'exclusion value'. To determine the appropriate value for k_p reference is made to diagrams similar to Fig. 2.2 that give the area under the normal distribution curve for the selected probability level. The area increases from 0.135% of the total area to 50% as k_p decreases from 3 to 0, representing a change in the probability level from 1 in 740 to 1 in 2.

What is not as straightforward is the choice of probability level to take account of the various stress conditions and the appropriate factors of safety to be assigned to these exclusion levels. Where overstress can lead to failure, it was the practice in the UK to take the 1 in 100 exclusion level. In the Codes of Practice before 1984 this applied to bending, tension, shear and compression parallel to the grain. Where overstress would not lead to other than local failure (as with compression perpendicular to the grain) the probability level was taken as 1 in 40. Probability levels of 1 in 100 and 1 in 40 have k_p values of 2.33 and 1.96 respectively.

These statistical basic stress values then had to be further modified to account for the dry condition (at that time 18% moisture content), size (conversion of the 20 mm size to 300 mm) and defects that set the particular grade. To this was added a further 'safety' factor by simply rounding the other factors into a practical number.

2.2.3 The methods in BS 5268-2: 1984

The testing of structural size members during the 1970s had shown a more realistic assessment of strength than small clear test pieces, therefore in BS 5268-2:1984 the strength values for visual grades were derived from full-sized tests, or, where there was insufficient data, assessments of full size values were made from the data for small clear specimens. Machine grades still relied on the relationship between the modulus of rupture and the modulus of elasticity to arrive at settings for grading machines. For the visual grades the exclusion level (as defined by the probability level) was set at 1 in 20 or the 5th percentile value for all properties. This value was, and still is, termed the *characteristic* value.

The characteristic bending strength and modulus of elasticity values were derived by test on SS grade timber from various common species. Other characteristic strength properties and other grade values were derived by proportion from these test values. Because the method of sampling produced statistically 'skewed' populations, the statistical analysis became complicated to the point where it ceased to be comprehensible to the wider audience to whom the previous normal distribution procedure had been fairly straightforward.

The characteristic grade strengths were then further modified for duration of load (converting the relatively short-term test value to a long-term value), for size (converting the full-size test value for, say, a 150 mm deep joist to the standard 300 mm depth) and a safety factor that included an allowance for the change from the 1 in 100 exclusion level in the previous Code to 1 in 20 in 1984.

2.2.4 The methods in BS 5268-2: 1996

The 1996 revision of BS 5268-2 brought in the European standards for solid timber. This included the statistical analytical methods given in BS EN 384. By comparison with all previous methods, including the normal distribution, the ranking method described in BS EN 384 for modulus of rupture and density, is simple: place the results in ascending order with the 5th percentile being the value for which 5% of the test values are lower. Adjustment is then made for size, moisture content and duration of load in a manner similar to the previous editions of CP 112 and BS 5268. The mean value of modulus of elasticity is determined by test. From these three test values the characteristic values for other properties are determined as follows:

tensile strength parallel to grain	$f_{t,0,k}$	$= 0.6 f_{m,k}$
compressive strength parallel to grain	$f_{c,0,k}$	$= 5(f_{m,k})^{0.45}$
shear strength	$f_{v,k}$	$= 0.2(f_{m,k})^{0.8}$
tension strength perpendicular to grain	$f_{t,90,k}$	$= 0.001\rho_k$
compressive strength perpendicular to grain	$f_{c,90,k}$	$= 0.015\rho_k$
modulus of elasticity parallel to grain for softwood species	$E_{0,05}$	$= 0.67E_{0,mean}$
modulus of elasticity parallel to grain for hardwood species	$E_{0,05}$	$= 0.84E_{0,mean}$
mean modulus of elasticity perpendicular to grain – softwoods	$E_{90,mean}$	$= E_{0,mean}/30$
mean modulus of elasticity perpendicular to grain – hardwoods	$E_{90,mean}$	$= E_{0,mean}/15$
shear modulus	G_{mean}	$= E_{0,mean}/16$

2.3 MODULUS OF ELASTICITY AND SHEAR MODULUS

There is no factor of safety on the modulus of elasticity. In BS 5268-2 two values of E are tabulated for the strength grade of each species and for each strength class. These are E_{mean} and the characteristic value $E_{minimum}$ ($E_{0,05}$). They are 'true' values of E (i.e. they do not contain an element of shear modulus) therefore deflection calculations must include shear deflection as well as bending deflection.

E_{mean} is the arithmetic mean from a number of test results, so BS 5268-2 permits this value to be used in a load-sharing situation such as wall studs or floor joists where four or more members act together to support a common load. By and large, the use of E_{mean} in these circumstances leads to acceptable components. It can be demonstrated statistically that a far greater number of pieces would have to act together to justify the use of E_{mean}. However, authoritative span tables for joists and rafters have for many years been based on E_{mean} without undue or adverse effects.

E_{min} is a statistical minimum value which, for the purposes of BS 5268, is set at the lower 5th percentile level. BS 5268-2 requires E_{min} to be used in designs where a single piece of solid timber acts alone to support a load and in stability calculations, for example, the buckling of a column.

Where 2, 3 or 4 or more pieces are joined together to act as one, E_{min} may be increased by the factor K_9 for sawn softwood timber that is nailed, screwed or bolted together and K_{28} for vertically laminated softwood timber which for 4 or more pieces have a value of 1.24. It can be shown that

$$E_N = E_{mean} - \frac{1.645s}{\sqrt{N}}$$

where s is the standard deviation and 1.645 is k_p for the 5th percentile value. Hence, knowing that $E_{min} = 0.67E_{mean}$, i.e. when $N = 1$, then

$$s = \frac{0.33E_{mean}}{1.645} = 0.201E_{mean}$$

For $N = 2$, then

$$E_N = E_{mean} - \frac{1.645 \times 0.201E_{mean}}{\sqrt{2}} = 0.766E_{mean} \quad \text{or} \quad \frac{0.766}{0.67}E_{min}$$

Thus, K_9 and $K_{28} = 1.143$ compared with the BS 5268-2 value of 1.14. Likewise for $N = 4$ the calculated value of K_9 and K_{28} by this method is 1.243 and the tabulated value 1.24. While K_9 has a maximum value for 4 pieces, K_{28} is tabulated for up to 8 laminates. It can be shown by calculation that the K_{28} value is 1.32, similar to the BS 5268-2 tabulated value.

2.4 GRADE STRESS

2.4.1 General

Grade stress is defined as 'the stress which can safely be permanently sustained by material of a specific section size and of a particular strength class, or species and (strength) grade'. The reference to section size is necessary because BS 5268-2 has a depth factor for solid timber in bending and a width factor for solid timber in tension. The tabulated grade stresses for bending and tension are based on a depth of 300 mm for bending and the largest dimension of a tension member also being 300 mm. The strength values for other sizes are modified by K_7 and K_{14} for bending and tension. These factors apply to both solid timber and glulam.

2.4.2 North American timbers

The depth and width factors described in section 2.4.1 also apply in BS 5268-2: 1996 to North American timbers where traditionally the strength values of framing grades were based on an 89 mm dimension. Design using these timbers is now similar in procedure to European timbers. One aspect of this standardization that should be used with extreme caution is the grade stress values in tension for the lower North American grades such as Structural Light Framing No. 3, Light Framing Utility and Stud where previously no tension value was given. The North American rules for these grades are related more to compression than to flexural or tension members.

2.4.3 Grade stresses for compression perpendicular to the grain

Two values are given for compression perpendicular to the grain in the strength class table (Table 7) of BS 5268-2. The first assumes no wane while the second gives a reduced value allowing for the deduction in the bearing width approximately equal to the maximum wane allowed. All the other grade stress tables give a bearing strength that includes the effects of wane and allows the designer to increase the bearing strength value by 33% if it is known that wane will not be present at the bearing. This makes bearing design simple in that the only decision is whether wane will or will not be present. The bearing capacity is then calculated on the target size of the bearing surface with the appropriate compression perpendicular to the grain grade stress.

2.4.4 Grade shear stress

The usual argument with regard to shear stress is the presence of splits and fissures at the position of maximum shear. It is unlikely that a timber member will be loaded in such a manner that failure in shear could occur. In simple terms, if a beam is split horizontally in half by a fissure, then the shear stress in the full section would be

$$\text{shear force} \times 1.5/(b \times h)$$

and in the two beams of depth $0.5h$ would be

$$0.5 \times \text{shear force} \times 1.5/(b \times 0.5h)$$

which gives identical shear stress values. Obviously the method of support has to allow the load applied to the whole beam to reach the two smaller members, which would occur with load applied to the top surface of the whole beam and bearing on the underside. Taking a very pessimistic view that the shear force is carried on a breadth equivalent to the beam breadth less the depth of fissure, it will rarely exceed the allowable shear stress.

2.5 LOAD SHARING

2.5.1 General

Load sharing comprises two distinctly separate processes. Firstly, the statistical basis discussed in section 2.3 allows improvement in strength for a number of pieces of timber acting together. Secondly, in such a group where there is means of distributing load between individual members by, say, a head binder in a wall, then the effects of distribution will be to transfer a higher proportion of load to the stronger and presumed stiffer members.

It is recognized in loading codes that even though loadings are quoted as the load per unit area, this uniformly distributed load is far from uniform in its distribution. There are statistical analysis techniques that allow consideration of the probability of loading occurring. Load sharing implies loads coming from a larger area than that on a single element of the structure and hence the probability that the average load per element will be lower.

Many sophisticated theories have been put forward in an attempt to quantify the enhancement of performance under load. Some give values as low as 2 or 3% increase and others as much as 40%. What none can assess are the secondary effects in a loaded structure of continuities and fixities where simply supported spans and pinned joints are assumed for calculation purposes.

Load sharing does occur and the simplest method for the designer is the 10% increase provided by the K_8 factor.

2.6 MOISTURE CONTENT

The European standards introduced in 1996 achieved a major breakthrough with regard to the moisture content used as a basis for determining strength properties and dimensions of solid timber. Before 1996 the dry strength values given in BS 5268-2 were at 18% moisture content, the North American dry values were at 19% and the UK standards for measuring dimensions of timber were set at 20%. The introduction of service classes has rationalized the most widely used service class 2 at 20%, the same as the measurement standard BS EN 336. The North American dry condition thus remains at 19%, which is of little consequence. At least consistency has been reached with European timber.

As described in section 1.2.5.2, there is a lower service class 1 where the equilibrium moisture content is likely to be 12%. This is about equivalent to the environment created in a well-heated and thermally insulated dwelling. With the construction of four-, five- and six-storey timber-framed buildings attention is being paid to limiting shrinkage movement of the timber by using super dried timber for floor joists in particular at a moisture content of 12%. There is some trade off against the extra cost for drying this timber by increased strength values.

The relationship of strength values and E to moisture content (m.c.) is given by

$$\log_n \sigma_1 = \log_n \sigma_s + C(M_s - M_1) \quad \text{and} \quad \log_n E_1 = \log_n E_s + C(M_s - M_1)$$

where σ_1 and E_1 are the strength and E values at the required moisture content M_1
σ_s and E_s are the strength and moisture content at fibre saturation point M_s (say 27% m.c.)
C is a constant found by substituting known values of σ and E at, say, 20% m.c.

So, for C24 timber with σ_m at 20% (= 7.5 N/mm^2) and σ_m at 27% (= 7.5 × 0.8 (K_2) = 6.0 N/mm^2):

$$C = \frac{\log_n 7.5 - \log_n 6.0}{(27 - 20)} = 0.032$$

For M_1 = 12%

$$\log_n \sigma_1 = \log_n 6.0 + 0.032\,(27 - 12) = 2.272$$

so σ_m at 12% = 9.70 N/mm^2, which is an increase of 29%. This procedure is worth pursuing where it is known that moisture contents will be equivalent to service class 1.

Only dry stresses are dealt with in this manual, because the designer is invariably concerned with design in conditions where the timber will be at less than 20% equilibrium moisture content. The stresses and moduli for wet conditions derived from BS 5268 by multiplying dry stresses and moduli by K_2 are those appropriate for the fibre saturation point.

Chapter 3
Loading

3.1 TYPES OF LOADING

The types of loading normally taken by a timber component or structure are the same as those taken by components or structures of other materials. These are:

- self-weight of the components
- other permanent loads such as finishes, walls, tanks and contents, partitions, etc.
- imposed loading such as stored materials
- imposed loading caused by people either as part of an overall uniformly distributed loading or as an individual concentrated load representing one person
- imposed loading such as snow on roofs
- wind loading, either vertical, horizontal, inclined, external, internal pressure or suction and drag.

In the UK, dead and imposed loads for various building usages including dynamic effects, parapets, balustrades and barriers (including vehicle barriers) are given in BS 6399: Part 1. Wind loadings are given in BS 6399: Part 2, while imposed loadings on roofs, including snow loading, are given in BS 6399: Part 3.

For agricultural buildings the designer is referred to BS 5502 'Buildings and structures for agriculture' and in particular BS 5502: Part 22 'Code of practice for design, construction and loading'.

It is beyond the scope of this book to consider seismic or explosive actions.

3.2 LOAD DURATION

Timber has the property of being able to withstand higher stresses for short periods of time than those it can withstand for longer periods or permanently. It is therefore necessary to know whether a particular type of loading is 'long term', 'medium term', 'short term' or 'very short term'. BS 5268: Part 2 gives values for the load–duration factor, K_3, which varies from 1.00 to 1.75 for solid timber and glued laminated timber. The corresponding load–duration factors for other materials and fasteners are derived from the relevant K_3 value.

Dead loading due to self-weight, finishes, partitions, etc., is obviously in place all the time and is *long-term* loading with a K_3 value of 1.00.

Imposed uniformly distributed floor loading is taken in the UK as *long term*, even though observations and measurements of different environments and applications have shown that only part of this uniformly distributed load is permanently in place. The reason that all this imposed load is considered to be permanent is that the defined loading provides an input to the calculation procedures that will generally result in a construction of adequate strength and stiffness. In other words, it is a device that can often save the designer going into the detailed assessment of creep deflection, the vibration characteristics of the construction, etc. There are situations, however, where this simplistic approach is not entirely satisfactory (see section 4.15.7 regarding the vibration and deflection limit of long span domestic floor joists).

Snow occurs in the UK for periods of a few weeks each year or less frequently, and falls into the category of *medium-term* loading having a K_3 value of 1.25. In some countries snow may have to be treated as a long-term load.

The spatial compass (or dimension) of a wind gust gives an indication of the duration of such a gust. Where the dimension of the wind gust exceeds 50 m, the duration of load is assumed to be 15 seconds and is taken as *short term* with $K_3 = 1.5$. For dimensions smaller than 50 m the duration of loading is taken as either 5 seconds or 3 seconds and both are described as *very short term* with $K_3 = 1.75$.

It is always worth while to compare the duration of load for the various K_3 values with the time for the tests to establish the mechanical properties of the material. Typically the test duration factor is equivalent to 1.78. The factor K_3 can only be applied to strength calculations – it is not applicable to the modulus of elasticity or shear modulus.

3.3 CONCENTRATED LOADINGS

In the original UK loading codes joists and beams less than 2.4 m in span had to be capable of supporting a uniformly distributed load equivalent in magnitude to 2.4 m of the specifed uniformly distributed loading. This procedure is often described as the 'equivalent slab loading' and can still be a useful device in some circumstances. The present-day loading codes require consideration of a concentrated load as an alternative to the uniformly distributed load. Not surprisingly perhaps the magnitude of the concentrated load is often equivalent to the effect of 2.4 m of uniformly distributed load. This concentrated load is intended to simulate the loading from a person (particularly for roof structures) or the foot of a piece of furniture or equipment or a trolley wheel.

Each application of the concentrated load can therefore have a different duration of load. For example, the 0.9 kN load applied at either rafter or ceiling level (but never both at the same time) represents a person on or in the roof for purposes of construction, repair or access to store or retrieve goods. This loading is therefore taken as *short-term* duration. Concentrated floor loads on domestic floors are *medium-term* duration while the loadings on other floors are of *long-term* duration except where they occur in corridors or on stairs of residential and institutional type buildings where they are of *short-term* duration as they represent transient loading in areas predominantly occupied by people.

Concentrated loads are assumed to act over an area 'appropriate to their cause'. Typically this has been taken as a 125 mm square for footfall loads and 50 mm square for furniture and equipment. Concentrated loads can also be a rung load on a ladder, a load at specified centres for catwalks or a load per metre for balconies.

When concentrated imposed loads were first introduced in place of the equivalent slab loading, there were concerns that only constructions such as insitu concrete floor slabs would be able to distribute the load through the construction. A number of tests were conducted at the Forest Products Research Laboratory, Princes Risborough, on typical timber floors comprising joists and various floor and roof deck sheathings, e.g. tongue and grooved boards and plywood. These tests showed that the sheathings were capable of providing significant distribution of the concentrated load from the point of application to the adjacent joists. Very conservatively the joist immediately beneath the applied load takes no more than 50% of the total with the balance going equally (say 25%) to the adjacent joists. Using this assumption for the distribution of concentrated load will usually give a less onerous design condition than that for the corresponding uniformly distributed load. Distribution of concentrated loads in this manner is not permitted for ceiling constructions unless there is a decking similar to that applied to normal floors.

3.4 DEAD LOADING

The dead load of materials in a construction can be established with a fair degree of accuracy. However, in some circumstances (e.g. investigating an unexplained deflection on site) it is important to realize that quoted dead loadings have a tolerance and the density of materials can vary. The anticipated thickness can vary, thus affecting weight. Even changes in moisture content of a material can affect weight. The designer has to base his design on the best information available at the time of preparing the design. With regard to strength, the factor of safety used in the derivation of the grade stress takes account of any normal variation or tolerance in the weight of known materials, but with regard to deflection, any increase or decrease in the weight of materials can produce a noticeable effect on deflections. In one instance a bituminous felt and chippings roof covering was replaced at a late stage in the design with a single plastic sheet covering – a change in total dead load from 0.5 to 0.25 kN/m^2. As a consequence the camber built into the supporting glulam beams (usually taken as twice the dead load deflection) that were already manufactured when the change was made, gave the unfortunate impression of a vaulted rather than a flat ceiling.

Section 28.1 lists the generally accepted weight of the more common building materials. When the manufacturer of an individual item is known, the manufacturer's quoted weights should be used in preference to these guide weights. Further reference can be made to BS 648:1964 'Schedule of weights of building materials' which is surprisingly comprehensive in its content despite its age.

3.5 IMPOSED LOADINGS FOR FLOORS

BS 6399: Part 1 'Code of practice for dead and imposed loads', gives the imposed loadings to be used on floors. For beams and joists, in addition to the dead load, one of two alternative imposed loads has to be considered to determine the maximum stresses and deflections: a uniformly distributed load, or a concentrated load.

Where there is an adequate means of load sharing, the mean modulus of elasticity and shear modulus may be used in estimating deformations of floor and roof joists. Where the imposed loading is mechanical plant or storage loads, the minimum modulus of elasticity and shear modulus should be used as this assumption will to some extent allow for creep deflection with the passage of time. Where the imposed loading could cause vibration, e.g. gymnasia and dance floors, then the minimum moduli should be used thus providing a 'stiffer' floor that will be less affected by the dynamic actions.

3.6 IMPOSED LOADINGS FOR ROOFS

BS 6399: Part 3 'Code of practice for imposed roof loads' gives the imposed loadings to be used on flat, pitched or curved roofs.

On flat roofs and sloping roofs up to and including 30° for which no access is provided to the roof (other than that necessary for cleaning and repair), the imposed uniformly distributed loading excluding snow is 0.60 kN/m^2 measured on true plan taken as *medium-term* duration. An alternative uniformly distributed load representing the effects of snow (also *medium term*) has to be considered or a separate *short-term* concentrated load of 0.9 kN on a 125 mm square. Where there is access to the roof, the uniformly distributed load increases to 1.5 kN/m^2, and the concentrated load becomes 1.8 kN with no change in the snow loading.

For a roof slope, $\alpha°$, between 30° and 60° the roof loading is given by $0.6[(60 - \alpha)/30]$ kN/m^2. For roof pitches steeper than 60° the imposed loading is zero. BS 5268-7 discounts the concentrated load where the roof slope exceeds 30° but it is still necessary to consider the normal imposed uniformly distributed loading or the alternative snow loading.

3.7 SNOW LOADING

Freshly fallen snow weighs approximately 0.8 kN/m^3 but compacted snow can weigh as much as 2.0 kN/m^3. The increasing thermal insulation requirements for modern roof constructions means that heat from within a building may not clear deposited snow so that accumulations behind, say, a high parapet can partially melt without draining, freeze, receive more snow and continue the process until perhaps 0.5 m depth of ice builds up, equivalent to about 5.0 kN/m^2, whereas the design loading may have been taken as the minimum of 0.6 kN/m^2.

Snow loading may be a uniform deposit over the entire flat, pitched or curved roof surface but this symmetrical snow load can be redistributed by wind as an asymmetrical loading on pitched and curved roofs. Both the uniform and asymmetrical snow distributions are of *medium-term* duration.

Where snow can be wind driven to accumulate in valleys or against upstands it is necessary to calculate the 'drift' load. The magnitude of the 'drift' load is

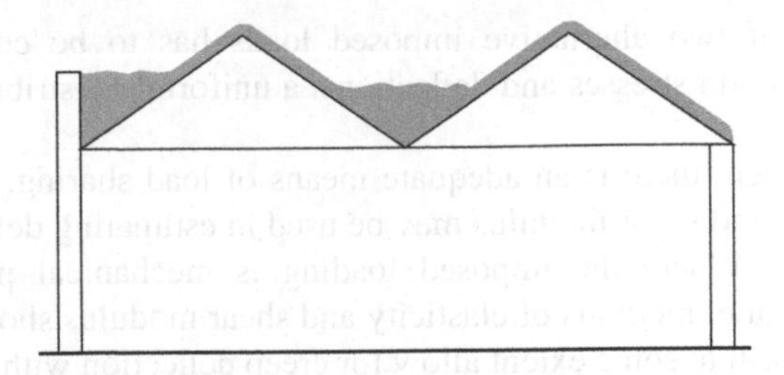

Fig. 3.1 Build-up of snow behind deep fascia and in a valley.

calculated by bringing the volume of previously deposited uniformly distributed snow into a valley or against an upstand. In the latter case there is a limiting slope for the snow of about 12°. In certain circumstances the snow may never fill an available space while in others any excess is considered to be blown away. The probability of drifting to give the calculated maximum load is likely to be a rare event so the loading condition is classed as an *accidental loading.* In this situation the permissible long-term strength values given in BS 5268 can be doubled (equivalent to a K_3 value of 2.00).

The basic snow loading is obtained from Figure 1 of BS 6399: Part 3. This gives the likely distribution of fresh snow across the United Kingdom varying from 1.0 kN/m^2 in the Cairngorms to 0.3 kN/m^2 on the south coast of England. Allowance has to be made for the altitude of the site where this exceeds 100 m. Including this altitude allowance gives the site snow load. As a generality, the site snow load will not exceed 0.6 kN/m^2 south of a line drawn between Bristol and the Wash.

The site snow load is then modified by the snow load shape coefficient, μ_1, to give the snow load on the roof s_d. For example, a site in Aberdeen has a basic snow load of 0.8 kN/m^2 and, assuming the altitude of the site does not exceed 100 m, the site snow load is also 0.8 kN/m^2. With a symmetrical 30° duo pitch roof the value of μ_1 for uniform load is 0.8, giving the uniform snow load on the roof as $0.8 \times 0.8 = 0.64$ kN/m^2 and the value of μ_1 for the asymmetric load is 1.2, giving the snow load on the roof as $1.2 \times 0.8 = 0.96$ kN/m^2. This latter load would be the critical value for the imposed roof load on one half of the pitched roof and would be used in the design of rafters and purlins in a traditional form of roof. For the design of trusses spanning the full width of the roof the uniform load gives a higher total load on the roof and is likely to control the design.

3.8 ROOF LOADINGS ON SMALL BUILDINGS

The time expended to make a comprehensive assessment of snow loading effects can be as lengthy for a dwelling as for a large warehouse. BS 6399: Part 2 therefore offers the option of considering as a special case 'small buildings' where the roof is pitched and is no more than 10 m wide, has no parapets and the total area does not exceed 200 m^2, of which not more than 35 m^2 can be at a lower level. This latter condition limits the size of any drift accumulating on the lower roof. There can be no abrupt changes in height of more than 1 m and the areas of projections such as chimney stacks and dormer windows are limited to 1 m^2.

In these circumstances the imposed load, i.e. normal loading and snow loading, is limited to the worst effect from either 0.75 kN/m^2 or 1.25 times the site snow load, s_0, or a concentrated load of 0.9 kN. The previously given rule for reducing distributed loads for roof pitches steeper than 30° also applies. It is unlikely that consideration of (1.25 × site snow load) would be necessary for sites located south of a line from Bristol to the Wash.

Up to the publication of BS 6399-3 in 1988, 0.75 kN/m^2 had been the required uniformly distributed loading for roofs with access only for maintenace and repair for 50 years or more. In checking roofs known to be built before 1988 this would be the design value used. Many designers still use this value for all roofs, particularly where similar roofs have been shown to be satisfactory in the past.

3.9 WIND LOADING

Timber is frequently associated with buildings of lightweight construction, and the general lack of mass usually requires attention to be paid to wind loading as it affects not only the stresses in individual members and the overall stability but connections and anchorages. The wind loading for which structures must be designed in the UK is calculated from BS 6399-2.

BS 6399-2 and the previous wind code CP3: ChV: Part 2 both predict the maximum gust load on a component or part of a building or the entire structure that is then used in static design. Both codes require the derivation of a design or effective wind speed that is converted to a dynamic wind pressure which with the relevant pressure coefficients, gives the wind forces acting on the building, part or component.

For the majority of sites in the UK the designer, when using CP3: ChV: Part 2, had only to assess the site exposure (S_2), as the other factors relating to topography, statistical return period and wind direction (S_1, S_3 and S_4) were generally assumed to be unity. The pressure coefficients for buildings were defined from the geometry of simple rectangular plan shapes.

BS 6399: Part 2, on the other hand, provides alternative methods for deriving both the effective wind speed and the pressure coefficients for a building part or component. This leads to computational procedures that range from the simple (hand calculator) to the complex (programmable calculator or computer based).

To arrive at the effective wind speed, BS 6399: Part 2 offers three basic methods:

- *Method 1* The Standard method
- *Method 2* The Directional method for effective wind speed, with Standard method pressure coefficients (as described in clause 3.4.2 of the code)
- *Method 3* The Directional method for effective wind speed, with Directional method pressure coefficients.

For hand calculations the designer will find Method 1 the simplest for buildings located in country areas, while Method 2 will be more advantageous in towns. Method 3 essentially requires a software driven solution.

All three methods use the same procedure to determine the site wind speed (V_s) which is derived from the basic wind speed (Figure 6 of the Code) and the effects of wind direction, seasonal or duration effect and the probability of occurrence. They all have an 'altitude factor' – Method 1 includes altitude with topography whereas Methods 2 and 3 take into account only altitude (with the effects of topography for Methods 2 and 3 being included in the factor S_b – see below).

The site wind speed is then factored by S_b to allow for terrain and building effects. The value of S_b is determined by the distance of the site from the sea, whether the site is located in the country or in a town and the effective height of the building.

The effective height of a building is derived from clause 1.7.3.3 and is related to the upwind shielding effects of other buildings or permanent obstructions. In country terrain there is usually little benefit available from upwind shielding but in towns the effects are very significant. The Note to clause 1.7.3.3 gives guidance on the height of surrounding buildings that may be assumed in the absence of specific measurements. With this assumed height the dimension X may be taken as 20 m where no plan dimensions exist.

For Method 1 the factor S_b is given in Table 4 of the Code. When using Methods 2 and 3, S_b is calculated from the equations and tables in clause 3.2.3.2. With Method 2, the gust factor (g_t) is taken as a fixed value of 3.44 for a diagonal dimension a of 5 m whereas in the more complex Method 3, g_t is a variable factor depending upon the effective height and the dimension a.

Methods 1 and 2 then return to clause 2.1.2 for the derivation of dynamic pressure and wind forces using relatively simple external and internal pressure coefficients. Method 3 proceeds to clause 3.1.2 using more comprehensive but likewise more complicated pressure coefficients.

Division of the height of a building into parts, e.g. storey heights – a procedure often followed in timber designs when using CP3: ChV: Part 2 – is not permitted in BS 6399: Part 2 unless the building height exceeds the horizontal dimension of the elevation facing the wind. This is a very unlikely geometrical proportion for a timber building. It is therefore essential that the designer takes full account of upwind shielding and uses Method 2 where possible. In these particular circumstances the resulting wind forces, particularly in a town, will in most cases be less than those derived from CP3: ChV: Part 2.

Consider as an example a typical pair of semidetached two-storey flat-roofed houses located in the city of Leeds at an altitude of 120 m. The distance from the site to the edge of the city to the north, south and west is 5 km and the distance to the east is 1.5 km. It is assumed that there are no topographical features such as an escarpment affecting the wind calculation. The eaves height is 6 m (2 storeys × 3 m) with plan dimensions of the building 10 m × 8 m. The houses are surrounded by similar buildings. As the orientation of such a dwelling relative to the cardinal points is not definable in a large development, the maximum likely site wind speed is calculated and the wind forces calculated from the corresponding pressure. To simplify the calculation only the wind forces acting on the long elevation will be considered.

Example

Using CP3: ChV: Part 2

Basic wind speed 45 m/s S_2 (3B) = 0.60 (3 m) and 0.67 (6 m) $S_1 = S_3 = S_4 = 1.00$
for the top storey $V_s = 0.67 \times 45 = 30.00$ m/s
for the bottom storey $V_s = 0.60 \times 45 = 27.00$ m/s
the dynamic pressure $q = 0.613 \times 30^2 = 552$ N/m² for the top storey
$q = 0.613 \times 27^2 = 447$ N/m² for the bottom storey

Combined c_{pe} on the windward and leeward elevations is [+0.7 − (−0.25)] = 0.95
Total wind force on long elevation is 3 × 10 × 0.95(552 + 447)/1000 = 28.47 kN

Using BS 6399: Part 2, Method 1
Basic wind speed V_b = 22.5 m/s (*note*: the wind speed is for 1 hour's duration whereas the previous Code was on a 3 second basis)
The effective height is 6(H_r) − 1.2 × 6(H_0) + 0.2 × 20(X) = 2.8 m
Using the worst parameters with the altitude factor $S_a = 1.12$ and $S_d = S_s = S_p = 1.0$ gives

$$V_b = 1.12 \times 22.5 = 25.20\,\text{m/s}$$

The minimum distance to the sea is 95 km, and as the distance into the town is less than 2 km, country conditions have to be assumed; hence, from Table 4,

$S_b = 1.31$ for $H_e = 2.8$ m, in the country, 100 km from the sea

$$V_e = 1.31 \times 25.20 = 33.01\,\text{m/s} \quad \text{and} \quad q_s = 0.613 \times 33.01^2 = 668\,\text{N/m}^2$$

(*Note*: if the effects of upwind shielding are ignored, i.e. $H_e = 6.0$ m, then $S_b = 1.48$, $V_e = 37.30$ m/s and $q_s = 853$ N/m²)
The c_{pe} values for a ratio of (plan dimension parallel to wind/eaves height) = 1.33 are

$$0.778 - (-0.278) = 1.056$$

and to arrive at the wind forces it is necessary to calculate the diagonal dimension a of the elevation under consideration. This dimension a gives the factor c_a which is a size effect for the size of the wind gust (smaller dimension gusts have higher pressures in simple terms):

$$a = \sqrt{10^2 + 8^2} = 12.8\,\text{m} \quad \text{and hence from Figure 4,} \quad c_a = 0.93$$

For alternative B in this figure, the wind force on the elevation is then,

$$P = 0.85 \times 10 \times 6 \times 0.93 \times 1.056 \times 668(1 + C_r)/1000$$

where C_r, the dynamic augmentation factor, is 0.01 approximately and can be ignored for timber structures. Hence

$P = 33.5$ kN, which is 17% higher than the CP3: ChV: Part 2 value.

Using BS 6399: Part 2, Method 2
Consider for this example only the cardinal points for the alternative wind directions in Table 3.1. (In a full calculation the wind directions should be in 30° increments, not 90°.)

Table 3.1 Cardinal points vs. wind directions

		North 0°	East 90°	South 180°	West 240°*
Altitude	S_a	1.12	1.12	1.12	1.12
Direction	S_d	0.78	0.74	0.85	1.0
Seasonal	S_s	1.0	1.0	1.0	1.0
Probability	S_p	1.0	1.0	1.0	1.0
$V_s = S_a \cdot S_d \cdot S_s \cdot S_p$		19.66	18.65	21.42	25.20
Distance, sea	(km)	90	90	330	80
Distance, town	(km)	5.0	1.5	5.0	5.0
	S_c	0.771	0.854	0.765	0.777
	T_c	0.624	0.649	0.624	0.624
	g_t	3.44	3.44	3.44	3.44
	S_t	0.209	0.209	0.209	0.209
	T_t	1.85	1.85	1.85	1.85
	S_h	0	0	0	0
$S_b = S_c T_c(1 + (g_t S_t T_t) + S_h)$		1.21	1.29	0.95	1.13
$V_e = S_b V_s$		23.79	24.06	20.35	28.48
$q_s = 0.613 \times V_e^2$		347 N/m²	355 N/m²	254 N/m²	497 N/m²

*The value of S_d is 1.0 at 240° and 0.99 at true west (270°), so the worst condition has been taken.

The critical design pressure is 497 N/m². It can be seen that the maximum value of S_b does not now give the maximum design wind pressure because this maximum S_b is used with a lower value of V_s due to the direction factor. By making these refinements the value of q_s is now 34% lower than with Method 1.

The wind force on the long elevation is calculated as for Method 1 but using the lower value of q_s, i.e.

$$P = 0.85 \times 10 \times 6 \times 0.93 \times 1.056 \times 497/1000 = 24.9\,\text{kN}$$

This is some 12.5% less than the CP3: ChV: Part 2 method but the effort required is considerably greater. To obtain the most economical return from BS 6399: Part 2 tabulated data, 'ready reckoners' or a software solution are a necessity if only to determine the dynamic wind pressure.

The above example demonstrates some of the intricacies of BS 6399-2 in relation to the effects of wind on the walls of a building. The calculation for the effects of wind on roof structures is more complex because the number of pressure zones on a duopitch roof is at least 6 and with pitches in the range 15° to 45° there can be alternative +ve and −ve pressure coefficients for a particular zone. This certainly complicates stability calculations and makes the stress calculations for a roof truss close to impossible without making overall simplifying assumptions such as taking uniform pressures on the windward and leeward slopes equal to the maximum values derived from BS 6399-2.

Where there is a net wind uplift or a net overturning moment due to wind BS 5268-3 for trussed rafters calls for a factor of safety of 1.4, i.e. the wind action must be multiplied by 1.4 and the result must not be greater than the dead load or permanent load value or restoring moment (including the contribution from holding down straps, etc.). This principle should be applied to all timber structures.

3.10 UNBALANCED LOADING

The critical condition of loading in the overall design of a component is usually with the total loading in position. However, particularly with lattice constructions, the maximum loading in some of the internal members occurs with only part of the imposed loading in position.

The vertical shear in diagonal *X* in Fig. 3.2 is zero with the balanced loading. The maximum force in diagonal *X* occurs with the unbalanced loading shown in Fig. 3.3. The vertical shear and hence the force in diagonal *Y* is less with the balanced loading shown in Fig. 3.4 than with the unbalanced loading shown in Fig. 3.5.

Obviously the positioning of the dead loading cannot vary, but the designer

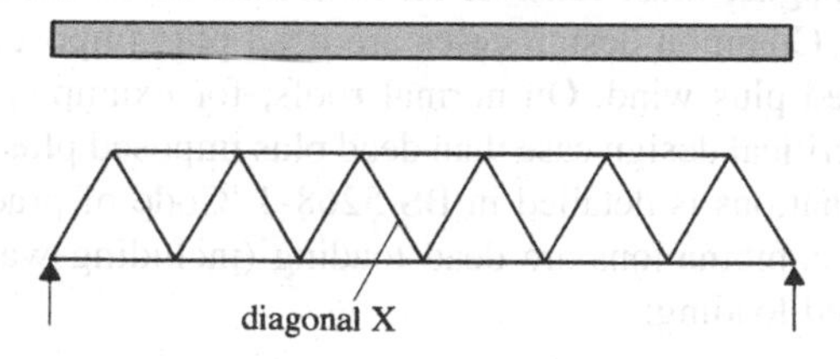

Fig. 3.2

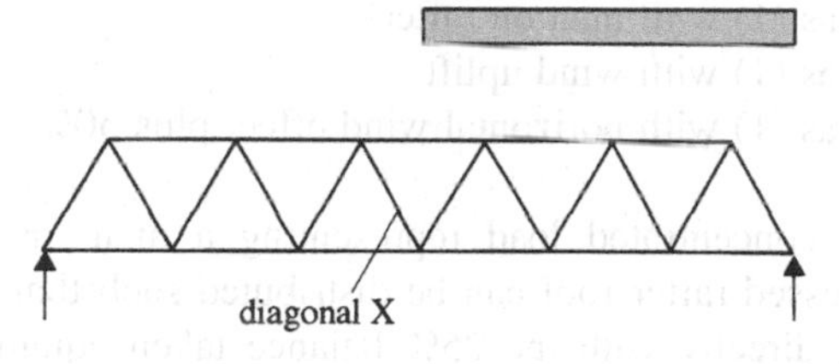

Fig. 3.3

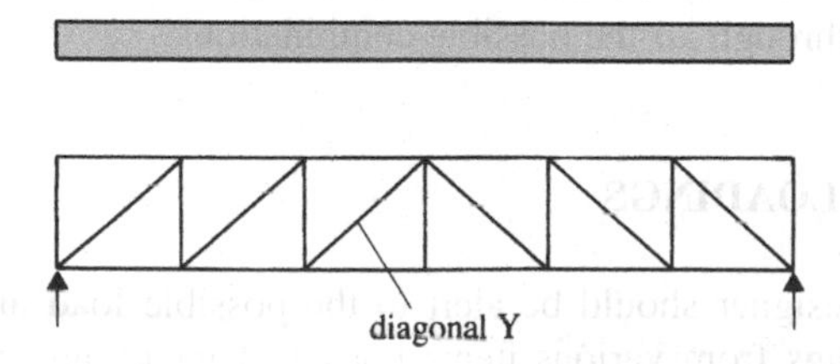

Fig. 3.4

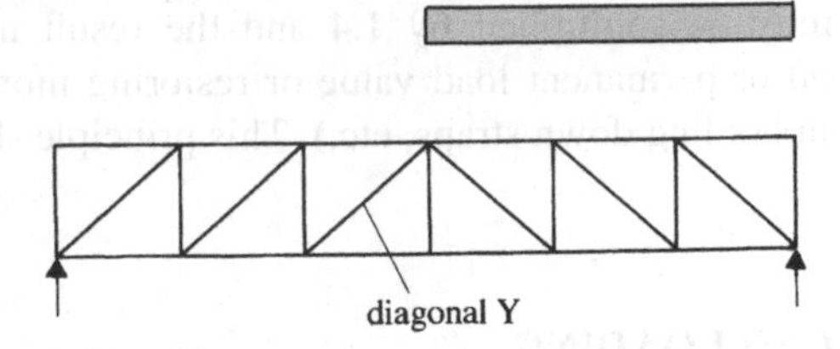

Fig. 3.5

should be alert to the effect on certain members (particularly internal members in frameworks) caused by unbalanced imposed loadings. It is most important to consider this when the balanced loading leads to zero force in any particular member. In particular care should be taken in adopting realistic forces in joints. The effects of unbalanced loading can occur during erection.

3.11 COMBINATIONS OF LOADING

The designer must consider all combinations of the loadings of which he is made aware. Common design cases are dead plus imposed, dead plus wind, dead plus imposed plus wind. On normal roofs, for example, dead plus wind is usually a more critical design case than dead plus imposed plus wind. An example of loading combinations is detailed in BS 5268-3 'Code of practice for trussed rafter roofs'. These combinations are dead loading (including water tanks) plus the following imposed loading:

(1) dead UDL on rafter and ceiling tie plus water tanks	(long term)
(2) as (1) with snow UDL	(medium term)
(3) as (1) with man on ceiling tie	(short term)
(4) as (1) with man on rafter	(short term)
(5) as (1) with wind uplift	(very short term)
(6) as (1) with horizontal wind effect plus 50% snow	(very short term)

The concentrated load representing a man on the ceiling tie or the rafter in a trussed rafter roof can be distributed such that 75% is taken by the member loaded directly with the 25% balance taken equally by the members on either side.

With experience of specific framework configurations the designer should be able to identify the two or three critical load combinations and may not have to work through all the possible combinations.

3.12 SPECIAL LOADINGS

The designer should be alert to the possible load and deflection implications in buildings from various items not all of which are within his control. A few are discussed below.

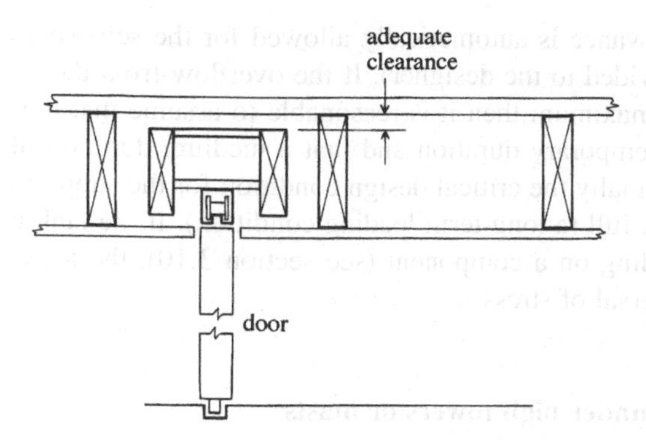

Fig. 3.6

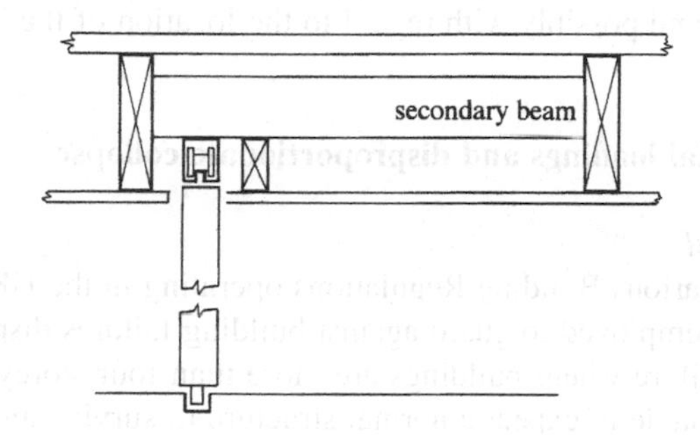

Fig. 3.7

3.12.1 Sliding doors

A sliding door can run on a bottom track, in which case the designer has to ensure that the deflection of any component above the door will not prevent the door from opening or closing. Often the extent of vertical adjustment or tolerance built into door runners is unrealistically small.

A sliding door can be top hung, in which case one can either provide a separate support beam or beams (Fig. 3.6) so that deflection of the main support beams will not cause the door to jam, or provide short beams between the main beams (Fig. 3.7), so that if it is found that too little adjustment has been provided in the door mechanism, adjustment to the secondary beams can be carried out without major structural alterations. It is not advisable to hang a door directly under a main beam (also see section 12.6.5).

3.12.2 Water tanks

Tanks are usually quoted as having a nominal capacity together with an actual dimensional size say 300 litres for a tank 0.60 m × 0.45 m × 0.50 m deep. These actual dimensions give the 'overflowing' capacity (in this example it is 450 litres). By taking the weight in design calculations as the 'overflowing' capacity, an allowance is automatically allowed for the self-weight of the tank (a load rarely provided to the designer). If the overflow from the tank fails and the tank fills to its maximum then it is reasonable to assume that this 'flood' overloading is only of temporary duration and that a medium duration of load factor is appropriate. Normally the critical design condition for the supporting structure occurs with the tank full (a long-term loading condition). If the tank is placed to give unbalanced loading on a component (see section 3.10), the act of emptying it could lead to reversal of stress.

3.12.3 Roofs under high towers or masts

Roofs under high towers or tall masts have been penetrated by icicles which form then break off. This potential problem should be discussed in the design of such roofs and possibly with regard to the location of the building itself.

3.12.4 Accidental loadings and disproportionate collapse

3.12.4.1 General

The various Building Regulations operating in the UK call for special procedures to be employed to guard against building failures disproportionate to the cause of the failure where buildings are more than four storeys in height. It would be unreasonable to expect a normal structure to survive an atomic blast but a localized internal explosion or a vehicular impact would come within this consideration. Until the early 1990s timber-framed buildings were limited to three storeys in height by fire requirements. Changes in the UK Building Regulations now allow timber constructions of five storeys in Scotland. In England, Wales and Northern Ireland the practical limit is eight storeys. This topic is therefore an essential design consideration for the taller buildings.

The designer's tasks have also been made more onerous with the introduction of *The Construction (Design and Management) Regulations 1994* which requires consideration of the safety of personnel not only during construction but while routine maintenance is being carried out.

3.12.4.2 Disproportionate collapse

The Building Regulations for the UK offer three ways of satisfying this statutory requirement:

(1) by providing adequate horizontal and vertical ties through the structure; these forces are given in the relevant design code of practice;
(2) with horizontal ties being effective, but not vertical ties, to remove the vertical supports one at a time to check by calculation that the remaining structure can span over the removed member;
(3) where neither horizontal nor vertical tying can be achieved, each support member should be removed one at a time and on its removal the area at risk of collapse should not exceed 15% of the area of the storey or 70 m^2, whichever is the less.

Options (1) and (2) are applicable to framed buildings, i.e. one relying on the strength and stiffness of beams and columns. It is unlikely that a timber building of five storeys or more will be a framed structure as the racking resistance required for normal loading conditions will generate a panellized, platform frame type of construction. Vertical, storey height diaphragm walls with continuous edge floor members providing the flanges of a vertical beam have been shown to perform exceptionally well in disproportionate collapse tests giving compliance with option (3).

3.12.4.3 *Safety during construction and maintenance*

The kind of problem that has to be considered is best illustrated by an example of a problem that occurred when CDM was not even a twinkle in the eye of the Health and Safety Executive (indeed the HSE did not exist). A very large (5000 m^2) single-storey, open plan, flat-roofed building was queried by the Factory Inspectorate with regard to stability if at any time during construction or subsequently one of the internal, free-standing columns was removed. The Inspectorate had in mind, particularly, a dumper truck colliding with a column during the construction stage.

A timber-framed building has little 'intentional' continuity. For example, there are considerable continuity effects with insitu concrete that may be enhanced by the introduction of reinforcing bars and the bolted joints in steel construction afford horizontal and vertical ties. Precast concrete, on the other hand, can have the same problems as timber and this was the probable reason why the Inspectorate took such an interest in the timber building described above.

A timber-framed building, on the other hand, has considerable 'coincidental' continuity through overlapping timbers and sheet materials as well as the nailed, screwed and bolted joints. These elements can be developed as often none will normally have been considered as contributing to the overall stability. The solution to the particular problem described was to arrange for the sheet of plywood in the roof construction immediately over a beam to 'straddle' the beam and for this sheet to be site fixed. This meant that the ends of the roof joists bearing on the beam were exposed and could be spliced by working from above. The plywood decking was then completed which further added to the continuity of the roof construction. If a column was removed it could be shown that the remaining roof would hang as a catenary from the adjoining beams. The duration of load in this condition would be short term and provided the dumper driver did not panic and remove other columns then the building would survive this accidental loading.

By careful consideration of the layout of the component parts of a timber building it is possible to insert continuity ties between joists and beams and also to make sure that sheathing can be arranged to bridge discontinuities in the supporting structure.

Chapter 4
The Design of Beams: General Notes

4.1 RELATED CHAPTERS

This chapter deals in detail with the considerations necessary for the design of beams. As such it is a reference chapter. Several other chapters deal with the actual design of beams, each one being devoted to one main type:

Chapter 5	Beams of solid timber. Principal and load-sharing members.
Chapter 6	Multiple section beams. Composites of two or three rectangular sections.
Chapter 7	Glulam beams. Vertical and horizontal glulam, parallel or tapered beams.
Chapter 8	Thin web beams.
Chapter 9	Lateral stability of beams. Full or partial lateral restraint.
Chapter 10	Stress skin panels. Single and double skin.
Chapter 11	Solid timber decking.
Chapter 12	Deflection. Practical and special considerations.

4.2 DESIGN CONSIDERATIONS

The principal considerations in the design of beams are:

- bending stress
- prevention of lateral buckling
- shear stress
- deflection: both bending deflection and shear deflection
- bearing at supports and under any point loads
- prevention of web buckling (with ply web beams).

The size of timber beams may be governed by the requirement of the section modulus to limit the bending stress without lateral buckling of the compression edge or fracture of the tension edge, or of the section inertia to limit deflection, or the requirement to limit the shear stress.

Generally, bending is critical for medium-span beams, deflection for long-span beams and shear only for heavily loaded short-span beams. In most cases it is advisable however to check for all three conditions. The relevant strengths (permissible stresses) are computed by modifying the grade stresses by the factors from BS 5268-2, which are discussed in this chapter.

4.3 EFFECTIVE DESIGN SPAN

In simple beam design it is normal to assume that the bearing pressure at the end of a simply supported beam is uniformly distributed over the bearing area and no account need be taken of the difference in intensity due to the rotation of the ends of the beam (clause 2.10.2 of BS 5268-2). If, therefore, only sufficient bearing area is provided to limit the actual bearing pressure to the permissible value, the effective design span should be taken as the distance between the centres of bearing. In many cases, however, the bearing area provided is much larger than that required to satisfy the bearing pressure requirement. In these circumstances the calculations for the effective design span need be taken only as the distance between the bearing areas which would be required by the limitation of maximum bearing stress, rather than the centres of the bearing areas actually provided (Fig. 4.1).

In determining the effective design span from the clear span, it is usually acceptable to assume an addition of around 0.05 m for solid timber joists and 0.1 m for built-up beams on spans up to 12 m, but more care is needed with beams of longer span or with heavily loaded beams. The effective design span is given the symbol L_e in BS 5268-2. L is used in most of the design examples in this manual for convenience. Even if a beam is built-in at the ends so that the loading can occur only over the clear span, it is usual to assume in designs that the loading occurs over the effective design span.

4.4 LOAD-SHARING SYSTEMS

4.4.1 Lateral distribution of load

When a beam system has adequate provision for lateral distribution of loads, e.g. by purlins, binders, boarding, battens, etc., it can be considered to be a load-sharing system. In such a system a member less stiff than the rest will tend to deflect more than the other members. As it does, the load on it will be transferred laterally to the adjoining stiffer members. Rather than give load-relieving factors, this load sharing is dealt with in BS 5268-2 by permitting higher values of E than E_{min} and increases in grade stresses in the design of certain components.

In the special case of four or more members occurring in a load-sharing system and being spaced not further apart than 610 mm (as with a domestic floor joist system) the appropriate grade stresses may be increased by 10% (factor K_8 of 1.1), and E_{mean} (and G_{mean}) may be used in calculations of deflection (although BS 5268-2 calls for the use of E_{min} if a flooring system is supporting an area intended for mechanical plant and equipment, or for storage, or for floors subject to considerable vibration such as gymnasia and ballrooms).

Factor K_8 applies to bearing stresses as well as bending, shear and, where appropriate, to compression or tension stresses parallel to the grain.

In the case of two or more members fastened together securely to form a trimmer joist or lintel, the grade stresses in bending and shear parallel to the grain and compression perpendicular to the grain (i.e. bearing) may also be increased by 10% (factor K_8). Also, E_{min} may be increased by factor K_9. See section 2.3.

Where members are glued together to form a vertically laminated member having the strong axis of the pieces in the vertical plane of bending, the bending, tension and shear parallel to grain stresses may be increased by factor K_{27} while E_{min} and compression parallel to grain may be increased by factor K_{28} (see section 7.3.2). Compression perpendicular to grain stresses may be increased by $K_{29} = 1.10$.

There are some inconsistencies when comparing K_8 with K_{27} and K_{28} (for compression parallel to the grain). For example, if a trimmer consists of two softwood members nailed together, the grade stresses for bending and shear parallel to the grain may be increased by $K_8 = 1.10$, whereas if the members were glued together the relevant factor is K_{27} with a value of 1.11. If a compression member comprises two pieces nailed together then no increase may be taken for grade stress for compression parallel to the grain but if the pieces are glued together then a K_{28} value of 1.14 may be incorporated. With compression perpendicular to the grain, at an end bearing for example, factor $K_8 = 1.10$ may be used for two or more members acting together, i.e. they may be either constrained to act together by a strong transverse bearer or nailed together or glued together.

This should not be a serious matter for a designer; although, if designing a two-piece column which also takes a bending moment and is supported at one end on an inclined bearing surface, it may seem odd to increase the bending strength parallel to grain for load sharing but not the compression strength parallel to grain. Also, when considering an end bearing, the bearing strength perpendicular to grain component may be increased for load sharing, but not that for the bearing strength parallel to grain, even though the bearing stresses being considered are at the same position!

It is not normal to design glulam members (horizontally laminated) as part of a load-sharing system. However BS 5268-2 recognizes the statistical effect within each member of four or more laminates by giving various coefficients K_{15} to K_{20} (see Table 7.1).

In the case of a built-up beam such as a ply web beam, BS 5268-2 recognizes the load-sharing effect by permitting the use in designs of factors K_{27}, K_{28} and K_{29} as for vertically glued members. In applying K_{27} and K_{28} to the grade stresses it is

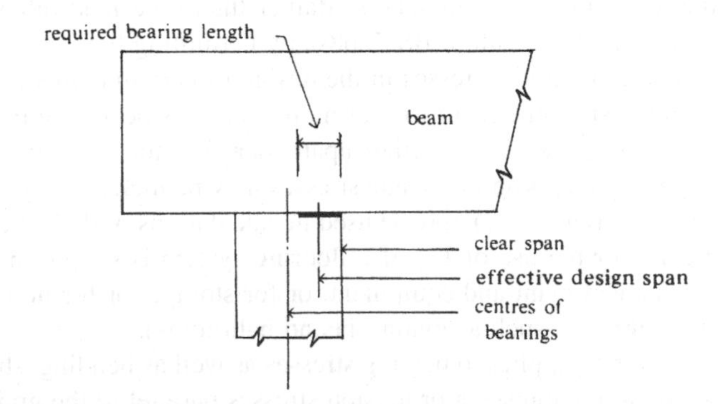

Fig. 4.1

accepted that the number of laminates, N, is appropriate to the number of members in each flange. In applying K_{28} to E_{min} the value of N is appropriate to the total number of solid timber members in the cross-section. Because K_{29} is the factor which relates to bearing, the same comment given above for vertically laminated members also applies to ply web beams.

Very often, ply web beams are placed at 1.200 m centres with secondary noggings fixed at regular intervals between them at the level of the top and bottom flanges. As such, many practising engineers have considered this to be a load-sharing system. Some engineers have used E_{mean} in calculating deflections while some have used E derived from $E_{min} \times K_{28}$. The extent to which a ply web beam system can be taken as a load-sharing system is discussed in Chapter 8 in relation to E and G values and in section 19.7.3 in relation to finger jointing.

4.4.2 Concentrated load. Load-sharing system

When a concentrated load such as that specified in BS 6399-1:1996 acts on a floor supported by a load-sharing system, clause 4.1 of BS 6399-1 permits the concentrated load to be disregarded in the design of the supporting members. When the concentrated load acts on a flat roof or sloping roof (sections 3.6 and 3.8) it must be considered, particularly for short-span beams.

In a load-sharing system lateral distribution of the load will reduce the effect on any one member, particularly if the decking is quite thick and the effect can 'spread' (sideways and along the span) through the thickness of the decking. A conservative assumption with a typical domestic or similar floor construction is that 50% of the concentrated load will act on the member directly under the load with the adjacent members taking 25% each.

For domestic floors BS 5268-2 states that the concentrated load defined in BS 6399-1 should be taken as medium term. BS 5268-3 allows the concentrated load acting at ceiling or rafter level to be taken as short-term duration ($K_3 = 1.5$) as this is taken to be a person walking on or in the roof.

Other concentrated loads (as from partitions) must be given special consideration. One case which is open to discussion is illustrated in Fig. 4.2. The usual questions are whether or not the joists B supporting the partition should be designed using E_{mean} on the assumption that they are part of a load-sharing system and

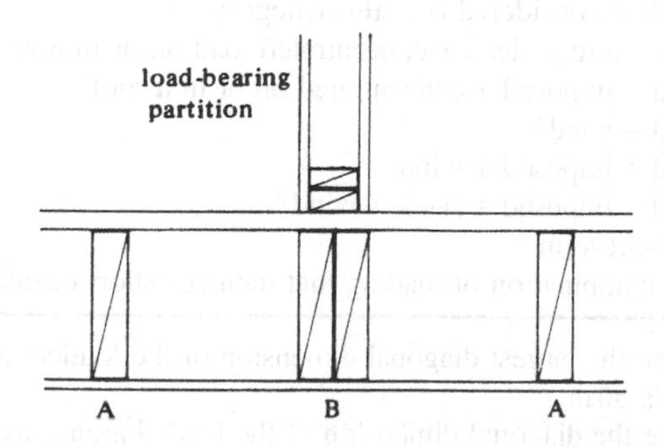

Fig. 4.2

whether or not the two joists A will carry a share of the load from the partition. Providing it will not lead to significant differential deflection between joists A and B it is probably better to assume that joists B support the total weight of the partition and a share of the floor loading, excluding any imposed loading under the 'footprint' of the partition, with an E value of E_N. It is prudent to carry out an additional deflection check to ensure that under dead loading only the deflections of joists A and B are similar and that any difference is unlikely to be noticeable.

4.5 LOAD–DURATION FACTOR

Timber, other wood-based board materials and fasteners can sustain a very much greater load for a short period of time than they can for a longer period of time, or permanently. BS 5268-2 takes account of this by quoting duration of loading factors for long-term, medium-term, short-term and very short-term loadings.

The K_3 modification factors for solid timber and glulam given in Table 4.1 are applicable to all strength properties including bearing, but may not be applied to moduli of elasticity or shear moduli. Modulus of elasticity and shear modulus values for solid timber and glulam are assumed to be constant irrespective of the duration of the loading.

A timber beam does not deflect by the full calculated amount until several hours after the application of a load. The testing clause of BS 5268-2 requires that only 80% of the calculated deflection should have taken place after 24 hours. There is therefore a case for disregarding or modifying deflection calculations in some cases when wind gust loading becomes the design criteria. It should be reasonable to

Table 4.1 Modification factors K_3 for duration of loading

Duration of loading	Value of K_3
Long term (e.g. dead + permanent imposed)	
domestic uniformly distributed loads to be considered in this category	1.00
Medium term (e.g. roof dead + imposed and dead + temporary imposed load)	
for domestic floors the concentrated load given in BS 6399-1 when considered is in this category	1.25
Short term (e.g. dead + concentrated load on or in a roof)	
dead + imposed + concentrated on or in a roof	
dead + wind*	
dead + imposed + wind*	
dead + imposed + snow + wind*	1.5
Very short term	
any combination of loading that includes short duration wind†	1.75

* Where the largest diagonal dimension of the loaded area a as defined in BS 6399-2 exceeds 50 m.

† Where the diagonal dimension of the loaded area a as defined in BS 6399-2 is less than 50 m.

assume that only 80% of the deflection calculated by assuming the full gust loading will take place during the few seconds the load is assumed to act.

The design must be checked to ensure that the strength values are not exceeded for any relevant condition of loading. Because of the duration of loading factors it is possible, for example, for the design case of dead loading only to be more critical than the case of dead plus imposed roof loading.

Other values for load–duration factors are given in BS 5268-2 with reference to

- plywood (including E and G) K_{36}
- tempered hardboard (including E and G) K_{38}, K_{39}, K_{40} and K_{41}
- wood particleboard (including E and G) K_{81}, K_{82}, K_{83} and K_{84}
- fasteners K_{48}, K_{52}, K_{58}, K_{62} and K_{66}

4.6 LATERAL STABILITY

There are several methods of linking the degree of lateral stability of a timber beam to the permissible stress in bending. Those from BS 5268-2 are described below and special methods for designing for conditions of partial restraint are covered in Chapter 9.

4.6.1 Maximum depth-to-breadth ratios (solid and laminated members)

BS 5268-2 gives limiting values of depth-to-breadth ratios which, if complied with for solid or laminated members, justify use of the full grade bending stress (increased by the load–duration or load-sharing factor if applicable) in calculations. These ratios are given in Table 19 of BS 5268-2 and are presented here in Table 4.2. When the depth-to-breadth ratio exceeds the value corresponding to the appropriate degree of lateral support, BS 5268-2 permits the designer to check to ensure

Table 4.2

Degree of lateral support	Maximum depth-to-breadth ratio
No lateral support	2
Ends held in position	3
Ends held in position and members held in line, as by purlins or tie rods, maximum centres will be defined in an amendment to BS 5268	4
Ends held in position and compression edge held in line, as by direct connection of sheathing, deck or joists	5
Ends held in position and compression edge held in line, as by direct connection of sheathing, deck or joists, together with adequate bridging or blocking spaced at intervals not exceeding six times the depth	6
Ends held in position and both edges firmly held in line	7

that there is no risk of lateral buckling under the design loading. This usually requires the use of an appropriately reduced bending stress. One method determined by the author is discussed in section 9.2.

4.6.2 Maximum second moment of area ratios (built-up beams)

BS 5268-2 requires built-up beams such as ply web beams to be checked to ensure that there is no risk of buckling (of the compression flange) under design load. This checking can be by special calculation, or by considering the compression flange as a column, or by providing lateral and/or end restraint as detailed in clause 2.10.10 of BS 5268-2 which is summarized in Table 4.3 of this manual and is related to the ratio of second moments of area.

Designers often find that it is easier and more acceptable to match a restraint condition appropriate to a higher fixity ratio than the one actually occurring, and it must be obvious that this should be acceptable to any approving authority. For example, with beams having an I_x / I_y ratio between 31 and 40, it is more likely that the designer will provide full restraint to the compression flanges rather than a brace at 2.4 m centres, particularly if the beams are exposed. Although this does not satisfy a pedantic interpretation of BS 5268-2, it provides more restraint and more than satisfies the intention of the clause.

As stated above, BS 5268-2 permits the compression flange of a ply web beam to be considered as a horizontal column which will tend to buckle laterally between points of restraint. Such an approach is conservative, it makes no recognition of the stabilizing influence of the web and tension flange or the location of the load in relationship to the depth of the beam. (See section 9.4.)

4.7 MOISTURE CONTENT

It is considered that solid timber of a thickness of about 100 mm or less (and glulam of any thickness) in a covered building in the UK will have an equilibrium mois-

Table 4.3 Maximum second moment of areas related to degree of restraint

$\frac{I_x}{I_y}$	Degree of restraint
Up to 5	No lateral support required
5–10	The ends of beams should be held in position at the bottom flange at the supports
10–20	Beams should be held in line at the ends
20–30	One edge should be held in line (N.B. not necessarily the compression edge)
30–40	The beam should be restrained by bridging or bracing at intervals of not more than 2.4 m
More than 40	Compression flange should be fully restrained

ture content of approximately 20% or less. BS 5268-2 gives stresses and E values for timber for two appropriate service conditions which are:

- *Service class 1*: dry exposure conditions where the average moisture content will not exceed 12% (internal use in continuously heated buildings).
- *Service class 2*: dry exposure conditions where the average moisture content will not exceed 20% for any significant period of time. (Covered and generally unheated, giving an average moisture content of 18%, or covered and generally heated, giving an average moisture content of 15%.)

Timbers which are fully exposed to external use are defined as service class 3 characterized as wet exposure conditions where the moisture content of solid timber exceeds 20% for significant periods.

For solid sections (not glulam) exceeding 100 mm in thickness, BS 5268-2 clause 2.6.1 recommends that service class 3 (wet exposure) stresses should be used in structural design because it is considered that the natural drying of these thick sections is a very slow process and load may be applied before the moisture content reaches the 'dry' threshold. Obviously if the sections are properly dried before installation this restriction does not apply. Also, it is as well to be aware that sections of 100 mm or thicker are liable to contain fissures.

BS 5268-2 does not tabulate stresses for wet exposure conditions but gives factors K_2 by which dry values should be multiplied. Note the former factor K_1 for the increase in section properties due to increase in dimension on wetting has been deleted from BS 5268-2.

Although it is preferable for timber to be dried before installation to a moisture content within a few percent of the level it will attain in service, it is essential, if dry stresses have been used in the design, for the timber to have dried to around 20% moisture content before a high percentage of the total design load is applied. If it is found that beams are still well over 20% moisture content when the time comes to apply, say, more than about one half of the total dead load, then the loading should be delayed or the beams should be propped until the moisture content is brought down to 20% or less, otherwise large creep deflections may take place (see section 4.15). If beams are erected at a moisture content of 22% or less in a situation where drying can continue freely, and only part of the dead load is applied, there is very little likelihood of serious trouble from creep deflection.

4.8 BENDING STRESSES

At the stress levels permitted by BS 5268-2 it is acceptable to assume a straight-line distribution of bending stress in solid beams between the maximum value in the outer compression fibres and the outer tension fibres (Fig. 4.3). See Chapters 8 and 10 for the variations applicable with ply web beams and stress skin panels.

The symbols used for bending parallel to the grain in BS 5268 are:

$\sigma_{m,g,\parallel}$ grade bending stress parallel to the grain
$\sigma_{m,adm,\parallel}$ permissible bending strength parallel to the grain
$\sigma_{m,a,\parallel}$ applied or actual bending stress parallel to the grain

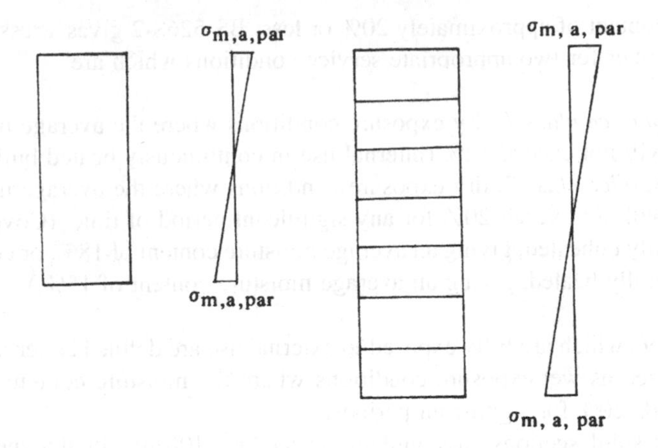

Fig. 4.3

BS 5268-2 permits some of the subscripts to be omitted if there is no chance of misunderstanding.

Bending strengths are influenced by depth (see section 4.9). Load-sharing and load–duration factors are applicable to bending stresses.

4.9 DEPTH AND FORM FACTORS

4.9.1 Depth factor for flexural members

All other things being equal, the greater the depth, h, of a timber beam the less strong is it. The depth factor, K_7, is in two parts:

for depths between 72 and 300 mm $K_7 = \left(\frac{300}{h}\right)^{0.11}$, giving a maximum value at 72 mm of 1.17

for depths over 300 mm $K_7 = 0.81\frac{h^2 + 92300}{h^2 + 56800}$

This factor is now applicable to all timbers in BS 5268-2, including the North American framing grades.

Although the intention of this chapter is to cover beams, it is convenient to discuss the width factor for tension members while discussing the depth factor for beams. 'Width' is the greater dimension of a rectangular section in tension and the width modification factor is K_{14}. For widths, h, of solid timber and glulam tension members over 72 mm there is a single formula, $K_{14} = (300/h)^{0.11}$. K_{14} is applied to the grade tension stresses tabulated in BS 5268-2 (note the reservation given in section 2.4.2 for lower grades of North American timbers in tension.

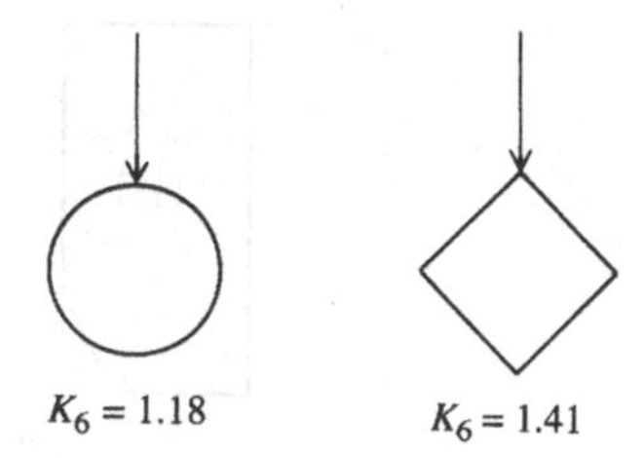

Fig. 4.4

4.9.2 Form factor for flexural members

The grade stresses in BS 5268-2 for flexural members apply to solid and laminated timber of a rectangular cross-section (including square sections with the load parallel to one of the principal axes).

For solid circular sections and solid square sections with the load in the direction of a diagonal, the grade stresses should be multiplied by the factor K_6, as indicated in Fig. 4.4.

4.10 BEARING

4.10.1 Bearing across and along the grain

The grade stress for bearing across the grain of timber is the grade stress for compression perpendicular to the grain. The grade stress for bearing on end grain is the grade stress for compression parallel to the grain (at a slenderness ratio of zero). Load-sharing and load–duration factors apply (but see section 4.4.1). The symbols used are:

- parallel to the grain
  - $\sigma_{c,g,\parallel}$ grade bearing stress parallel to the grain
  - $\sigma_{c,adm,\parallel}$ permissible bearing strength parallel to the grain
  - $\sigma_{c,a,\parallel}$ applied or actual bearing stress parallel to the grain
- perpendicular to the grain
  - $\sigma_{c,g,\perp}$ grade bearing stress perpendicular to the grain
  - $\sigma_{c,adm,\perp}$ permissible bearing strength perpendicular to the grain
  - $\sigma_{c,a,\perp}$ applied or actual bearing stress perpendicular to the grain.

BS 5268-2 permits some of the subscripts to be omitted if there is no chance of misunderstanding.

The grade stresses tabulated in BS 5268-2 for compression (or bearing) perpendicular to the grain have been reduced to take account of wane (see Fig. 4.5) as described in section 2.4.3. Where the specification specifically excludes wane at the bearing position the grade stress may be increased as follows:

- For strength class grade the higher of the two listed values for compression perpendicular to grain is applicable.

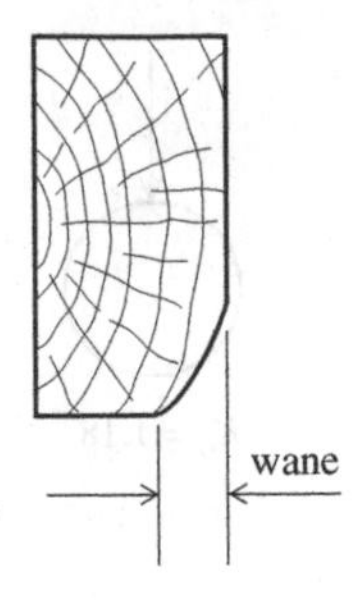

Fig. 4.5

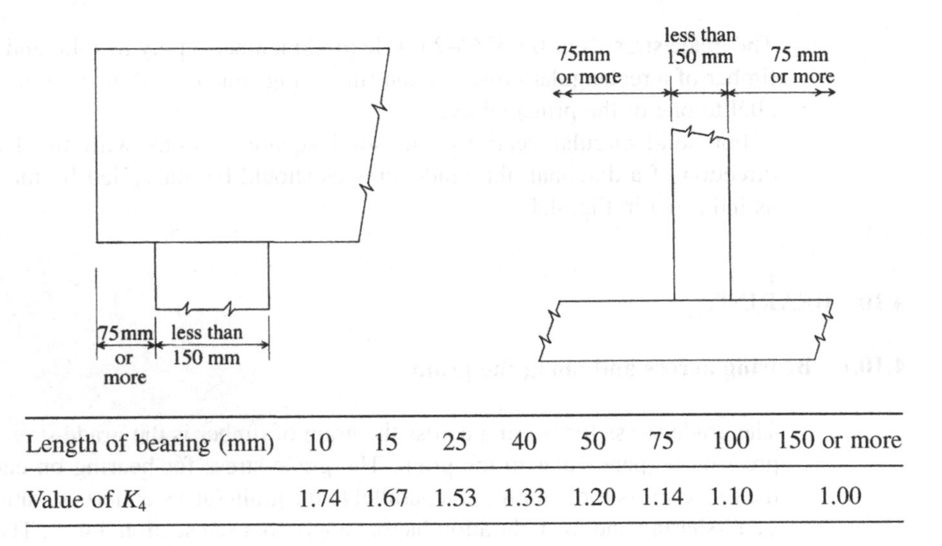

Length of bearing (mm)	10	15	25	40	50	75	100	150 or more
Value of K_4	1.74	1.67	1.53	1.33	1.20	1.14	1.10	1.00

Fig. 4.6

- For SS and GS grades the SS grade compression perpendicular to grain stress $\times 1.33$ is applicable to both grades.
- For Canadian softwoods to NLGA rules, USA softwoods to NGRDL rules and North American softwoods to North American MSR rules the SS grade compression perpendicular to grain stress $\times$ 1.33 is applicable to all grades.

The grade stresses tabulated in BS 5268-2 for compression (or bearing) parallel to the grain have not been reduced to take account of wane. If the end in question is bearing on timber of the same species this need not be considered because the critical bearing stress will be perpendicular to the grain of the other piece of timber. If, however, the piece being considered is bearing directly onto steel, for example, and wane is permitted, a reduction in the permissible stress (or the area of bearing) should strictly speaking be taken. BS 5268-2 does not give guidance on this point but it would make sense to calculate the reduced end grain area which would remain if wane occurred up to the grade limit. By reference to BS 4978 it can be calculated that the reduced areas for GS and SS rectangular sections are

94% and 97% respectively. As can be seen, even with GS, the reduction in area is not dramatic and can usually be disregarded.

When the bearing length is less than 150 mm and the bearing occurs 75 mm or more from the end of a member (as shown in Fig. 4.6) the permissible bearing stress may be increased by the factor K_4 tabulated in Fig. 4.6. Interpolation is permitted.

4.10.2 Plywood. Bearing on the face

BS 5268-2 gives values in Tables 34 to 47 for bearing on the face. The load–duration (K_3) factor applies.

4.10.3 Plywood. Bearing on edge

BS 5268 does not give values for edge bearing of plywood. It is not usual to require plywood to take edge bearing, and if such cases arise it will be necessary to test the strength for the particular circumstances.

4.11 SHEAR

4.11.1 Solid sections

BS 5268-2 gives permissible shear parallel to grain stresses which relate to maximum shear stresses taking into account the parabolic shear stress distribution over the depth of the section and not the average value (as assumed in steelwork design). Load–duration and load-sharing factors apply. The formula for calculating shear stress at any level of a built-up section is:

$$\tau = \frac{F_v A_u \bar{y}}{b I_x} \tag{4.1}$$

where τ = the shear parallel to grain stress at the level being considered
F_v = the vertical external shear
A_u = the area of the beam above the level at which τ is being calculated
$\bar{y}$ = the distance from the neutral axis of the beam to the centre of the area A_u
I_x = the complete second moment of area of the beam at the cross-section being considered
b = the breadth of the beam at the level at which τ is being calculated.

If $F_v A_u \bar{y}/I_x$ is evaluated, this gives the total shear force parallel to grain above the level being considered per unit length of beam.

In the special case of a rectangular beam (Fig. 4.7) the maximum shear stress parallel to grain occurring at the neutral axis becomes

$$\tau = \frac{3F_v}{2A} \tag{4.2}$$

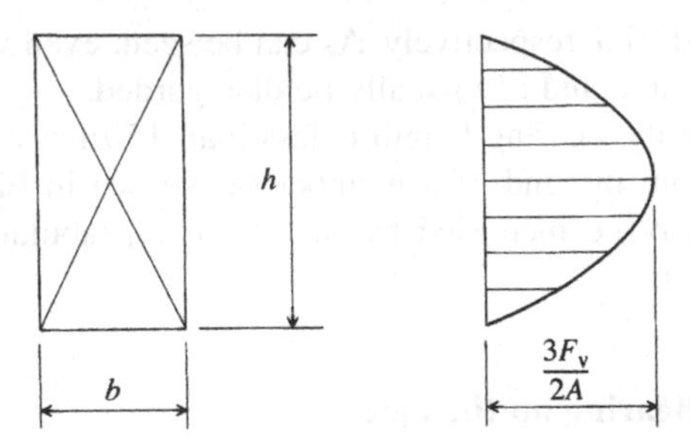

Fig. 4.7

where A is the total cross-sectional area of the beam. (For the derivation of this simplified formula, see section 4.11.3.)

4.11.2 Ply web beams. Panel shear stress

Panel shear stress is the term used in timber engineering for the horizontal shear stress in a plywood web. For a symmetrical ply web beam the maximum panel shear stress occurs at the x–x axis. The general formula for horizontal shear stress must be modified to allow for the effect of the material in the flanges and web having differing E values, thus:

$$\tau_p = \frac{F_v(E_{fN}S_{Xf} + E_w S_{Xw})}{tEI_x} \qquad (4.3)$$

where τ_p = the panel shear stress
F_v = the vertical external shear
E_{fN} = the E value of the flanges taking account of the number of pieces N acting together in the whole beam, not just one flange
E_w = the E value of the web
S_{Xf} = the first moment of area of the flange elements above (or below) the x–x axis
S_{Xw} = the first moment of area of the web above (or below) the x–x axis
t = the sum of the thicknesses of the web
EI_x = the sum of the EI values of the separate parts.

Panel shear stress is discussed further in Chapter 8 on ply web beams, as is 'rolling shear stress' on the glue lines between flanges and web.

4.11.3 Glulam

The horizontal shear stress is a maximum on the neutral axis and is also the maximum stress on the glue line if a glue line occurs at the neutral axis.

The horizontal shear stress at the neutral axis (Fig. 4.8) is as for solid beams:

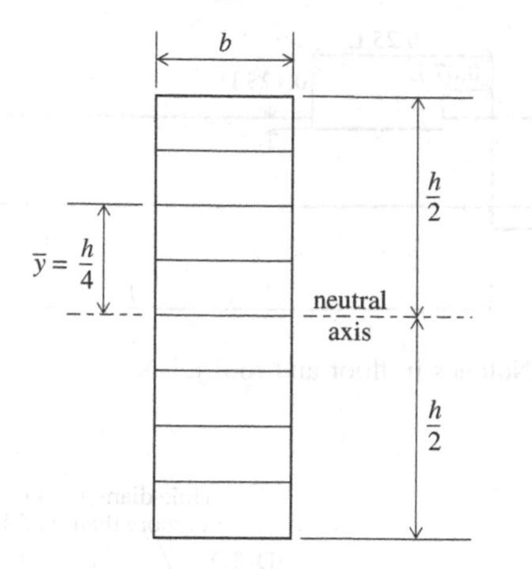

Fig. 4.8

$$\tau = \frac{F_v(A/2)\bar{y}}{bI_x} = \frac{F_v(bh/2)(h/4)}{b(bh^3/12)} = \frac{3F_v}{2bh} = \frac{3F_v}{2A}$$

where A is the total cross-sectional area of the beam.

Shear stress is not usually critical in solid or glulam rectangular section unless the span is small and the loading is heavy.

4.12 THE EFFECT OF NOTCHES AND HOLES

BS 5268-2 requires that allowance should be made in calculations for the effects of notches or holes.

Notwithstanding this, BS 5268-2 states that the effect of notches and holes need not be calculated in simply supported floor or roof joists not more than 250 mm deep where notches not exceeding 0.125 of the joist depth are located between 0.07 and 0.25 of the span from the support, and holes drilled at the neutral axis with a diameter not exceeding 0.25 of the joist depth (and not less than 3 diameters apart centre-to-centre) are located between 0.25 and 0.4 of the span from the support. In other circumstances, where it is necessary to drill or notch a member, allowance for the hole or notch should be made in the design. These are illustrated in Figs 4.9 and 4.10.

In considering the effect of round holes on the bending stress in solid joists it is usually not necessary to consider any effects of stress concentrations. Deflection will hardly be affected, as it is a function of *EI* over the full joist span. Holes in ply web beams are discussed in Chapter 8.

Shear stress for an un-notched solid rectangular section shall be limited so that:

$$\tau = \frac{3F_v}{2bh} \leq \tau_{adm} \quad \text{(modified as appropriate by } K_3, K_8 \text{ or } K_{27}, K_{19} \text{ and } K_{27}\text{)}$$

where h = total depth of section.

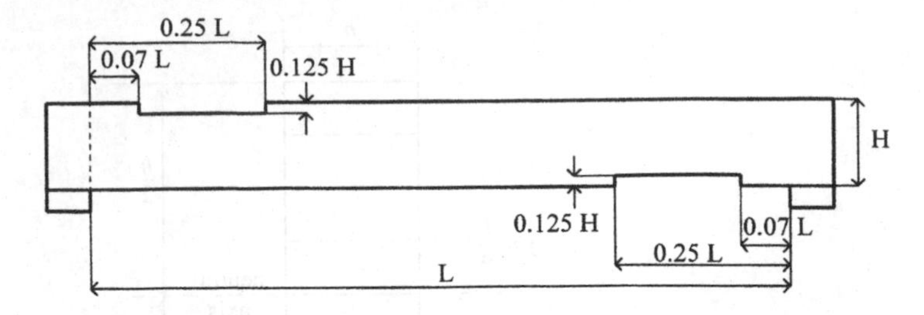

Fig. 4.9 Notches in floor and roof joists.

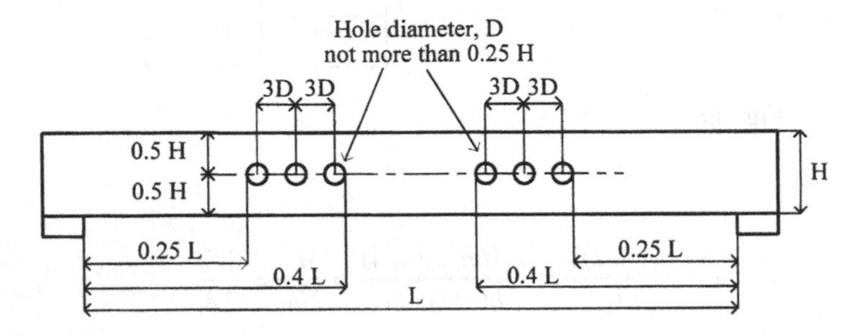

Fig. 4.10 Holes in floor and roof joists.

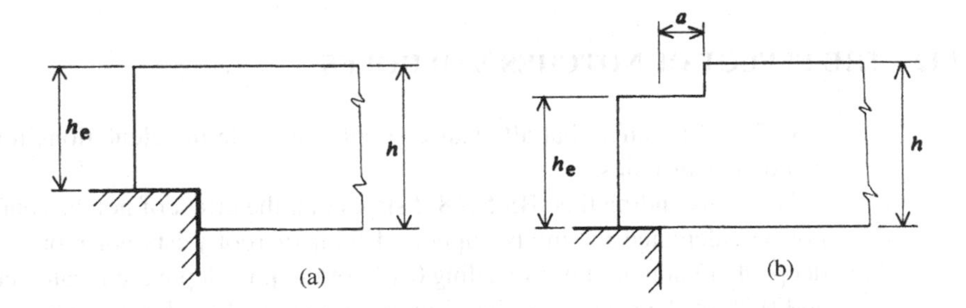

Fig. 4.11

Shear stress for a notched solid rectangular section is:

$$\tau = \frac{3F_v}{2bh_e} \leq \tau_{adm} \times K_5 \quad \text{(modified as appropriate by } K_3, K_8 \text{ or } K_{27}, K_{19} \text{ and } K_{27}\text{)}$$

where h_e = effective depth of section (see Fig. 4.11).
For a notch on the lower edge, see Fig. 4.11(a):

$$K_5 = \frac{h_e}{h}$$

For a notch on the top edge, see Fig. 4.11(b):

$$K_5 = \frac{h(h_e - a) + ah_e}{h_e^2} \quad \text{for } a \leq h_e$$

$$\text{and } K_5 = 1.0 \quad \text{for } a > h_e$$

4.12.1 Shear stress concentrations

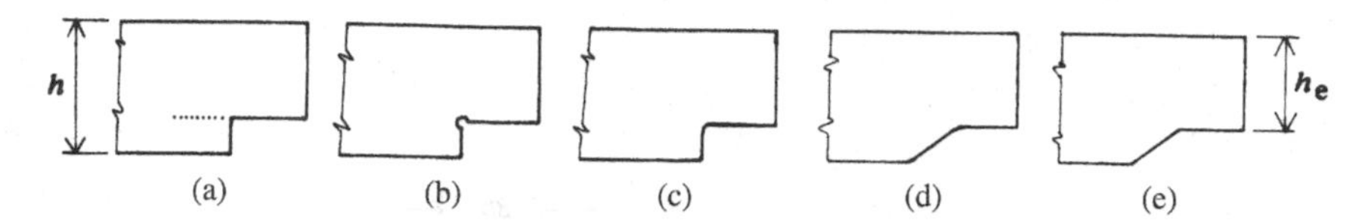

Fig. 4.12

Where beams or joists are notched at the end to suit the support conditions, there is a stress concentration resulting from the rapid change of section.

Square cut notches should be avoided wherever possible. Poor workmanship can lead to over-sawing the notch with a tendency to propagate an early shear failure along the dotted line shown in Fig. 4.12(a). One method of avoiding this problem is to specify all notches to be cut to a pre-drilled hole of, say, 8 mm diameter as shown in Fig. 4.12(b), or to form a generous fillet or taper to the notch as shown in Figs 4.12(c), (d) or (e).

4.12.2 Shear capacity of a notch in the bottom edge

The shear stresses (formula (4.2)) can be translated to shear capacities, e.g. for an un-notched rectangular section

$$\bar{F}_v = \frac{2A\tau}{3}$$

for a notch in the lower edge with $h_e = \alpha h$ (see Fig. 4.11(a)),

$$\tau_{adm} = \alpha\,\tau \quad \text{and} \quad \bar{F}_v = \frac{2}{3}\alpha^2 A\tau$$

It can be seen from the above formula that the shear capacity of the notched section is equal to the shear capacity of the un-notched section multiplied by α^2 and therefore reduces rapidly with increased depth of notch (e.g. with half the full depth remaining, the shear capacity is a quarter that of the full section). BS 5268-2 limits such notches to 50% of the full beam depth.

4.12.3 Shear capacity of a notch in the top edge

Notches may be required in the top edge (Figs 4.13 to 4.15) to suit fixing details for gutters, reduced fascias, etc. Such notches must not exceed 50% of the full

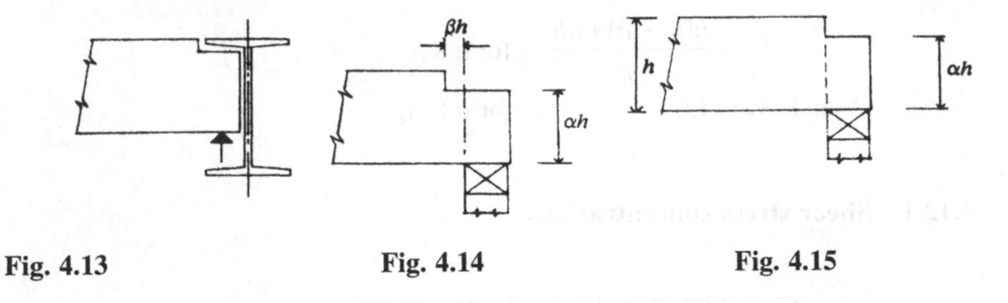

Fig. 4.13 Fig. 4.14 Fig. 4.15

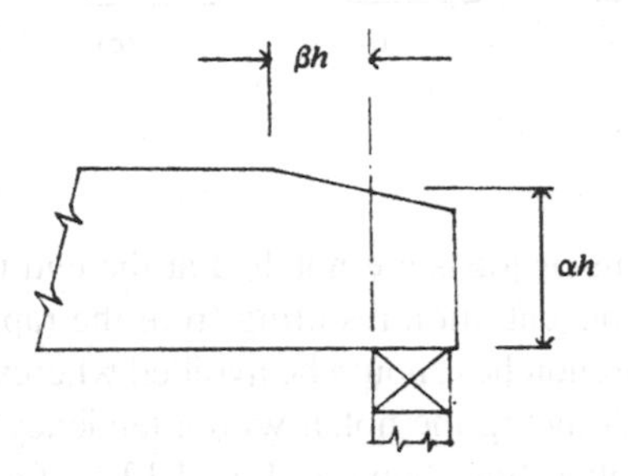

Fig. 4.16

beam depth. The projection of the notch, βh, beyond the inside edge of the bearing line influences the shear capacity of the reduced section.

Figure 4.15 is a special condition where $\beta h = 0$ and the end of the notch coincides with the inside edge of the bearing. Provided that α is at least 50% of the full section the shear capacity of the full depth section is realized.

For a notch in the top edge, two conditions exist:

when $\beta < \alpha$ (see Fig. 4.14) $$\bar{F}_v = \frac{2}{3}\left(1+\beta-\frac{\beta}{\alpha}\right)A\tau$$

when $\beta > \alpha$ the section is evaluated on a depth of (αh) $$\bar{F}_v = \frac{2}{3}\alpha A\tau$$

When the top edge is bevelled (Fig. 4.15) the two formulae above apply with αh measured at the inner face of the bearing and βh measured horizontally from the point of effective depth.

A summary of these formulae is presented in Table 4.4.

4.13 SHEAR IN BEAMS SUPPORTED BY FASTENINGS AND IN ECCENTRIC JOINTS

When a beam is supported along its length, as indicated in Figs 4.17 and 4.18, the force F being transmitted locally into the beam should be resisted by the net

Table 4.4

Detail	Shear capacity
h; h	$\overline{F}_{vu}$
h; αh	$\alpha^2 \overline{F}_{vu}$
h; αh	$\overline{F}_{vu}$, but limit α to 0.5 (see text)
βh; h; αh	* $\left(1 + \beta - \dfrac{\beta}{\alpha}\right) \overline{F}_{vu}$
αh or more; h; αh; 45° or less	$\alpha \overline{F}_{vu}$

$\overline{F}_{vu}$ is the full shear capacity for the un-notched section.

Although presented in a different way to the notes relating to Fig. 4.11, the formulae in this table give the same results as the formulae in BS 5268.

*This formula applies only when βh does not exceed αh.

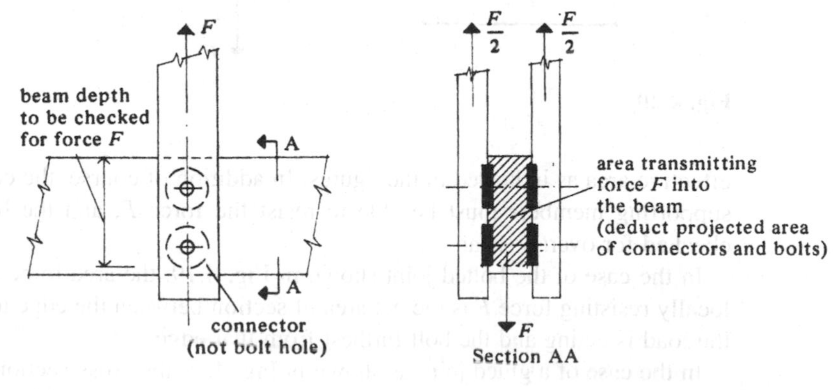

Fig. 4.17

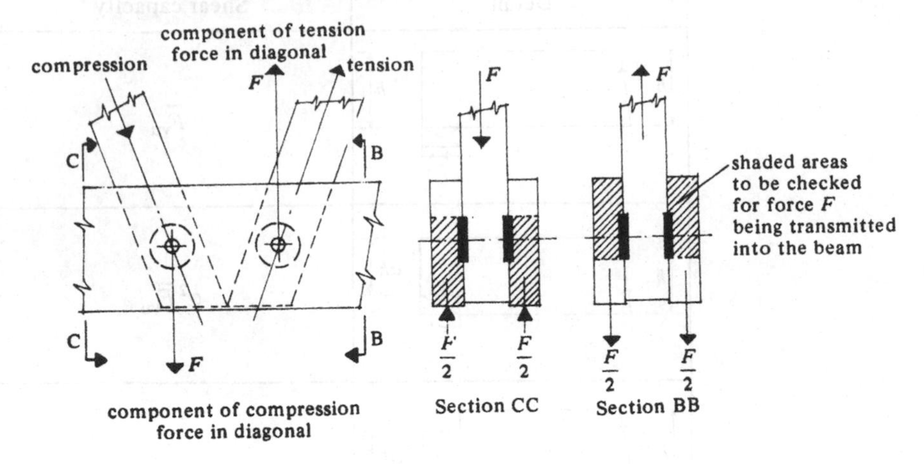

Fig. 4.18

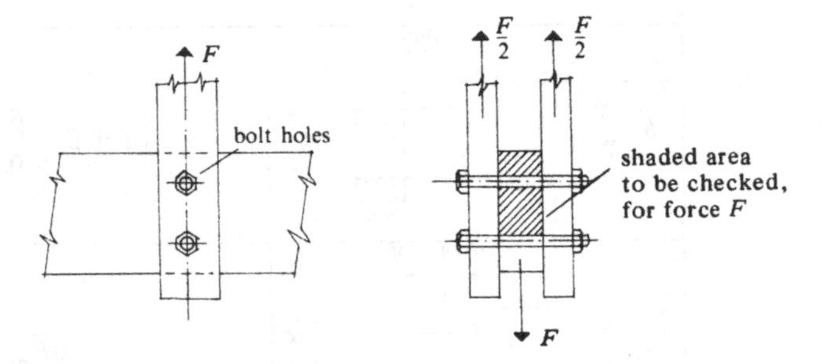

Fig. 4.19

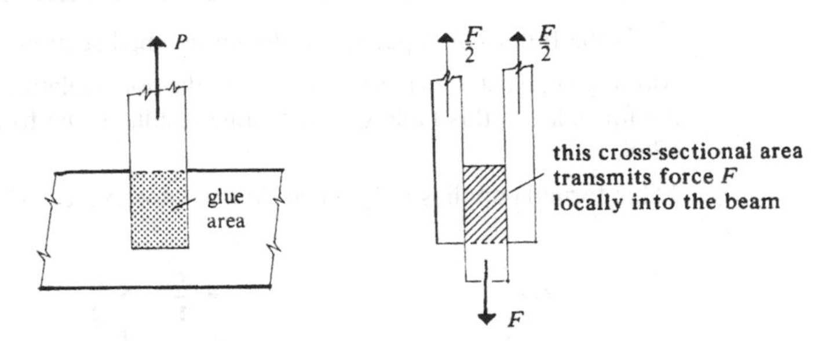

Fig. 4.20

effective area as indicated in the figures. In addition of course, the connectors and supporting members must be able to resist the force F, and the beam must be checked for overall shear.

In the case of the bolted joint shown in Fig. 4.19, the area to be considered as locally resisting force F is the net area of section between the edge towards which the load is acting and the bolt furthest from that edge.

In the case of a glued joint as shown in Fig. 4.20 the cross-sectional area of the beam transmitting force F locally into the beam is the shaded area. However, in

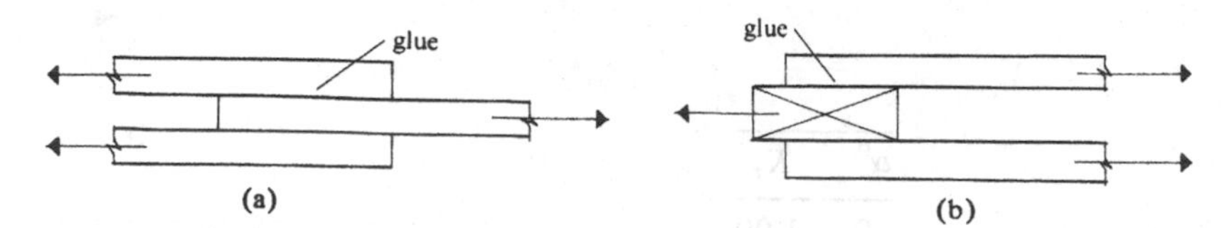

Fig. 4.21

normal circumstances, the critical factor at this position would be the rolling shear at the glue interface.

4.14 GLUE-LINE STRESSES

Providing that one of the recommended glue types is used correctly (see Chapter 19 on glue joints) the glue is always stronger than the surrounding timber. The strength of glue lines is therefore calculated as the permissible shear on the face of the timbers or plywood being glued. Load–duration and load-sharing factors apply because the stresses are timber stresses.

Glue has little strength in tension, therefore connections should be designed to prevent any tension effect on the glue line. The aim should be to achieve pure shear. The type of overlap joint shown in Figs 4.21(a) and (b) should be symmetrical if glued.

When the forces stress the glue along the grain of timber, Fig. 4.21(a), the permissible stress is the permissible stress for the timber grade for shear parallel to the grain (reduced by a factor K_{70} of 0.9 if the glue-line pressure during curing is obtained by nails or staples rather than by positive pressure from clamps or similar items (clause 6.10.1.4 of BS 5268-2).

When the forces are stressing the glue across the grain of timber, Fig. 4.21(b), the permissible stress is one-third of the permissible stress for the timber grade for shear parallel to the grain. This shear across the grain is referred to as 'rolling shear' as the tendency is for the outer fibres to roll like a pile of logs being pushed sideways. If there are likely to be frequent large changes in moisture content (say ±5%, it is not good practice to glue pieces of solid timber together with the angle between the grain directions of the pieces approaching 90°, due to the different moisture movement characteristics in swelling and shrinking along and across the grain.

For forces acting at angles other than 0° and 90° to the grain (e.g. area *A* in Fig. 4.22 but not area *B*) the permissible stress is given by

$$\tau_{\alpha} = \tau_{adm,\parallel}(1 - 0.67\sin\alpha) = \tau_{adm,\parallel}K_{\tau}$$

where $t_{adm,\parallel}$ = the permissible shear parallel to the grain stress for the timber
α = the angle between the direction of the load and the longitudinal axis of the piece of timber (see Fig. 4.22).

Note that when $\alpha = 90°$, $\sin\alpha = 1.0$ and τ_{α} is one-third of $\tau_{adm,\parallel}$, which is the value for rolling shear. Values for $1 - 0.67\sin\alpha$ are tabulated and shown in graph form in Fig. 4.22.

α°	K_τ
0	1.00
5	0.94
10	0.88
15	0.83
20	0.77
25	0.72
30	0.67
35	0.62
40	0.57
45	0.53
50	0.49
55	0.45
60	0.42
65	0.40
70	0.37
75	0.36
80	0.34
85	0.34
90	0.33

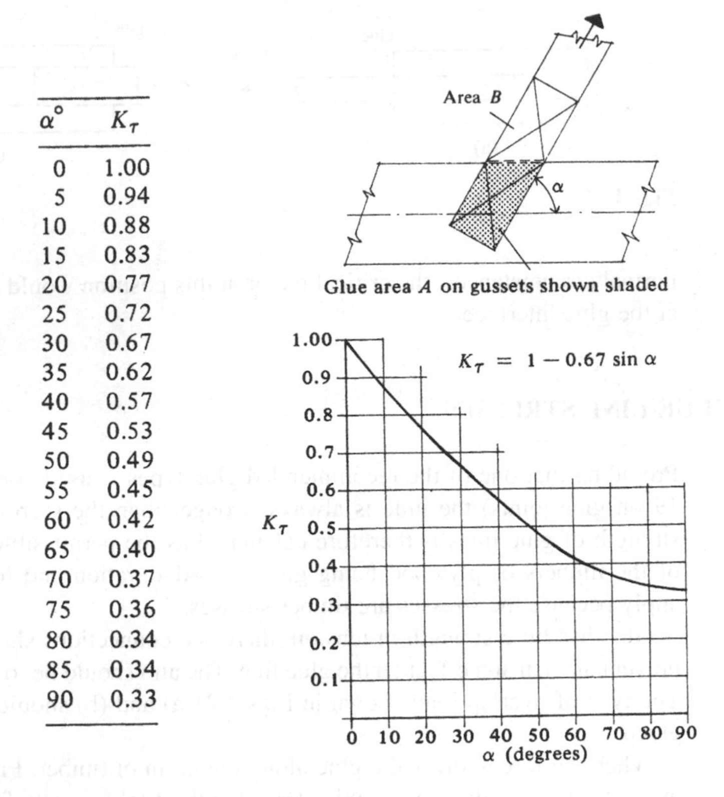

Fig. 4.22 Graph of K_τ for α from 0° to 90°.

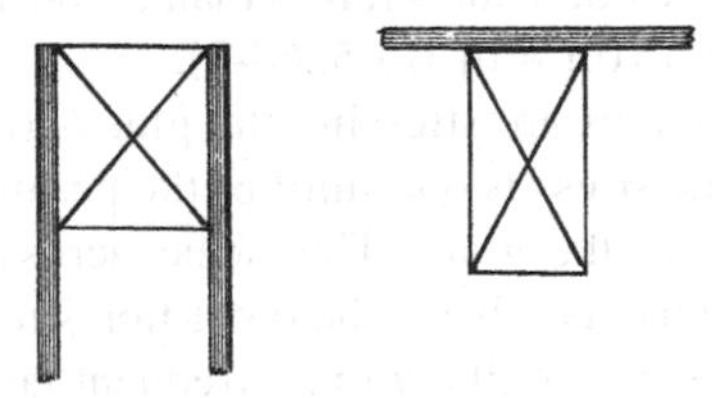

Fig. 4.23

When the forces are stressing the face of plywood either across or along the face veneer (Fig. 4.23) the permissible stress is the permissible 'rolling shear' stress for the plywood grade. Even if the stress is parallel to the face grain it is considered that the next veneer face is so close to the surface that 'rolling' of the fibres in this veneer could occur before the full shear strength of the face veneer is generated.

In the special case of the flange-to-web connection of a ply web beam and the connection of the plywood (or other board) to the outermost joist of a glued stress skin panel (Fig. 4.24), Clause 4.7 of BS 5268-2 requires that the permissible shear stress at the glue line be multiplied by $0.5(K_{37})$. This is an arbitrary factor to take account of likely stress distribution along the glue line. The maximum stress depends on the geometry of the beam section, the combination of shear force and moment and the loading on the beam. The range of possible K_{37} values is from just

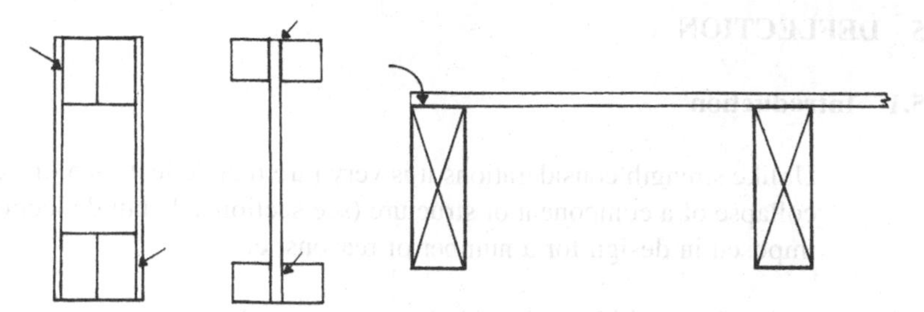

Fig. 4.24

under 0.5 to just under 1.0. From a practical design viewpoint the derivation of K_{37} for a particular situation would be unduly onerous unless some simplification for a software solution could be found.

This may also explain why some tests on ply web beams have achieved factors of safety above 10. Normally, however, a designer would be expected to work to the lower rolling shear value to which must be added the effect of $K_{70}(0.90)$ if the bonding pressure is provided by nails or staples as required by clause 6.10.1.3 of BS 5268-2.

BS 5268-2 (clause 6.10.1.1) puts a limit of 50 mm on the individual pieces of solid timber being glued together and also sets a limit of 29 mm thickness for plywood glued to timber of any thickness. The special case for plywood sandwiched between two timbers, as shown in the I beam in Fig. 4.24, is not specifically covered and a 50 mm limit for timber thickness is recommended by the author. It is prudent to take note of this clause, because if thick pieces are distorted in any way it is extremely difficult to hold them in line during assembly and, if they are forced into place before gluing, subsequent release of the built-in fabrication stresses could damage glue lines.

BS 6446 'Specification for the manufacture of glued structural components of timber and wood based panel products', as its title implies, covers in greater detail the requirements for producing glued joints between timber and essentially plywood. Table 4.5 shows the recommended sizes and maximum spacing of nails to give an adequate glue line pressure between timber and plywood.

Table 4.5 Sizes and maximum spacing of nails for pressure on glue line

Headside thickness (mm)	Nail diameter (mm)	Minimum pointside penetration (mm)	Maximum spacing* (mm)	Maximum edge distance (mm)	Average glue line area per nail (mm^2)
Under 10	2.65	30	75	30	7500
10 to 20	2.65	40	100	30	7500
Over 20	3.35	40	100	30	7500

* Where the headside thickness is 30 mm or over and the gluing surfaces are brought into close contact by cramping only while nail pressure is applied, the maximum spacing may be increased by 50%.

4.15 DEFLECTION

4.15.1 Introduction

Unlike strength considerations it is very rare that deflection alone could lead to the collapse of a component or structure (see section 2.1) but deflection limits are still imposed in design for a number of reasons, e.g.

- to keep within acceptable limits the visual appearance and functional requirements of the component or structure
- to prevent damage to brittle finishes such as plaster
- to prevent undue deflection of roof structures so that rainwater ponds and cannot reach outlets – in the worst situation the accumulation of water could cause overloading of the structure and failure
- to limit the effects of creep
- to provide sufficient stiffness in the construction that vibrations and other dynamic effects are not a problem.

4.15.2 Deflection limits and appropriate *E* values

Any limit which is placed on the magnitude of deflection is purely arbitrary, usual practice or good practice and may be relaxed by the specifying authority in certain cases if this will lead to acceptable economy; or tightened, perhaps for appearance or if a particularly brittle component is being supported. For example, it used to be common practice to limit deflection to span/500 when large cast-iron water tanks were being supported, whereas on storage racks where men can walk, span/240 is often used, or span/180 where there is storage with no access for personnel.

In normal building construction deflection of the timber elements will not threaten the safety of the occupants and in these circumstances deflection and deformation do not come within the scope of the Building Regulations. It may, however, come within the scope of professional indemnity insurers.

'True' *E* values are tabulated in BS 5268-2, therefore the designer should also include a calculation of shear deflection in determining the stiffness of a beam, even for simple floor joists.

All beams carry a dead loading which is in place throughout the life of the member. Most beams also carry an imposed loading which, in the case of a roof beam carrying snow, will only be in place for a small percentage of the life of the beam and may never reach the amount specified in BS 6399-1. In the case of a floor beam, part of the imposed loading will be in place for most of the time (e.g. furniture) whereas the rest of the design-imposed load (e.g. people) will only be in place for part of the time and may never reach the amount specified. In the case of a ceiling joist in a roof, the imposed loading could be in place for years at a time (e.g. suitcases and other storage items).

Although BS 5268-2 (as with similar documents for other materials) only puts a limit on the deflection position below the horizontal with the total load in place, this total load may never occur or may occur only for brief occasional periods and often the more important consideration is the deflection under the permanent or dead loading. When a beam is built up from several pieces (e.g. a ply web beam or glulam) it is possible to build in a camber ('subject to consideration being given to the effect of excessive deformation') so that under the action of the dead loading the beam settles approximately to a horizontal position. In these cases, the deflection limit of $0.003L$ applies to deflection under the imposed loading only. If a beam is overcambered, the deflection limit of $0.003L$ still applies to the total imposed loading, no reduction being allowed to this because of the overcambering.

In the case of beams of solid sawn timber, there is no opportunity to provide camber, however, the span of such beams is limited to around 6 m, at which span the usual deflection limit under total loading is 18 mm (or 14 mm if one is working to the arbitrary limit in BS 5268 for domestic floor joists). Therefore, by proportion, the deflection under permanent loading in most cases is unlikely to be more than around 8–10 mm, which is hardly likely to be noticed unless a beam lines up with a horizontal feature, or the joists are supported by a trimmer which itself deflects 8–10 mm under permanent loading (see Fig. 12.12).

The decision of which E value (and G value) to use in calculating the deflection of components in various situations has already been covered in sections 2.3, 3.4 and 4.4.1 and are summarized below. In addition, practical and special considerations of deflection are discussed in Chapter 12. Where a value for E is stated below (e.g. E_{min}) it should be understood that the appropriate value of shear modulus (e.g. G_{min}) is also intended.

- *Single member principal beam.* A beam having one piece of timber in cross section, acting in isolation, must be regarded as a principal member with deflection calculated using E_{min} (and G_{min}).
- *Trimmer joist or lintel.* If a trimmer joist or lintel is constructed from two or more pieces solidly fixed together by mechanical fasteners to be able to share the loading, the E_{min} and G_{min} values may be increased by K_9 having a stop-off point at four or more pieces.
- *Vertically glued laminated beam.* Where a vertically glued laminated beam is constructed from two or more pieces the E_{min} and G_{min} values may be increased by K_{28} having a stop-off point of eight or more pieces (unless used as part of a load-sharing system with members at centres not exceeding 610 mm, in which case mean values may be used).
- *Built-up beams* (such as ply web beams). The E and G values to use in calculating deflection are the minimum values modified by factor K_{28} for the number of pieces of timber in the cross section. (Clause 2.10.10 of BS 5268-2.)
- *Load-sharing system at close centres.* Where a load-sharing system of four or more members (such as joists or studs) spaced at centres not in excess of 610 mm has adequate provision for lateral distribution of loads, the mean values of moduli may be used in calculations of deflection (unless the area is to support mechanical plant and equipment, or is used for storage, or is subject to vibrations as with a gymnasium or ballroom, in which case minimum values of moduli are to be used).
- *Horizontally laminated glulam beams.* Deflection is calculated by multiplying the mean value of the moduli for the highest strength class used in the manufacture of the beam by factor K_{20}.

4.15.3 Bending deflection and shear deflection

The deflection of any beam is a combination of bending deflection and shear deflection. In addition, if a beam is installed at a high moisture content and is permitted to dry out with a high percentage of the load in place, 'creep' deflection will take place during the drying-out process. This is one reason why it is important to install beams at a moisture content reasonably close to the moisture content they will attain in service. Shear deflection is usually a fairly small percentage of the total deflection of solid sections, but BS 5268-2 requires it to be taken into account in the design of solid timber joists and glulam (see sections 4.15.5 and 4.15.6). Shear deflection is likely to be significant in the design of ply web beams (see section 8.6).

4.15.4 Bending deflection

The formulae for bending deflection of simply supported beams for various conditions of loading are well established. Several are given in Chapter 27 of this manual for central deflection, and for maximum deflection when this is different to central deflection. It can be shown with a simply supported beam that no matter what the loading system is, if the central bending deflection is calculated instead of the maximum deflection, the error is never more than 2.57%. Also the maximum bending deflection always occurs within $0.0774L$ of the centre of span. For complicated loading conditions it can take a considerable time to calculate the position of the maximum deflection even before calculating its magnitude. In this situation it is invariably quite satisfactory to calculate the central deflection as the difference between this value and the true maximum is likely to be less than the variability of the E value of the timber.

In all the formulae below, L should strictly be taken as L_e (see section 4.3).

With a complicated system of loading, one can calculate deflection for the exact loading, but this can be time consuming and an easier way is to calculate the deflection from the equivalent uniformly distributed loading which would give the same deflection as the actual loading.

The deflection δ_m at midspan is calculated as:

$$\delta_m = \frac{5F_eL^3}{384EI}$$

The equivalent uniform load may be determined for most loading conditions as:

$$F_e = FK_m$$

where F = actual load

K_m = a coefficient taken from Tables 4.10–4.17 according to the nature of the actual load.

Where more than one type of load occurs on a span, F_e is the summation of the individual FK_m values. Tables 4.10–4.17 give values of K_m for several loading conditions, and the following example shows how the tables are used.

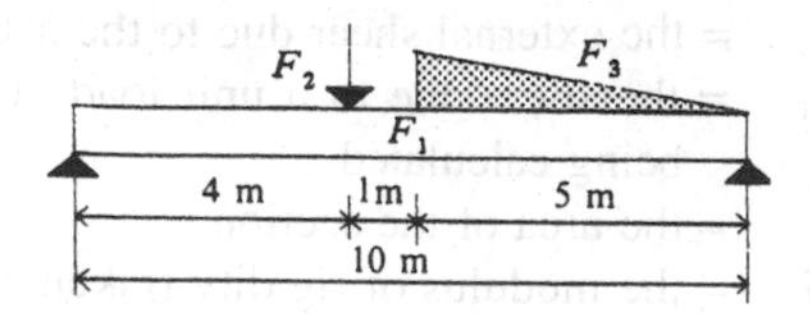

Fig. 4.25

Table 4.6

Load	Load (kN) F	Refer to	n	K_m	FK_m
F_1	2.0	Table 4.11	1.0	1.00	2.00
F_2	6.0	Table 4.17	0.4	1.51	9.06
F_3	8.0	Table 4.12	0.5	1.28	10.24
					$F_e = 21.30$

Coefficient n is a ratio of span (see Tables 4.10–4.17).

Example

Calculate the bending deflection at midspan for the beam shown in Fig. 4.25. Assume an EI capacity of 10100 kN m^2.

Calculate F_e in tabular form as shown in Table 4.6.

$$\text{Bending deflection at midspan} = \frac{5F_eL^3}{384EI} = \frac{5 \times 21.3 \times 10^3}{384 \times 10\,000}$$

$$= 0.0275\,\text{m} = 27.5\,\text{mm}$$

4.15.5 Shear deflection

In addition to deflection caused by bending stresses, there is a further deflection caused by shear stresses (except in the special case of a beam subject to pure bending which is free from shear stresses). Shear deflection is disregarded in most structural materials, although it should certainly be calculated, for example, in the case of a deep heavily loaded steel plate girder.

Timber beams are frequently deep in relation to their span and have a low G/E value (where G is the modulus of rigidity) taken in BS 5268-2 as $\frac{1}{16}$ (0.0625) compared to 0.4 for mild steel.

By the method of unit loads:

$$\text{Shear deflection} = K_{\text{form}} \int_0^L \frac{F_v F_i}{AG} dx$$

where K_{form} = a form factor dependent on the cross-sectional shape of the beam (1.2 for a rectangle)

F_v = the external shear due to the actual loading
F_i = the shear due to a unit load at the point where the deflection is being calculated
A = the area of the section
G = the modulus of rigidity (taken as $E/16$).

The shear deflection is normally added to the centre span bending deflection, therefore it is the centre span shear deflection in which one is interested. With the unit load placed at centre span, $F_i = 0.5$, and it can be shown that:

$$\text{Shear deflection at midspan} = \frac{K_{\text{form}} \times \text{area of product of shear force diagrams to midspan}}{AG}$$

$$= \frac{K_{\text{form}} \times M_0}{AG}$$

where M_0 is the moment at midspan irrespective of the distribution of the applied loading.

To reduce design work, M_0 for a simple span may be calculated as $F_0 L/8$, where F_0 is the equivalent total uniform load to produce the moment M_0

$$M_0 = FK_v$$

where F = actual load
K_v = a coefficient taken from Tables 4.10–4.17 according to the nature of the actual load.

Where more than one type of load occurs on a span, F_0 is the summation of the individual FK_v values.

Example

Calculate the shear deflection at midspan for the beam shown in Fig. 4.25. Assume the AG capacity to be 34 660 kN and $K_{\text{form}} = 1.2$ (for a solid rectangular beam). Calculate F_0 in tabular form (Table 4.7).

$$M_0 = \frac{F_0 L}{8} = \frac{22.27 \times 10}{8} = 27.8\,\text{kN m}$$

Table 4.7

Load	Load (kN) F	Refer to	n	K_v	FK_m
F_1	2.0	Table 4.11	1.0	1.00	2.00
F_2	6.0	Table 4.17	0.4	1.60	9.60
F_3	8.0	Table 4.12	0.5	1.33	10.67
					$F_e = 22.27$

Coefficient n is a ratio of span (see Tables 4.10–4.17).

$$\text{Shear deflection at midspan} = \frac{K_{\text{form}} \times M_0}{AG} = \frac{1.2 \times 27.8}{34\,660}$$
$$= 0.000\,96\,\text{m} = 0.96\,\text{mm}$$

This represents not only a very small actual value but only 3.5% of the bending deflection. It will be seen later in this handbook that in the design of ply web beams shear deflection can be in excess of 15% of the bending deflection and is very significant, but it is usually only a small deflection in real terms with beams of solid timber in the span range where deflection is critical.

4.15.6 Magnitude of shear deflection of solid rectangular sections

To give an indication of the proportion of beams where shear deflection is important, the effect of shear deflection on solid uniform rectangular sections subject to uniformly distributed loading is illustrated in Fig. 4.26.

$$\text{Total deflection} = \delta_t = \delta_m + \delta_v$$

where δ_m = bending deflection
δ_v = shear deflection.

Therefore (with p = load per unit length):

$$\delta_t = \frac{5pL^3}{384EI} + \frac{K_{\text{form}} \times M_0}{GA}$$

$\frac{L}{h}$	$\frac{\delta_t}{\delta_m}$
12	1.106
13	1.091
14	1.078
15	1.068
16	1.060
17	1.053
18	1.047
19	1.042
20	1.038
21	1.035
22	1.032
23	1.029
24	1.027
25	1.025

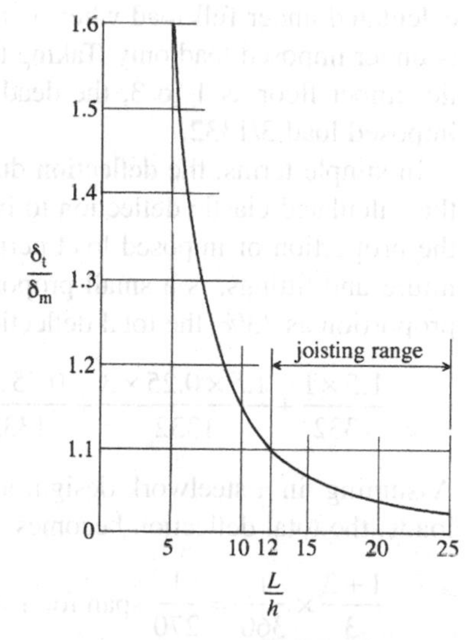

Fig. 4.26

but $GA = Ebh/16$ and $EI = Ebh^3/12$, therefore

$$GA = \frac{3EI}{4h^2}$$

so that, with $K_{form} = 1.2$

$$\delta_t = \frac{5pL^3}{384EI} + \frac{1.2pL^2}{8} \times \frac{4h^2}{3EI}$$

The ratio δ_t/δ_m indicates the relative importance of shear deflection:

$$\frac{\delta_t}{\delta_m} = 1 + \frac{\delta_v}{\delta_m} = 1 + 15.36\left(\frac{h}{L}\right)^2$$

The ratio δ_t/δ_m for values of L/h is plotted in Fig. 4.26

In the normal range of joisting, shear deflection can be expected to add between 2.5 and 10% to the value of bending deflection which, for spans where deflection is critical, represents a small amount in real terms. The designer can judge quickly from the graph in Fig. 4.26 whether or not shear deflection is likely to be significant (but note that BS 5268-2 requires a calculation of shear deflection even for floor or roof joists of solid timber).

Taking the case of a 250 mm deep joist over a 4.6 m span with a total permissible deflection (0.003L) of 13.8 mm for an uncambered beam, shear deflection would be around 0.8 mm for $L/h = 18.4$.

4.15.7 Creep and dynamic effects

The deflection limits given in BS 5268-2 are often criticized for being too simplistic but in reality they achieve the required purposes without being too obvious in their execution. For example, the limit of 0.003 × span (1/333 of the span) is calculated under full load whereas in structural steelwork the limit of 1/360 span is under imposed load only. Taking the ratio of dead to imposed load for a domestic timber floor as 1 to 3, the dead load deflection limit is then 1/1332 and the imposed load 3/1332.

In simple terms, the deflection due to creep under a permanent load can cause the calculated elastic deflection to increase by about 50%. It is also accepted that the proportion of imposed load permanently on a domestic floor, e.g. due to furniture and fittings, is a small proportion of the design-imposed load. Taking this proportion as 33%, the total deflection (including creep) under full load becomes:

$$\frac{1.5 \times 1}{1332} + \frac{1.5 \times 0.25 \times 3}{1332} + \frac{0.75 \times 3}{1332} = \frac{4.875}{1332} = \frac{1}{273} \text{ span for a timber floor}$$

Assuming, in a steelwork design, a similar ratio of 1 to 3 for dead and imposed loads, the total deflection becomes

$$\frac{1+3}{3} \times \frac{1}{360} = \frac{1}{270} \text{ span for a steel (no creep) construction under full load.}$$

This demonstrates that creep has not been entirely ignored in the BS 5268-2 limit.

Creep is likely to be a major problem where a large part of the imposed load is in place for long periods of time, e.g. floors used for storage. In these circumstances BS 5268-2 recommends the use of E_{min} even where the construction is load-sharing. The ratio of $E_{mean}/E_{min} = 1.5$ (section 2.2.5) can be seen to provide some allowance for creep in the deflection calculation.

Seviceability requires consideration of the 'comfort' of the user, and into this comes dynamic effects such as vibration as a person walks across a floor. This is not a simple topic to resolve for it relates to the mass of the floor construction and the actual arrangement of structural members in the floor providing lateral distribution of applied dynamic load and damping of the generated vibration. The problem became manifest in the 1970s as domestic floor spans tended to extend beyond the then typical 3.6 m. Analysis of the relationship between mass and stiffness showed that a finite deflection limit would provide an adequate control of the dynamic effects for domestic floors. Experience has shown that the threshold of human tolerance requires the natural frequency of the floor to be greater than 8 Hz. A simple analogy of floor vibration to that of a spring gives

$$n_0 = \frac{1}{2\pi}\sqrt{\frac{g}{\delta_G}}$$

where n_0 = natural frequency (Hz)
g = acceleration due to gravity (9810 mm/s^2)
δ_G = dynamic deflection of the floor under its own weight.

The dynamic deflection is calculated using the dynamic E value which, for simplicity, can be taken as $1.33E_{mean}$, so taking 75% of the normal dead load deflection, calculation gives the dynamic dead load deflection and the above expression becomes

$$n_0 = \frac{1}{2\pi}\sqrt{\frac{9810}{0.75\delta_G}} = \frac{18.2}{\sqrt{\delta_G}} \quad \text{or} \quad \frac{18}{\sqrt{\delta_G}} \quad \text{in simple terms}$$

After much discussion due to the range of possible values for mass, stiffness and damping characteristics a limit of 14 mm was set in BS 5268-2. Taking the range of typical dead loads for domestic floors as 0.25 kN/m^2 (25% of the total) to 0.75 kN/m^2 (33%) then with the 14 mm total deflection limit, the natural frequencies are 9.6 and 8.4 Hz respectively. The success of this limit can also be judged by the very few complaints that have been recorded since its introduction! BS 5268-2 also requires the use of E_{min} for floors subject to vibration such as gymnasia and dance floors. This provides an increase in both dead load mass and stiffness.

A development of the 14 mm philosophy is to say that an adequately stiff floor can usually be achieved where the dead load deflection does not exceed 3.5 mm (the 25% example above) irrespective of the imposed load category. Care must be exercised in using this approach where rythmic applications of load can occur, e.g. a dance floor or marching soldiers on a bridge.

In present-day terminology these effects are grouped under the heading 'Serviceability'. The approach given in EN 1990 'Basis of design' (otherwise known as Eurocode 0) leads to a more transparent assessment of the serviceability criteria. Verification of these criteria should consider the following:

- deformations that affect the appearance, the comfort of users or the functioning of the structure (including the functioning of machines or services) or that cause damage to finishes or non-structural elements
- vibrations that cause discomfort to people, or that limit the functional effectiveness of the structure
- damage that is likely to adversely affect appearance, durability or the functioning of the structure.

The applications of these serviceability principles are set out in Eurocode 5 DD ENV1995-1: 1994. The modulus of elasticity to be used in all deflection calculations is the mean value. In particular the procedures give a more reasoned approach where more than one type of imposed load is applied to a member, e.g. a beam supporting both roof and floor loads.

For deflection there are three basic requirements:

(1) instantaneous deflection under imposed loads $$u_{\text{inst Q},1} + \sum_{j>1} u_{\text{inst Q},j}\psi_{0,j} \leq \frac{\text{span}}{350}$$

(2) final deflection under total load, including creep $$u_{\text{fin G}} + u_{\text{fin Q}} - u_{\text{c}} \leq \frac{\text{span}}{200}$$

and for domestic floor joists

(3) total instantaneous deflection $$\sum_{i>0} u_{\text{inst G},i} + \sum_{j>0} u_{\text{inst Q},j} \leq 14\,\text{mm}$$

where $u_{\text{inst G},i}$ = instantaneous deflection under dead load i
$u_{\text{inst Q},j}$ = instantaneous deflection under imposed load j
u_{c} = in-built camber

and $$u_{\text{fin G}} = \sum_{i>0} u_{\text{inst G},i}(1 + k_{\text{def}})$$

$$u_{\text{fin Q}} = u_{\text{inst Q},1}(1 + \psi_{2,1}k_{\text{def}}) + \sum_{j>1} u_{\text{inst Q},j}(\psi_{0,j} + \psi_{2,j}k_{\text{def}})$$

with $\psi_{0,j}$ = load combination factor (see Table 4.8)
$\psi_{2,i}$ = load combination factor (see Table 4.8)
k_{def} = factor for creep deflection (see Table 4.9)

Where a timber member having a moisture content greater than 20% is loaded and then dries out under load to a service class 2 condition, the k_{def} value for service class 3 should be increased by 1.00, e.g. the long-term value becomes 2.00 + 1.00 = 3.00. This reflects the well-recognized phenomenon when timber dries while under load and is a very good reason for ensuring that structural timbers intended for use in service class 2 are at the correct moisture content when installed.

Consider the domestic floor with dead load 25% of the total and the span less than 4.666 m. BS 5268-2 gives the deflection limit as span/333 using E_{mean}. This

Table 4.8 Load combination factors

Imposed load category	ψ_0	ψ_2
Residential, institutional, educational	0.7	0.3
Offices and banks	0.7	0.3
Public assembly	0.7	0.6
Retail	0.7	0.6
Storage	1.0	0.8
Snow loads	0.6	0
Wind loads	0.6	0

Table 4.9 Deflection factors for solid timber and glulam k_{def}

	Service class		
	1	2	3
Long term	0.60	0.80	2.00
Medium term	0.25	0.25	0.75
Short and very short term	0	0	0

has been shown previously to give span/270 with creep. EC5 gives the long-term deflection, including creep, in service class 2 as:

$$[0.25 \times \text{total load} \times (1 + 0.8)] + [0.75 \times \text{total load} \times (1 + 0.3 \times 0.8)]$$

$$= 1.38 \times \text{total load} \leq \frac{\text{span}}{200}\text{, which is equivalent under total load to } \frac{\text{span}}{276}$$

There is therefore very little difference between the two methods for the simple condition of a single imposed load.

4.16 BENDING AND SHEAR DEFLECTION COEFFICIENTS

Tables 4.10–4.17 give coefficients K_m and K_V which can be used to expedite the calculations of bending and shear deflections in the centre of simply supported beams. The notes and examples in sections 4.15.4 and 4.15.5 explain how they should be used. Coefficient n in these tables is a ratio of the span.

F is total load in all cases

Table 4.10

n	K_m	K_v
0	–	–
0.05	0.120	0.1
0.1	0.238	0.2
0.15	0.355	0.3
0.2	0.467	0.4
0.25	0.575	0.5
0.3	0.677	0.6
0.333	0.740	0.667
0.35	0.771	0.7
0.4	0.858	0.8
0.45	0.934	0.9
0.5	1.000	1.000
0.55	1.054	1.082
0.60	1.095	1.133
0.65	1.123	1.161
0.667	1.130	1.167
0.7	1.138	1.171
0.75	1.141	1.166
0.8	1.133	1.150
0.85	1.114	1.123
0.9	1.080	1.089
0.95	1.046	1.047
1.000	1.000	1.000

Table 4.11

n	K_m	K_v
0	1.600	2.0
0.05	1.598	1.95
0.1	1.592	1.9
0.15	1.583	1.85
0.2	1.570	1.8
0.25	1.553	1.75
0.3	1.533	1.7
0.333	1.518	1.667
0.35	1.510	1.65
0.4	1.485	1.6
0.45	1.456	1.55
0.5	1.425	1.5
0.55	1.391	1.45
0.6	1.355	1.4
0.65	1.325	1.35
0.667	1.311	1.333
0.7	1.277	1.3
0.75	1.234	1.25
0.8	1.190	1.2
0.85	1.145	1.15
0.9	1.098	1.1
0.95	1.049	1.05
1.000	1.000	1.000

Table 4.12

n	K_m	K_v
0	–	–
0.05	0.160	0.133
0.1	0.317	0.266
0.15	0.471	0.4
0.2	0.620	0.533
0.25	0.760	0.567
0.3	0.891	0.8
0.333	0.988	0.889
0.35	1.010	0.933
0.4	1.116	1.067
0.45	1.207	1.2
0.5	1.280	1.333
0.55	1.334	1.431
0.60	1.368	1.474
0.65	1.382	1.477
0.667	1.375	1.472
0.7	1.376	1.453
0.75	1.351	1.407
0.8	1.309	1.346
0.85	1.251	1.272
0.9	1.180	1.188
0.95	1.095	1.097
1.000	1.000	1.000

F is total load in all cases

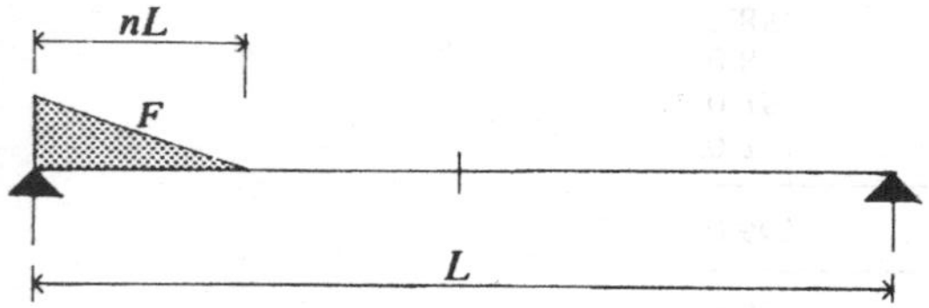

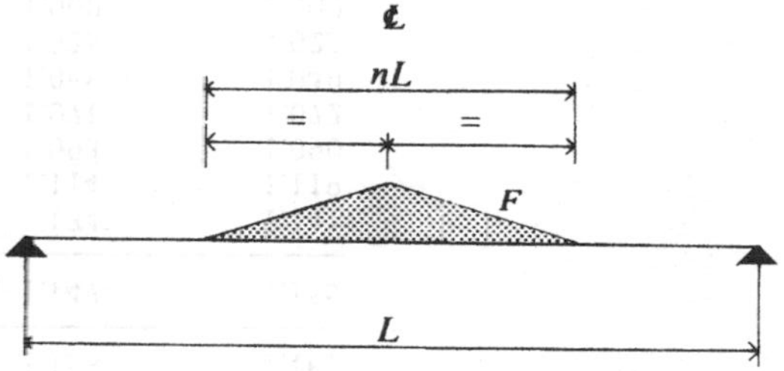

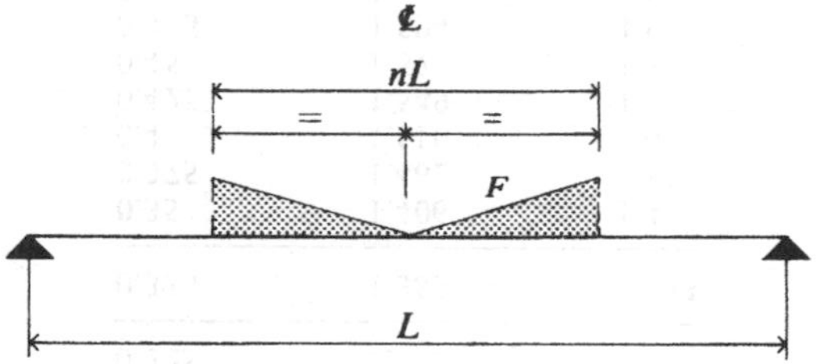

Table 4.13

n	K_m	K_v
0	–	–
0.05	0.080	0.067
0.1	0.159	0.133
0.15	0.238	0.2
0.2	0.315	0.267
0.25	0.370	0.333
0.3	0.463	0.4
0.333	0.510	0.444
0.35	0.532	0.467
0.4	0.600	0.533
0.45	0.662	0.6
0.5	0.720	0.667
0.55	0.774	0.732
0.6	0.822	0.793
0.65	0.864	0.845
0.667	0.877	0.861
0.7	0.901	0.890
0.75	0.932	0.925
0.8	0.957	0.954
0.85	0.976	0.975
0.9	0.990	0.989
0.95	0.997	0.997
1.000	1.000	1.000

Table 4.14

n	K_m	K_v
0	1.600	2.0
0.05	1.599	1.967
0.1	1.596	1.933
0.15	1.591	1.9
0.2	1.585	1.867
0.25	1.576	1.833
0.3	1.566	1.8
0.333	1.559	1.778
0.35	1.554	1.767
0.4	1.541	1.733
0.45	1.526	1.7
0.5	1.510	1.667
0.55	1.492	1.633
0.6	1.473	1.6
0.65	1.453	1.567
0.667	1.446	1.556
0.7	1.431	1.533
0.75	1.409	1.5
0.8	1.385	1.467
0.85	1.360	1.433
0.9	1.334	1.4
0.95	1.307	1.367
1.000	1.280	1.334

Table 4.15

n	K_m	K_v
0	1.600	2.0
0.05	1.597	1.933
0.1	1.588	1.867
0.15	1.574	1.8
0.2	1.555	1.733
0.25	1.530	1.667
0.3	1.500	1.6
0.333	1.479	1.556
0.35	1.467	1.533
0.4	1.428	1.467
0.45	1.386	1.4
0.5	1.340	1.333
0.55	1.290	1.267
0.6	1.237	1.2
0.65	1.181	1.133
0.667	1.161	1.111
0.7	1.122	1.067
0.75	1.060	1.0
0.8	0.996	0.933
0.85	0.929	0.867
0.9	0.861	0.8
0.95	0.791	0.733
1.000	0.72	0.667

F is total load in all cases

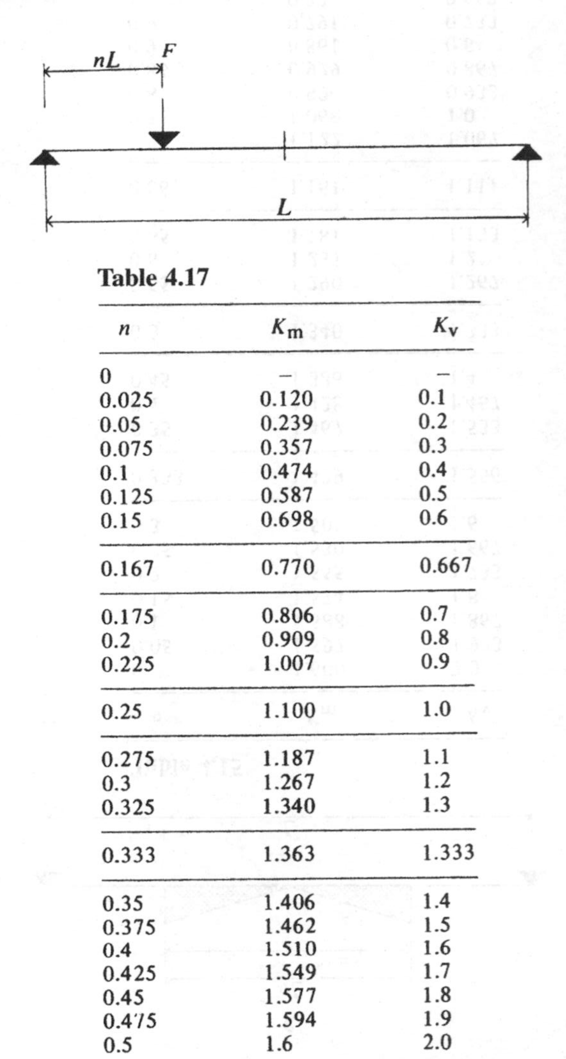

Table 4.16

n	K_m	K_v
0	1.28	1.333
0.05	1.279	1.332
0.1	1.277	1.327
0.15	1.273	1.320
0.2	1.267	1.311
0.25	1.260	1.300
0.3	1.252	1.287
0.333	1.245	1.277
0.35	1.242	1.273
0.4	1.230	1.257
0.45	1.211	1.240
0.5	1.203	1.222
0.55	1.188	1.203
0.6	1.171	1.183
0.65	1.153	1.162
0.667	1.147	1.155
0.7	1.134	1.141
0.75	1.114	1.119
0.8	1.093	1.090
0.85	1.071	1.072
0.9	1.048	1.049
0.95	1.025	1.025
1.000	1.000	1.000

Table 4.17

n	K_m	K_v
0	–	–
0.025	0.120	0.1
0.05	0.239	0.2
0.075	0.357	0.3
0.1	0.474	0.4
0.125	0.587	0.5
0.15	0.698	0.6
0.167	0.770	0.667
0.175	0.806	0.7
0.2	0.909	0.8
0.225	1.007	0.9
0.25	1.100	1.0
0.275	1.187	1.1
0.3	1.267	1.2
0.325	1.340	1.3
0.333	1.363	1.333
0.35	1.406	1.4
0.375	1.462	1.5
0.4	1.510	1.6
0.425	1.549	1.7
0.45	1.577	1.8
0.475	1.594	1.9
0.5	1.6	2.0

Chapter 5

Beams of Solid Timber

5.1 INTRODUCTION

For many of the uses of solid timber, such as floor or roof joists, the designer can refer to load/span tables in which the designs have already been carried out employing the principles set out in BS 5268-7. Such tables are given in Approved Document A of *The Building Regulations 1991* for England and Wales and in tables issued by TRADA.

The design of a timber beam or joist was once relatively straightforward but the introduction of a more comprehensive design approach, as introduced into clause 2.10.7 and Annex K of BS 5268-2, can extend the design work involved. It would appear that this comprehensive approach may be regarded as optional and to be considered generally when creep deflection is expected to be a problem. The designer should note that when Annex K is used the E value for deflection calculations is the mean value even for a principal member.

5.2 GENERAL DESIGN

Chapter 4 details the various factors and aspects which must be taken into account in the design of beams. They are not repeated at length in this chapter but when a factor is used in the calculations and it is thought that the designer might wish to check back, reference is made to the relevant paragraph of Chapter 4.

The designer is in one of two different situations in the design process. The first alternative is assessing a section and grade of timber that has already been decided, e.g. checking the design of a component or construction that has already been prepared by another designer. This is a straightforward situation, in which the section properties and the grade are known, so the designer can immediately calculate the actual stresses and compare them with the strength values. In the second case, the designer originates the design knowing only the span and load configuration. The process of assuming cross-section dimensions and timber grade and then checking for the various stresses and deflections can be time consuming, unless the designer has enough experience or good fortune to choose the correct section and grade at the first attempt.

Both situations are covered by typical examples in this chapter. To eliminate the time-consuming guesswork of the second situation, tables of beam capacities are introduced from which the required capacities for shear, moment and deflection performance can be compared easily with the required values.

5.3 PRINCIPAL BEAMS OF SOLID TIMBER

With principal members composed of one piece of timber there is obviously no load sharing and the minimum value of E must be used. The load duration factor, K_3, applies to bending, shear and bearing strength values and the depth factor, K_7, can be applied to bending values.

5.3.1 Example of checking a previously selected floor trimmer beam (principal member)

Check the suitability of a 72 × 220 strength class C24 for use within a domestic floor (i.e. service class 2).

The clear span is 2.80 m and the trimmer supports incoming joists spaced at 0.60 m centres.
Dead load = 0.35 kN/m² inclusive of trimmer self-weight.
Imposed load = 1.50 kN/m² or a concentrated load of 1.4 kN located anywhere on the span.
Incoming joists have a span of 4.20 m; they occur only on one side of the trimmer and provide lateral restraint for the trimmer.
Note 2 to clause 2.8 of BS 5268-2 allows the concentrated imposed load on a domestic floor to be taken as medium-term duration.

The incoming beams are at close enough centres to justify taking the load on the trimmer as being uniformly distributed. A bearing length of 50 mm has been provided at each end giving an effective design span of 2.85 m.

Loading

$$\text{Dead UDL} = 2.85 \times \frac{4.20}{2} \times 0.35 = 2.09\,\text{kN}$$

$$\text{Imposed UDL} = 2.85 \times \frac{4.20}{2} \times 1.50 = 8.98\,\text{kN}$$

$$\text{Total UDL long term} = 11.07\,\text{kN}$$

$$\text{End reaction} = \text{maximum end shear} = \frac{11.07}{2} = 5.54\,\text{kN}$$

End bearing

Applied end bearing stress (long term with 50 mm bearing length)

$$= \frac{5.54 \times 1000}{72 \times 50} = 1.54\,\text{N/mm}^2$$

Permissible bearing stress (wane permitted) = grade stress × K_3
= 1.9 × 1.0 = 1.9 N/mm² **O.K.**

Permissible bearing stress (no wane permitted) = grade stress × K_3 = 2.4 × 1.0
= 2.4 N/mm² **O.K.**

Notes: (1) Where the specification specifically excludes wane the higher value of compression perpendicular to grain stress is applicable.
(2) Enhanced bearing factor K_4 does not apply.

Shear

Shear force = F_v = 5.54 kN

Applied shear stress $= \dfrac{3 \times F_v}{2 \times b \times h} = \dfrac{3 \times 5.54 \times 1000}{2 \times 72 \times 220} = 0.52\,\text{N/mm}^2$

Permissible shear stress (long term) = grade stress × K_3 = 0.71 × 1.0
= 0.71 N/mm² **O.K.**

Bending

Maximum bending moment $= M_L = \dfrac{11.07 \times 2.85}{8} = 3.95\,\text{kNm}$

Section modulus $= Z_x = \dfrac{72 \times 220^2}{6} = 0.581 \times 10^6\,\text{mm}^3$

Applied bending stress $= \dfrac{M_L}{Z_x} = \dfrac{3.95 \times 10^6}{0.581 \times 10^6} = 6.8\,\text{N/mm}^2$

Permissible bending stress = grade stress × K_3 × K_7 = 7.50 × 1.0 × 1.032
= 7.74 N/mm² **O.K.**

Note: See Table 5.3 for values of K_7.

Deflection

For a principal member $E = E_{min}$ = 7200 N/mm² and $G = E_{min}/16$ = 450 N/mm².

$I_x = 72 \times 220^3/12 = 63.9 \times 10^6\,\text{mm}$

$EI_x = E_{min} \times I_x = 460\,\text{kN m}^2$

$AG = A \times G = (72 \times 220) \times 450/1000 = 7128\,\text{kN}$

Bending deflection $= \dfrac{5 \times F \times L_e^3}{384 \times EI_x} = \dfrac{5 \times 11.07 \times 2.85^3}{384 \times 460} = 0.00725\,\text{m}$

Shear deflection $= \dfrac{K_{form} \times M}{AG} = \dfrac{1.2 \times 3.95}{7128} = 0.00066\,\text{m}$

Total deflection = 0.00791 m

Alternatively, total deflection may be calculated as:

Total deflection = bending deflection × K_V

where $K_V = 1 + 15.36(h/L)^2 = 1 + 15.36(220/2850)^2 = 1.0915$

Total deflection = 0.00725 × 1.0915 = 0.00791 m

Permissible deflection = 0.003 × 2.85 = 0.00855 m **O.K.**

The second loading condition to be checked is a medium-term condition. By inspection the 1.8 kN concentrated imposed load can be seen to be less critical than the uniformly imposed loading condition, i.e.

$$\text{Moment} = \frac{2.09 \times 2.85}{8} + \frac{1.4 \times 2.85}{4} = 1.74\,\text{kNm} < 3.95\,\text{kNm}$$

and no further design check is needed.

Prior to the introduction of Annex K the deflection assessment would stop at this point with the trimmer shown to be satisfactory.

Clause 2.10.7 of BS 5268-2 is now amended to offer 'more comprehensive procedures covering various load types and durations . . . given in Annex K'. If the designer considers that such a procedure is required the deflection assessments would continue as follows:

From section 4.15.7 or Annex K for residential category,

$\psi_0 = 0.7$, $\psi_2 = 0.3$ and $k_{def} = 0.8$

By proportion of dead to imposed load,

Dead load instantaneous deflection $= \delta_G = 0.0015\,\text{m}$
Imposed load instantaneous deflection $= \delta_Q = 0.0064\,\text{m}$

Case (a) Instantaneous deflection under imposed load

Deflection $= \delta_Q = 0.0064\,\text{m}$
Allowable deflection = span/350 = 2.85/350 = 0.008 m **O.K.**

Case (b) Final deflection under total load including creep

$$\begin{aligned}\text{Deflection} &= \delta_G(1 + k_{def}) + \delta_Q(1 + \psi_2 k_{def})\\ &= \delta_G(1 + 0.8) + \delta_Q(1 + 0.3 \times 0.8)\\ &= 1.8\delta_G + 1.24\delta_Q\end{aligned}$$

Therefore

$$\begin{aligned}\text{Deflection} &= (1.8 \times 0.0015) + (1.24 \times 0.0064)\\ &= 0.0027 + 0.0079 = 0.0106\,\text{m}\end{aligned}$$

Allowable deflection = span/200 = 2.85/200 = 0.0142 m **O.K.**

The deflections calculated for case (a) and case (b) are conservative having been based on E_{min} in lieu of the permitted E_{mean}.

5.3.2 Example where the section, species/grade are to be determined

Find a section, species/grade of timber to meet the span and loading conditions of section 5.3.1.

From section 5.3.1 it can be seen that any section to meet the design requirements must have the following capacities:

Long-term shear capacity = 5.54 kN
Long-term moment capacity = 3.95 kN m

Because shear deflection must be taken into account it is not possible to calculate an accurate stiffness capacity but a useful method of estimating the required *EI* capacity required to limit deflection to 0.003 of span involves a transposition of the conventional formula for a uniformly distributed load.

With a total UDL of *F* on span *L*, at a maximum permissible deflection of 0.003*L* and taking account of the ratio δ_t/δ_m in Fig. 4.26:

$$\delta_{total} = \frac{5FL^3}{384EI} \times \frac{\delta_t}{\delta_m} = 0.003L$$

from which

$$EI = 4.34FL^2 \times \frac{\delta_t}{\delta_m}$$

Without knowing the depth of the joist which will be used, the L/h ratio cannot be calculated and hence the ratio of δ_t/δ_m is not known. However, it can be anticipated that L/h is likely to be between 12 and 15. If a value of 12 is adopted, this gives δ_t/δ_m at 1.106 from Fig. 4.26.

Therefore an initial estimate for EI may be taken as $EI = 4.8FL^2$

For the general case of any loading condition this formula becomes $EI = 4.8F_eL^2$ where F_e is the 'equivalent UDL' which will give the same bending deflection at midspan as the actual loading (see section 4.15.6).

For a total UDL of 11.1 kN the estimated EI capacity required is
$4.8 \times 11.1 \times 2.85^2 = 391\,\text{kN}\,\text{m}^2$
For a dead UDL of 2.09 kN plus a concentrated central load of 1.8 kN
$F_e = 2.09 + (1.8 \times 1.6) = 4.97\,\text{kN}$ (1.6 is K_m from Table 4.17 with $n = 0.5$)
and the estimated EI capacity required is
$4.8 \times 2.85^2 \times 4.97 = 194\,\text{kN}\,\text{m}^2$
The total UDL condition governs the EI requirement.

From Table 5.3 it can be seen that a 72 × 220 C24 section is satisfactory for shear and bending. The EI capacity of 460 kN m² indicates that the deflection is satisfactory, subject perhaps to a final design check now that L/h is established.

5.4 LOAD-SHARING SYSTEMS OF SOLID TIMBER

With a load-sharing system of four or more beams of solid timber spaced at not more than 610 mm centres, the load-sharing factor ($K_8 = 1.1$) discussed in section 4.4 applies to all stresses. E_{mean} is usually applicable subject to the limitation explained in section 4.15.1.

5.4.1 Example of checking a previously selected load-sharing floor joist system

Check the suitability of 38 × 235 C16 joisting spaced at 0.6 m centres over a clear span of 3.3 m for use in a domestic floor (service class 2).

Dead loading (including the self-weight of joists) is 0.35 kN/m², the imposed loading is 1.5 kN/m² (all of which must be considered as being permanently in place); the concentrated load may be neglected (see section 3.3). A bearing length of 50 mm has been provided at each end giving an effective design span of 3.35 m.

Loading

Dead UDL = 3.35 × 0.6 × 0.35 = 0.70 kN
Imposed UDL = 3.35 × 0.6 × 1.5 = 3.01 kN
Total UDL long term = 0.70 + 3.01 = 3.71 kN
End reaction = maximum end shear = 3.71/2 = 1.85 kN

End bearing

Applied end bearing stress (long term with 50 mm bearing length)
= 1850/(38 × 50) = 0.97 N/mm^2
Permissible bearing stresses:
(wane permitted) = grade stress × K_3 × K_8 = 1.7 × 1.0 × 1.1 = 1.87 N/mm^2
O.K.
(no wane permitted) = grade stress × K_3 × K_8 = 2.2 × 1.0 × 1.1 = 2.42 N/mm^2
O.K.

Notes: (1) Where the specification specifically excludes wane the higher value of compression perpendicular to grain stress is applicable.
(2) Enhanced bearing factor K_4 does not apply.

Shear

Shear force = F_v = 1.85 kN
End reaction = maximum end shear = 1.85 kN

$$\text{Applied shear stress} = \frac{3 \times 1850}{2 \times 38 \times 235} = 0.31\,\text{N/mm}^2$$

Permissible shear stress (long term) = grade stress × K_3 × K_8 = 0.67 × 1.0 × 1.1
= 0.74 N/mm^2 **O.K.**

Bending

$$\text{Long-term bending moment} = M_L = \frac{3.71 \times 3.35}{8} = 1.55\,\text{kN m}$$

$$Z_x = 0.350 \times 10^6\,\text{mm}^3$$

$$\text{Applied bending moment} = \frac{M_L}{Z_x} = \frac{1.55 \times 10^6}{0.350 \times 10^6} = 4.43\,\text{N/mm}^2$$

Permissible bending stress = grade stress × K_3 × K_7 × K_8
= 5.3 × 1.0 × 1.03 × 1.1
= 6.00 N/mm^2 **O.K.**

Note: See Table 5.2 for values of K_7.

Deflection

For a load-sharing member E = 8800 N/mm^2 and $G = E_{min}/16$ = 550 N/mm^2.

I_x = 38 × 235^3/12 = 41.1 × 10^6 mm
$EI_x = E_{mean} \times I_x$ = 362 kN m^2
$AG = A \times G$ = (38 × 235) × 550/1000 = 4911 kN

$$\text{Bending deflection} = \frac{5 \times F \times L^3}{384 \times EI_x} = \frac{5 \times 3.71 \times 3.35^3}{384 \times 362} = 0.00501\,\text{m}$$

$$\text{Shear deflection} = \frac{K_{form} \times M}{AG} = \frac{1.2 \times 1.55}{4911} = 0.00038\,\text{m}$$

Total deflection = 0.005 01 + 0.000 38 = 0.0054 m

Alternatively, total deflection may be calculated as:

Total deflection = bending deflection × K_V
where $K_V = 1 + 15.36(h/L)^2 = 1 + 15.36\,(235/3350)^2 = 1.0756$
Total deflection = 0.005 01 × 1.0756 = 0.0054 m
Permissible deflection = 0.003 × 3.35 = 0.010 m **O.K.**

5.4.2 Example where the section, species/grade are to be determined

Find a range of sections/grades of timber to meet the span and loading conditions of section 5.4.1.

From section 5.4.1 it can be seen that any section to meet the design requirements must have the following capacities:

Long-term shear capacity = 1.85 kN

Long-term moment capacity = 1.55 kN m

Approximate EI_x required = $4.80FL^2 = 4.80 \times 3.71 \times 3.35^2 = 200\,\text{kN m}^2$

Refer to section 5.5 for shear, moment and *EI* capacities for various species/grades and choose from

38 × 235 C16 from Table 5.2
38 × 184 C24 from Table 5.4
35 × 195 TR26 from Table 5.6
44 × 170 TR26 from Table 5.6

Where the *EI* capacity is close to the required value a final deflection check should be made, i.e. the 44 × 170 TR26 with an *EI* capacity of 198 kN m² should be checked as follows:

$$\text{Bending deflection} = \frac{5 \times F \times L_e^3}{384 \times EI_x} = \frac{5 \times 3.71 \times 3.35^3}{384 \times 198} = 0.00917\,\text{m}$$

Total deflection = bending deflection × K_V
where $K_V = 1 + 15.36(h/L)^2 = 1 + 15.36(170/3350)^2 = 1.0025$
Total deflection = 0.009 17 × 1.0025 = 0.0092 m
Permissible deflection = 0.003 × 3.35 = 0.010 m **O.K.**

Consideration of the list of possible sections illustrates that shear is seldom critical and the choice depends in most cases on bending and deflection considerations. Deflection tends to be the limit in load-sharing systems and bending may be the limit in principal members.

5.5 GEOMETRICAL PROPERTIES OF SOLID TIMBER SECTIONS IN SERVICE CLASSES 1 AND 2

Tables 5.1 to 5.6 offer section capacities for timber sizes/grades that are generally commercially available. C16 is more available than C24 so that C16 may be preferable for load-sharing systems requiring large quantities of a particular section, whereas C24 may be prefered for principal members. The range of sections is also available in TR26 material, which although initially produced for the trussed rafter market, is becoming increasingly popular as an alternative to C16 and C24 material.

5.6 PRINCIPAL MEMBERS BENDING ABOUT BOTH THE *x–x* AND *y–y* AXES

When the direction of load does not coincide with one of the principal axes of a section there is bending about both axes and the design procedure is usually one of trial and error to find a section which will resist the combined bending and deflection of the section.

If a load F acts at an angle θ to the y–y axis (Fig. 5.1), then the components of the load acting about the x–x and y–y axes are $F\cos\theta$ and $F\sin\theta$ respectively. The axis v–v is at an inclination ϕ to the x–x axis, where:

$$\tan\phi = \frac{I_x}{I_y}\tan\theta = \left(\frac{h}{b}\right)^2 \tan\theta$$

The critical points for bending stress are at A and B which are the fibres furthest from the v–v axis and at these points the bending stress is:

$$\sigma_{m,a,par} = \left[\frac{M_x}{Z_x} + \frac{M_y}{Z_y}\right] \quad \text{tension or compression}$$

The direction of the deflection is normal to the v–v axis. The total deflection δ_t is the geometrical sum of the deflections δ_x and δ_y

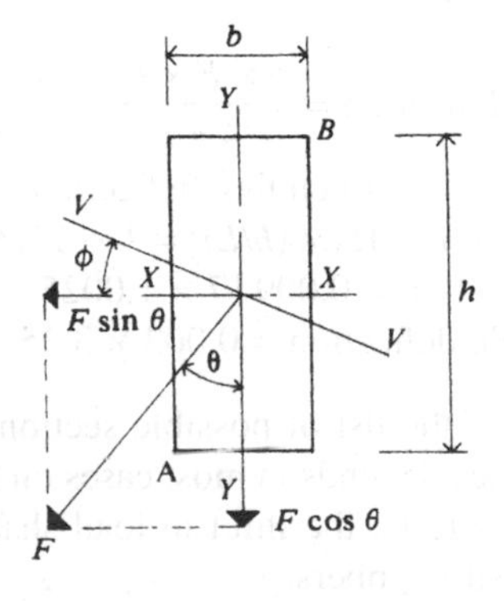

Fig. 5.1

Table 5.1 Geometrical properties: principal members, strength class C16, service classes 1 and 2

Section				Section properties		Long term		Medium term		
b (mm)	h (mm)	h/b	Depth factor K_7	Z ($\times 10^6$ mm^3)	I ($\times 10^6$ mm^4)	Shear (kN)	Moment (kN m)	Shear (kN)	Moment (kN m)	EI (kN m^2)
35	72	2.1	1.17	0.030	1.1	1.13	0.19	1.41	0.23	6
38	63	1.7	1.19	0.025	0.8	1.07	0.16	1.34	0.20	5
38	89	2.3	1.14	0.050	2.2	1.51	0.30	1.89	0.38	13
38	140	3.7	1.09	0.124	8.7	2.38	0.72	2.97	0.89	50
38	235	6.2	1.03	0.350	41.1	3.99	1.90	4.99	2.38	238
45	70	1.6	1.17	0.037	1.3	1.41	0.23	1.76	0.29	7
45	95	2.1	1.13	0.068	3.2	1.91	0.41	2.39	0.51	19
45	122	2.7	1.10	0.112	6.8	2.45	0.65	3.07	0.82	39
45	147	3.3	1.08	0.162	11.9	2.95	0.93	3.69	1.16	69
45	170	3.8	1.06	0.217	18.4	3.42	1.22	4.27	1.53	107
45	195	4.3	1.05	0.285	27.8	3.92	1.58	4.90	1.98	161
45	220	4.9	1.03	0.363	39.9	4.42	1.99	5.53	2.49	232
60	195	3.3	1.05	0.380	37.1	5.23	2.11	6.53	2.64	215
60	220	3.7	1.03	0.484	53.2	5.90	2.65	7.37	3.32	309
72	97	1.3	1.13	0.113	5.5	3.12	0.68	3.90	0.85	32
72	122	1.7	1.10	0.179	10.9	3.92	1.05	4.90	1.31	63
72	147	2.0	1.08	0.259	19.1	4.73	1.49	5.91	1.86	111
72	195	2.7	1.05	0.456	44.5	6.27	2.54	7.84	3.17	258
72	220	3.1	1.03	0.581	63.9	7.08	3.19	8.84	3.98	371
97	97	1.0	1.13	0.152	7.4	4.20	0.91	5.25	1.14	43
97	147	1.5	1.08	0.349	25.7	6.37	2.00	7.96	2.50	149
97	195	2.0	1.05	0.615	59.9	8.45	3.42	10.56	4.27	348
97	220	2.3	1.03	0.782	86.1	9.53	4.29	11.91	5.36	499
97	295	3.0	1.00	1.407	207.5	12.78	7.47	15.98	9.34	1204

Coefficients

Duration of load (clause 2.8, Table 14 of BS 5268-2)

Long term $K_3 = 1.0$

Medium term $K_3 = 1.25$

Depth factor (clause 2.10.6)

$K_7 = (300/h)^{0.11}$ for $72 < h < 300$

$K_7 = 0.81\,(h^2 + 92\,300)/(h^2 + 56\,800)$ for $h > 300$

Derivation of capacities

Shear

Grade stress = 0.67 N/mm^2

Capacity = $(2bh)/3 \times$ grade stress $\times K_3$

Moment

Grade stress = 5.3 N/mm^2

Capacity = $Z \times$ grade stress $\times K_3 \times K_7$

Deflection

$E_{min} = 5800$ N/mm^2

Table 5.2 Geometrical properties: load-sharing members, strength class C16, service classes 1 and 2

Section			No. of units	Depth factor	Load-sharing factor		Section properties		Long term		Medium term		
b (mm)	h (mm)	h/b	N	K_7	K_8	K_9	Z ($\times 10^6$ mm^3)	I ($\times 10^6$ mm^4)	Shear (kN)	Moment (kN m)	Shear (kN)	Moment (kN m)	EI (kN m^2)
35	72	2.1	1	1.17	1.10	1.00	0.030	1.1	1.24	0.21	1.55	0.26	10
38	63	1.7	1	1.19	1.10	1.00	0.025	0.8	1.18	0.17	1.47	0.22	7
38	89	2.3	1	1.14	1.10	1.00	0.050	2.2	1.66	0.33	2.08	0.42	20
38	140	3.7	1	1.09	1.10	1.00	0.124	8.7	2.61	0.79	3.27	0.98	76
38	235	6.2	1	1.03	1.10	1.00	0.350	41.1	4.39	2.09	5.48	2.62	362
45	70	1.6	1	1.17	1.10	1.00	0.037	1.3	1.55	0.25	1.93	0.31	11
45	95	2.1	1	1.13	1.10	1.00	0.068	3.2	2.10	0.45	2.63	0.56	28
45	122	2.7	1	1.10	1.10	1.00	0.112	6.8	2.70	0.72	3.37	0.90	60
45	147	3.3	1	1.08	1.10	1.00	0.162	11.9	3.25	1.02	4.06	1.28	105
45	170	3.8	1	1.06	1.10	1.00	0.217	18.4	3.76	1.35	4.70	1.68	162
45	195	4.3	1	1.05	1.10	1.00	0.285	27.8	4.31	1.74	5.39	2.18	245
45	220	4.9	1	1.03	1.10	1.00	0.363	39.9	4.86	2.19	6.08	2.74	351
60	195	3.3	1	1.05	1.10	1.00	0.380	37.1	5.75	2.32	7.19	2.91	326
60	220	3.7	1	1.03	1.10	1.00	0.484	53.2	6.49	2.92	8.11	3.65	469
72	97	1.3	1	1.13	1.10	1.00	0.113	5.5	3.43	0.75	4.29	0.93	48
72	122	1.7	1	1.10	1.10	1.00	0.179	10.9	4.32	1.15	5.39	1.44	96
72	147	2.0	1	1.08	1.10	1.00	0.259	19.1	5.20	1.64	6.50	2.04	168
72	195	2.7	1	1.05	1.10	1.00	0.456	44.5	6.90	2.79	8.62	3.49	392
72	220	3.1	1	1.03	1.10	1.00	0.581	63.9	7.78	3.50	9.73	4.38	562
97	97	1.0	1	1.13	1.10	1.00	0.152	7.4	4.62	1.00	5.78	1.26	65
97	147	1.5	1	1.08	1.10	1.00	0.349	25.7	7.01	2.20	8.76	2.75	226
97	195	2.0	1	1.05	1.10	1.00	0.615	59.9	9.29	3.76	11.62	4.70	527
97	220	2.3	1	1.03	1.10	1.00	0.782	86.1	10.49	4.72	13.11	5.90	757
97	295	3.0	1	1.00	1.10	1.00	1.407	207.5	14.06	8.22	17.57	10.27	1826

Coefficients

Duration of load (clause 2.8, Table 14 of BS 5268-2)
Long term $K_3 = 1.00$
Medium term $K_3 = 1.25$

Depth factor (clause 2.10.6)
$K_7 = (300/h)^{0.11}$ for $72 < h < 300$
$K_7 = 0.81\ (h^2 + 92300)/(h^2 + 56800)$ for $h > 300$

Load-sharing factors (clause 2.10.11)
$K_8 = 1.10$ when $N > 1$
$K_9 = 1.00$ when $N = 1$
$K_9 = 1.14$ when $N = 2$
$K_9 = 1.21$ when $N = 3$
$K_9 = 1.24$ when $N > 3$

Derivation of capacities

Shear
Grade stress = 0.67 N/mm^2
Capacity = $(2bh)/3 \times$ grade stress $\times K_3 \times K_8$

Moment
Grade stress = 5.3 N/mm^2
Capacity = $Z \times$ grade stress $\times K_3 \times K_7 \times K_8$

Deflection
$E = E_{mean} = 8800$ N/mm^2

Table 5.3 Geometrical properties: principal members, strength class C24, service classes 1 and 2

Section				Section properties		Long term		Medium term		
b (mm)	h (mm)	h/b	Depth factor K_7	Z ($\times10^6$ mm^3)	I ($\times10^6$ mm^4)	Shear (kN)	Moment (kN m)	Shear (kN)	Moment (kN m)	EI (kN m^2)
35	72	2.1	1.17	0.030	1.1	1.19	0.27	1.49	0.33	8
38	63	1.7	1.19	0.025	0.8	1.13	0.22	1.42	0.28	6
38	89	2.3	1.14	0.050	2.2	1.60	0.43	2.00	0.54	16
38	140	3.7	1.09	0.124	8.7	2.52	1.01	3.15	1.27	63
38	235	6.2	1.03	0.350	41.1	4.23	2.69	5.28	3.37	296
45	70	1.6	1.17	0.037	1.3	1.49	0.32	1.86	0.40	9
45	95	2.1	1.13	0.068	3.2	2.02	0.58	2.53	0.72	23
45	122	2.7	1.10	0.112	6.8	2.60	0.92	3.25	1.16	49
45	147	3.3	1.08	0.162	11.9	3.13	1.31	3.91	1.64	86
45	170	3.8	1.06	0.217	18.4	3.62	1.73	4.53	2.16	133
45	195	4.3	1.05	0.285	27.8	4.15	2.24	5.19	2.80	200
45	220	4.9	1.03	0.363	39.9	4.69	2.82	5.86	3.52	287
60	195	3.3	1.05	0.380	37.1	5.54	2.99	6.92	3.74	267
60	220	3.7	1.03	0.484	53.2	6.25	3.76	7.81	4.69	383
72	97	1.3	1.13	0.113	5.5	3.31	0.96	4.13	1.20	39
72	122	1.7	1.10	0.179	10.9	4.16	1.48	5.20	1.85	78
72	147	2.0	1.08	0.259	19.1	5.01	2.10	6.26	2.63	137
72	195	2.7	1.05	0.456	44.5	6.65	3.59	8.31	4.49	320
72	220	3.1	1.03	0.581	63.9	7.50	4.51	9.37	5.63	460
97	97	1.0	1.13	0.152	7.4	4.45	1.29	5.57	1.61	53
97	147	1.5	1.08	0.349	25.7	6.75	2.83	8.44	3.54	185
97	195	2.0	1.05	0.615	59.9	8.95	4.83	11.19	6.04	432
97	220	2.3	1.03	0.782	86.1	10.10	6.07	12.63	7.59	620
97	295	3.0	1.00	1.407	207.5	13.54	10.57	16.93	13.21	1494

Coefficients

Duration of load (clause 2.8, Table 14 of BS 5268-2)
Long term $K_3 = 1.0$
Medium term $K_3 = 1.25$

Depth factor (clause 2.10.6)
$K_7 = (300/h)^{0.11}$ for $72 < h < 300$
$K_7 = 0.81(h^2 + 92300)/(h^2 + 56800)$ for $h > 300$

Derivation of capacities

Shear
Grade stress = 0.71 N/mm^2
Capacity = $(2bh)/3 \times$ grade stress $\times K_3$

Moment
Grade stress = 7.5 N/mm^2
Capacity = $Z \times$ grade stress $\times K_3 \times K_7$

Deflection
E_{min} = 7200 N/mm^2

Table 5.4 Geometrical properties: load-sharing members, strength class C24, service classes 1 and 2

Section			No. of units	Depth factor	Load-sharing factor		Section properties		Long term		Medium term		
b (mm)	h (mm)	h/b	N	K_7	K_8	K_9	Z ($\times 10^6$ mm^3)	I ($\times 10^6$ mm^4)	Shear (kN)	Moment (kN m)	Shear (kN)	Moment (kN m)	EI (kN m^2)
38	89	2.3	1	1.14	1.10	1.00	0.050	2.2	1.76	0.47	2.20	0.59	24
38	140	3.7	1	1.09	1.10	1.00	0.124	8.7	2.77	1.11	3.46	1.39	94
38	235	6.2	1	1.03	1.10	1.00	0.350	41.1	4.65	2.96	5.81	3.71	444
45	70	1.6	1	1.17	1.10	1.00	0.037	1.3	1.64	0.36	2.05	0.44	14
45	95	2.1	1	1.13	1.10	1.00	0.068	3.2	2.23	0.63	2.78	0.79	35
45	122	2.7	1	1.10	1.10	1.00	0.112	6.8	2.86	1.02	3.57	1.27	74
45	147	3.3	1	1.08	1.10	1.00	0.162	11.9	3.44	1.45	4.31	1.81	129
45	170	3.8	1	1.06	1.10	1.00	0.217	18.4	3.98	1.90	4.98	2.38	199
45	195	4.3	1	1.05	1.10	1.00	0.285	27.8	4.57	2.47	5.71	3.08	300
45	220	4.9	1	1.03	1.10	1.00	0.363	39.9	5.15	3.10	6.44	3.87	431
60	195	3.3	1	1.05	1.10	1.00	0.380	37.1	6.09	3.29	7.61	4.11	400
60	220	3.7	1	1.03	1.10	1.00	0.484	53.2	6.87	4.13	8.59	5.16	575
72	97	1.3	1	1.13	1.10	1.00	0.113	5.5	3.64	1.05	4.55	1.32	59
72	122	1.7	1	1.10	1.10	1.00	0.179	10.9	4.57	1.63	5.72	2.03	118
72	147	2.0	1	1.08	1.10	1.00	0.259	19.1	5.51	2.31	6.89	2.89	206
72	195	2.7	1	1.05	1.10	1.00	0.456	44.5	7.31	3.95	9.14	4.93	480
72	220	3.1	1	1.03	1.10	1.00	0.581	63.9	8.25	4.96	10.31	6.20	690
97	97	1.0	1	1.13	1.10	1.00	0.152	7.4	4.90	1.42	6.12	1.78	80
97	147	1.5	1	1.08	1.10	1.00	0.349	25.7	7.42	3.12	9.28	3.90	277
97	195	2.0	1	1.05	1.10	1.00	0.615	59.9	9.85	5.32	12.31	6.65	647
97	220	2.3	1	1.03	1.10	1.00	0.782	86.1	11.11	6.68	13.89	8.35	930
97	295	3.0	1	1.00	1.10	1.00	1.407	207.5	14.90	11.63	18.62	14.54	2241

Coefficients

Duration of load (clause 2.8, Table 14 of BS 5268-2)
Long term $K_3 = 1.00$
Medium term $K_3 = 1.25$

Depth factor (clause 2.10.6)
$K_7 = (300/h)^{0.11}$ for $72 < h < 300$
$K_7 = 0.81(h^2 + 92300)/(h^2 + 56800)$ for $h > 300$

Load-sharing factors (clause 2.10.11)
$K_8 = 1.10$ when $N > 1$
$K_9 = 1.00$ when $N = 1$
$K_9 = 1.14$ when $N = 2$
$K_9 = 1.21$ when $N = 3$
$K_9 = 1.24$ when $N > 3$

Derivation of capacities

Shear
Grade stress = 0.71 N/mm^2
Capacity = $(2bh)/3 \times$ grade stress $\times K_3 \times K_8$

Moment
Grade stress = 7.5 N/mm^2
Capacity = $Z \times$ grade stress $\times K_3 \times K_7 \times K_8$

Deflection
$E = E_{mean} = 10800$ N/mm^2

Table 5.5 Geometrical properties: principal members, strength class TR26, service classes 1 and 2

Section				Section properties		Long term		Medium term		
b (mm)	h (mm)	h/b	Depth factor K_7	Z ($\times 10^6$ mm^3)	I ($\times 10^6$ mm^4)	Shear (kN)	Moment (kN m)	Shear (kN)	Moment (kN m)	EI (kN m^2)
35	60	1.7	1.19	0.021	0.6	1.54	0.25	1.93	0.31	5
35	72	2.1	1.17	0.030	1.1	1.85	0.35	2.31	0.44	8
35	84	2.4	1.15	0.041	1.7	2.16	0.47	2.70	0.59	13
35	97	2.8	1.13	0.055	2.7	2.49	0.62	3.11	0.78	20
35	122	3.5	1.10	0.087	5.3	3.13	0.96	3.91	1.20	39
35	147	4.2	1.08	0.126	9.3	3.77	1.36	4.72	1.70	69
35	170	4.9	1.06	0.169	14.3	4.36	1.79	5.45	2.24	106
35	195	5.6	1.05	0.222	21.6	5.01	2.33	6.26	2.91	160
44	72	1.6	1.17	0.038	1.4	2.32	0.44	2.90	0.56	10
44	97	2.2	1.13	0.069	3.3	3.13	0.78	3.91	0.98	25
44	122	2.8	1.10	0.109	6.7	3.94	1.21	4.92	1.51	49
44	147	3.3	1.08	0.158	11.6	4.74	1.71	5.93	2.14	86
44	170	3.9	1.06	0.212	18.0	5.49	2.26	6.86	2.82	133
44	195	4.4	1.05	0.279	27.2	6.29	2.92	7.87	3.65	201
44	220	5.0	1.03	0.355	39.0	7.10	3.67	8.87	4.59	289

Coefficients

Duration of load (clause 2.8, Table 14 of BS 5268-2)

Long term $K_3 = 1.00$

Medium term $K_3 = 1.25$

Depth factor (clause 2.10.6)

$K_7 = (300/h)^{0.11}$ for $72 < h < 300$

$K_7 = 0.81(h^2 + 92300)/(h^2 + 56800)$ for $h > 300$

Derivation of capacities

Shear

Grade stress = 1.1 N/mm^2

Capacity = $(2bh)/3 \times$ grade stress $\times K_3 \times K_8$

Moment

Grade stress = 10 N/mm^2

Capacity = $Z \times$ grade stress $\times K_3 \times K_7$

Deflection

$E_{min} = 7400$ N/mm^2

Table 5.6 Geometrical properties: load-sharing members, strength class TR26, service classes 1 and 2

Section		No. of units	Depth factor	Load-sharing factor		Section properties		Long term		Medium term		
b (mm)	h (mm)	N	K_7	K_8	K_9	Z ($\times 10^6$ mm^3)	I ($\times 10^6$ mm^4)	Shear (kN)	Moment (kN m)	Shear (kN)	Moment (kN m)	EI (kN m^2)
35	60	1	1.19	1.10	1.00	0.021	0.6	1.69	0.28	2.12	0.34	7
35	72	1	1.17	1.10	1.00	0.030	1.1	2.03	0.39	2.54	0.49	12
35	84	1	1.15	1.10	1.00	0.041	1.7	2.37	0.52	2.96	0.65	19
35	97	1	1.13	1.10	1.00	0.055	2.7	2.74	0.68	3.42	0.85	29
35	122	1	1.10	1.10	1.00	0.087	5.3	3.44	1.05	4.31	1.32	58
35	147	1	1.08	1.10	1.00	0.126	9.3	4.15	1.50	5.19	1.87	102
35	170	1	1.06	1.10	1.00	0.169	14.3	4.80	1.97	6.00	2.47	158
35	195	1	1.05	1.10	1.00	0.222	21.6	5.51	2.56	6.88	3.20	238
44	72	1	1.17	1.10	1.00	0.038	1.4	2.56	0.49	3.19	0.61	15
44	97	1	1.13	1.10	1.00	0.069	3.3	3.44	0.86	4.30	1.07	37
44	122	1	1.10	1.10	1.00	0.109	6.7	4.33	1.33	5.41	1.66	73
44	147	1	1.08	1.10	1.00	0.158	11.6	5.22	1.89	6.52	2.36	128
44	170	1	1.06	1.10	1.00	0.212	18.0	6.03	2.48	7.54	3.10	198
44	195	1	1.05	1.10	1.00	0.279	27.2	6.92	3.22	8.65	4.02	299
44	220	1	1.03	1.10	1.00	0.355	39.0	7.81	4.04	9.76	5.05	429

Coefficients

Duration of load (clause 2.8, Table 14 of BS 5268-2)

Long term $K_3 = 1.00$

Medium term $K_3 = 1.25$

Depth factor (clause 2.10.6)

$K_7 = (300/h)^{0.11}$ for $72 < h < 300$

$K_7 = 0.81(h^2 + 92300)/(h^2 + 56800)$ for $h > 300$

Load-sharing factors (clause 2.9)

$K_8 = 1.10$

Derivation of capacities

Shear

Grade stress = 1.1 N/mm^2

Capacity = $(2bh)/3 \times$ grade stress $\times K_3 \times K_8$

Moment

Grade stress = 10 N/mm^2

Capacity = $Z \times$ grade stress $\times K_3 \times K_7 \times K_8$

Deflection

$E = E_{mean} = 11000$ N/mm^2

$$\delta_t = \sqrt{(\delta_x)^2 + (\delta_y)^2}$$

where δ_x = deflection about x–x axis
δ_y = deflection about y–y axis.

Bending about both the x–x and y–y axes greatly increases the dimensions of the required rectangular section, consequently the designer should endeavour where possible to place the section so that the load acts directly in the plane of maximum stiffness. For example, a purlin between trusses should preferably be placed in a vertical plane either by employing tapered end blocking pieces or by being supported vertically in metal hangers. Frequently, however, this is not possible. In traditional house construction, a purlin is often placed normal to the roof slope to simplify the fixing of secondary rafters, and the force parallel to the roof slope can be resisted by the secondary system with its roof sarking, triangulation with ceiling joists, etc. Where such advantageous assumptions are justified, the purlin need only be designed for the force $F\cos\theta$ acting about the x–x axis, but attention must be paid to the adequacy of end fixings, horizontal cross ties, etc. to ensure that the force $F\sin\theta$ is also adequately resisted or balanced.

Where the permissible bending stress about both axes is the same, the equation for combining bending stresses can be conveniently modified as follows to avoid the need to determine sectional properties about the y–y axis. (When the permissible bending stresses differ, the method of combining stress ratios as shown in section 7.3.3 for glulam beams must be adopted.)

$$\sigma_{m,a,par} = \frac{M_x}{Z_x} + \frac{M_y}{Z_y} = \frac{F\cos\theta\, L^2}{8Z_x} + \frac{F\sin\theta\, L^2}{8Z_y}$$

$$= \frac{FL^2}{8Z_x}\left(\cos\theta + \frac{Z_x}{Z_y}\sin\theta\right)$$

$$= \frac{M}{Z_x}\left(\cos\theta + \frac{h}{b}\sin\theta\right)$$

where M is the full bending moment applied to the section at θ to the y–y axis.

It can be seen that $\sigma_{m,a,par}$ is a function of h/b and that slender sections with a high h/b ratio have larger bending stresses, as would be expected.

If h/b is known, one can calculate the value required for $\overline{M}_x$ so that $\sigma_{m,a,par}$ does not exceed the permissible stress:

$$\text{required } \overline{M}_x = M\left[\cos\theta + \frac{h}{b}\sin\theta\right]$$

Similarly the equation for total deflection can be simplified to avoid the need to determine sectional properties about the y–y axis.

$$\delta_t = \sqrt{(\delta_x)^2 + (\delta_y)^2}$$

which, for a uniform load (disregarding shear deflection), is:

$$\delta_t = \sqrt{\left(\frac{5F\cos\theta\, L^3}{384EI_x}\right) + \left(\frac{5F\sin\theta\, L^3}{384EI_y}\right)}$$

$$= \frac{5FL^3}{384EI_x}\sqrt{\left[\cos^2\theta + \left(\frac{I_x\sin\theta}{I_y}\right)^2\right]}$$

$$= \frac{5FL^3}{384EI_x}\sqrt{1 + \left[\left(\frac{h}{b}\right)^4 - 1\right]\sin^2\theta}$$

If h/b is known, one can calculate the value required for EI_x to limit the deflection $\delta_{t,m}$ to a given amount, e.g. for δ_t not to exceed 0.003 of span. By transposing with $\delta_t = 0.003L$ the formula (disregarding shear deflection) becomes:

$$\text{required } EI_x = 4.34FL^2\sqrt{1 + \left[\left(\frac{h}{b}\right)^4 - 1\right]\sin^2\theta}$$

In considering shear deflection, rather than calculating about both axes, increase the required EI_x capacity given above by the ratio shown in Fig. 4.26.

The section with the smallest cross section for a given loading will be realized if the deflection about both axes can be made the same.

To give equal deflection about the x–x and y–y axes, one can equate $\delta_x = \delta_y$ and show that h/b must be equal to $\sqrt{\cot\theta}$. Values of $\sqrt{\cot\theta}$ for values of θ from 2.5° to 45° in 2.5° increments are given in Table 5.7.

With the resultant preferred value for h/b, the required value of EI_x becomes:

Table 5.7

θ (degrees)	$\sqrt{\cot\theta}$
2½	4.79
5	3.38
7½	2.76
10	2.38
12½	2.12
15	1.93
17½	1.78
20	1.65
22½	1.55
25	1.46
27½	1.38
30	1.32
32½	1.25
35	1.19
37½	1.14
40	1.09
42½	1.04
45	1.00

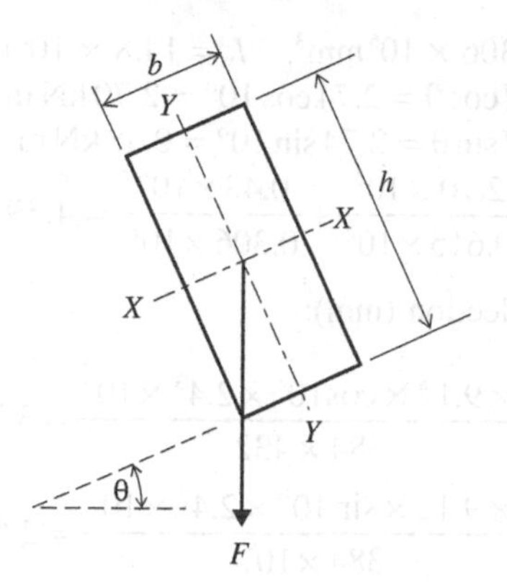

Fig. 5.2

$$\text{required } EI_x = 4.34FL^2\sqrt{[1+(\cos^2\theta-1)\sin^2\theta]}$$
$$= 4.34FL^2\sqrt{2\cos^2\theta}$$

with $h/b \leq \sqrt{\cot\theta}$, $EI_x = 6.14FL^2\cos\theta$, to which an allowance for shear deflection should be made.

Example of design of beam bending about x–x and y–y axes

Determine an economic purlin size (Fig. 5.2) to suit a span of 2.4 m. The medium-term vertical load is 3.8 kN/m run and the angle $\theta = 10°$.

Total vertical load $F = 9.12\,\text{kN}$

Total vertical bending moment $M = \dfrac{9.12 \times 2.4}{8} = 2.74\,\text{kN m}$

For bending deflection, the required EI_x for economical section is

$$6.14 \times 9.12 \times 2.4^2 \times \cos 10° = 318\,\text{kN m}^2$$

to which an allowance for shear deflection must be added.

The required EI capacity should be produced by a section having $h/b \leq \sqrt{\cot 10°}$ (i.e. $h/b \leq 2.38$).

From Table 5.3 check the suitability of 97 × 195 C24.

$EI_x = 432\,\text{kN m}^2$ with $h/b = 2.0$
$L/h = 2400/195 = 12.3$ which gives $\delta_t/\delta_m = 1.10$ from Fig. 4.26
Required $EI_x = 318 \times 1.10 = 350\,\text{kN m}^2$
Required $\overline{M}_x = 2.74\,(\cos 10° + 2\sin 10°) = 3.65\,\text{kN m}$

From Table 5.3, medium-term moment capacity = 6.04 kN m

The section meets the deflection and bending criteria. Check this simplified method by the calculations below.

For 97 × 195 C24,

permissible medium-term bending stress = $7.5 \times 1.25 \times 1.05 = 9.84\,\text{N/mm}^2$
and $E_{min} = 7200\,\text{N/mm}^2$
$Z_x = 0.615 \times 10^6\,\text{mm}^3$, $I_x = 59.9 \times 10^6\,\text{mm}^4$ and $EI_x = 432\,\text{kN m}^2$

$Z_y = 0.306 \times 10^6\,\text{mm}^3$, $I_y = 14.8 \times 10^6\,\text{mm}^4$ and $EI_y = 107\,\text{kN m}^2$
$M_x = M\cos\theta = 2.74\cos 10° = 2.70\,\text{kN m}$
$M_y = M\sin\theta = 2.74\sin 10° = 0.48\,\text{kN m}$

$$\sigma_{m,a} = \frac{2.70\times10^6}{0.615\times10^6} + \frac{0.48\times10^6}{0.306\times10^6} = 4.39 + 1.57 = 5.96\,\text{N/mm}^2 \quad \textbf{O.K.}$$

Bending deflection (mm):

$$\delta_x = \frac{5\times9.12\times\cos10°\times2.4^3\times10^3}{384\times432} = 3.74\,\text{mm}$$

$$\delta_x = \frac{5\times9.12\times\sin10°\times2.4^3\times10^3}{384\times107} = 2.66\,\text{mm}$$

$$\delta_t = \sqrt{3.74^2 + 2.66^2} = 4.57\,\text{mm}$$

Shear deflection (mm): $G = 7200/16 = 450\,\text{N/mm}^2$

$$\delta_x = \frac{K_{\text{form}}M_x}{AG} = \frac{1.2\times2.70\times10^6}{97\times195\times450} = 0.38\,\text{mm}$$

$$\delta_y = \frac{1.2\times0.48\times10^6}{97\times195\times450} = 0.068\,\text{mm}$$

$$\delta_t = \sqrt{0.38^2 + 0.068^2} = 0.39\,\text{mm}$$

Total bending + shear deflection = 4.57 + 0.39 = 4.96 mm
Allowable deflection at 0.003 × span = 0.003 × 2400 = 7.2 mm **O.K.**
Alternatively:

$$\text{Bending stress} = \frac{M}{Z_x}\left(\cos\theta + \frac{h}{b}\sin\theta\right) = \frac{2.74\times10^6}{0.615\times10^6}[0.985 + (2.0\times0.174)]$$
$$= 5.94\,\text{N/mm}^2$$

compared to 5.96 N/mm² calculated previously.

$$\text{Bending deflection} = \delta_t = \frac{5FL^3}{384EI_x}\sqrt{1+\left[\left(\frac{h}{b}\right)^4 - 1\right]\sin^2\theta}$$
$$= \frac{5\times9.12\times2.4^3}{384\times432}\sqrt{1+15\times0.0301}$$
$$= 0.004\,58\,\text{m}$$

which agrees with the value of 4.57 mm previously calculated and to which shear deflection is to be added.

Chapter 6
Multiple Section Beams

6.1 INTRODUCTION

Timbers are frequently grouped together to form 2-member, 3-member or 4-member sections (Fig. 6.1) to act as lintels or trimmers. The members are suitably connected together in parallel to support a common load and should be regarded as principal members with limited enhancement to allow for improved permissible shear stress, bending stress and E value.

6.2 MODIFICATION FACTORS

The shear and bending stresses parallel to grain should be multiplied by the load-sharing modification factor $K_8 = 1.1$ and the minimum modulus of elasticity should be modified by the factor K_9 for deflection calculations. For softwoods K_9 has the value 1.14 for two members, 1.21 for three members and 1.24 for four or more members. All other modification factors are as for principal members.

6.3 CONNECTION OF MEMBERS

Individual members should be suitably connected together, i.e. by nailing or bolting. The type of connection will depend upon the manner in which the load is applied to the beam, as illustrated in Fig. 6.2. Nailing is preferred to bolting. Bolting of trimmers may result in bolt heads clashing with joist hangers and some loss of section which may on occasions reduce the strength of the section.

Referring to Fig. 6.2, the loads are applied as follows:

- *(a) Direct load.* The load is applied through a spreader member so that each element immediately takes its proportion of load. In this instance only nominal nail fixings are required.
- *(b) Load applied to one side of member.* Sufficient fixings must be provided to transfer the assumed load being taken by each component. For example, in a 2-member section half the loading must be transferred between members. For a 3-member section two-thirds of the loading on the loaded face is to be transferred into the central member and one-third of the loading on the loaded face is to be transferred to the member furthest from the loaded face. Four-member sections are normally loaded on both sides.

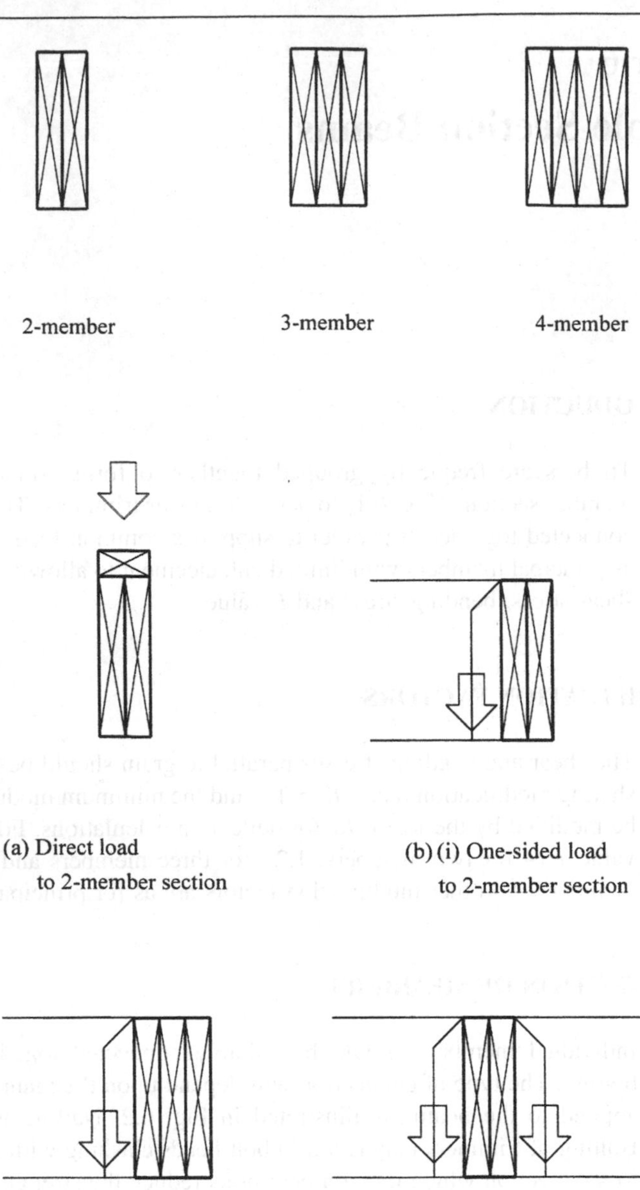

Fig. 6.1

Fig. 6.2

- *(c) Load applied to both faces of member.* The fixing arrangement between members will depend on the relative loading on each side of the beam and should be such as to transfer the appropriate variation in loadings into each component of the member.

Table 6.1 Section capacities: TR26, two members (BS 5268: Part 2: 1996)

	Section				Load-sharing factor		Section properties		Long term		Medium term		
Code	b (mm)	h (mm)	No. of units N	Depth factor K_7	K_8	K_9	Z ($\times 10^6$ mm^3)	I ($\times 10^6$ mm^4)	Shear (kN)	Moment (kN m)	Shear (kN)	Moment (kN m)	EI (kN m^2)
DJ60	70	60	2	1.19	1.10	1.14	0.042	1.3	3.39	0.55	4.24	0.69	11
DJ72	70	72	2	1.17	1.10	1.14	0.060	2.2	4.07	0.78	5.08	0.97	18
DJ84	70	84	2	1.15	1.10	1.14	0.082	3.5	4.74	1.04	5.93	1.30	29
DJ97	70	97	2	1.13	1.10	1.14	0.110	5.3	5.48	1.37	6.85	1.71	45
DJ122	70	122	2	1.10	1.10	1.14	0.174	10.6	6.89	2.11	8.61	2.64	89
DJ147	70	147	2	1.08	1.10	1.14	0.252	18.5	8.30	3.00	10.38	3.75	156
DJ170	70	170	2	1.06	1.10	1.14	0.337	28.7	9.60	3.95	12.00	4.93	242
DJ195	70	195	2	1.05	1.10	1.14	0.444	43.3	11.01	5.12	13.76	6.40	365
DJ72	88	72	2	1.17	1.10	1.14	0.076	2.7	5.11	0.98	6.39	1.22	23
DJ97	88	97	2	1.13	1.10	1.14	0.138	6.7	6.89	1.72	8.61	2.15	56
DJ122	88	122	2	1.10	1.10	1.14	0.218	13.3	8.66	2.65	10.83	3.31	112
DJ147	88	147	2	1.08	1.10	1.14	0.317	23.3	10.44	3.77	13.04	4.71	197
DJ170	88	170	2	1.06	1.10	1.14	0.424	36.0	12.07	4.96	15.08	6.20	304
DJ195	88	195	2	1.05	1.10	1.14	0.558	54.4	13.84	6.43	17.30	8.04	459
DJ220	88	220	2	1.03	1.10	1.14	0.710	78.1	15.62	8.08	19.52	10.10	659

Coefficients

Duration of load (clause 2.8, Table 14 of BS 5268-2)

Long term $K_3 = 1.00$

Medium term $K_3 = 1.25$

Depth factor (clause 2.10.6)

$K_7 = (300/h)^{0.11}$ for $72 < h < 300$

$K_7 = 0.81(h^2 + 92300)/(h^2 + 56800)$ for $h > 300$

Load-sharing factors (clause 2.10.11)

$K_8 = 1.10$ when $N > 1$

$K_9 = 1.00$ when $N = 1$

$K_9 = 1.14$ when $N = 2$

$K_9 = 1.21$ when $N = 3$

$K_9 = 1.24$ when $N > 3$

Derivation of capacities

Shear

Grade stress = 1.1 N/mm^2

Capacity = $(2bh)/3 \times$ grade stress $\times K_3 \times K_8$

Moment

Grade stress = 10 N/mm^2

Capacity = $Z \times$ grade stress $\times K_3 \times K_7 \times K_8$

Deflection

E_{min} = 7400 N/mm^2

$E = E_{min} \times K_9$

Table 6.2 Section capacities: TR26, three members (BS 5268: Part 2: 1996)

Code	Section		No. of units	Depth factor	Load-sharing factor		Section properties		Long term		Medium term		
	b (mm)	h (mm)	N	K_7	K_8	K_9	Z ($\times 10^6$ mm^3)	I ($\times 10^6$ mm^4)	Shear (kN)	Moment (kN m)	Shear (kN)	Moment (kN m)	EI (kN m^2)
TJ60	105	60	3	1.19	1.10	1.21	0.063	1.9	5.08	0.83	6.35	1.03	17
TJ72	105	72	3	1.17	1.10	1.21	0.091	3.3	6.10	1.17	7.62	1.46	29
TJ84	105	84	3	1.15	1.10	1.21	0.123	5.2	7.11	1.56	8.89	1.95	46
TJ97	105	97	3	1.13	1.10	1.21	0.165	8.0	8.22	2.05	10.27	2.56	72
TJ122	105	122	3	1.10	1.10	1.21	0.260	15.9	10.33	3.16	12.92	3.95	142
TJ147	105	147	3	1.08	1.10	1.21	0.378	27.8	12.45	4.50	15.56	5.62	249
TJ170	105	170	3	1.06	1.10	1.21	0.506	43.0	14.40	5.92	18.00	7.40	385
TJ195	105	195	3	1.05	1.10	1.21	0.665	64.9	16.52	7.68	20.65	9.59	581
TJ72	132	72	3	1.17	1.10	1.21	0.114	4.1	7.67	1.47	9.58	1.83	37
TJ97	132	97	3	1.13	1.10	1.21	0.207	10.0	10.33	2.58	12.91	3.22	90
TJ122	132	122	3	1.10	1.10	1.21	0.327	20.0	12.99	3.98	16.24	4.97	179
TJ147	132	147	3	1.08	1.10	1.21	0.475	34.9	15.65	5.66	19.57	7.07	313
TJ170	132	170	3	1.06	1.10	1.21	0.636	54.0	18.10	7.44	22.63	9.31	484
TJ195	132	195	3	1.05	1.10	1.21	0.837	81.6	20.76	9.65	25.95	12.06	730
TJ220	132	220	3	1.03	1.10	1.21	1.065	117.1	23.43	12.12	29.28	15.15	1049

Coefficients

Duration of load (clause 2.8, Table 14 of BS 5268-2)

Long term $K_3 = 1.00$
Medium term $K_3 = 1.25$

Depth factor (clause 2.10.6)

$K_7 = (300/h)^{0.11}$ for $72 < h < 300$
$K_7 = 0.81(h^2 + 92\,300)/(h^2 + 56\,800)$ for $h > 300$

Load-sharing factors (clause 2.10.11)

$K_8 = 1.10$ when $N > 1$
$K_9 = 1.00$ when $N = 1$
$K_9 = 1.14$ when $N = 2$
$K_9 = 1.21$ when $N = 3$
$K_9 = 1.24$ when $N > 3$

Derivation of capacities

Shear

Grade stress = 1.1 N/mm^2
Capacity = $(2bh)/3 \times$ grade stress $\times K_3 \times K_8$

Moment

Grade stress = 10 N/mm^2
Capacity = $Z \times$ grade stress $\times K_3 \times K_7 \times K_8$

Deflection

$E_{min} = 7400$ N/mm^2
$E = E_{min} \times K_9$

Table 6.3 Section capacities: TR26, four members (BS 5268: Part 2: 1996)

Code	Section		No. of units	Depth factor	Load-sharing factor		Section properties		Long term		Medium term		EI
	b (mm)	h (mm)	N	K_7	K_8	K_9	Z ($\times 10^6$ mm^3)	I ($\times 10^6$ mm^4)	Shear (kN)	Moment (kN m)	Shear (kN)	Moment (kN m)	(kN m^2)
QJ60	140	60	4	1.19	1.10	1.24	0.084	2.5	6.78	1.10	8.47	1.38	23
QJ72	140	72	4	1.17	1.10	1.24	0.121	4.4	8.13	1.56	10.16	1.95	40
QJ84	140	84	4	1.15	1.10	1.24	0.165	6.9	9.49	2.08	11.86	2.60	63
QJ97	140	97	4	1.13	1.10	1.24	0.220	10.6	10.95	2.73	13.69	3.42	98
QJ122	140	122	4	1.10	1.10	1.24	0.347	21.2	13.78	4.22	17.22	5.27	194
QJ147	140	147	4	1.08	1.10	1.24	0.504	37.1	16.60	6.00	20.75	7.50	340
QJ170	140	170	4	1.06	1.10	1.24	0.674	57.3	19.20	7.90	24.00	9.87	526
QJ195	140	195	4	1.05	1.10	1.24	0.887	86.5	22.02	10.23	27.53	12.79	794
QJ72	176	72	4	1.17	1.10	1.24	0.152	5.5	10.22	1.96	12.78	2.45	50
QJ97	176	97	4	1.13	1.10	1.24	0.276	13.4	13.77	3.44	17.21	4.30	123
QJ122	176	122	4	1.10	1.10	1.24	0.437	26.6	17.32	5.30	21.65	6.63	244
QJ147	176	147	4	1.08	1.10	1.24	0.634	46.6	20.87	7.54	26.09	9.43	428
QJ170	176	170	4	1.06	1.10	1.24	0.848	72.1	24.14	9.93	30.17	12.41	661
QJ195	176	195	4	1.05	1.10	1.24	1.115	108.8	27.68	12.86	34.61	16.08	998
QJ220	176	220	4	1.03	1.10	1.24	1.420	156.2	31.23	16.16	39.04	20.20	1433

Coefficients

Duration of load (clause 2.8, Table 14 of BS 5268-2)

Long term $K_3 = 1.00$

Medium term $K_3 = 1.25$

Depth factor (clause 2.10.6)

$K_7 = (300/h)^{0.11}$ for $72 < h < 300$

$K_7 = 0.81(h^2 + 92300)/(h^2 + 56800)$ for $h > 300$

Load-sharing factors (clause 2.10.11)

$K_8 = 1.10$ when $N > 1$

$K_9 = 1.00$ when $N = 1$

$K_9 = 1.14$ when $N = 2$

$K_9 = 1.21$ when $N = 3$

$K_9 = 1.24$ when $N > 3$

Derivation of capacities

Shear

Grade stress = 1.1 N/mm^2

Capacity = $(2bh)/3$ × grade stress × K_3 × K_8

Moment

Grade stress = 10 N/mm^2

Capacity = Z × grade stress × K_3 × K_7 × K_8

Deflection

$E_{min} = 7400$ N/mm^2

$E = E_{min} \times K_9$

6.4 STANDARD TABLES

Tables 6.1 to 6.3 provide section capacites for 2-member, 3-member and 4-member sections of TR26 grade. These represent the most popular sections.

6.5 DESIGN EXAMPLE

Determine a suitable 2-member section to take a uniform dead load (including self-weight) of 3.95 kN/m over an effective span of 3.0 m.

$L = 3.0\,\text{m}$
$w = 3.95\,\text{kN/m}$
Total load $= W = 3.95 \times 3 = 11.85\,\text{kN}$
Maximum shear $= 11.85/2 = 5.92\,\text{kN}$
Maximum moment $= WL/8 = 11.85 \times 3.0/8 = 4.44\,\text{kN m}$
Approximate EI required $= 4.8WL^2 = 4.8 \times 11.85 \times 3^2 = 511\,\text{kN m}^2$

From Table 6.1 try 88/220 TR26

Long-term shear capacity = 15.62 kN > 5.92 kN, therefore **O.K.**
Long-term moment capacity = 8.08 kN m > 4.44 kN m, therefore **O.K.**

$$\text{Bending deflection at midspan} = \frac{5wL^4}{384EI}$$

$$= \frac{5 \times 3.95 \times 3^4}{384 \times 659} = 0.00632\,\text{m}$$

Shear deflection factor $= 1 + 15.36\,(\text{depth/span})^2$

$$= 1 + 15.36\left(\frac{220}{3000}\right)^2 = 1.0826$$

Total deflection $= 0.00632 \times 1.0826 = 0.00684\,\text{m}$
Allowable deflection at 0.003 of span $= 0.009\,\text{m}$ **O.K.**

Section satisfactory.

Chapter 7
Glulam Beams

7.1 INTRODUCTION

A glulam section is one manufactured by gluing together laminations with their grain essentially parallel. In the UK a horizontally glued laminated member is defined as having the laminations parallel to the neutral plane while a vertically laminated member has the laminates at right angles to the neutral plane. This leads to different strength values for horizontally and vertically laminated members made from the same species and grade of timber. Because Eurocode 5 does not differentiate between the orientation of the laminates in ascribing strength values to glulam, the European definitions are simpler – horizontal glulam is glued laminated timber with the glue line plane perpendicular to the long side of the glulam section while with vertical glulam the glue line plane is perpendicular to the short side of the section.

There are four basic reasons for laminating timber:

- sections can be produced very much larger than can be obtained from a single piece of timber
- large defects such as knots can be distributed throughout a glulam section by converting the solid timber section into laminates and forming a glulam section
- structural members of tapered and curved profiles can be produced easily (by laminating a previously two-dimensional curved glulam a three-dimensional portal or arch frame can be made, albeit at a cost!)
- members can be cambered to offset deflections due, say, to the self-weight of the structure.

The production requirements for glulam are now given in BS EN 386 'Glued laminated timber – Performance requirements and minimum production requirements'. It is necessary to have a further three standards, BS EN 390 'Glued laminated timber – Sizes – Permissible deviations', BS EN 391 'Glued laminated timber – Delamination test of gluelines' and BS EN 392 'Glued laminated timber – Shear test of gluelines' in order to achieve comparability with the now withdrawn BS 4169 'Glued laminated timber structures'.

In selecting the species of timber for laminating the two prime characteristics are the 'gluability' and dimensional stability with changing atmospheric conditions. Many tropical hardwoods are difficult to glue because of natural oils and resins contaminating the gluing surface. Timber with large movement values such as Ekki have to be used with caution and special care taken in the orientation of laminates to reduce the risk of glue line splitting. The most common species used in Europe is whitewood.

It is essential to control the moisture content of the timber at the time of manufacture. BS EN 386 specifies that the moisture content of the laminates should lie in the range 8% to 15% with the range between the highest and lowest laminates in the member being 4%. It is also prudent to have the moisture content at manufacture as close as possible to the likely equilibrium moisture in service, particularly in heated environments, to avoid glue line splitting.

BS EN 386 gives the maximum finished thicknesses for laminates forming members to be used in the three service class environments. For the conifer species, in service class 1 the maximum thickness is 45 mm and the cross-sectional area 10 000 mm^2 (say 45 × 220 mm); in service class 2 the thickness is 45 mm and 9000 mm^2; while the limitations in service class 3 are 35 mm and 7000 mm^2. Where the cross-sectional area is greater than 7500 mm^2 it is recommended that a longitudinal saw kerf is run down the centre of the laminate to reduce the risk of excessive cupping. The requirements for the broadleaved species are more stringent.

The laminates have to be kiln dried to the required moisture content and it follows that there must be adequate storage space available to store the timber at all stages of the production cycle so that the costs of kilning are not thrown away. The dried laminates will still be in the sawn condition so they are then planed on the surfaces to be bonded. These planed surfaces must be flat and parallel. It is normal practice to glue up the section within 24 hours of planing as there is a risk of timber distortion (cupping) and, in even the most benign environments, chemical contamination of the surfaces.

The storage, mixing and application of the adhesive may appear to be straightforward. Far from it. Getting the adhesive on to the laminates and laying on the next laminate (open assembly time), and then applying the clamping pressure within a further time limit (closed assembly time) can be an interesting logistical problem bearing in mind that there is also a finite time period from the time of mixing the adhesive to it becoming unuseable (pot life). These various time periods are influenced by the ambient conditions – for example, with a relative humidity of 30% (60%+ is the norm) the open assembly time can be well under 10 minutes due to the rapid evaporation of the volatile constituents of the adhesive whereas the laminating team would be expecting at least 20 minutes in normal circumstances.

Pressure has to be applied to the glue lines while the adhesive is setting to obtain a close contact between laminates (the thicker the glue line the lower the glue line strength). There is also an element of flattening of the laminates required to remove minor cupping distortion. For laminates of the maximum thickness this glue line pressure is often taken as 0.7 N/mm^2 (100 lbf/in^2). Pressure can be applied in a number of ways – the most usual being either clamps or air bags. The need to maintain the gluing surfaces in close contact during the setting of the adhesive rules out simply nailing the laminates together. There will tend to be some movement of the timber during the setting period and the withdrawal characteristics of even annular ring shank nails will not contain this movement. As a consequence the glue line is stressed perpendicular to the plane of adhesion, which is the worst possible direction for an adhesive whether partially set or fully set and cured.

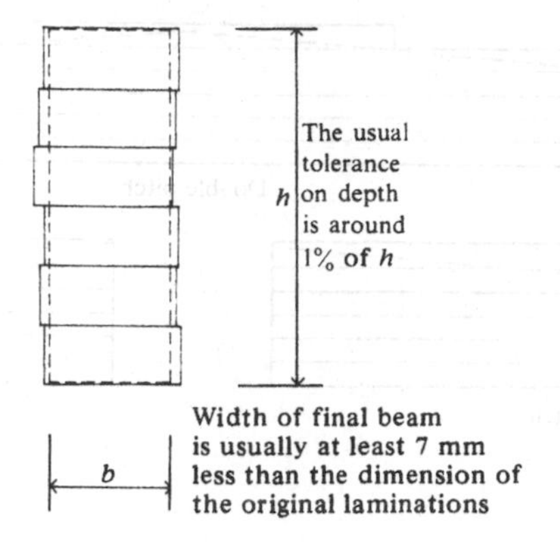

Fig. 7.1

The fabricated member has to remain under pressure for a period of time depending on the glue line temperature. Bearing in mind that the rate of setting of most adhesives used for glulam production is very slow below 15 °C and effectively ceases at 10 °C, it follows that the laminates themselves should be at least 15 °C at the time of gluing. There is no possibility of taking timber from an unheated storage shed in winter with the temperature at −5 °C and running it through the laminating process in an hour or two.

All the foregoing problems of control of the gluing process apply equally to any gluing on site which is why site gluing is rarely carried out. In one situation the site heating to achieve the requisite temperature only managed to set fire to the enclosing, protecting structure with dire consequences!

With temperatures raised to 30 °C it is possible to release the clamping pressure after 4 hours or so and gently move the laminated member to a storage area for the adhesive to cure and harden further. After perhaps 24 hours curing the section can be machined to remove any deviations caused by the different widths and straightness of the individual laminates and the relative slippage sideways during laminating. In this machining process the squeeze out of adhesive at each glue line is removed. A consistent squeeze out for each glue line along the length of the member is a good indicator of uniform spread of adhesive and adequate glue line pressure. By the time this machining process takes place this adhesive will be very hard and will damage the cutters of the planer or thicknesser. For this reason these cutters cannot be used to plane the laminates before gluing as any chips in the edge of the blades will result in 'tramlines' up to 1.5 mm high in the intended gluing surface.

It is normal for the standard glulam sections which can be obtained from stockists to be manufactured without a camber. Incorporating a camber into bespoke glulam simply requires a former to gently bend the laminates. The art in this procedure comes in guessing the spring back on release of clamping pressure.

Glulam sections are built up from thin members, and it is therefore possible to manufacture complicated shapes, curving laminates within limits. The designs in this chapter (indeed in this manual) are limited to beams with parallel, mono-pitch or duo-pitch profiles. The method of building up laminations for beams with sloping tops is indicated in Fig. 7.2. Although the grain of the laminations which

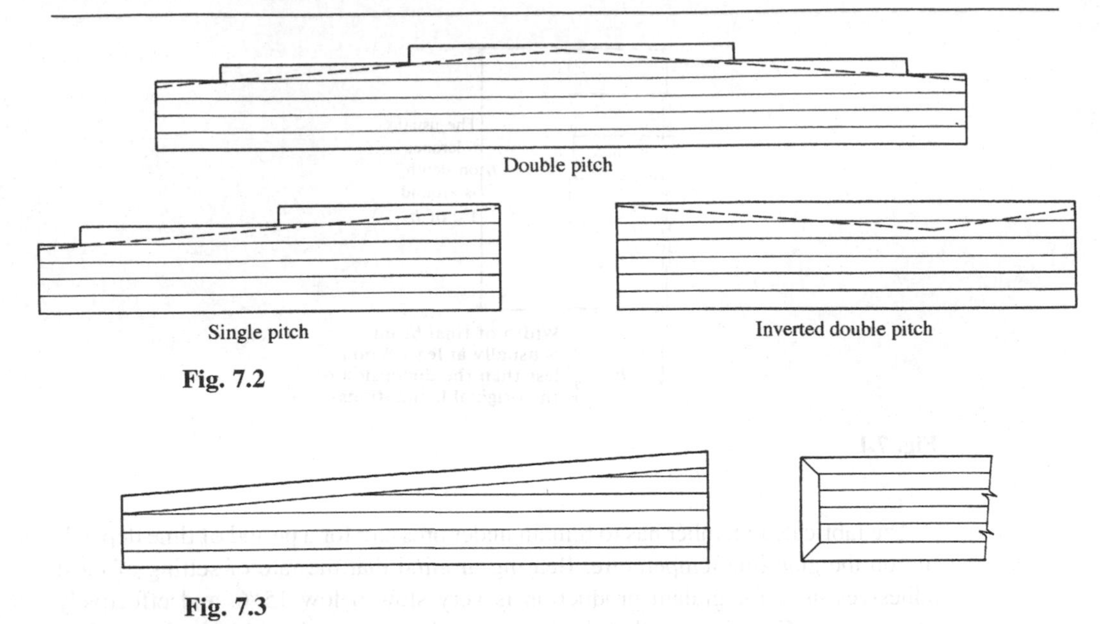

Fig. 7.2

Fig. 7.3

occur near the sloping surface is not parallel to this surface, it is permissible to consider these laminations as full strength in the design.

When it is necessary, for appearance, to fit a lamination parallel to the sloping surface or the end of the beam (Fig. 7.3), provided it is correctly glued into place, it may be considered to add to the strength of the beam.

In no circumstances should two prefabricated part-sections which have been manufactured separately be glued together as sketched in Fig. 7.4 and treated as a fully composite beam. Each part is almost certain to have too much inertia to be held by one glue line once the clamps are removed. It has been proved by experience, however, that a section which, for example, is too deep for the sanding machine, may be manufactured initially with a 'dry joint', the two pieces being separated after the glue has been cured; then each piece is sanded (not on the dry joint), glued, re-assembled and clamped. Where feasible, the method of adding to a glulam section should be by one laminate at a time, because great care and manufacturing expertise are required to obtain a successful member by the method explained above.

7.2 TIMBER STRESS GRADES FOR GLULAM

The concept of special laminating timber grades for glulam no longer exists. Timber graded to BS EN 518, BS EN 519 or BS 5756 is used for both horizontal and vertical glulam with the appropriate modification factors to the grade strength values for orientation and number of laminates. This change in the grading rules allows both visual and machine grading to be used for glulam. The modification factors for horizontal glulam depend on the timber grade and number of laminates. It is permissible to use either the species/strength grade or the corresponding

Fig. 7.4

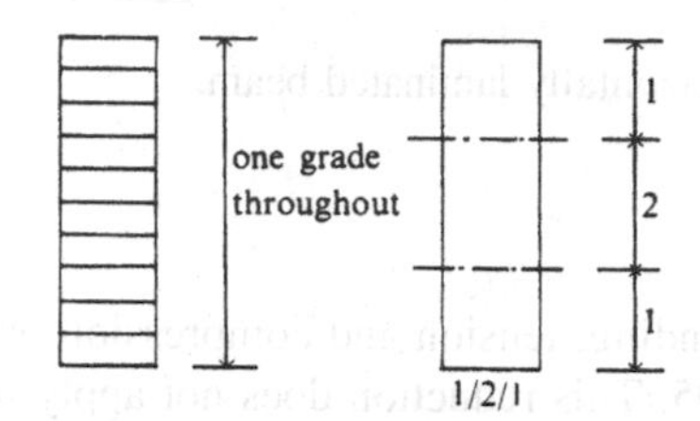

The higher grade must occupy at least the outer zones (i.e. 25% each of depth)

Fig. 7.5

strength class with the modification factors in both cases related to the strength class. For example, consider a horizontal glulam member to be made from North American Douglas fir–larch, J&P grade No. 1 so from Table 3 of BS 5268-2 this species/grade lies in strength class C16. The bending strength for C16 is 5.3 N/mm^2 whereas the species/grade value is 5.6 N/mm^2. Either of these strength values may be used but the modification factors in both cases are based on the C16 grade given in Table 21 of the standard. For vertically laminated members the modification factors are related only to the number of laminations irrespective of the grade strength.

Horizontal glulam may be made with a single grade of laminate throughout or from two grades provided they are not more than three grades apart in strength class terms, e.g. C24 and C16 may be used but not C24 with C14. The stronger grade should occupy the two outer zones with each zone forming at least 25% of the depth of the member, as shown in Fig. 7.5.

7.3 STRENGTH VALUES FOR HORIZONTALLY OR VERTICALLY LAMINATED BEAMS

7.3.1 Horizontally laminated beams

With a horizontally laminated beam (see Fig. 7.6) the method of deriving the grade strength is to multiply the relevant strength class or species/grade value for bending, compression, etc., by the appropriate laminating factors given in Table 7.1 and by the modification factors K_2 to K_{12} as for straight beams of solid timber and K_{33} to K_{35} for curved profiles. Where the glulam section is made from two grades, the modification factors K_{15} to K_{20} are taken for the higher grade but the

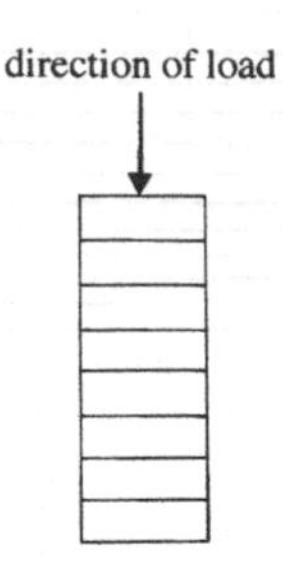

Fig. 7.6 Horizontally laminated beam.

resulting bending, tension and compression parallel to the grain values are multiplied by 0.95. This reduction does not apply to compression perpendicular to the grain, shear or modulus of elasticity. For both principal and load-sharing members the grade modulus of elasticity to be used is $E_{mean} \times K_{20}$. It should be noted that the grade modulus of elasticity value is constant irrespective of the number of laminates. This is also the value to be used when calculating K_{12} for the slenderness reduction of compression members.

Wane is not permitted with glulam, therefore the grade strength value for compression perpendicular to the grain is either the higher value given for the strength class or the species/grade value multiplied by 1.33.

The laminating factors given in Table 7.1 for softwood laminations were originally derived in the early 1960s from a statistical analysis of hypothetical knot distributions in laminates and then in the built-up member. This analysis was supplemented by a number of tests on particular layouts for the knots. The UK continues to use this approach whereas Eurocode 5 considers laminated timber to be a special product called 'glulam' and assigns strength values to the material irrespective of the direction of the applied loading in relation to the plane of the glue lines. There are rules in EN 386 relating to the production of horizontally and vertically laminated timber but glulam is then designed in the same way as solid timber but with strength values given in BS EN 1194.

7.3.2 Vertically laminated beams

The grade strength values for bending. compression, shear parallel to the grain and modulus of elasticity are calculated by multiplying the relevant strength class or species/grade value by the relevant factor, K_{27} to K_{29}, given in Table 7.2 for the number of laminations in the member. The factor K_{28} is applied to E_{min} rather than using E_{mean} as for horizontally laminated members. The value of $K_8 = 1.1$ can only be used with K_{27} and K_{29} where a load-sharing system of four or more vertically laminated members exists.

Lateral restraint is unlikely to be a problem but should be checked, remembering that any beam tends to buckle only at right angles to the direction of bending.

Table 7.1 Modification factors K_{15} to K_{20} for single-grade softwood horizontally glued laminated members

Strength class	Number of laminations*	Bending parallel to grain K_{15}	Tension parallel to grain K_{16}	Compression parallel to grain K_{17}	Compression perpendicular to grain† K_{18}	Shear parallel to grain K_{19}	Modulus of elasticity‡ K_{20}
C16 and C18	4	1.05	1.05	1.07	1.69	2.73	1.17
	5	1.16	1.16				
	7	1.29	1.29				
	10	1.39	1.39				
	15	1.49	1.49				
	20+	1.57	1.57				
C22 and C24	4	1.26	1.26	1.04	1.55	2.34	1.07
	5	1.34	1.34				
	7	1.39	1.39				
	10	1.43	1.43				
	15	1.48	1.48				
	20+	1.52	1.52				
C27 and C30	4+	1.39	1.39	1.11	1.49	1.49	1.03

* Interpolation is permitted for intermediate number of laminations.
† K_{18} should be applied to the lower value given for compression perpendicular to grain.
‡ K_{20} should be applied to the mean modulus of elasticity.

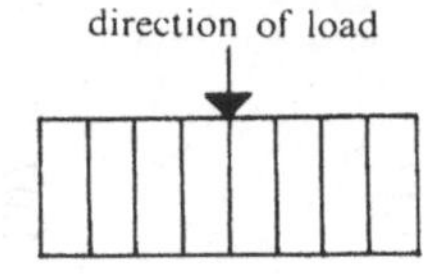

Fig. 7.7 Vertically laminated beam.

Table 7.2 Modification factors K_{27} to K_{29} for softwood vertically glued laminated members

Number of laminations	Bending, tension and shear parallel to grain K_{27}	Modulus of elasticity and compression parallel to grain* K_{28}	Compression perpendicular to grain† K_{29}
2	1.11	1.14	1.1
3	1.16	1.21	
4	1.19	1.24	
5	1.21	1.27	
6	1.23	1.29	
7	1.24	1.3	
8 or more	1.25	1.32	

* When applied to the modulus of elasticity, K_{21} is applicable to the minimum value of E.
† If no wane is present, K_{29} should have the value of 1.33 and, regardless of the grade of timber used, should be applied to the SS grade stress for the species.

7.3.3 Glulam beams with bending about both axes

With bending about both axes the distribution of bending stresses is as shown in Fig. 7.8. At point A the two compression bending stress ratios are additive. At point C the two tensile bending stress ratios are additive and at points B and D the two stress ratios are combined algebraically.

The permissible bending stress about x–x = $\sigma_{mx,adm,par}$
The permissible bending stress about y–y = $\sigma_{my,adm,par}$
At point A the bending stresses are combined thus:

$$\frac{\sigma_{mx,a,par}}{\sigma_{mx,adm,par}} + \frac{\sigma_{my,a,par}}{\sigma_{my,adm,par}} \leq 1.0$$

The deflection of the beam $\delta = \sqrt{\delta_x^2 + \delta_y^2}$ and will take place at an angle θ to the direction of loading on the x–x axis where $\tan\theta = \delta_y/\delta_x$.

7.4 APPEARANCE GRADES FOR GLULAM MEMBERS

BS 4169 had three appearance grades: regularized, planed and sanded. There is no comparable grading in BS EN 386 so the specifier or purchaser has to provide in detail his own requirements. The definitions given in BS 4169 are given below for information. They have the advantage of being finishes which the glulam industry can provide.

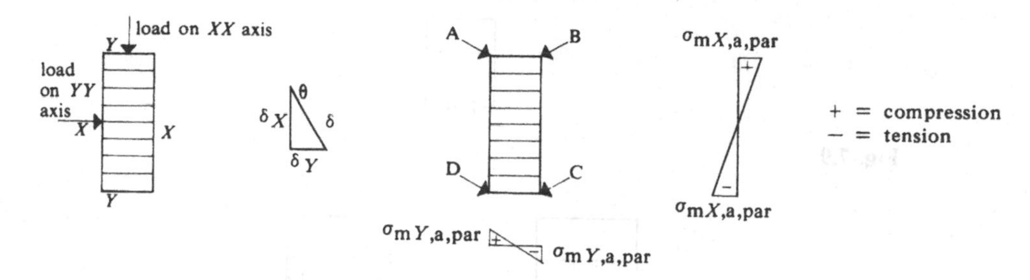

Fig. 7.8

The term 'regularized' has a different definition, although a similar intention, to the same word used in relation to solid timber. The definitions of the three classifications are:

Regularized. Not less than 50% of the surface sawn or planed to remove the protruding laminations. Surface defects not made good or filled.

This classification is suitable for use in industrial buildings or in similar utilitarian situations where appearance is not of prime importance. Finishing treatment, when specified, would usually be of the opaque or pigmented type.

Planed. Fully planed surface free from glue stains. Significant knot holes, fissures, skips in planing, voids and similar defects on exposed surfaces filled or made good with glued inserts.

This classification is suitable for most applications other than where varnish or similar non-reflective finish is specified.

Sanded. Exposed surfaces fully planed, with knot holes, fissures, voids and similar defects filled or made good with glued inserts, and sanded. Normal secondary sanding marks are acceptable. Outside laminations selected with reasonable care to match grain and colour at end joints where practicable, and free from loose knots and open knot holes. Reasonable care to be exercised in matching the direction of grain and colour of glued inserts.

This classification is recommended for use where appearance is a prime consideration and where it is desired to apply a varnish or similar finish.

Although these classifications make reference to finishing, the manufacturer will not varnish, etc., or provide protective covering unless such requirements are added to the specification. Laminations will contain the natural characteristics of the species within the grading limits for the individual laminations, including fissures on the edges and faces of laminations. The corners of members may be 'pencil rounded' (Fig. 7.9) to avoid chipped edges.

The limiting sizes of permitted characteristics and defects for individual laminations are given in BS 4978 for visual graded timber. However, no limits are stated for fissures which may appear on the sides of completed glulam members. From an appearance point of view, this can be a matter for individual specification or discussion. From a structural aspect, such fissures are unlikely to be a worry and the effect on the shear parallel to grain stress can be calculated by normal formulae, substituting b_{net} for b in the formula

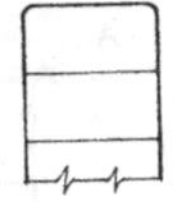

Fig. 7.9

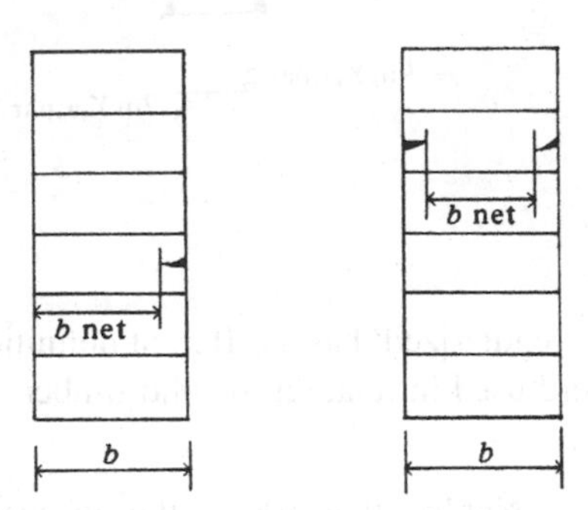

Fig. 7.10

$$\tau = \frac{F_v S_x}{b I_x} \quad \text{(see Fig. 7.10).}$$

The above principle can also be applied where there is a split in a glue line. Shear stress is not normally the design criterion so there is usually a reasonable margin between applied shear stress and the permissible shear stress as enhanced by factor K_{19}.

7.5 JOINTS IN LAMINATIONS

7.5.1 Joints in horizontal laminations

With structural glulam the end jointing of individual laminations is carried out almost certainly by finger jointing or scarf joints. If a butt joint is used, then the laminate in which it occurs has to be disregarded in the stress calculations, and this usually makes the member uneconomical. Finger joints and scarf joints may be considered not to reduce the *EI* value of the member. Finger joints can be used on either axis of laminates and have equal strength whichever way they are cut.

There are two ways of establishing if the strength of an end joint in a lamination is adequate for its function. The simple way is to provide a joint with an efficiency in bending which compares with the efficiency in bending of the laminate. The second way is to calculate the strength required of the joint in bending, tension or compression (or combinations of bending and tension, or bending and compression), and to provide a joint of adequate strength.

When a horizontally laminated member is made from C24 or C16 laminates, providing the end joint has an efficiency in bending of at least:

70% for C24 laminates, or
55% for C16 laminates,

Table 7.3 Finger joint profile efficiency ratings

Finger profile			Efficiency ratings (%)	
Length l (mm)	Pitch p (mm)	Tip width t (mm)	Bending and tension parallel to grain	Compression parallel to grain
55	12.5	1.5	75	88
50	12	2	75	83
40	9	1	65	89
32	6.2	0.5	75	92
30	6.5	1.5	55	77
30	11	2.7	50	75
20	6.2	1	65	84
15	3.8	0.5	75	87
12.5	4	0.7	65	82
12.5	3	0.5	65	83
10	3.7	0.6	65	84
10	3.8	0.6	65	84
7.5	2.5	0.2	65	92

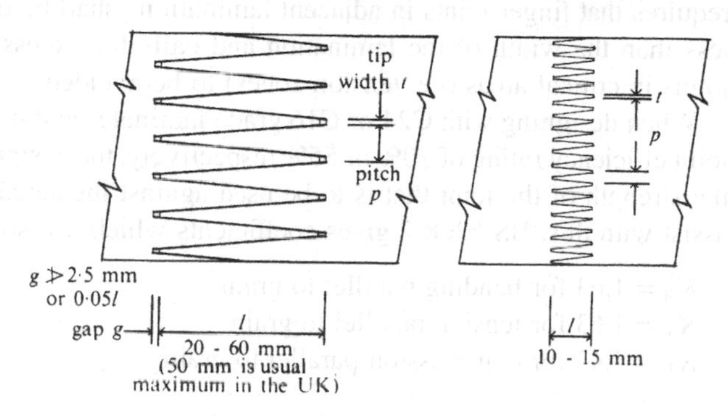

Fig. 7.11 Types of structural finger joint.

then BS 5268-2 (clause 3.4 and Table 96) permits the joint to be used without a further design check.

Guidance on the efficiency in bending, tension and compression of various finger joints is given in BS 5268-2 Annex C, reproduced here as Table 7.3 with length l, pitch p and tip width t, as illustrated in Fig. 7.11.

Note that, although the efficiency in tension is the same as in bending, the efficiency in compression is higher. The efficiencies in bending and tension have been established by test. Tests in compression tend to indicate an efficiency of 100% but, rather than quote 100%, the following formula is used:

$$\text{Efficiency rating in compression} = [(p - t)/p] \times 100\%$$

End joints should be staggered in adjacent laminations as illustrated in Fig. 7.12. This will ensure that the actual joint efficiency is larger than the quoted value because, even in the outer lamination, as well as the strength of the joint in isolation, the adjacent lamination will act as a splice plate which will increase the

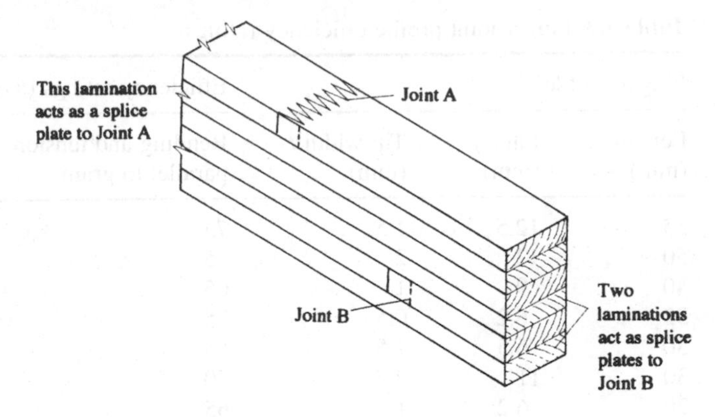

Fig. 7.12

strength of the joint. Finger joints are usually randomly spaced but BS 4169 requires that finger joints in adjacent laminations shall be offset by a distance not less than the width of the lamination and calls for excessive grouping of finger joints in critical areas (i.e. tension zones) to be avoided.

When designing with C24 or C16 grade laminates and unable to provide an end joint efficiency rating of 70% or 55% respectively, the designer is required to check the strength of the joint that is to be used against the actual strength required. To assist with this, BS 5268-2 gives coefficients which, for softwood, are:

$K_{30} = 1.63$ for bending parallel to grain
$K_{31} = 1.63$ for tension parallel to grain
$K_{32} = 1.43$ for compression parallel to grain.

Example

Consider a horizontally laminated beam 540 mm deep of 12 laminates and calculate the permissible extreme fibre stress in bending for the outer lamination under medium-term loading. The beam is C24 grade throughout and the finger jointing to be used has an efficiency rating in bending of 55% (i.e. less than the 70% required for 'blanket approval' with C24 grade).

The permissible bending stress in the lamination, as limited by the joint being used, is calculated as:

The C24 grade stress for the species (7.50 N/mm^2)
× the relevant moisture content factor K_2 (1.00 for dry exposure)
× the relevant load–duration factor K_3 (1.25)
× the modification factor for depth of *member* K_7 (0.893 for 540 mm)
× the ratio for efficiency in bending of the joint (0.55 in this case)
× factor K_{30} (1.63)

Permissible bending stress in laminate $= 7.50 \times 1.00 \times 1.25 \times 0.893 \times 0.55 \times 1.63$
$= 7.50\,N/mm^2$

If the laminates were free of end joints, the permissible bending stress for this member would be:

The C24 grade stress for the species ($7.50\,N/mm^2$)
× the relevant moisture content factor K_2 (1.00 for dry exposure)
× the relevant load–duration factor K_3 (1.25)
× the modification factor for depth of member K_7 (0.893 for 540 mm)
× the laminating factor K_{15} (1.45 in this case)

Permissible bending stress $= 7.50 \times 1.00 \times 1.25 \times 0.893 \times 1.45 = 12.1\,N/mm^2$

The design check is extended in the case of finger joints in a member taking axial tension as well as bending.

Example

Consider a horizontally laminated beam 540 mm deep of 12 laminates and calculate the permissible extreme fibre stress in bending for the outer lamination under medium-term loading. The beam is C24 grade throughout and the finger jointing to be used has an efficiency rating in bending of 55% (i.e. less than the 70% required for 'blanket approval' with C24 grade). All as the previous example.

As before, the permissible bending stress is $7.5\,N/mm^2$.

The permissible tension stress in the lamination, as limited by the joint being used, is calculated as:

The C24 grade stress for the species ($4.50\,N/mm^2$)
× the relevant moisture content factor K_2 (1.00 for dry exposure)
× the relevant load–duration factor K_3 (1.25)
× the modification factor for width of member K_{14}, see section 4.9.1 (0.937 for 540 mm width)
× the ratio for efficiency in bending of the joint (0.55 in this case)
× factor K_{31} (1.63)

Permissible tension stress in laminate $= 4.5 \times 1.00 \times 1.25 \times 0.937 \times 0.55 \times 1.63$
$= 4.72\,N/mm^2$

For the purpose of this example assume applied bending stress is $4.0\,N/mm^2$ and applied tension stress is $1.5\,N/mm^2$.

$$\frac{\sigma_{m,a}}{\sigma_{m,adm}} + \frac{\sigma_{t,a}}{\sigma_{t,adm}} \le 1.0$$

$$\frac{4.0}{7.5} + \frac{1.5}{4.72} = 0.85 \le 1.0 \quad \textbf{O.K.}$$

Section is adequate despite the 55% efficiency rated finger joints. Combined bending and compression would be assessed in a similar fashion.

7.5.2 Joints in vertical laminations

With a member made up of vertical laminates the stress grades which apply are those for solid timber (e.g. C24, C16, etc.). As with joints in horizontal laminates the designer has the option of providing a joint of a certain efficiency for a particular stress grade or of comparing the strength of a joint of a certain efficiency with the strength required for the actual stresses in the member.

When a horizontally laminated member is made from C24 or C16 laminates, providing the end joint has an efficiency in bending of at least:

70% for C24 laminates, or
55% for C16 laminates,

then BS 5268-2 (clause 3.4 and Table 96) permits the joint to be used without a further design check.

Efficiency ratings for a range of commercial finger joint profiles is given in Annex C of BS 5268-2 and in Table 7.3.

When designing with C24 or C16 grade laminates and unable to provide an end joint efficiency rating of 70% or 55% efficiency respectively, clause 3.4 of BS 5268-2 requires the permissible stresses to be reduced accordingly.

7.6 CHOICE OF GLUE FOR GLULAM

BS 5268-2 in clause 6.10.1.2 requires the adhesive to be appropriate to the environment in which the member/joint will be used. Table 94 of BS 5268-2 gives requirements for permissible adhesive types for high/low hazard exposure catagories for internal/external intended use. If specifying MR (moisture resistant) adhesive complying with BS 1204, the designer should ensure that the particular formulation is suitable for the service condition and intended life of the component.

For further information refer to Chapter 19.

7.7 PRESERVATIVE TREATMENT

There are two stages at which preservation can be carried out. Either individual laminations can be preserved before assembly or the member can be treated after assembly and after all notching, etc., has been carried out. Usually a manufacturer prefers to adopt the latter method and to treat with an organic-solvent process rather than use a water-borne treatment. The specifier is encouraged to discuss each case with the manufacturer because space and size of preservation plant available, weight of components, etc., all have a bearing on the ability of a manufacturer to preserve members.

If individual laminations are preserved by a water-borne method (either before or after end jointing) the laminates will have to be kiln dried before being assembled into a member. Even if machined before being treated they will probably have to be re-machined to obtain a surface suitable for gluing and this will remove some of the preserved timber. The adhesive and the preservative must be compatible. The manufacturer of each must be consulted.

If individual laminations are preserved by an organic-solvent process after being machined this will have no effect on the moisture content, nor will it lead to deterioration of the surface. There is some doubt as to whether or not a water repellent or other additive can be included in the preservative. Certainly the preservative and the adhesive must be compatible.

If completed members are to be preserved, the specifier must ensure that equipment is available for the size and weight of members. If a water-borne process is used it is likely that a certain amount of surface deterioration will occur and either time must be allowed for air drying or cost must be included for kiln drying (which could also lead to further surface deterioration (fissures)). An organic-solvent process is usually preferable.

If there is a requirement to use certain fire-retardant treatments it may not be possible to glue at all after treatment. If treatment is to be carried out on completed members it will probably be necessary to delay treatment until the glue is fully cured (which may be seven days or more) and compatibility must be checked.

Whitewood is 'resistant' to water-borne preservatives whereas redwood is easy to preserve. Both can be treated by organic-solvent processes.

7.8 STANDARD SIZES

Several attempts have been made at both national and international level to agree a range of standard sizes. Some manufacturers prefer to use 45 mm finished laminations machined from 50 mm. Some prefer 33.3 mm from 38 mm. The latter allows 100 mm increments of depth, but requires more glue lines and machining. The tendency is for European (including Nordic) manufacturers to standardize on 45 mm increments of depth when manufacturing straight members.

Beams are usually supplied in widths of 90, 115, 135, 160 and 185 mm. Wider sections are possible but this involves special layup of twin members in each lamination with staggered vertical joints.

7.9 TABLES OF PROPERTIES AND CAPACITIES OF STANDARD SIZES IN C24 GRADE

7.9.1 Introduction

The most popular grade used by fabricators is C24, as this grade is compatible with the available economic commercial grades, and can be end jointed without loss of strength efficiency.

C16 grade is less popular, perhaps due to its less attractive appearance in a component which is often chosen for its appearance value, or perhaps because it requires the same expenditure of labour and an equal degree of quality control in manufacture as C24 grade, but has less strength capacity.

Beams are not normally precambered unless specifically requested.

It is unusual to find a manufacturer offering grades higher than C24 or adopting combined grade laminations unless specially requested to do so. Consequently the remainder of this discussion relates only to C24 single-grade sections.

A limited range of the smaller uncambered sections is frequently available from stock. These are usually of C24 grade.

The net laminate thickness for other than curved work (such as portal frames) is usually 45 mm for maximum economy in manufacture. This thickness would normally be used for straight members and those provided with a nominal camber to off-set all or part of the anticipated dead load deflection.

As the depth h of the section increases, the depth factor K_7 reduces, whereas K_{15}, the modification factor for number of laminations, increases. These two factors tend to counteract each other. The product of K_7 and K_{15} varies between 1.33 and 1.28, giving a reasonably consistent value for the permissible bending stress at all beam depths.

7.9.2 Horizontally laminated beams

Capacities of sections are given in Tables 7.4 to 7.8 for C24 horizontally laminated single-grade beams determined from the following considerations.

Modification factors: duration of load: long term $K_3 = 1.0$
medium term $K_3 = 1.25$

Depth factor: $K_7 = (300/h)^{0.11}$ for $72 < h < 300$
$K_7 = 0.81(h^2 + 92\,300)/(h^2 + 56\,800)$ for $h > 300$

where h = depth of section.

Lamination factors: K_{15} = as tabulated
$K_{19} = 2.34$
$K_{20} = 1.07$

Shear capacity

Grade stress = 0.71 N/mm^2
Shear capacity = $(2bh/3) \times$ grade stress $\times K_3 \times K_{19}$

where b = breadth of section and h = depth of section.

Moment capacity

Grade stress = 7.5 N/mm^2
Moment capacity = $Z \times$ grade stress $\times K_3 \times K_7 \times K_{15}$

where Z = section modulus = $bh^2/6$.

EI capacity

$E_{mean} = 10\,800$ N/mm^2
$E = E_{mean} \times K_{20} = 10\,800 \times 1.07 = 11\,556$ N/mm^2
I = second moment of area = $bh^3/12$
EI capacity = $E \times I$

Table 7.4 C24; horizontally laminated glulam, 90 mm wide

Section						Section properties		Long term		Medium term		
b (mm)	h (mm)	h/b	No. of lams L	Depth factor K_7	Factor K_{15}	Z ($\times 10^6$ mm^3)	I ($\times 10^6$ mm^4)	Shear (kN)	Moment (kN m)	Shear (kN)	Moment (kN m)	EI (kN m^2)
90	180	2.0	4	1.06	1.26	0.486	43.7	17.9	4.9	22.4	6.1	505
90	225	2.5	5	1.03	1.34	0.759	85.4	22.4	7.9	28.0	9.8	987
90	270	3.0	6	1.01	1.37	1.094	147.6	26.9	11.3	33.6	14.2	1706
90	315	3.5	7	0.99	1.39	1.488	234.4	31.4	15.4	39.3	19.3	2709
90	360	4.0	8	0.96	1.40	1.944	349.9	35.9	19.7	44.9	24.7	4044
90	405	4.5	9	0.94	1.42	2.460	498.2	40.4	24.6	50.5	30.7	5757
90	450	5.0	10	0.92	1.43	3.038	683.4	44.9	30.0	56.1	37.5	7898
90	495	5.5	11	0.91	1.44	3.675	909.7	49.3	35.9	61.7	44.9	10512
90	540	6.0	12	0.89	1.45	4.374	1181.0	53.8	42.5	67.3	53.1	13647
90	585	6.5	13	0.88	1.46	5.133	1501.5	58.3	49.6	72.9	62.0	17351
90	630	7.0	14	0.87	1.47	5.954	1875.4	62.8	57.3	78.5	71.7	21672
90	675	7.5	15	0.87	1.48	6.834	2306.6	67.3	65.7	84.1	82.1	26655
90	720	8.0	16	0.86	1.49	7.776	2799.4	71.8	74.6	89.7	93.3	32349

Table 7.5 C24; horizontally laminated glulam, 115 mm wide

Section						Section properties		Long term		Medium term		
b (mm)	h (mm)	h/b	No. of lams L	Depth factor K_7	Factor K_{15}	Z ($\times 10^6$ mm^3)	I ($\times 10^6$ mm^4)	Shear (kN)	Moment (kN m)	Shear (kN)	Moment (kN m)	EI (kN m^2)
115	180	1.6	4	1.06	1.26	0.621	55.9	22.9	6.2	28.7	7.8	646
115	225	2.0	5	1.03	1.34	0.970	109.2	28.7	10.1	35.8	12.6	1261
115	270	2.3	6	1.01	1.37	1.397	188.6	34.4	14.5	43.0	18.1	2180
115	315	2.7	7	0.99	1.39	1.902	299.5	40.1	19.7	50.2	24.6	3461
115	360	3.1	8	0.96	1.40	2.484	447.1	45.9	25.2	57.3	31.5	5167
115	405	3.5	9	0.94	1.42	3.144	636.6	51.6	31.4	64.5	39.3	7357
115	450	3.9	10	0.92	1.43	3.881	873.3	57.3	38.3	71.6	47.9	10092
115	495	4.3	11	0.91	1.44	4.696	1162.3	63.1	45.9	78.8	57.4	13432
115	540	4.7	12	0.89	1.45	5.589	1509.0	68.8	54.2	86.0	67.8	17438
115	585	5.1	13	0.88	1.46	6.559	1918.6	74.5	63.4	93.1	79.2	22171
115	630	5.5	14	0.87	1.47	7.607	2396.3	80.2	73.3	100.3	91.6	27691
115	675	5.9	15	0.87	1.48	8.733	2947.3	86.0	84.0	107.5	104.9	34059
115	720	6.3	16	0.86	1.49	9.936	3577.0	91.7	95.4	114.6	119.2	41335
115	765	6.7	17	0.85	1.50	11.217	4290.4	97.4	107.8	121.8	134.7	49580
115	810	7.0	18	0.85	1.50	12.575	5093.0	103.2	120.0	129.0	150.0	58854
115	855	7.4	19	0.85	1.51	14.011	5989.8	108.9	134.4	136.1	168.0	69219
115	900	7.8	20	0.84	1.52	15.525	6986.3	114.6	149.2	143.3	186.5	80733

Table 7.6 C24; horizontally laminated glulam, 135 mm wide

Section			No. of	Depth		Section properties		Long term		Medium term		
b (mm)	h (mm)	h/b	lams L	factor K_7	Factor K_{15}	Z ($\times 10^6$ mm^3)	I ($\times 10^6$ mm^4)	Shear (kN)	Moment (kN m)	Shear (kN)	Moment (kN m)	EI (kN m^2)
135	180	1.3	4	1.06	1.26	0.729	65.6	26.9	7.3	33.6	9.1	758
135	225	1.7	5	1.03	1.34	1.139	128.1	33.6	11.8	42.1	14.8	1481
135	270	2.0	6	1.01	1.37	1.640	221.4	40.4	17.0	50.5	21.2	2559
135	315	2.3	7	0.99	1.39	2.233	351.6	47.1	23.1	58.9	28.9	4063
135	360	2.7	8	0.96	1.40	2.916	524.9	53.8	29.6	67.3	37.0	6066
135	405	3.0	9	0.94	1.42	3.691	747.3	60.6	36.9	75.7	46.1	8636
135	450	3.3	10	0.92	1.43	4.556	1025.2	67.3	45.0	84.1	56.3	11847
135	495	3.7	11	0.91	1.44	5.513	1364.5	74.0	53.9	92.5	67.4	15768
135	540	4.0	12	0.89	1.45	6.561	1771.5	80.7	63.7	100.9	79.6	20471
135	585	4.3	13	0.88	1.46	7.700	2252.3	87.5	74.4	109.3	93.0	26027
135	630	4.7	14	0.87	1.47	8.930	2813.0	94.2	86.0	117.8	107.5	32507
135	675	5.0	15	0.87	1.48	10.252	3459.9	100.9	98.6	126.2	123.2	39983
135	720	5.3	16	0.86	1.49	11.664	4199.0	107.7	111.9	134.6	139.9	48524
135	765	5.7	17	0.85	1.50	13.168	5036.6	114.4	126.5	143.0	158.1	58203
135	810	6.0	18	0.85	1.50	14.762	5978.7	121.1	140.9	151.4	176.1	69090
135	855	6.3	19	0.85	1.51	16.448	7031.5	127.8	157.8	159.8	197.2	81257
135	900	6.7	20	0.84	1.52	18.225	8201.3	134.6	175.2	168.2	219.0	94774

Table 7.7 C24; horizontally laminated glulam, 160 mm wide

Section						Section properties		Long term		Medium term		
b (mm)	h (mm)	h/b	No. of lams L	Depth factor K_7	Factor K_{15}	Z ($\times 10^6$ mm^3)	I ($\times 10^6$ mm^4)	Shear (kN)	Moment (kN m)	Shear (kN)	Moment (kN m)	EI (kN m^2)
160	180	1.1	4	1.06	1.26	0.864	77.8	31.9	8.6	39.9	10.8	899
160	225	1.4	5	1.03	1.34	1.350	151.9	39.9	14.0	49.8	17.5	1755
160	270	1.7	6	1.01	1.37	1.944	262.4	47.8	20.1	59.8	25.2	3033
160	315	2.0	7	0.99	1.39	2.646	416.7	55.8	27.4	69.8	34.3	4816
160	360	2.3	8	0.96	1.40	3.456	622.1	63.8	35.1	79.7	43.8	7189
160	405	2.5	9	0.94	1.42	4.374	885.7	71.8	43.7	89.7	54.6	10236
160	450	2.8	10	0.92	1.43	5.400	1215.0	79.7	53.3	99.7	66.7	14041
160	495	3.1	11	0.91	1.44	6.534	1617.2	87.7	63.9	109.7	79.9	18688
160	540	3.4	12	0.89	1.45	7.776	2099.5	95.7	75.5	119.6	94.3	24262
160	585	3.7	13	0.88	1.46	9.126	2669.4	103.7	88.1	129.6	110.2	30847
160	630	3.9	14	0.87	1.47	10.584	3334.0	111.6	101.9	139.6	127.4	38527
160	675	4.2	15	0.87	1.48	12.150	4100.6	119.6	116.8	149.5	146.0	47387
160	720	4.5	16	0.86	1.49	13.824	4976.6	127.6	132.7	159.5	165.8	57510
160	765	4.8	17	0.85	1.50	15.606	5969.3	135.6	150.0	169.5	187.4	68981
160	810	5.1	18	0.85	1.50	17.496	7085.9	143.5	167.0	179.4	208.7	81884
160	855	5.3	19	0.85	1.51	19.494	8333.7	151.5	187.0	189.4	233.7	96304
160	900	5.6	20	0.84	1.52	21.600	9720.0	159.5	207.6	199.4	259.5	112324

Table 7.8 C24; horizontally laminated glulam, 185 mm wide

Section						Section properties		Long term		Medium term		
b (mm)	h (mm)	h/b	No. of lams L	Depth factor K_7	Factor K_{15}	Z ($\times 10^6$ mm^3)	I ($\times 10^6$ mm^4)	Shear (kN)	Moment (kN m)	Shear (kN)	Moment (kN m)	EI (kN m^2)
185	180	1.0	4	1.06	1.26	0.999	89.9	36.9	10.0	46.1	12.5	1039
185	225	1.2	5	1.03	1.34	1.561	175.6	46.1	16.2	57.6	20.2	2029
185	270	1.5	6	1.01	1.37	2.248	303.4	55.3	23.3	69.2	29.1	3507
185	315	1.7	7	0.99	1.39	3.059	481.9	64.5	31.7	80.7	39.6	5568
185	360	1.9	8	0.96	1.40	3.996	719.3	73.8	40.6	92.2	50.7	8312
185	405	2.2	9	0.94	1.42	5.057	1024.1	83.0	50.5	103.7	63.2	11835
185	450	2.4	10	0.92	1.43	6.244	1404.8	92.2	61.7	115.3	77.1	16234
185	495	2.7	11	0.91	1.44	7.555	1869.8	101.4	73.9	126.8	92.3	21608
185	540	2.9	12	0.89	1.45	8.991	2427.6	110.6	87.3	138.3	109.1	28053
185	585	3.2	13	0.88	1.46	10.552	3086.4	119.9	101.9	149.8	127.4	35667
185	630	3.4	14	0.87	1.47	12.238	3854.9	129.1	117.8	161.4	147.3	44547
185	675	3.6	15	0.87	1.48	14.048	4741.3	138.3	135.1	172.9	168.8	54791
185	720	3.9	16	0.86	1.49	15.984	5754.2	147.5	153.4	184.4	191.8	66496
185	765	4.1	17	0.85	1.50	18.044	6902.0	156.8	173.3	195.9	216.7	79759
185	810	4.4	18	0.85	1.50	20.230	8193.0	166.0	193.1	207.5	241.4	94679
185	855	4.6	19	0.85	1.51	22.540	9635.8	175.2	216.2	219.0	270.2	111352
185	900	4.9	20	0.84	1.52	24.975	11238.8	184.4	240.1	230.5	300.1	129875

7.9.3 Vertically laminated beams

Capacities of sections are given in Tables 7.9 to 7.13 for C24 vertically laminated beams determined from the following considerations.

Modification factors: duration of load: long term $K_3 = 1.0$
medium term $K_3 = 1.25$
Depth factor: $K_7 = (300/b)^{0.11}$ for $72 < b < 300$
$K_7 = 0.81(b^2 + 92\,300)/(b^2 + 56\,800)$ for $b > 300$

where b = depth of section.

Lamination factors: K_{27} = as tabulated
K_{28} = as tabulated

Shear capacity

Grade stress = 0.71 N/mm^2
Shear capacity = $(2bh) \times$ grade stress $\times K_3 \times K_{27}$

where b = depth of section and h = breadth of section.

Moment capacity

Grade stress = 7.5 N/mm^2
Moment capacity = $Z \times$ grade stress $\times K_3 \times K_7 \times K_{15}$

where Z = section modulus = $hb^2/6$.

EI capacity

E_{min} = 7200 N/mm^2
$E = E_{min} \times K_{28}$ N/mm^2
I = second moment of area = $hb^3/12$
EI capacity = $E \times I$

Table 7.9 C24; vertically laminated glulam, 90 mm deep

Section		No. of	Depth	Factor		Section properties		Long term		Medium term		
b (mm)	h (mm)	lams L	factor K_7	K_{27}	K_{28}	Z ($\times 10^6$ mm^3)	I ($\times 10^6$ mm^4)	Shear (kN)	Moment (kN m)	Shear (kN)	Moment (kN m)	EI (kN m^2)
90	180	4	1.14	1.19	1.24	0.243	10.9	9.1	2.5	11.4	3.1	98
90	225	5	1.14	1.21	1.27	0.304	13.7	11.6	3.1	14.5	3.9	125
90	270	6	1.14	1.23	1.29	0.365	16.4	14.1	3.8	17.7	4.8	152
90	315	7	1.14	1.24	1.30	0.425	19.1	16.6	4.5	20.8	5.6	179
90	360	8	1.14	1.25	1.32	0.486	21.9	19.2	5.2	24.0	6.5	208
90	405	9	1.14	1.25	1.32	0.547	24.6	21.6	5.9	27.0	7.3	234
90	450	10	1.14	1.25	1.32	0.608	27.3	24.0	6.5	30.0	8.1	260
90	495	11	1.14	1.25	1.32	0.668	30.1	26.4	7.2	32.9	8.9	286
90	540	12	1.14	1.25	1.32	0.729	32.8	28.8	7.8	35.9	9.8	312
90	585	13	1.14	1.25	1.32	0.790	35.5	31.2	8.5	38.9	10.6	338
90	630	14	1.14	1.25	1.32	0.851	38.3	33.5	9.1	41.9	11.4	364
90	675	15	1.14	1.25	1.32	0.911	41.0	35.9	9.8	44.9	12.2	390
90	720	16	1.14	1.25	1.32	0.972	43.7	38.3	10.4	47.9	13.0	416

Table 7.10 C24; vertically laminated glulam, 115 mm deep

Section				Factor		Section properties		Long term		Medium term		
b (mm)	h (mm)	No. of lams L	Depth factor K_7	K_{27}	K_{28}	Z ($\times 10^6 mm^3$)	I ($\times 10^6 mm^4$)	Shear (kN)	Moment (kN m)	Shear (kN)	Moment (kN m)	EI (kN m^2)
115	180	4	1.11	1.19	1.24	0.397	22.8	11.7	3.9	14.6	4.9	204
115	225	5	1.11	1.21	1.27	0.496	28.5	14.8	5.0	18.5	6.3	261
115	270	6	1.11	1.23	1.29	0.595	34.2	18.1	6.1	22.6	7.6	318
115	315	7	1.11	1.24	1.30	0.694	39.9	21.3	7.2	26.6	9.0	374
115	360	8	1.11	1.25	1.32	0.794	45.6	24.5	8.3	30.6	10.3	434
115	405	9	1.11	1.25	1.32	0.893	51.3	27.6	9.3	34.4	11.6	488
115	450	10	1.11	1.25	1.32	0.992	57.0	30.6	10.3	38.3	12.9	542
115	495	11	1.11	1.25	1.32	1.091	62.7	33.7	11.4	42.1	14.2	596
115	540	12	1.11	1.25	1.32	1.190	68.4	36.7	12.4	45.9	15.5	650
115	585	13	1.11	1.25	1.32	1.289	74.1	39.8	13.4	49.8	16.8	705
115	630	14	1.11	1.25	1.32	1.389	79.8	42.9	14.5	53.6	18.1	759
115	675	15	1.11	1.25	1.32	1.488	85.5	45.9	15.5	57.4	19.4	813
115	720	16	1.11	1.25	1.32	1.587	91.3	49.0	16.5	61.2	20.7	867
115	765	17	1.11	1.25	1.32	1.686	97.0	52.1	17.6	65.1	22.0	921
115	810	18	1.11	1.25	1.32	1.785	102.7	55.1	18.6	68.9	23.2	976
115	855	19	1.11	1.25	1.32	1.885	108.4	58.2	19.6	72.7	24.5	1030
115	900	20	1.11	1.25	1.32	1.984	114.1	61.2	20.7	76.5	25.8	1084

Table 7.11 C24; vertically laminated glulam, 135 mm deep

Section		No. of lams	Depth factor	Factor		Section properties		Long term		Medium term		
b (mm)	h (mm)	L	K_7	K_{27}	K_{28}	Z ($\times 10^6$ mm^3)	I ($\times 10^6$ mm^4)	Shear (kN)	Moment (kN m)	Shear (kN)	Moment (kN m)	EI (kN m^2)
135	180	4	1.09	1.19	1.24	0.547	36.9	13.7	5.3	17.1	6.7	329
135	225	5	1.09	1.21	1.27	0.683	46.1	17.4	6.8	21.7	8.5	422
135	270	6	1.09	1.23	1.29	0.820	55.4	21.2	8.3	26.5	10.3	514
135	315	7	1.09	1.24	1.30	0.957	64.6	25.0	9.7	31.2	12.1	605
135	360	8	1.09	1.25	1.32	1.094	73.8	28.8	11.2	35.9	14.0	702
135	405	9	1.09	1.25	1.32	1.230	83.0	32.3	12.6	40.4	15.7	789
135	450	10	1.09	1.25	1.32	1.367	92.3	35.9	14.0	44.9	17.5	877
135	495	11	1.09	1.25	1.32	1.504	101.5	39.5	15.4	49.4	19.2	965
135	540	12	1.09	1.25	1.32	1.640	110.7	43.1	16.8	53.9	21.0	1052
135	585	13	1.09	1.25	1.32	1.777	119.9	46.7	18.2	58.4	22.7	1140
135	630	14	1.09	1.25	1.32	1.914	129.2	50.3	19.6	62.9	24.5	1228
135	675	15	1.09	1.25	1.32	2.050	138.4	53.9	21.0	67.4	26.2	1315
135	720	16	1.09	1.25	1.32	2.187	147.6	57.5	22.4	71.9	28.0	1403
135	765	17	1.09	1.25	1.32	2.324	156.8	61.1	23.8	76.4	29.7	1491
135	810	18	1.09	1.25	1.32	2.460	166.1	64.7	25.2	80.9	31.5	1578
135	855	19	1.09	1.25	1.32	2.597	175.3	68.3	26.6	85.4	33.2	1666
135	900	20	1.09	1.25	1.32	2.734	184.5	71.9	28.0	89.9	35.0	1754

Table 7.12 C24; vertically laminated glulam, 160 mm deep

Section		No. of	Depth	Factor		Section properties		Long term		Medium term		
b (mm)	h (mm)	lams L	factor K_7	K_{27}	K_{28}	Z ($\times 10^6$ mm^3)	I ($\times 10^6$ mm^4)	Shear (kN)	Moment (kN m)	Shear (kN)	Moment (kN m)	EI (kN m^2)
160	180	4	1.07	1.19	1.24	0.768	61.4	16.2	7.3	20.3	9.2	549
160	225	5	1.07	1.21	1.27	0.960	76.8	20.6	9.3	25.8	11.7	702
160	270	6	1.07	1.23	1.29	1.152	92.2	25.2	11.4	31.4	14.2	856
160	315	7	1.07	1.24	1.30	1.344	107.5	29.6	13.4	37.0	16.7	1006
160	360	8	1.07	1.25	1.32	1.536	122.9	34.1	15.4	42.6	19.3	1168
160	405	9	1.07	1.25	1.32	1.728	138.2	38.3	17.4	47.9	21.7	1314
160	450	10	1.07	1.25	1.32	1.920	153.6	42.6	19.3	53.3	24.1	1460
160	495	11	1.07	1.25	1.32	2.112	169.0	46.9	21.2	58.6	26.5	1606
160	540	12	1.07	1.25	1.32	2.304	184.3	51.1	23.1	63.9	28.9	1752
160	585	13	1.07	1.25	1.32	2.496	199.7	55.4	25.1	69.2	31.3	1898
160	630	14	1.07	1.25	1.32	2.688	215.0	59.6	27.0	74.6	33.8	2044
160	675	15	1.07	1.25	1.32	2.880	230.4	63.9	28.9	79.9	36.2	2190
160	720	16	1.07	1.25	1.32	3.072	245.8	68.2	30.9	85.2	38.6	2336
160	765	17	1.07	1.25	1.32	3.264	261.1	72.4	32.8	90.5	41.0	2482
160	810	18	1.07	1.25	1.32	3.456	276.5	76.7	34.7	95.9	43.4	2628
160	855	19	1.07	1.25	1.32	3.648	291.8	80.9	36.6	101.2	45.8	2774
160	900	20	1.07	1.25	1.32	3.840	307.2	85.2	38.6	106.5	48.2	2920

Table 7.13 C24; vertically laminated glulam, 185 mm deep

Section		No. of lams	Depth factor	Factor		Section properties		Long term		Medium term		
b (mm)	h (mm)	L	K_7	K_{27}	K_{28}	Z ($\times10^6\,mm^3$)	I ($\times10^6\,mm^4$)	Shear (kN)	Moment (kN m)	Shear (kN)	Moment (kN m)	EI ($kN\,m^2$)
185	180	4	1.05	1.19	1.24	1.027	95.0	18.8	9.7	23.4	12.1	848
185	225	5	1.05	1.21	1.27	1.283	118.7	23.8	12.3	29.8	15.4	1086
185	270	6	1.05	1.23	1.29	1.540	142.5	29.1	15.0	36.4	18.7	1323
185	315	7	1.05	1.24	1.30	1.797	166.2	34.2	17.6	42.8	22.0	1556
185	360	8	1.05	1.25	1.32	2.054	189.9	39.4	20.3	49.3	25.4	1805
185	405	9	1.05	1.25	1.32	2.310	213.7	44.3	22.8	55.4	28.6	2031
185	450	10	1.05	1.25	1.32	2.567	237.4	49.3	25.4	61.6	31.7	2257
185	495	11	1.05	1.25	1.32	2.824	261.2	54.2	27.9	67.7	34.9	2482
185	540	12	1.05	1.25	1.32	3.080	284.9	59.1	30.5	73.9	38.1	2708
185	585	13	1.05	1.25	1.32	3.337	308.7	64.0	33.0	80.0	41.2	2934
185	630	14	1.05	1.25	1.32	3.594	332.4	69.0	35.5	86.2	44.4	3159
185	675	15	1.05	1.25	1.32	3.850	356.2	73.9	38.1	92.4	47.6	3385
185	720	16	1.05	1.25	1.32	4.107	379.9	78.8	40.6	98.5	50.8	3611
185	765	17	1.05	1.25	1.32	4.364	403.6	83.7	43.1	104.7	53.9	3836
185	810	18	1.05	1.25	1.32	4.620	427.4	88.7	45.7	110.8	57.1	4062
185	855	19	1.05	1.25	1.32	4.877	451.1	93.6	48.2	117.0	60.3	4288
185	900	20	1.05	1.25	1.32	5.134	474.9	98.5	50.8	123.1	63.4	4513

7.10 TYPICAL DESIGNS

7.10.1 Typical design of C24 grade laminated beam loaded about major axis

Determine a suitable section in C24 grade glulam to support a uniformly distributed load of 4.0 kN/m dead load (including self-weight) and 2.4 kN/m medium term imposed over an effective span of $L_e = 6.0$ m. Assume full lateral restraint to the section.

Long-term load = 6.0 × 4.0 = 24 kN
Medium-term load = F = 6.0 × (4.0 + 2.4) = 38.4 kN

Shear

Long-term shear = 24/2 = 12 kN
Medium-term shear = 38.4/2 = 19.2 kN

Moment

Long-term moment = 24 × 6/8 = 18 kN m
Medium-term moment = 38.4 × 6/8 = 28.8 kN m

Deflection

For deflection under medium-term loading to be limited to 0.003 of span:

Estimated EI required = $4.8\,WL^2 = 4.8 \times 38.4 \times 6.0^2 = 6635$ kN m^2

From Table 7.4, try 90 × 450 C24 glulam:

Long-term shear capacity = 44.9 kN > 12 kN
Medium-term shear capacity = 56.1 kN > 19.2 kN

Long-term moment capacity = 30.0 kN m > 18.0 kN m
Medium-term moment capacity = 37.5 kN m > 28.8 kN m
EI_x = 7898 kN m^2 > 6635 kN m^2

The EI_x provided is 19% higher than the estimated requirement and would normally be taken as adequate. If the EI_x provided were to be only slightly less, or slightly more, than the estimated value a final check on deflection should be undertaken to confirm the influence of shear deflection.

$$\text{Bending deflection} = \frac{5 \times F \times L_e^3}{384 \times EI_x} = \frac{5 \times 38.4 \times 6.0^3 \times 10^3}{384 \times 7898} = 0.0137\,\text{m}$$

Total deflection = bending deflection × K_v

where $K_v = 1 + 15.36\,(h/L)^2 = 1 + 15.63\,(450/6000)^2 = 1.088$

Total deflection = 0.0137 × 1.088 = 0.0149 m

Allowable deflection = 0.003 × 6.0 = 0.018 m, which is satisfactory.

7.10.2 Typical design of C24 grade laminated beam loaded about both the major *x–x* and minor *y–y* axes

Using the capacity tables, determine a suitable section in C24 grade glulam to suit a span of L_e= 5.0 m. The beam is to support a medium-term vertical load of 4.0 kN/m (including self-weight) and the angle θ = 15° (see Fig. 7.13).

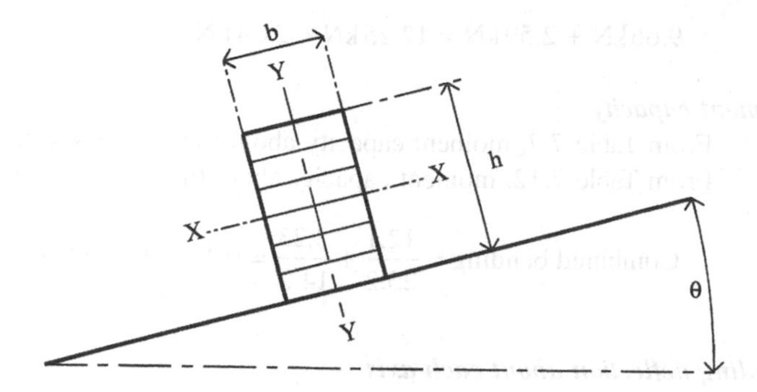

Fig. 7.13

Total vertical load = $F = 5.0 \times 4.0 = 20\,\text{kN}$
Load about x–x axis = $F_x = F\cos\theta = 19.3\,\text{kN}$
Load about y–y axis = $F_y = F\sin\theta = 5.18\,\text{kN}$

Maximum vertical shear = $V = 20/2 = 10\,\text{kN}$
Shear about x–x axis = $V\cos\theta = 9.66\,\text{kN}$
Shear about y–y axis = $V\sin\theta = 2.59\,\text{kN}$

Maximum vertical bending moment = $M = FL_e/8 = 20 \times 5.0/8 = 12.5\,\text{kN m}$
Moment about x–x axis = $M\cos\theta = 12.1\,\text{kN m}$
Moment about y–y axis = $M\sin\theta = 3.23\,\text{kN m}$

For deflection, referring to section 5.6,

$$\text{Required } EI_x \text{ (for economic section)} = 6.14\,FL_e^2 = 6.14 \times 20 \times 5.0^2 = 3070\,\text{kN m}^2$$

This EI should be provided by a section having $h/b \le \sqrt{\cot 15°} = 1.93$ (see Table 5.1 for values of $\sqrt{\cot\theta}$). The section should have $EI_x \cong 3070\,\text{kN m}^2$ coinciding with $h/b \cong 1.93$.

From Table 7.4, a 90 × 360 gives $EI_x = 4044\,\text{kN m}^2$ but $h/b = 4 > 1.93$
From Table 7.5, a 115 × 315 gives $EI_x = 3461\,\text{kN m}^2$ but $h/b = 2.7 > 1.93$
From Table 7.6, a 135 × 315 gives $EI_x = 4063\,\text{kN m}^2$ but $h/b = 2.3 > 1.93$
From Table 7.7, a 160 × 270 gives $EI_x = 3033\,\text{kN m}^2$ with $h/b = 1.7 < 1.93$

From the above, check the suitability of 160 × 270 section.

Check shear

From Table 7.7, shear capacity about the x–x axis = $59.8\,\text{kN} > 9.66\,\text{kN}$
From Table 7.12, shear capacity about the y–y axis = $31.4\,\text{kN} > 2.59\,\text{kN}$

BS 5268-2 does not give any recommendation for the assessment of combined shear stress due to bi-axial loading. The shear stress in a rectangular section will be at its maximum along the y–y axis for loading about the x–x axis and similarly will be at a maximum along the x–x axis for loading about the y–y axis. Therefore it is only at the intersect of the x–x axis and y–y axis that the combined shear stress reaches its maximum value. It will be conservative to combine the applied shear forces about the x–x and y–y axes and limit this total to the lower shear capacity given for the y–y axis, i.e.

$$9.66\,\text{kN} + 2.59\,\text{kN} = 12.25\,\text{kN} < 31.4\,\text{kN}$$

Check moment capacity

From Table 7.7, moment capacity about the x–x axis = 25.2 kN m > 12.1 kN m
From Table 7.12, moment capacity about the y–y axis = 14.2 kN m > 3.23 kN m

$$\text{Combined bending} = \frac{12.1}{25.2} + \frac{3.23}{14.2} = 0.48 + 0.23 = 0.17 < 1.0 \quad \textbf{O.K.}$$

Check bending deflection about each axis

Total deflection about the x–x axis = bending deflection × K_v
where $K_v = 1 + 15.36(h/L)^2 = 1 + 15.63(270/5000)^2 = 1.046$

$$\delta_x = \frac{5 \times K_v \times F_x \times L_e^3}{384 \times EI_x} = \frac{5 \times 1.046 \times 19.3 \times 5.0^3}{384 \times 3033} = 0.0108\,\text{m}$$

Total deflection about y–y axis = bending deflection × K_v, where

$$K_v = 1 + 15.36(b/L)^2 = 1 + 15.63(160/5000)^2 = 1.016$$

$$\delta_y = \frac{5 \times K_v \times F_y \times L_e^3}{384 \times EI_y} = \frac{5 \times 1.016 \times 5.18 \times 5.0^3}{384 \times 856} = 0.010\,\text{m}$$

Total deflection = $\sqrt{(\delta_x)^2 + (\delta_y)^2} = 0.0147\,\text{m}$
Allowable deflection = 0.003 × 5.0 = 0.018 m, which is satisfactory.

7.11 THE CALCULATION OF DEFLECTION AND BENDING STRESS OF GLULAM BEAMS WITH TAPERED PROFILES

7.11.1 Introduction

The calculation of bending deflection and shear deflection for glulam beams with tapering profiles by the normal stain energy method is extremely time-consuming, as is the calculation of the position and value of the maximum bending stress. In sections 7.11.2, 7.11.4 and 7.11.6, formulae and coefficients are given for simply supported beams of symmetrical duo-pitch, mono-pitch and inverted duo-pitch to cover most cases of normal loading. These coefficients facilitate the calculation of bending deflection and shear deflection at midspan, and maximum bending stress.

The formulae for bending deflection at midspan are established from the general formula:

$$\delta_m = \int_0^L \frac{M_x m_x \mathrm{d}x}{EI_x}$$

where M_x = the moment at x due to the applied loading
m_x = the moment at x due to unit load applied at midspan.

In addition, the varying nature of the second moment of area is taken into account by introducing a ratio

$$\frac{H}{h} = n$$

where H = the midspan depth
h = the minimum end depth.

The formulae for shear deflection at midspan are established from the general formula:

$$\delta_v = K_{form} \int_0^L \frac{F_{vx} F_{vu} dx}{(AG)_x}$$

where K_{form} = a form factor equal to 1.2 for a solid rectangle
F_{vx} = the vertical shear at x due to the applied loading
F_{vu} = the vertical shear at x due to unit load applied at midspan.

In addition, the varying nature of the area is taken into account by introducing a ratio $H/h = n$, where H = the midspan depth and h = the minimum end depth.

7.11.2 Formulae for calculating bending deflection of glulam beams with tapered profiles

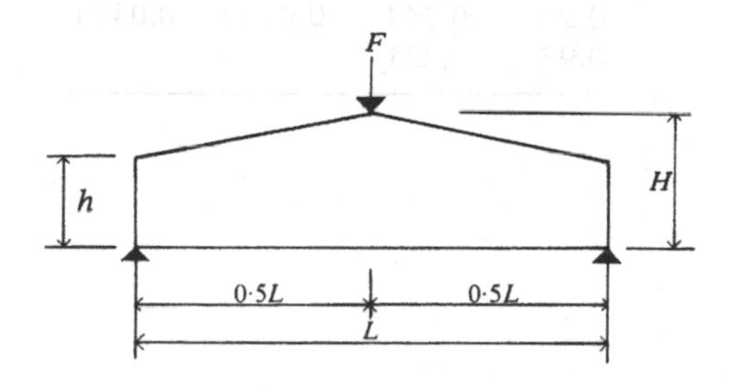

Central bending deflection is:

$$\delta_{m1} = \frac{5FL^3 \Delta_{m1}}{384EI_h}$$

where $\Delta_{m1} = 4.8\left[\frac{1}{n-1}\right]^3 \left[\log_e n - \frac{3n^2 - 4n + 1}{2n^2}\right]$ as tabulated

I_h = the second moment of area at depth h
n = H/h.

Values of Δ_{m1} for values of n from 1.0 to 3.9

n	1.0	2.0	3.0
0	1.6	0.327	0.126
0.05	1.433		
0.10	1.291	0.292	0.116
0.15	1.167		
0.20	1.060	0.262	0.108
0.25	0.966		
0.30	0.883	0.236	0.100
0.35	0.810		
0.40	0.746	0.213	0.0933
0.45	0.688		
0.50	0.637	0.194	0.0870
0.55	0.590		
0.60	0.549	0.177	0.0813
0.65	0.511		
0.70	0.477	0.162	0.0762
0.75	0.446		
0.80	0.418	0.149	0.0714
0.85	0.392		
0.90	0.369	0.137	0.0670
0.95	0.347		

Example: for $n = 1.90$, $\Delta_{m1} = 0.369$

Fig. 7.14

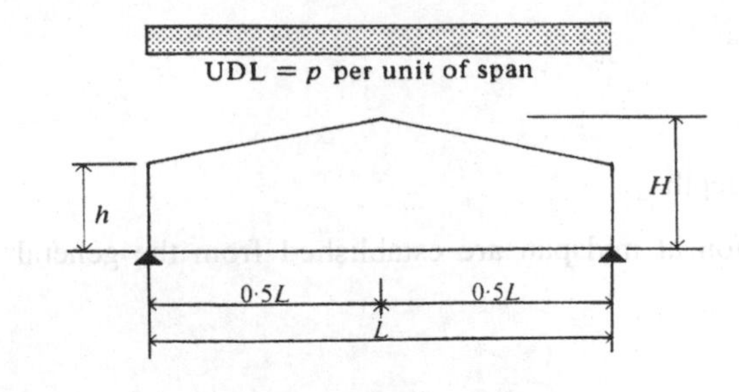

Central bending deflection is:

$$\delta_{m2} = \frac{5pL^4\Delta_{m2}}{384EI_h}$$

where $\Delta_{m2} = 2.4\left[\frac{1}{n-1}\right]^3 \left[\frac{2n+1}{n-1}\log_e n - \frac{8n^2-3n+1}{2n^2}\right]$ as tabulated

I_h = the second moment of area at depth h

n = H/h.

Fig. 7.15

Values of Δ_{m2} for values of n from 1.0 to 3.9

n	1.0	2.0	3.0
0	1.000	0.218	0.0869
0.05	0.900		
0.10	0.814	0.195	0.0806
0.15	0.739		
0.20	0.673	0.176	0.0749
0.25	0.616		
0.30	0.565	0.159	0.0698
0.35	0.520		
0.40	0.481	0.144	0.0651
0.45	0.445		
0.50	0.413	0.132	0.0609
0.55	0.384		
0.60	0.358	0.120	0.0571
0.65	0.334		
0.70	0.313	0.111	0.0535
0.75	0.293		
0.80	0.276	0.102	0.0503
0.85	0.259		
0.90	0.244	0.0939	0.0474
0.95	0.231		

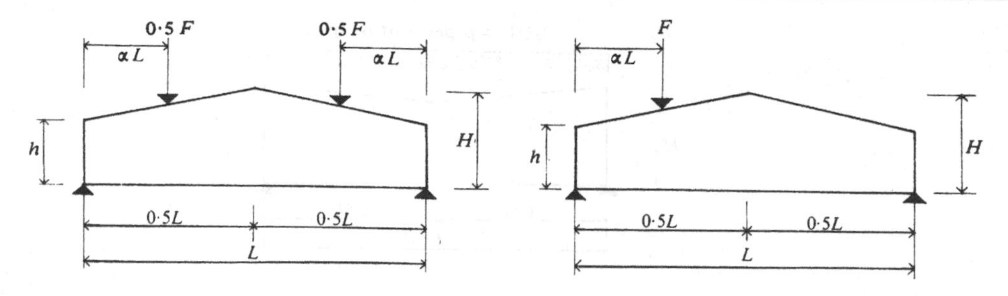

Central bending deflection is:

$$\delta_{m3} = \frac{5FL^3\Delta_{m3}}{384EI_h}$$

where $\Delta_{m3} = 4.8\left[\frac{1}{n-1}\right]\left[\frac{\log_e N}{n-1} - \frac{\alpha}{N} - \frac{\alpha(2n-1)}{n^2}\right]$ as tabulated

I_h = the second moment of area at depth h
$n = H/h$
$N = 2\alpha(n-1) + 1.$

Values of Δ_{m3} for values of n from 1.0 to 3.9

	α = 0.1			α = 0.2			α = 0.3			α = 0.4		
n	1.0	2.0	3.0	1.0	2.0	3.0	1.0	2.0	3.0	1.0	2.0	3.0
0	0.474	0.115	0.0495	0.909	0.209	0.0860	1.27	0.276	0.109	1.51	0.315	0.122
0.10	0.390	0.104	0.0462	0.745	0.188	0.0799	1.03	0.247	0.101	1.22	0.281	0.113
0.20	0.327	0.0946	0.0432	0.621	0.170	0.0745	0.854	0.223	0.0942	1.01	0.252	0.105
0.30	0.278	0.0862	0.0405	0.525	0.154	0.0695	0.717	0.201	0.0877	0.840	0.228	0.0973
0.40	0.239	0.0789	0.0380	0.449	0.141	0.0650	0.610	0.183	0.0818	0.711	0.206	0.0907
0.50	0.208	0.0725	0.0357	0.388	0.129	0.0609	0.524	0.166	0.0765	0.608	0.187	0.0846
0.60	0.182	0.0668	0.0337	0.338	0.118	0.0571	0.454	0.152	0.0716	0.525	0.171	0.0791
0.70	0.161	0.0617	0.0318	0.297	0.109	0.0537	0.397	0.139	0.0671	0.457	0.156	0.0741
0.80	0.143	0.0570	0.0300	0.263	0.100	0.0506	0.349	0.128	0.0631	0.401	0.144	0.0695
0.90	0.128	0.0531	0.0284	0.234	0.0927	0.0477	0.310	0.118	0.0593	0.354	0.132	0.0653
N	0.2*n* + 0.8			0.4*n* + 0.6			0.6*n* + 0.4			0.8*n* + 0.2		

Fig. 7.16

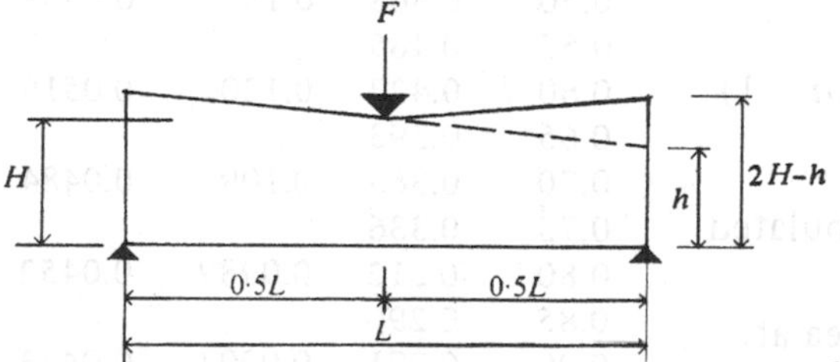

As this case is not particularly common values of n for values of Δ_{m4} are not tabulated. The same applies to the next case.

Central bending deflection is: $\delta_{m4} = \frac{FL^3}{16EI_h}\left[\frac{1}{n-1}\right]^3\left[\log_e \frac{2n-1}{n} - \frac{n^2-1}{2n^2}\right]$

Fig. 7.17

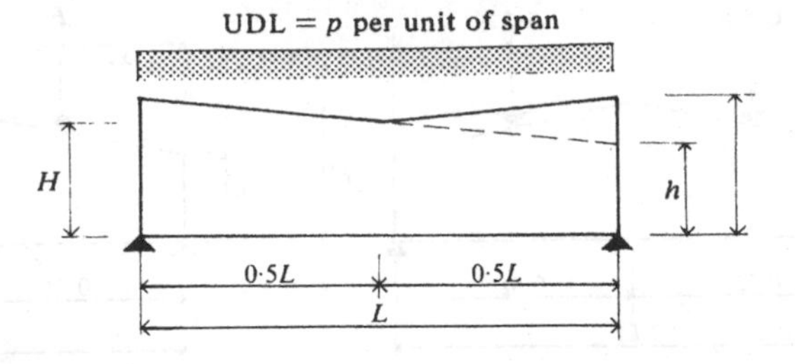

Central bending deflection is:

$$\delta_{m5} = \frac{pL^4}{32EI_h}\left[\frac{1}{n-1}\right]^3 \Big/ \left[\frac{6n^2-n+1}{2n^2} - \frac{4n-1}{n-1} \times \log_e \frac{2n-1}{n}\right]$$

Fig. 7.18

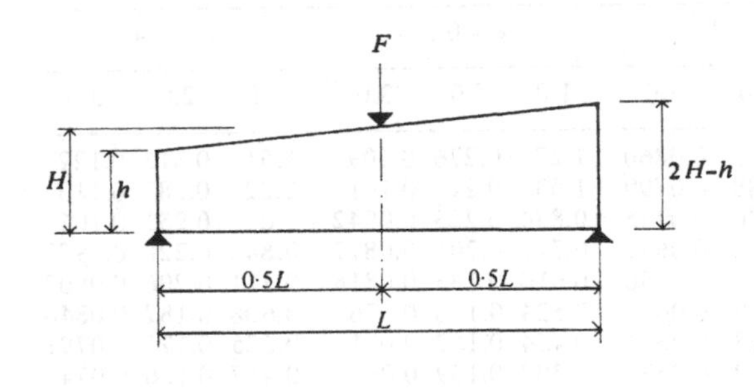

Central bending deflection is:

$$\delta_{m6} = \frac{5FL^3\Delta_{m6}}{384EI_h}$$

where $\Delta_{m6} = 2.4\left[\frac{1}{n-1}\right]^3\left[\log_e(2n-1) - \frac{2(n-1)}{n}\right]$ as tabulated

I_h = the second moment of area at depth h

$n = H/h$.

Fig. 7.19

Values of Δ_{m6} for values of n from 1.0 to 3.9

n	1.0	2.0	3.0
0	1.60	0.237	0.0828
0.05	1.38		
0.10	1.21	0.208	0.0761
0.15	1.06		
0.20	0.942	0.184	0.0702
0.25	0.839		
0.30	0.752	0.164	0.0649
0.35	0.678		
0.40	0.613	0.147	0.0601
0.45	0.557		
0.50	0.508	0.132	0.0558
0.55	0.465		
0.60	0.427	0.120	0.0519
0.65	0.393		
0.70	0.363	0.109	0.0484
0.75	0.336		
0.80	0.312	0.0989	0.0452
0.85	0.290		
0.90	0.271	0.0904	0.0423
0.95	0.253		

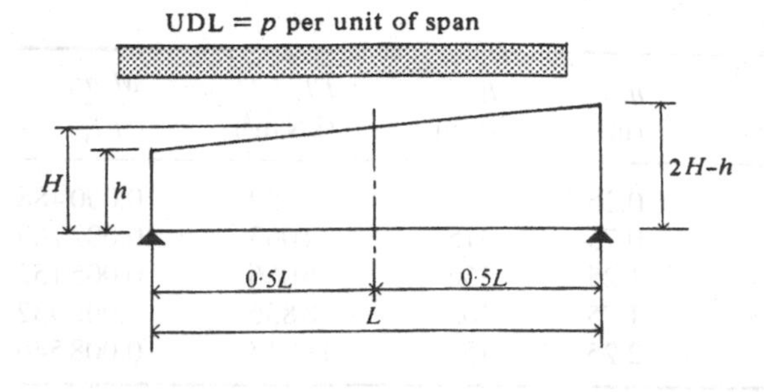

Central bending deflection is:

$$\delta_{m7} = \frac{5pL^4\Delta_{m7}}{384EI_h}$$

where $\Delta_{m7} = 1.2\left[\frac{1}{n-1}\right]^4\left[6n\log_e n - (4n-1)\log_e(2n-1) - \frac{(n-1)^2}{n}\right]$

I_h = the second moment of area at depth h

$n = H/h$.

Values of Δ_{m7} for values of n from 1.0 to 3.9			
n	1.0	2.0	3.0
0	1.000	0.153	0.0553
0.05	0.865		
0.10	0.756	0.135	0.0510
0.15	0.666		
0.20	0.590	0.120	0.0472
0.25	0.527		
0.30	0.473	0.107	0.0437
0.35	0.427		
0.40	0.387	0.0965	0.0406
0.45	0.353		
0.50	0.322	0.0872	0.0378
0.55	0.296		
0.60	0.272	0.0790	0.0352
0.65	0.251		
0.70	0.232	0.0719	0.0329
0.75	0.215		
0.80	0.200	0.0657	0.0308
0.85	0.187		
0.90	0.174	0.0602	0.0289
0.95	0.163		

Fig. 7.20

7.11.3 Example of calculating bending deflection of a tapered glulam beam

Consider a 135 mm wide C24 laminated beam tapered and loaded as in Fig. 7.21. Calculate the midspan bending deflection (a) by considering elemental strips and using the strain energy method, and (b) by using the coefficients from Fig. 7.14.

Method (a)

Divide the beam into a convenient number of elements, in this case 10 equal elements of 1 m.

$$\text{Deflection} = 2 \times \sum_1^5 \frac{M_x m_x \delta_s}{EI_x}$$

where M = the moment at midpoints of elements 1 to 5 due to the 20 kN point load

m = the unit moment at midpoints of elements 1 to 5 due to unit load at midspan

EI = EI at midpoints of elements 1 to 5

δ_s = increment of span (equal to 1 m in this particular example).

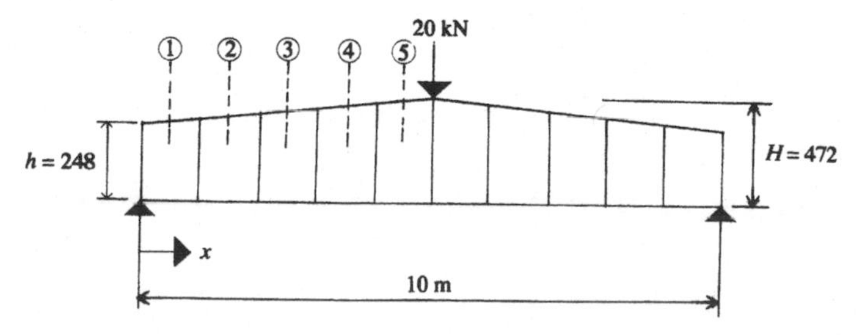

Fig. 7.21

Table 7.14

Element	x (m)	M_x (kN m)	m_x (m)	h_x (mm)	EI_x (kN m²)	$\frac{M_x m_x}{EI_x}$
1	0.5	5	0.25	270	2559	0.000488
2	1.5	15	0.75	315	4063	0.002769
3	2.5	25	1.25	360	6066	0.005152
4	3.5	35	1.75	405	8836	0.006932
5	4.5	45	2.25	450	11847	0.008546
						$\Sigma = 0.023887$

At a distance x from the left-hand support up to midspan,

$$M_x = 10x, \quad m_x = 0.5x \quad \text{and} \quad \delta_s = 1.0\,\text{m}.$$

By tabulation, consider one half of the beam, taking the EI value of each element from Fig. 7.21 appropriate to the mid-position depth of each element.

From Table 7.14 for full span:

$$\text{Deflection} = 2 \times 0.0239 \times 1.0 = 0.0478\,\text{m}$$

Method (b)

Use coefficients from Fig. 7.14 and the EI value at the end support:

$n = H/h = 472/248 = 1.903$, and interpolating from Fig. 7.14, $\Delta_{m1} = 0.368$

$I_h = 135 \times 248^3/12 = 171.6 \times 10^6\,\text{mm}^4$

$E = E_{\text{mean}} \times K_{20} = 10\,800 \times 1.07 = 11\,556\,\text{N/mm}^2$

$EI = 11\,556 \times 171.6 \times 10^6/10^9 = 1983\,\text{kN m}^2$

$$\text{Deflection (see Fig. 7.14)} = \frac{5 \times 20 \times 10^3 \times 0.368}{384 \times 1983}$$

$$= 0.0483\,\text{m}$$

In method (a) the end and midspan depths were chosen to give hx, in Table 7.9, equal to standard depths, and thereby simplify the tabular work.

In practice, method (b) is further simplified by choosing a standard depth at d.

7.11.4 Formulae for calculating shear deflection of glulam beams with tapered profiles

The formulae for calculating the shear deflection of glulam beams with tapered profiles are shown on the next three pages.

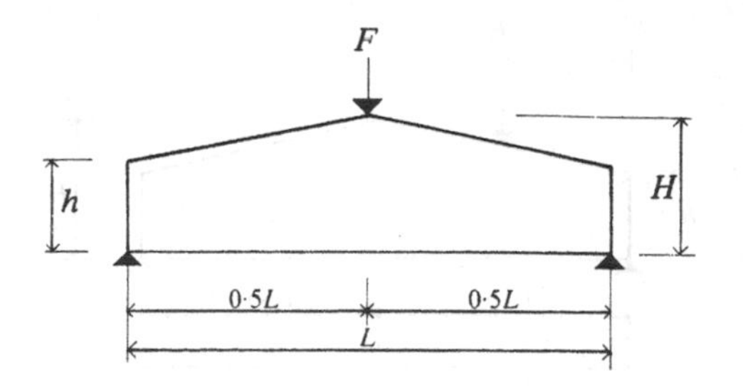

Central shear deflection is:

$$\delta_{v1} = \frac{K_{form} M_0 \Delta_{v1}}{GA_h}$$

where $M_0 = FL/4$

$\Delta_{v1} = \dfrac{\log_e n}{n-1}$ as tabulated

A_h = the area at depth h

K_{form} = the form factor (1.2 for a rectangular section)

$n = H/h$.

Fig. 7.22

Values of Δ_{v1} for values of n from 1.0 to 3.9

n	1.0	2.0	3.0
0	1.000	0.693	0.549
0.05	0.976		
0.10	0.953	0.674	0.539
0.15	0.932		
0.20	0.912	0.657	0.529
0.25	0.893		
0.30	0.874	0.641	0.519
0.35	0.857		
0.40	0.841	0.625	0.510
0.45	0.826		
0.50	0.811	0.611	0.501
0.55	0.797		
0.60	0.783	0.597	0.493
0.65	0.770		
0.70	0.758	0.584	0.485
0.75	0.746		
0.80	0.735	0.572	0.477
0.85	0.724		
0.90	0.713	0.560	0.469
0.95	0.703		

Example: for $n = 2.50$, $\Delta_{v1} = 0.611$.

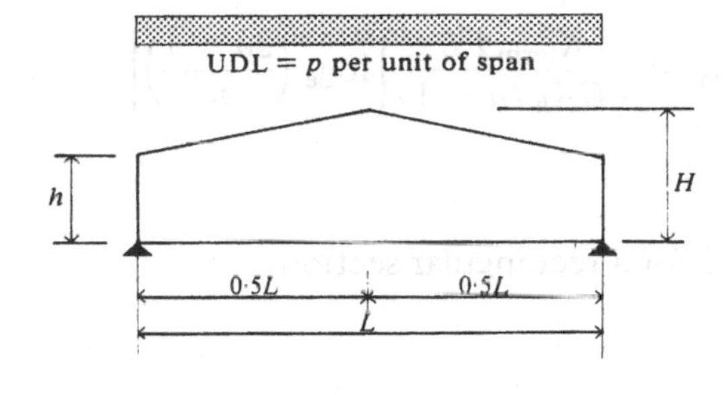

Central shear deflection is:

$$\delta_{v2} = \frac{K_{form} M_0 \Delta_{v2}}{GA_h}$$

where $M_0 = pL^2/8$

$$\Delta_{v2} = \left[\frac{2}{n-1}\right]\left[\left(\frac{n}{n-1}\right)\log_e n - 1\right]$$

A_h = the area at depth h

K_{form} = the form factor (1.2 for a rectangular section)

$n = H/h$.

Fig. 7.23

Values of Δ_{v2} for values of n from 1.0 to 3.9

n	1.0	2.0	3.0
0	1.000	0.773	0.648
0.05	0.984		
0.10	0.968	0.757	0.638
0.15	0.953		
0.20	0.939	0.742	0.629
0.25	0.926		
0.30	0.913	0.729	0.620
0.35	0.900		
0.40	0.888	0.715	0.611
0.45	0.877		
0.50	0.866	0.703	0.603
0.55	0.855		
0.60	0.844	0.691	0.595
0.65	0.834		
0.70	0.825	0.679	0.587
0.75	0.815		
0.80	0.806	0.668	0.580
0.85	0.797		
0.90	0.789	0.658	0.573
0.95	0.781		

Example: for $n = 2.60$, $\Delta_{v2} = 0.691$.

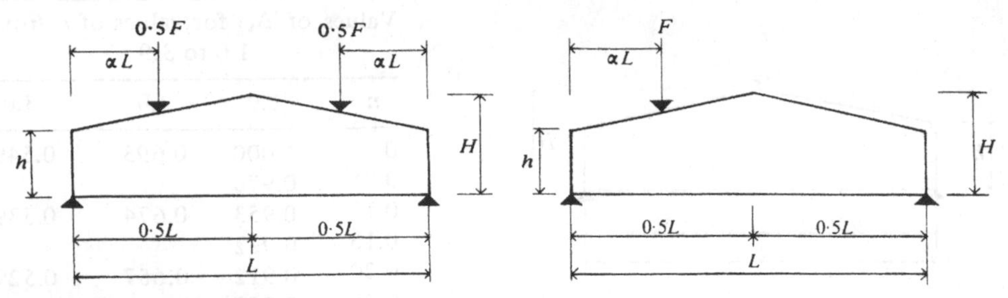

Central shear deflection is: $\delta_{v3} = \dfrac{K_{form} FL}{GA_h}\left[\dfrac{\log_e N}{n-1}\right]$

where $N = 2\alpha(n-1)+1$
$n = H/h$
A_h = the area at depth h
K_{form} = the form factor (1.2 for a rectangular section).

Fig. 7.24

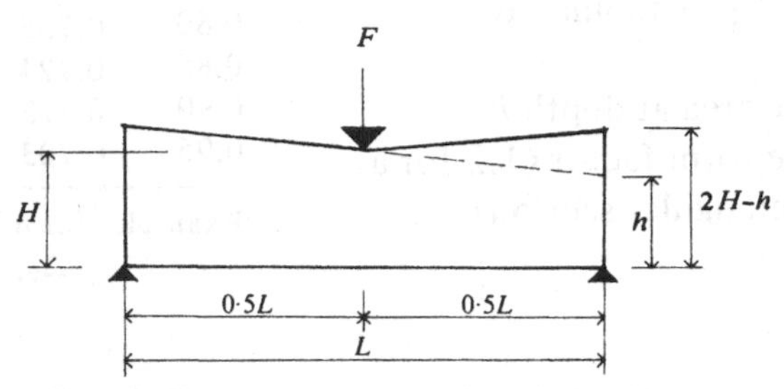

Central shear deflection is: $\delta_{v4} = \dfrac{K_{form} FL}{4GA_h(n-1)}\left[\log_e\left(\dfrac{2n-1}{n}\right)\right]$

where $n = H/h$
A_h = the area at depth h
K_{form} = the form factor (1.2 for a rectangular section).

Fig. 7.25

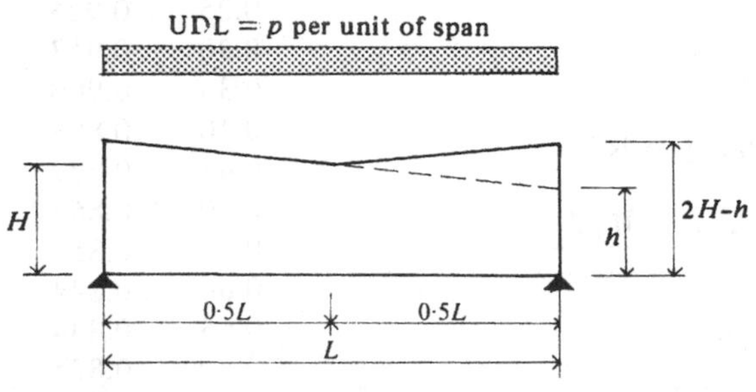

Central shear deflection is: $\delta_{v5} = \dfrac{K_{form} pL^2}{4GA_h(n-1)}\left[1 - \dfrac{n}{n-1}\log_e \dfrac{2n-1}{n}\right]$

where $n = H/h$
A_h = the area at depth h
K_{form} = the form factor (1.2 for a rectangular section).

Fig. 7.26

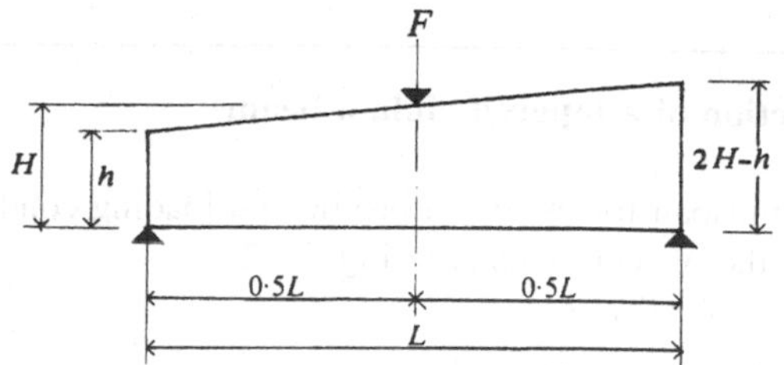

Central shear deflection is:

$$\delta_{v6} = \frac{K_{form} M_0 \Delta_{v6}}{GA_h}$$

where $\Delta_{v6} = \frac{1}{2(n-1)} \log_e (2n-1)$

as tabulated

M_0 = the moment at mid span

A_h = the area at depth h

K_{form} = the form factor (1.2 for a rectangular section)

$n = H/h$.

Values of Δ_{v6} for values of n from 1.0 to 3.9

n	1.0	2.0	3.0
0	1.000	0.549	0.402
0.05	0.953		
0.10	0.912	0.529	0.392
0.15	0.874		
0.20	0.841	0.510	0.383
0.25	0.811		
0.30	0.783	0.493	0.374
0.35	0.758		
0.40	0.735	0.477	0.366
0.45	0.713		
0.50	0.693	0.462	0.358
0.55	0.674		
0.60	0.657	0.448	0.351
0.65	0.641		
0.70	0.625	0.436	0.344
0.75	0.611		
0.80	0.597	0.424	0.337
0.85	0.584		
0.90	0.572	0.413	0.330
0.95	0.560		

Fig. 7.27

UDL = p per unit of span

Central shear deflection is:

$$\delta_{v7} = \frac{K_{form} M_0 \Delta_{v7}}{GA_h}$$

where $\Delta_{v7} = \frac{n}{(n-1)^2} \log_e \frac{n^2}{2n-1}$

as tabulated

M_0 = the moment at mid span

A_h = the area at depth h

K_{form} = the form factor (1.2 for a rectangular section)

$n = H/h$.

Values of Δ_{v7} for values of n from 1.0 to 3.9

n	1.0	2.0	3.0
0	1.000	0.575	0.441
0.05	0.953		
0.10	0.913	0.557	0.432
0.15	0.877		
0.20	0.845	0.539	0.423
0.25	0.816		
0.30	0.790	0.524	0.415
0.35	0.767		
0.40	0.745	0.509	0.407
0.45	0.725		
0.50	0.707	0.496	0.400
0.55	0.689		
0.60	0.673	0.483	0.393
0.65	0.658		
0.70	0.645	0.472	0.386
0.75	0.631		
0.80	0.619	0.461	0.379
0.85	0.607		
0.90	0.596	0.450	0.373
0.95	0.585		

Fig. 7.28

7.11.5 Example of calculating shear deflection of a tapered glulam beam

Calculate the shear deflection at midspan for the beam profile and loading condition illustrated in Fig. 7.21, using the Δ_{v1} coefficients in Fig. 7.22.

From Fig. 7.22, with $n = 1.903$, $\Delta_{v1} = 0.712$

$$E = E_{mean} \times K_{20} = 10\,800 \times 1.07 = 11\,556\,\text{N/mm}^2$$

The modulus of rigidity G is taken as

$$\frac{E}{16} = \frac{11556}{16} = 722\,\text{N/mm}^2$$

At depth $h = 248$ mm, shear rigidity is

$$GA_h = 722 \times 135 \times 248 = 24\,172\,000\,\text{N} = 24\,172\,\text{kN}$$

and the midspan moment $M_0 = 50$ kN m.

Shear deflection is:

$$\delta_{v1} = \frac{K_{form} \times M_0 \times \Delta_{v1}}{GA_h} = \frac{1.2 \times 50 \times 0.713}{24172} = 0.0017\,\text{m} = 1.7\,\text{mm}$$

7.11.6 Formulae for calculating position and value of maximum bending stress of glulam beams with tapered profiles

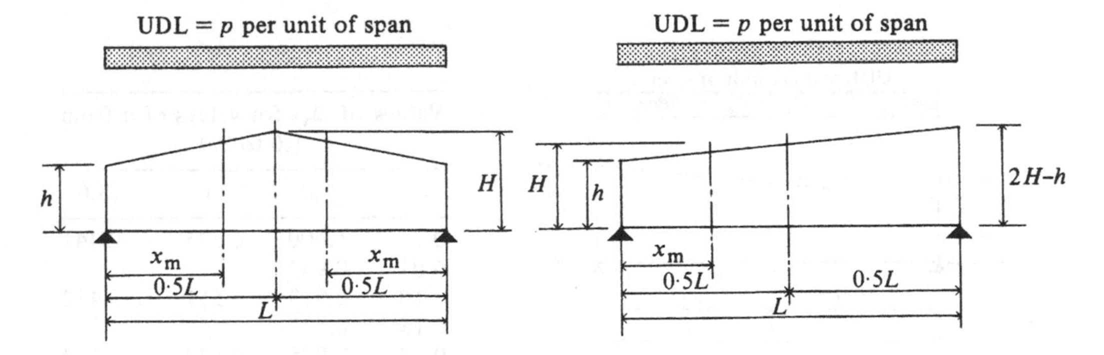

Fig. 7.29

Tapered beam with uniform loading

For a simply supported beam of the tapering profiles as shown in Fig. 7.29 with uniform loading, it can be shown that the position of the maximum bending stress occurs at:

$$x_m = \frac{Lh}{2H}$$

where x_m, L, h and H are as shown in Fig. 7.29.

It can be shown that:

$$\text{Maximum bending stress} = \frac{M_0}{Z_h(2n-1)}$$

where M_0 = the moment at midspan
Z_h = the section modulus at depth h
$n = H/h$.

By transposing, one arrives at an equation which will permit the designer to avoid time-consuming trial and error in the design of taper beams:

$$n = \frac{1}{2}\left[\frac{M_0}{\overline{m}_h} + 1\right]$$

where $\overline{m}_h$ is the moment of resistance at depth h.

Tapered beam with central point load

For a simply supported beam of the tapering profile shown in Fig. 7.30 with a central point load, it can be shown that the position of the maximum bending stress occurs at:

$$x_m = \frac{L}{2(n-1)}$$

where x, L, h and H are as shown in Fig. 7.30 and $n = H/h$.

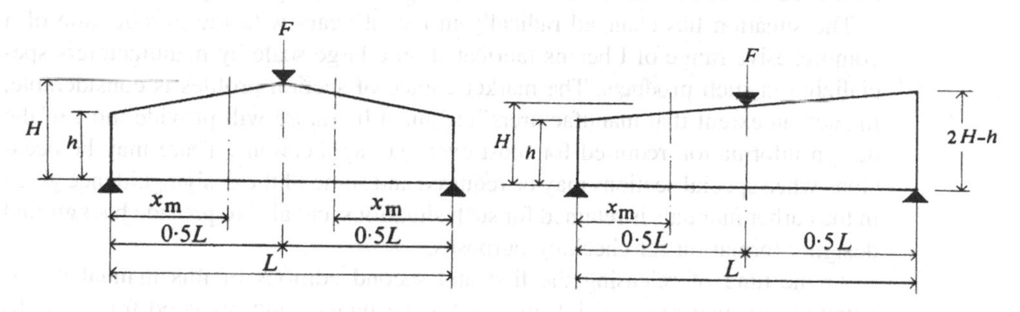

Fig. 7.30

The value of x_m is for values of n greater than 2. For cases with n less than 2 the maximum bending stress occurs at midspan. It can be shown that:

$$\text{Maximum bending stress} = \frac{M_0}{4Z_h(n-1)}$$

where M_0 = the moment at midspan
Z_h = the section modulus at depth h
$n = H/h$.

By transposing one arrives at an equation which will permit the designer to avoid time-consuming trial and error in the design of taper beams:

$$n = \frac{M_0}{4\overline{m}_h} + 1$$

where $\overline{m}_h$ is the moment of resistance at depth h.

Chapter 8

Thin Web Beams

8.1 INTRODUCTION

The design of thin web built-up beams was dealt with at some length in the first and second editions of this manual. At that time it was common practice for timber engineering companies in the UK to fabricate 'I' and 'box' profiles either to their own chosen standard sections or to special profiles designed by others.

The situation has changed radically in recent years with the introduction of a comprehesive range of I beams fabricated on a large scale by manufacturers specializing in such products. The market choice of section profiles is considerable, to such an extent that manufacturers' technical literature will provide most of the design information required for most everyday applications. There may be occasions when special sections may be required and some of the design guidance given in the earlier manuals is retained for such situations and also to provide background design information for checking purposes.

At the time of releasing the first and second editions of this manual it was common practice to use solid timbers for the flanges and plywood for the webs and to use nail/gluing to give the required interface contact between flanges and webs. Although there may still be a limited number of UK companies fabricating I beams to such a specification, the market is now dominated by a limited number of specialists who fabricate economically on a large scale and to a high quality.

For most applications of roof and floor joisting the reader is referred to technical information issued by manufacturers such as Trus Joist MacMillan, Boisse Cascade, Jones Joists, etc.

Sections and properties for a range of proprietary I beams are given in section 8.7.

8.2 PRIMARY DESIGN CONSIDERATIONS

To illustrate the design concepts solid timber flanges and plywood webs will be adopted.

8.2.1 Bending deflection

Thin web beams are a composite section with flanges and webs of different materials, and consequently, differing E values. On a few occasions, the E values of the flange and web may be close enough to justify the designer taking the section as monolithic and therefore calculating geometrical properties without any modification for differing E values. However, the more normal situation involves differing E values, and therefore the use of straightforward geometrical properties may not lead to a correct or sufficiently accurate design.

It will be seen from clause 2.9 of BS 5268-2 that the load-sharing K_8 factor of 1.1 together with the use of E_{mean} for deflection calculations do not apply to built-up beams. Special provisions are given in clause 2.10.10 of BS 5268-2. For deflection the total number of pieces of solid timber in both flanges should be taken to determine factor K_{28} to be applied to the minimum modulus of elasticity. These factors are summarized in Table 8.1

The full cross section of the plywood should be adopted and as the bending of the beam is about an axis perpendicular to the plane of the board (i.e. edge loaded) the E value for plywood webs should be taken as the modulus of elasticity in tension or compression. Stresses and moduli may be modified for duration of load using the factor K_{36} given in Table 39 of BS 5268-2 and summarized here in Table 8.2 for service classes 1 and 2.

The EI capacity is determined as follows:

Referring to Fig. 8.1(a) for a full depth web,

$$I_{web} = \frac{th^3}{12} \tag{8.1}$$

$$I_{flange} = \frac{b(h^3 - h_w^3)}{12} \tag{8.2}$$

$$EI_x = E_{flange} I_{flange} + E_{web} I_{web} \tag{8.3}$$

Table 8.1 Modification factor K_{28} for deflection

Number of pieces	Value of K_{28}
2	1.14
4	1.24
6	1.29
8 or more	1.32

Table 8.2 Modification factor K_{36} by which the grade stresses and moduli for long-term duration should be multiplied to obtain values for other durations

	Value of K_{36}	
Duration of loading	Stress	Modulus
Long term	1	1
Medium term	1.33	1.54
Short/very-short term	1.5	2

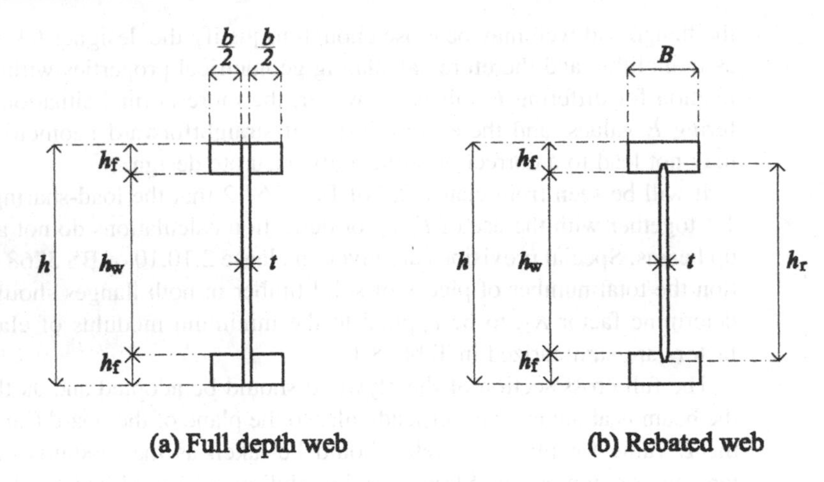

Fig. 8.1

Referring to Fig. 8.1(b) for a rebated web,

$$I_{\text{web}} = \frac{th_r^3}{12} \tag{8.4}$$

$$I_{\text{flange}} = \left[\frac{B(h^3 - h_w^3)}{12}\right] - I_{\text{web}} \tag{8.5}$$

$$EI_x = E_{\text{flange}} I_{\text{flange}} + E_{\text{web}} I_{\text{web}} \tag{8.6}$$

8.2.2 Shear deflection

The shear deflection will be a significant proportion of the total deflection (frequently 20% or more), and must be taken into account in determining deflection of thin web beams.

For simple spans the shear deflection at midspan for a section of constant EI_x is given in Section 4.15.5 as:

$$\delta_V = \frac{K_{\text{form}} \times M_0}{AG}$$

where K_{form} = a form factor
M_0 = the midspan bending moment
AG = the shear rigidity.

For an I or box section having flanges and webs of uniform thickness throughout the span, Roark's *Formulas for Stress & Strain* gives the formula for the section constant K_{form} (the form factor) as:

$$K_{\text{form}} = \left[1 + \frac{3(D_2^2 - D_1^2)D_1}{2D_2^3}\left(\frac{t_2}{t_1} - 1\right)\right]\frac{4D_2^2}{10r^2}$$

where D_1 = distance from neutral axis to the nearest surface of the flange
D_2 = distance from neutral axis to extreme fibre

t_1 = thickness of web (or webs, in box beams)
t_2 = width of flange (including web thickness)
r = radius of gyration of section with respect to neutral axis = $\sqrt{I_x/A}$.

When transposed into the terms of Fig. 8.1(a) this becomes:

$$K_{\text{form}} = \frac{Ah^2}{10I_x}\left[1+\frac{3(h^2-h_w^2)h_w}{2h^3}\left(\frac{b}{t}\right)\right]$$

where A = the area of the full section
b = the total flange width (excluding thickness of web).

If α is made equal to h_f/h so that $h_W = h(1 - 2\alpha)$ then:

$$K_{\text{form}} = \frac{Ah^2}{10I_x}\left[1+6\alpha(1-\alpha)(1-2\alpha)\left(\frac{b}{t}\right)\right]$$

or

$$K_{\text{form}} = \frac{Ah^2K_v}{10I_x}$$

where $K_v = 1 + 6\alpha(1 - \alpha)(1 - 2\alpha)(b/t)$.

Hence shear deflection is:

$$\delta_v = \frac{K_{\text{form}}M_0}{GA} = \frac{Ah^2K_vM_0}{10I_xGA} = \frac{h^2K_vM_0}{10GI_x}$$

For a solid section $\alpha = 0.5$, hence $K_v = 1.0$ and

$$\delta_v = \frac{h^2M_0}{10G(bh^3/12)} = \frac{1.2M_0}{GA}$$

which agrees with the form factor $K_{\text{form}} = 1.2$ given in section 4.15.5 for solid sections.

There is little, if any, inaccuracy in adopting Roark's recommended approximation that K_{form} may be taken as unity if A is taken as the area of the web or webs only.

The shear deflection may then be simplified to:

$$\delta_v = \frac{M_0}{A_wG_w} \tag{8.7}$$

where M_0 = the bending moment at midspan
A_w = the area of the webs
G_W = the modulus of rigidity of the webs.

8.2.3 Bending

The method most usually adopted to determine the maximum bending stress at the extreme fibre of a section assumes that the web makes no contribution to the bending strength. This method has much to commend it, although it will underestimate the bending strength of the section.

Table 8.3 Modification factor K_{27} for bending

Number of pieces in one flange	Value of K_{27}
1	1
2	1.11
3	1.16
4	1.19
5	1.21
6	1.23
7	1.24
8 or more	1.25

Apart from the simplicity of this method, disregarding the contribution made by the web to bending strength reduces the design of web splices, if these should be required, to a consideration of shear forces only.

The applied bending stress is calculated as:

$$\sigma_m = \frac{M_0}{Z_{flange}} \tag{8.8}$$

where M_0 = applied bending moment
Z_{flange} = section modulus of flange only.

The section modulus is

$$Z_{flange} = \frac{I_{flange}}{y_h} \tag{8.9}$$

where I_{flange} = inertia of flange only
y_h = the distance from the neutral axis to the extreme fibre being considered = $h/2$.

The permissible bending stress is:

$$\sigma_{m,adm} = \sigma_{grade} \times K_3 \times K_{27} \tag{8.10}$$

where K_3 = factor for duration of load
K_{27} = factor for number of pieces of solid timber in one flange as shown in Table 8.3.

8.2.4 Panel shear

Panel shear stress is the traditional term used for horizontal shear stress in a thin web beam, not to be confused with 'rolling shear stress' as discussed in section 8.2.5.

The maximum panel shear stress (v_p) occurs at the x–x axis and is determined as:

$$v_p = \frac{VQ}{tI_g} \tag{8.11}$$

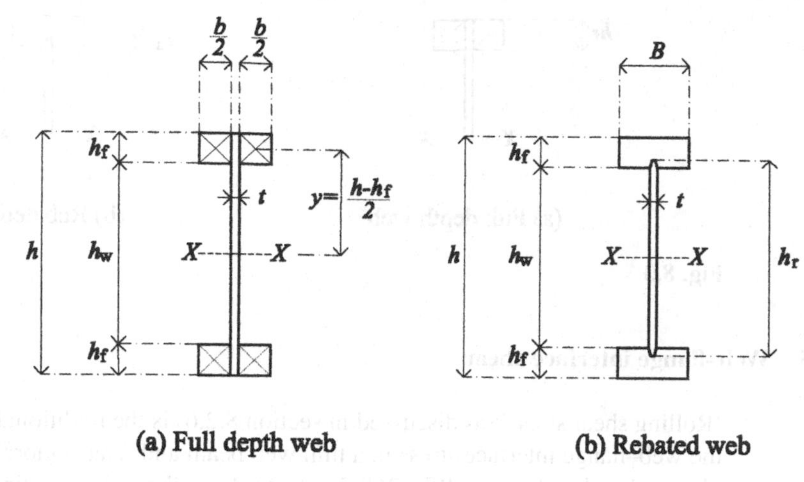

(a) Full depth web

(b) Rebated web

Fig. 8.2

where V = applied shear
Q = first moment of area of flanges and web or webs above the x–x axis
$= Q_{\text{flange}} + Q_{\text{web}}$
t = web thickness
I_g = second moment of area of the full section about the x–x axis.

Referring to Fig. 8.2(a) for a full depth web:

$$Q_{\text{flange}} = (bh_f)\left(\frac{h-h_f}{2}\right) \tag{8.12}$$

$$Q_{\text{web}} = \frac{th^2}{8} \tag{8.13}$$

Referring to Fig. 8.2(b) for a rebated web:

$$Q_{\text{flange}} = Q(\text{flange each side of web}) + Q(\text{flange over web})$$

$$= \left[(bh_f)\left(\frac{h-h_f}{2}\right)\right] + \left\{t\left(\frac{h-h_r}{2}\right)\left[\frac{h}{2} - \left(\frac{h-h_r}{2}\right)\right]\right\}$$

$$= \left[(bh_f)\left(\frac{h-h_f}{2}\right)\right] + \left\{\frac{t}{8}(h^2 - h_r^2)\right\} \tag{8.14}$$

$$Q_{\text{web}} = \frac{th_r^2}{8} \tag{8.15}$$

The applied panel shear stress should not exceed the value tabulated in Tables 40 to 56 of BS 5268-2 according to the type of plywood adopted. The listed grade stresses apply to long-term loading. For other durations of loading the stresses should be modified by K_{36} given in Table 39 of BS 5268-2, and repeated in this manual as Table 8.2.

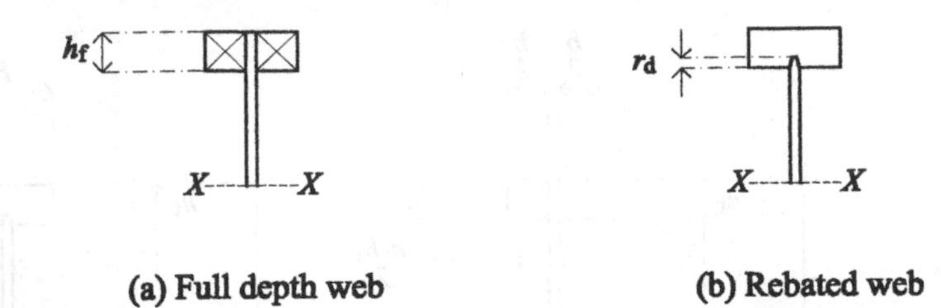

Fig. 8.3

8.2.5 Web–flange interface shear

'Rolling shear stress' (as discussed in section 8.2.6) is the traditional term used for the web–flange interface stress in a thin web beam and relates more specifically to plywood webs although BS 5268-2 retains the rolling shear definition for other web materials such as tempered hardboard and chipboard.

The applied rolling shear stress (v_r) is determined as:

$$v_r = \frac{VQ_f}{TI_g} \tag{8.16}$$

where V = applied shear
Q_f = first moment of area of flange only above the x–x axis
I_g = second moment of area of the full section about the x–x axis
T = total contact depth between webs and flanges above the x–x axis.

Referring to Fig. 8.3(a) for a full depth web:

$$Q_{flange} = Q_{flange} \quad \text{as eq. (8.12)}$$
$$T = 2 \times h_f$$

Referring to Fig. 8.3(b) for a rebated web:

$$Q_{flange} = Q_{flange} \quad \text{as eq. (8.14)}$$
$$T = 2 \times r_d$$

The applied rolling shear stress should not exceed the value tabulated in Tables 40 to 56 of BS 5268-2 according to the type of plywood adopted. The listed grade stresses apply to long-term loading. For other durations of loading the stresses should be modified by K_{36} given in Table 39 of BS 5268-2, and repeated in this manual as Table 8.2.

8.2.6 Rolling shear stress

The web–flange shear in a glued ply web beam is frequently referred to as 'rolling shear'. Figures 8.4 and 8.5 represent a plan on the top flange of an I and box beam, and Fig. 8.6 is an idealized magnification of the junction between web and flange.

The term 'rolling shear' is frequently used because it best describes the appearance of the failure which can result if the ultimate stress at the junction between

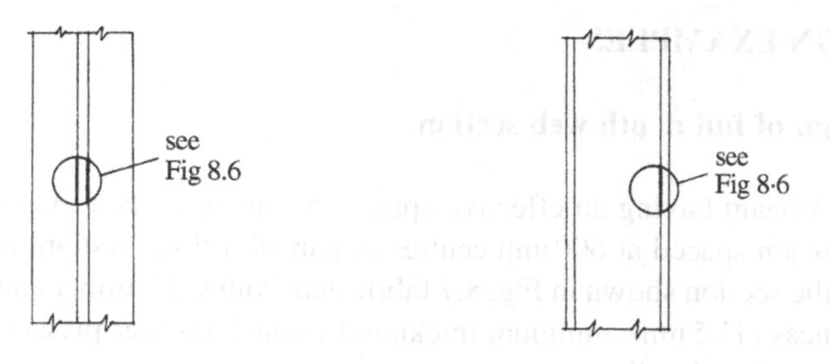

Fig. 8.4

Fig. 8.5

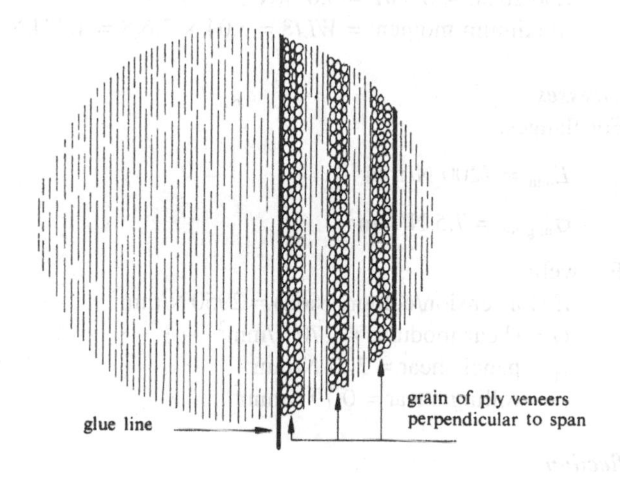

Fig. 8.6

web and flange is exceeded. In transferring horizontal shear forces from web to flange, a 'rolling action' takes place. If the face grain of the plywood runs perpendicular to the general grain direction of the timber in the flange, the rolling takes place at this interface. If the face grain of the plywood runs parallel to the general grain direction of the timber in the flange, the rolling will take place between the face veneer of the plywood and the next veneer into the plywood. To avoid this rolling action, the rolling shear must be limited by providing sufficient glued contact depth between flange and web.

Rolling shear in ply web beams applies only to the plywood, not to the solid timber. Any glue lines to the solid timber flanges in ply web beams are stressed parallel to grain.

Clause 4.7 of BS 5268-2 requires the permissible rolling shear stress to be multiplied by a stress concentration modification factor K_{37} which has a value of 0.5.

Clause 6.10.1.5 of BS 5268-2 requires the permissible rolling shear stress to be multiplied by modification factor $K_{70} = 0.9$ if the bonding pressure is provided by nails or staples.

8.3 DESIGN EXAMPLES

8.3.1 Design of full depth web section

A beam having an effective span of 5.6 m supports a long-term load of 1.8 kN/m^2 when spaced at 600 mm centres as part of a floor system. Determine suitability of the section shown in Fig. 8.7 fabricated from C24 timber and 12 mm nominal thickness (11.5 mm minimum thickness) Finnish conifer plywood using nails to develope the glue line pressure.

Span = L = 5.6 m
Total applied load = $W = 5.6 \times 0.6 \times 1.8 = 6.04$ kN
End shear = 6.04/2 = 3.02 kN
Maximum moment = $WL/8 = 6.04 \times 5.6/8 = 4.23$ kN m

Permissible stresses

For flanges:

$$E_{\min} = 7200 \text{ N/mm}^2$$

$$\sigma_{\text{m,grade}} = 7.5 \text{ N/mm}^2$$

For web:

E (for tension/compression) = 2950 N/mm^2
G = shear modulus = 270 N/mm^2
v_p = panel shear = 3.74 N/mm^2
v_r = rolling shear = 0.79 N/mm^2

Bending deflection

With nominal 30% allowance for shear deflection:

Anticipated EI required = $1.3 \times 4.34\, WL^2$
Approximate EI required = $5.64\, WL^2 = 5.64 \times 6.04 \times 5.6^2 = 1068$ kN m^2

From eq. (8.1),

$$I_{\text{web}} = 11.5 \times 300^3/12 = 25.8 \times 10^6 \text{ mm}^4$$

From eq. (8.2),

$$I_{\text{flange}} = 88(300^3 - 160^3)/12 = 168 \times 10^6 \text{ mm}^4$$

From Table 8.1, with 4 flange elements contributing to E value, $K_{28} = 1.24$

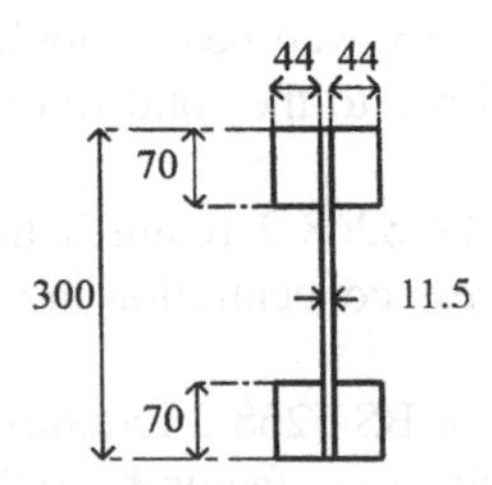

Fig. 8.7

$$E_{flange} = 7200 \times 1.24 = 8928 \text{ N/mm}^2$$

From eq. (8.3),

$$EI_x = (8928 \times 168 \times 10^6) + (2950 \times 25.8 \times 10^6) \text{ N mm}^2$$
$$= (1500 \times 10^9) + (76.1 \times 10^9) \text{ N mm}^2$$
$$= 1576 \text{ kN m}^2$$

$$\text{Bending deflection} = \frac{5WL^3}{384 \times EI_x} = \frac{5 \times 6.04 \times 5.6^3}{384 \times 1576} = 0.0088 \text{ m}$$

Shear deflection

$$A_w = 11.5 \times 300 = 3450 \text{ mm}^2$$
$$G_w = 270 \text{ N/mm}^2$$
$$A_w G_w = 3450 \times 270 = 931\,500 \text{ N} = 931.5 \text{ kN}$$

From eq. (8.7),

$$\text{Shear deflection} = \frac{4.23}{931.5} = 0.0045 \text{ m}$$

Total deflection

Total deflection = bending deflection + shear deflection, that is

$$0.0088 + 0.0045 = 0.0133 \text{ m}$$

Allowable deflection = 0.003 × span (= 0.0168 m) or 0.014 m which ever is the lesser. **O.K.**

Bending

From eq. (8.9),

$$Z_{flange} = \frac{168 \times 10^3}{150} = 1.12 \times 10^6 \text{ mm}^3$$

From eq. (8.8),

$$\text{Applied bending stress} = \frac{4.23 \times 10^3}{1.12 \times 10^6} = 3.78 \text{ N/mm}^2$$

From eq. (8.10),

$$\text{Allowable bending stress} = \sigma_{grade} \times K_3 \times K_{27}$$
$$= 7.5 \times 1.0 \times 1.11 = 8.25 \text{ N/mm}^2 \quad \textbf{O.K.}$$

Panel shear

From eq. (8.12),

$$Q_{flange} = (88 \times 70)[(300 - 70)/2] = 0.708 \times 10^6 \text{ mm}^3$$

From eq. (8.13),

$$Q_{web} = 11.5 \times 300^2/8 = 0.129 \times 10^6 \text{ mm}^3$$

$$Q = Q_{\text{flange}} + Q_{\text{web}} = 0.837 \times 10^6 \text{ mm}^3$$

$$I_g = I_{\text{flange}} + I_{\text{web}} = (168 \times 10^6) + (25.8 \times 10^6) = 193.8 \times 106 \text{ mm}^4$$

From eq. (8.11),

$$\text{Applied stress} = v_p = \frac{3020 \times 0.837 \times 10^6}{11.5 \times 193.8 \times 10^6} = 1.13 \text{ N/mm}^2$$

Allowable stress = 3.74 N/mm^2 **O.K.**

Web–flange interface shear

Total web–flange contact depth is

$$T = 2 \times h_f = 2 \times 70 = 140 \text{ mm}$$

From eq. (8.16),

$$\text{Applied interface stress} = v_r = \frac{3020 \times 0.708 \times 10^6}{140 \times 193.8 \times 10^6} = 0.08 \text{ N/mm}^2$$

Allowing for duration of load, stress concentration factor and bonding pressure provided by nailing:

$$\text{Allowable interface stress} = \text{grade stress} \times K_{36} \times K_{37} \times K_{70}$$
$$= 0.79 \times 1.0 \times 0.5 \times 0.9 = 0.36 \text{ N/mm}^2$$ **O.K.**

8.3.2 Design of rebated web section

A beam having an effective span of 5.0 m supports a long-term load of 1.8 kN/m^2 when spaced at 400 mm centres as part of a floor system. Determine the suitability of the section shown in Fig. 8.8 fabricated from C24 timber and 12 mm nominal thickness (11.5 mm minimum thickness) Finnish conifer plywood. Webs are factory pressure bonded to flanges.

Span = L = 5.0 m
Total applied load = W = 5.0 × 0.4 × 1.8 = 3.6 kN
End shear = 3.6/2 = 1.8 kN
Maximum moment = $WL/8$ = 3.6 × 5.0/8 = 2.25 kN m.

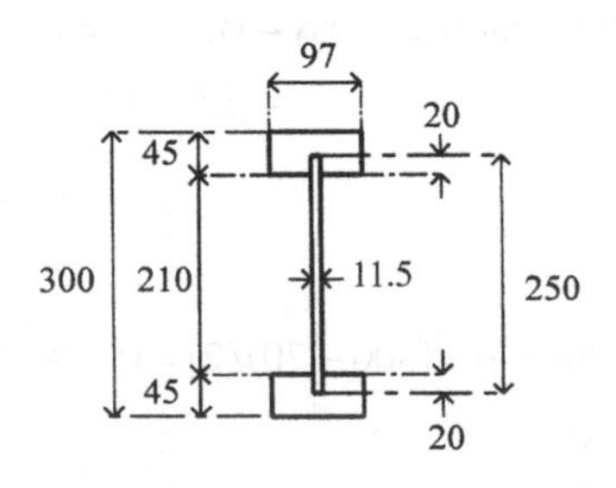

Fig. 8.8

Permissible stresses

For flanges:

$$E_{min} = 7200\,N/mm^2$$
$$\sigma_{m,grade} = 7.5\,N/mm^2$$

For web:

E (for tension/compression) = 2950 N/mm^2
G = shear modulus = 270 N/mm^2
v_p = panel shear = 3.74 N/mm^2
v_r = rolling shear = 0.79 N/mm^2

Bending deflection

With nominal 30% allowance for shear deflection:

Anticipated EI required = $1.3 \times 4.34\,WL^2$
Approximate EI required = $5.64\,WL^2 = 5.64 \times 3.6 \times 5.0^2 = 508\,kN\,m^2$

From eq. (8.4),

$$I_{web} = 11.5 \times 250^3/12 = 15.0 \times 10^6\,mm^4$$

From eq. (8.5),

$$I_{flange} = [97(300^3 - 200^3)/12] - I_{web}$$
$$= 146.3 \times 10^6 - 15.0 \times 10^6 = 131.3 \times 10^6\,mm^4$$

From Table 8.1, with two flange elements contributing to E value, $K_{28} = 1.14$

$$E_{flange} = 7200 \times 1.14 = 8208\,N/mm^2$$

From eq. (8.3),

$$EI_x = (8208 \times 131.3 \times 10^6) + (2950 \times 15.0 \times 10^6)\,N\,mm^2$$
$$= (1078 \times 10^9) + (44 \times 10^9)\,N\,mm^2$$
$$= 1122\,kN\,m^2$$

$$\text{Bending deflection} = \frac{5 \times W \times L^3}{384 \times EI_x} = \frac{5 \times 3.6 \times 5.0^3 \times 10^3}{384 \times 1122} = 0.0052\,m$$

Shear deflection

$A_w = 11.5 \times 250 = 2875\,mm^2$
$G_w = 270\,N/mm^2$
$A_wG_w = 2875 \times 270 = 776\,250\,N = 776.2\,kN$

From eq. (8.7),

$$\text{Shear deflection} = \frac{2.25}{776.2} = 0.0029\,m$$

Total deflection

Total deflection = bending deflection + shear deflection, that is,

$$0.0052 + 0.0029 = 0.0081\,m$$

Allowable deflection = 0.003 × span (= 0.015 m) or 0.014 m which ever is the lesser. **O.K.**

Bending

From eq. (8.9),

$$Z_{\text{flange}} = \frac{132.1 \times 10^3}{150} = 0.881 \times 10^6 \text{mm}^3$$

From eq. (8.8),

$$\text{Applied bending stress} = \frac{2.25 \times 10^6}{0.881 \times 10^6} = 2.25 \text{N/mm}^2$$

From eq. (8.10),

$$\begin{aligned}\text{Allowable bending stress} &= \sigma_{\text{grade}} \times K_3 \times K_{27} \\ &= 7.5 \times 1.0 \times 1.0 = 7.50 \text{N/mm}^2 \quad \textbf{O.K.}\end{aligned}$$

Panel shear

From eq. (8.12),

$$\begin{aligned}Q_{\text{flange}} &= \{86 \times 45[(300 - 45)/2]\} + \{10.9(300^2 - 250^2)/8\} \\ &= 0.493 \times 10^6 \text{mm}^3 + 0.037 \times 10^6 \text{mm}^3 \\ &= 0.53 \times 10^6 \text{mm}^3\end{aligned}$$

From eq. (8.13),

$$Q_{\text{web}} = \frac{11.5 \times 250^2}{8} = 0.090 \times 10^6 \text{mm}^3$$

$$Q = Q_{\text{flange}} + Q_{\text{web}} = 0.62 \times 10^6 \text{mm}^3$$

$$I_g = I_{\text{flange}} + I_{\text{web}} = (131.3 \times 10^6) + (15.0 \times 10^6) = 146.3 \times 10^6 \text{mm}^4$$

From eq. (8.11),

$$\text{Applied stress} = v_p = \frac{1800 \times 0.62 \times 10^6}{11.5 \times 146.3 \times 10^6} = 0.66 \text{N/mm}^2$$

Allowable stress = 3.74 N/mm² **O.K.**

Web–flange interface shear

Total web–flange contact depth = $T = 2 \times r_d = 2 \times 20 = 40$ mm
From eq. (8.16),

$$\text{Applied interface stress} = v_r = \frac{1800 \times 0.53 \times 10^6}{40 \times 146.3 \times 10^6} = 0.163 \text{N/mm}^2$$

Allowing for duration of load and stress concentration factor:

$$\begin{aligned}\text{Allowable interface stress} &= \text{grade stress} \times K_{36} \times K_{37} \\ &= 0.79 \times 1.0 \times 0.5 = 0.395 \text{N/mm}^2 \quad \textbf{O.K.}\end{aligned}$$

8.4 WEB SPLICES

8.4.1 Introduction

Plywood sheets usually have a longer side of 2.4 m and consequently web splices may be required at intervals along the beam length. The web splice is usually of the same specification as the web and for an I beam is placed on both sides of the web.

Proprietary beams referred to in section 8.1 usually have webs continuously bonded together so that splices are unnecessary.

8.4.2 Stresses in ply web splices

When a web splice is required only to transmit a pure shear F, on the assumption that the web is not transmitting a proportion of the bending moment (see section 8.2.3), the splice plate should be designed to transmit a force F, applied at the centroid of the area of the splice plate at each side of the joint. It is usual to resist the resulting stresses with a glued connection as shown in Fig. 8.9.

Shear F, must be resisted at an eccentricity $b_s/2$ resulting in a combination of shear and bending on the glue area.

The shear stress (τ_v) on the glue area due to vertical shear can be calculated on the assumption of an average distribution without concentration of stress:

$$\tau_v = \frac{F}{n\, b_s h_s}$$

where n is the number of glue lines.

The permissible stress is the rolling shear stress of the plywood modified by the load–duration factor K_3. If the bonding pressure is achieved by nails or staples, the permissible stress should be multiplied by factor $K_{70} = 0.9$.

The moment due to the eccentricity of shear force = $M_e = 0.5\,Fb_s$.

Referring to Fig. 8.10, the maximum stress due to the secondary moment on the glue area is:

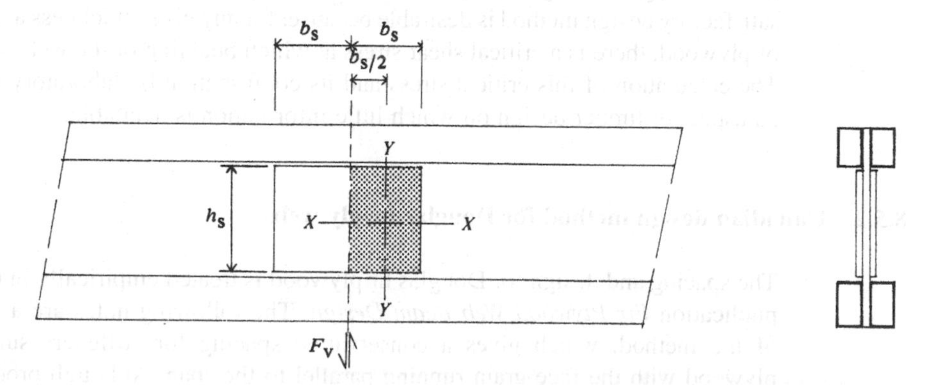

Fig. 8.9

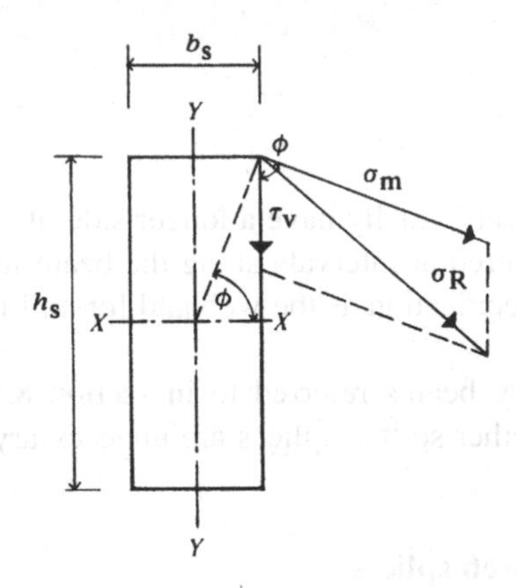

Fig. 8.10

$$\sigma_m = \frac{M_e a}{nI_p}$$

where n = number of glue lines
a = distance from the centre of the glued area to the extreme corner fibre $= 0.5\sqrt{b_s^2 + h_s^2}$
I_p = polar second moment of area $= I_x + I_y = (b_s h_s/12)(b^2 + h^2)$.

The maximum resulting stress σ_R acting on the glue area of the splice plate is the vector summation of τ_v and σ_m. This can be determined by plotting τ_v and σ_m to scale or calculated as:

$$\sigma_R = \sqrt{\sigma_m^2 + \tau_v^2 + 2a\tau_v \cos\phi}$$

8.5 WEB STIFFENERS

8.5.1 Introduction

BS 5268-2 gives no guidance on the design of stiffeners for thin web beams. A satisfactory design method is desirable because, for any given thickness and species of plywood, there is a critical shear stress at which buckling of the web will occur. The calculation of this critical stress and its confirmation by laboratory testing is an aspect of timber design on which little information is available.

8.5.2 Canadian design method for Douglas fir ply webs

The spacing and design for Douglas fir plywood is treated empirically in the COFI publication *Fir Plywood Web Beam Design*. The following notes are a summary of the method, which gives a conservative spacing for stiffeners suitable for plywood with the face grain running parallel to the span. Although produced for Douglas fir plywood it may be used for Douglas fir-faced plywood.

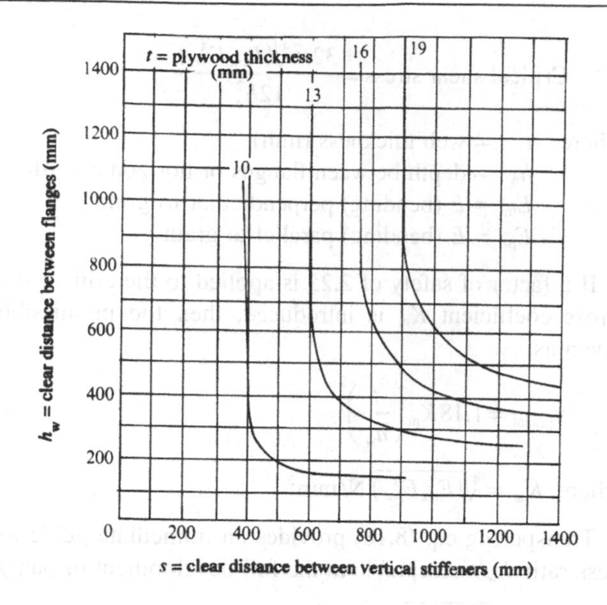

Fig. 8.11

The recommended basic clear distance between vertical stiffeners s for combinations of web thickness t and the clear distance between flanges h_w can be taken from Fig. 8.11. The basic spacing may be increased if the actual panel shear stress is less than the permissible according to eq. (8.17) except that the value of p must never be taken as less than 50% and the actual allowable improved clear spacing between stiffeners must not exceed $3s$.

$$s' = s\left(1 + \frac{100 - p}{25}\right) \tag{8.17}$$

where s' = the allowable improved clear stiffener spacing
p = percentage of allowable panel shear stress occurring in the panel.

p must not be taken as less than 50% and $s' > 3s$ (s from Fig. 8.11).

8.5.3 A method for Finnish birch-faced plywood

At present there are no recommendations for the spacing of stiffeners in Finnish birch-faced (or all-birch plywood) comparable to the method given in section 8.5.2 for Douglas fir, and the following proposals were given in earlier editions of this manual upon the more general recommendations outlined by Hanson in his book *Timber Engineers' Handbook* (Wiley, New York, 1948).

The procedure assumes that the plywood has the face grain perpendicular to span, in which circumstances experience shows that horizontal stiffeners are more effective in limiting web buckling than are the more conventional vertical stiffeners. This aspect is reflected in the method.

Hanson states that the critical panel shear stress for the commencement of web buckling is given by the equation:

$$\text{Critical shear stress} = \frac{32t^2\sqrt[4]{(E_{tra}E_{par}^3)}}{12h_w^2}$$

where t = web thickness (mm)
h_w = depth between flanges or horizontal stiffeners (mm)
E_{tra} = E (bending) perpendicular to grain
E_{par} = E (bending) parallel to grain.

If a factor of safety of 2.25 is applied to the critical shear stress and a material stress coefficient K_m is introduced, then the permissible panel shear stress is given as:

$$\tau_{p,adm} = 1.18K_m\left(\frac{t}{h_w}\right)^2 \tag{8.18}$$

where $K_m = \sqrt[4]{(E_{tra}E_{par}^3)}$ N/mm^2.

Transposing eq. (8.18) provides an immediate guide to the maximum slenderness ratio h_w/t acceptable to the full development of panel shear:

$$\frac{h_w}{t} = \sqrt{\frac{1.18K_m}{\tau_{p,adm}}}$$

Applying this formula to the range of Finnish plywoods listed in Tables 40–46 of BS 5268-2 gives a maximum slenderness ratio $h_w/t \cong 28$ for all grades and thicknesses.

8.6 HOLES OR SLOTS IN PLY WEB BEAMS

The passage of pipes, ducting or other services through the webs of beams requires a consideration of the resulting stresses in the beam at the position of the hole.

Consider the rectangle slot in the web of a beam as shown in Fig. 8.12. The bending stresses at the extreme fibres a and b at any section BB are given as:

$$\sigma_{ma} = \frac{M_A}{Z_A} + \frac{F_{vA}X(EI)_u}{Z_u[(EI)_u + (EI)_L]} \quad \text{and} \quad \sigma_{mb} = \frac{M_A}{Z_A} + \frac{F_{vA}X(EI)_L}{Z_L[(EI)_u + (EI)_L]}$$

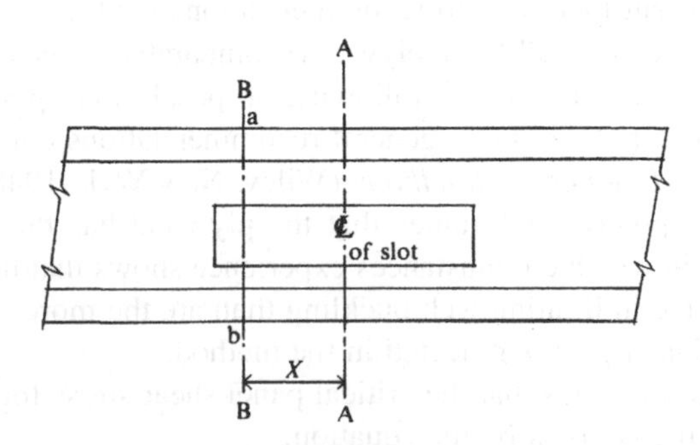

Fig. 8.12

where M_A = the moment at AA (mid length of slot)
F_{vA} = the shear at AA
X = the distance from section AA to BB
Z_A = the section modulus of full section at AA
Z_u = the section modulus of section at BB above slot
Z_L = the section modulus of section at BB below slot
$(EI)_u$ = the sectional rigidity of section at BB above slot
$(EI)_L$ = the sectional rigidity of section at BB below slot.

The formulae above are based on the assumption that the moment M_A produces a bending stress given by the first part of each formula, while the shear is divided between the upper and the lower portions of the beam in proportion to the sectional rigidity of the upper and lower portions of the beam.

For thin web beams it is usually convenient to form a slot symmetrical about the mid-depth of section, and then to reinforce the perimeter of the slot. With symmetry the above equations simplify to:

$$\sigma_{ma} = \sigma_{mb} = \frac{M_A}{Z_A} + \frac{F_{vA}X}{2Z_u}$$

In the case of simple span beams, a slot is frequently provided at midspan, where there is no shear under a symmetrical uniformly loading condition, and the slot may be of the clear depth h_w between flanges providing that the remaining flange section is adequate to resist the moment M_A. It is necessary, however, to check the slot for excessive stress due to unbalanced loadings.

8.7 PROPRIETARY SECTIONS

8.7.1 Range of sections and properties

There are a number of suppliers of timber I joists which the designer can specify directly from the manufacturer's literature or by involving the supplier's technical design adviser. Should the designer wish to make an assessment of a suitable section for a given situation then the information given in the remainder of this chapter may be of use.

Information is provided for the following manufacturers:

Manufacturers	*Available sections*	*Properties*
Trus Joist	Fig. 8.13	Table 8.4
Boise Cascade	Fig. 8.14	Table 8.5
James Jones	Figs 8.15 and 8.16	Table 8.6
Fillcrete Masonite	Figs 8.17 and 8.18	Table 8.7

The chapter concludes with a typical worked example using Table 8.4.

Fig. 8.13 Range of TJI® joists.

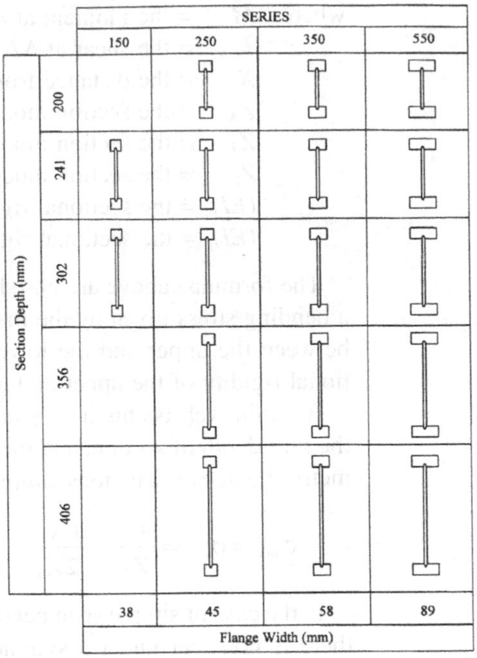

Table 8.4 TJI® joists

			Load sharing				Non-load sharing			
			Long		Medium		Long		Medium	
Type	*EI* (kN m²)	*AG* (kN)	*V* (kN)	*M* (kN m)	*V* (kN)	*M* (kN m)	*V* (kN)	*M* (kN m)	*V* (kN)	*M* (kN m)
150/241	438	1348	4.50	3.79	5.63	4.74	4.50	3.68	5.63	4.60
150/302	759	1846	5.70	5.06	7.13	6.33	5.70	4.91	7.13	6.14
250/160	180	686	3.10	2.50	3.88	3.13	3.10	2.43	3.88	3.04
250/200	318	1013	4.00	3.45	5.00	4.31	4.00	3.35	5.00	4.19
250/241	507	1348	5.00	4.46	6.25	5.58	5.00	4.33	6.25	5.41
250/302	875	1846	6.00	5.95	7.50	7.44	6.00	5.78	7.50	7.23
250/356	1303	2287	7.00	7.30	8.75	9.13	7.00	7.09	8.75	8.86
250/406	1796	2696	7.40	8.57	9.25	10.71	7.40	8.32	9.25	10.40
350/160	225	686	3.10	3.11	3.88	3.89	3.10	3.02	3.88	3.78
350/200	397	1013	4.00	4.31	5.00	5.39	4.00	4.18	5.00	5.23
350/241	631	1348	5.10	5.57	6.38	6.96	5.10	5.41	6.38	6.76
350/302	1083	1846	6.30	7.45	7.88	9.31	6.30	7.23	7.88	9.04
350/356	1603	2287	7.30	9.15	9.13	11.44	7.30	8.88	9.13	11.10
350/406	2199	2696	7.80	10.74	9.75	13.43	7.80	10.43	9.75	13.04
550/200	601	1184	5.20	6.60	6.50	8.25	5.20	6.41	6.50	8.01
550/241	952	1575	5.80	8.55	7.25	10.69	5.80	8.30	7.25	10.38
550/302	1624	2157	7.10	11.43	8.88	14.29	7.10	11.10	8.88	13.88
550/356	2393	2673	8.10	14.03	10.13	17.54	8.10	13.62	10.13	17.03
550/406	3269	3150	9.10	16.49	11.38	20.61	9.10	16.01	11.38	20.01

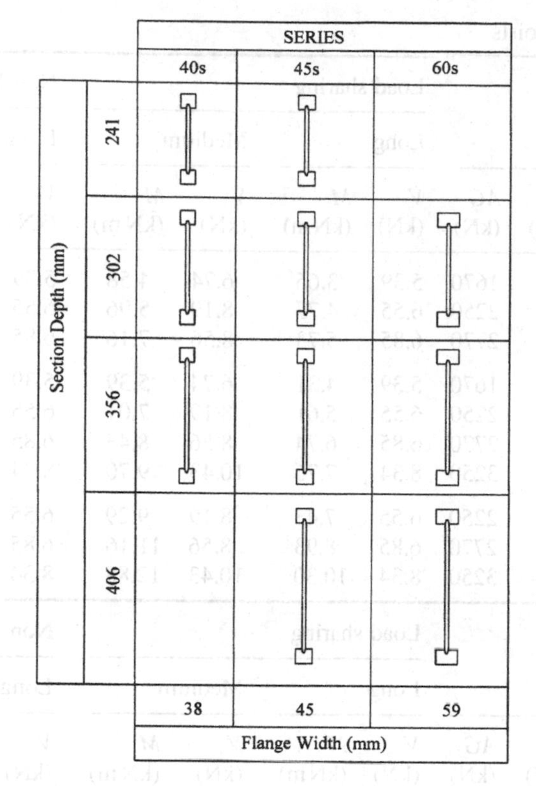

Fig. 8.14 Range of BCI® joists.

Table 8.5 BCI® joists

Service class 1			Load sharing				Non-load sharing			
			Long		Medium		Long		Medium	
Type	*EI* (kN m²)	*AG* (kN)	*V* (kN)	*M* (kN m)	*V* (kN)	*M* (kN m)	*V* (kN)	*M* (kN m)	*V* (kN)	*M* (kN m)
40/241	423	1670	5.39	3.66	6.74	4.58	5.39	3.51	6.74	4.39
40/302	726	2250	6.55	4.77	8.19	5.96	6.55	4.58	8.19	5.72
40/356	1069	2770	6.85	5.73	8.56	7.16	6.85	5.50	8.56	6.88
45/241	493	1670	5.39	4.31	6.74	5.39	5.39	4.14	6.74	5.17
45/302	843	2250	6.55	5.61	8.19	7.01	6.55	5.39	8.19	6.73
45/356	1240	2770	6.85	6.74	8.56	8.43	6.85	6.47	8.56	8.09
45/406	1681	3250	8.34	7.76	10.43	9.70	8.34	7.45	10.43	9.31
60/302	1099	2250	6.55	7.43	8.19	9.29	6.55	7.13	8.19	8.92
60/356	1610	2770	6.85	8.93	8.56	11.16	6.85	8.57	8.56	10.72
60/406	2177	3250	8.34	10.30	10.43	12.88	8.34	9.89	10.43	12.36

Service class 2			Load sharing				Non-load sharing			
			Long		Medium		Long		Medium	
Type	*EI* (kN m²)	*AG* (kN)	*V* (kN)	*M* (kN m)	*V* (kN)	*M* (kN m)	*V* (kN)	*M* (kN m)	*V* (kN)	*M* (kN m)
40/241	388	1380	5.39	3.30	6.74	4.13	5.39	3.17	6.74	3.96
40/302	666	1870	6.55	4.29	8.19	5.36	6.55	4.12	8.19	5.15
40/356	981	2300	6.85	5.15	8.56	6.44	6.85	4.94	8.56	6.18
45/241	452	1380	5.39	3.88	6.74	4.85	5.39	3.72	6.74	4.66
45/302	773	1870	6.55	5.05	8.19	6.31	6.55	4.85	8.19	6.06
45/356	1137	2300	6.85	6.06	8.56	7.58	6.85	5.82	8.56	7.27
45/406	1543	2700	8.34	6.99	10.43	8.74	8.34	6.71	10.43	8.39
60/302	1007	1870	6.55	6.69	8.19	8.36	6.55	6.42	8.19	8.03
60/356	1476	2300	6.85	8.04	8.56	10.05	6.85	7.72	8.56	9.65
60/406	1997	2700	8.34	9.27	10.43	11.59	8.34	8.90	10.43	11.12

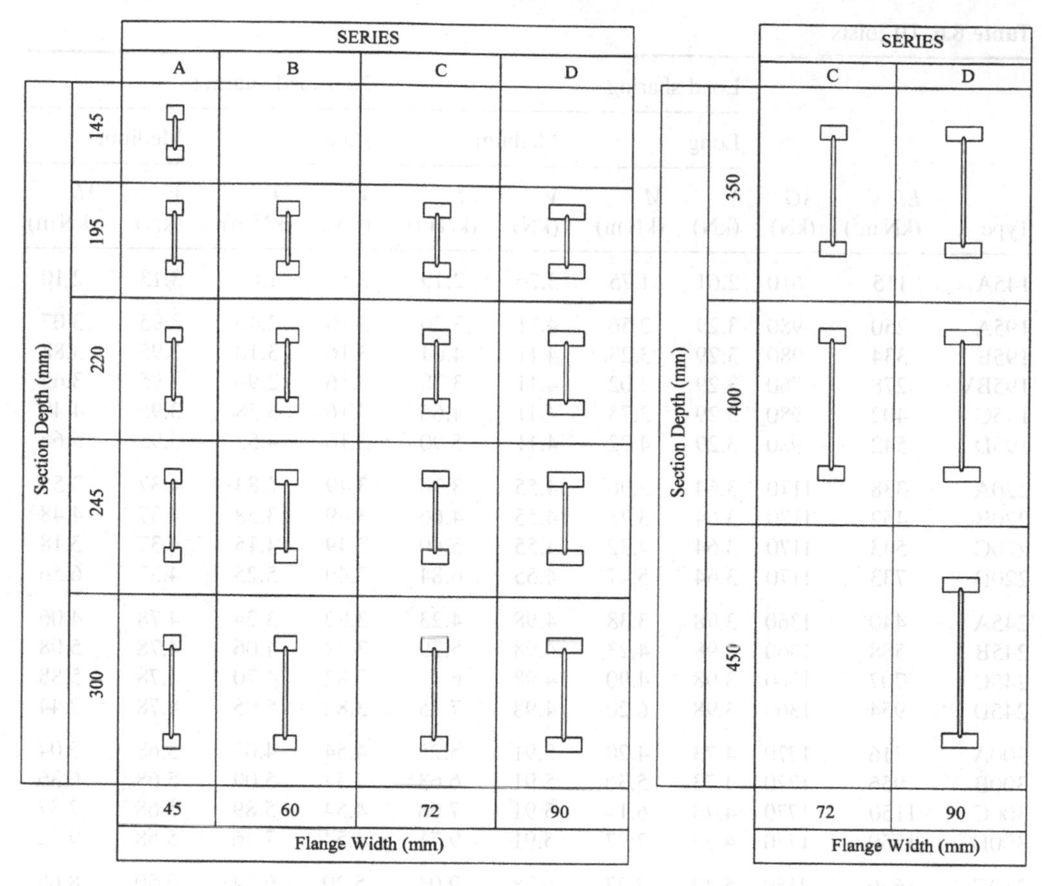

Fig. 8.15 Range of JJI joists.

Fig. 8.16 Range of JJI joists.

Table 8.6 JJI joists

			Load sharing				Non-load sharing			
			Long		Medium		Long		Medium	
Type	*EI* (kN m²)	*AG* (kN)	*V* (kN)	*M* (kN m)	*V* (kN)	*M* (kN m)	*V* (kN)	*M* (kN m)	*V* (kN)	*M* (kN m)
145A	115	610	2.61	1.75	3.26	2.19	2.51	1.68	3.13	2.10
195A	250	980	3.29	2.56	4.11	3.20	3.16	2.46	3.95	3.07
195B	334	980	3.29	3.23	4.11	4.04	3.16	3.10	3.95	3.88
195BV	278	760	3.29	3.02	4.11	3.78	3.16	2.90	3.95	3.62
195C	402	980	3.29	3.73	4.11	4.66	3.16	3.58	3.95	4.48
195D	542	980	3.29	4.72	4.11	5.90	3.16	4.53	3.95	5.66
220A	338	1170	3.64	2.96	4.55	3.70	3.49	2.84	4.37	3.55
220B	452	1170	3.64	3.73	4.55	4.66	3.49	3.58	4.37	4.48
220C	543	1170	3.64	4.32	4.55	5.40	3.49	4.15	4.37	5.18
220D	733	1170	3.64	5.47	4.55	6.84	3.49	5.25	4.37	6.56
245A	440	1360	3.98	3.38	4.98	4.23	3.82	3.24	4.78	4.06
245B	588	1360	3.98	4.23	4.98	5.29	3.82	4.06	4.78	5.08
245C	707	1360	3.98	4.90	4.98	6.13	3.82	4.70	4.78	5.88
245D	954	1360	3.98	6.20	4.98	7.75	3.82	5.95	4.78	7.44
300A	716	1770	4.73	4.20	5.91	5.25	4.54	4.03	5.68	5.04
300B	956	1770	4.73	5.30	5.91	6.63	4.54	5.09	5.68	6.36
300C	1150	1770	4.73	6.14	5.91	7.68	4.54	5.89	5.68	7.37
300D	1550	1770	4.73	7.77	5.91	9.71	4.54	7.46	5.68	9.32
350C	1640	2150	5.42	7.23	6.78	9.04	5.20	6.94	6.50	8.68
350D	2210	2150	5.42	9.16	6.78	11.45	5.20	8.79	6.50	10.99
400C	2240	2530	6.10	8.30	7.63	10.38	5.86	7.97	7.32	9.96
400D	3000	2530	6.10	10.52	7.63	13.15	5.86	10.10	7.32	12.62
450D	3920	2900	6.79	11.85	8.49	14.81	6.52	11.38	8.15	14.22

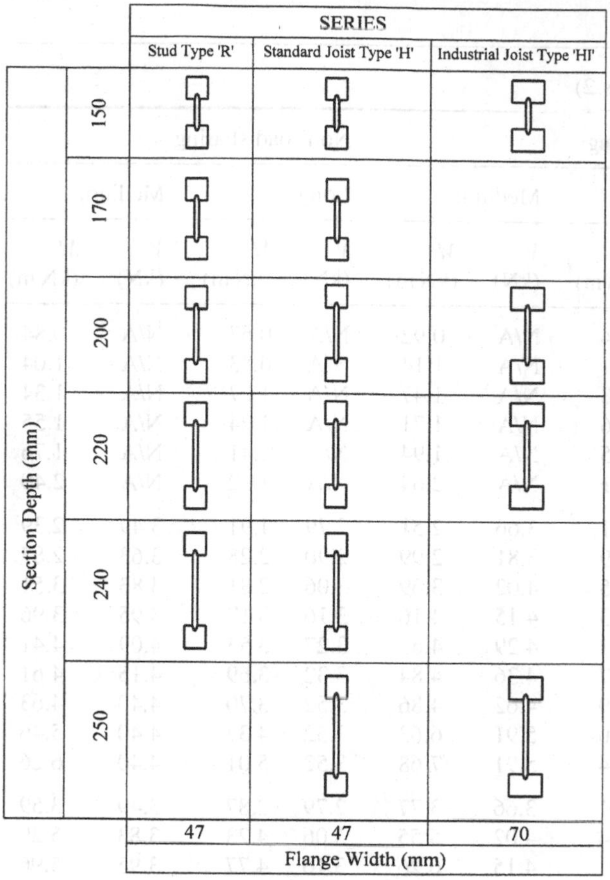

SERIES

Stud Type 'R' | Standard Joist Type 'H' | Industrial Joist Type 'HI'

Section Depth (mm): 150, 170, 200, 220, 240, 250

Flange Width (mm): 47, 47, 70

Fig. 8.17 Range of Masonite joists (1 of 2).

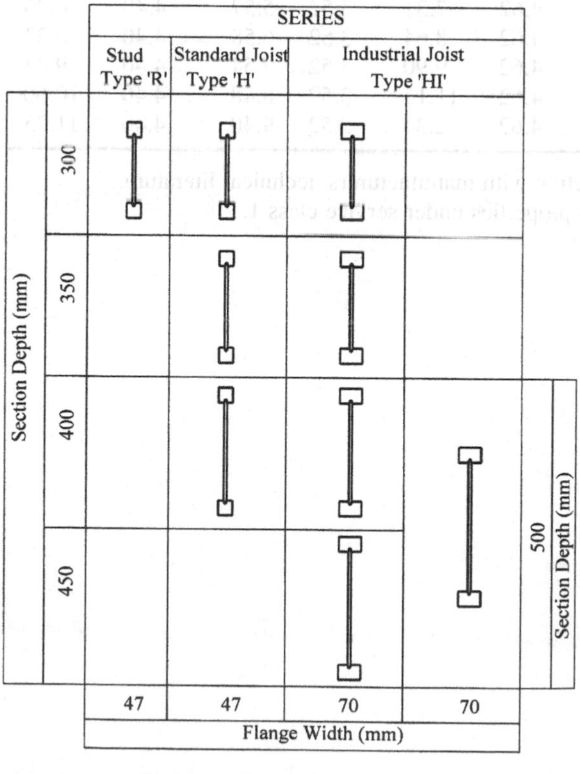

SERIES

Stud Type 'R' | Standard Joist Type 'H' | Industrial Joist Type 'HI'

Section Depth (mm): 300, 350, 400, 450; 500

Flange Width (mm): 47, 47, 70, 70

Fig. 8.18 Range of Masonite joists (2 of 2).

Table 8.7 Masonite joists (service class 2)

			Load sharing				Non-load sharing			
			Long		Medium		Long		Medium	
Type	*EI* (kN m²)	*AG* (kN)	*V* (kN)	*M* (kN m)	*V* (kN)	*M* (kN m)	*V* (kN)	*M* (kN m)	*V* (kN)	*M* (kN m)
150R	89	314	N/A	0.74	N/A	0.92	N/A	0.67	N/A	0.84
170R	125	408	N/A	0.91	N/A	1.14	N/A	0.83	N/A	1.04
200R	190	548	N/A	1.18	N/A	1.47	N/A	1.07	N/A	1.34
220R	242	642	N/A	1.36	N/A	1.71	N/A	1.24	N/A	1.55
240R	300	735	N/A	1.55	N/A	1.94	N/A	1.41	N/A	1.76
300R	513	1016	N/A	2.11	N/A	2.64	N/A	1.92	N/A	2.40
150H	127	442	2.93	2.01	3.66	2.51	2.79	1.91	3.49	2.39
170H	177	567	3.05	2.39	3.81	2.99	2.90	2.28	3.63	2.85
200H	268	754	3.21	2.95	4.02	3.69	3.06	2.81	3.83	3.51
220H	340	879	3.32	3.33	4.15	4.16	3.16	3.17	3.95	3.96
240H	422	1004	3.43	3.71	4.29	4.63	3.27	3.53	4.09	4.41
250H	465	1066	3.49	3.87	4.36	4.84	3.32	3.69	4.15	4.61
300H	608	1378	3.70	3.89	4.62	4.86	3.52	3.70	4.40	4.63
350H	870	1690	4.73	5.30	5.91	6.63	3.52	4.37	4.40	5.46
400H	1180	2002	4.73	6.14	5.91	7.68	3.52	5.01	4.40	6.26
150HI	191	442	2.93	3.01	3.66	3.77	2.79	2.87	3.49	3.59
200HI	403	754	3.21	4.44	4.02	5.55	3.06	4.23	3.83	5.29
220HI	512	879	3.32	5.01	4.15	6.26	3.16	4.77	3.95	5.96
250HI	700	1066	3.49	5.84	4.36	7.30	3.32	5.56	4.15	6.95
300HI	916	1378	3.70	5.85	4.62	7.31	3.52	5.57	4.40	6.96
350HI	1310	1690	3.70	6.91	4.62	8.64	3.52	6.58	4.40	8.23
400HI	1775	2002	3.70	7.92	4.62	9.90	3.52	7.54	4.40	9.43
450HI	2313	2314	3.70	8.90	4.62	11.13	3.52	8.48	4.40	10.60
500HI	2923	2626	3.70	9.87	4.62	12.34	3.52	9.40	4.40	11.75

These properties should be read in conjunction with manufacturers' technical literature.
See manufacturers' literature for improved properties under service class 1.

8.7.2 Worked example

A beam having an effective span of 5.0 m supports a long-term load of 1.8 kN/m^2 when spaced at 400 mm centres as part of a load-sharing floor system. Choose a suitable TJI section from Table 8.4 and verify the deflection.

Span = L = 5.0 m
Total applied load = $W = 5.0 \times 0.4 \times 1.8 = 3.6$ kN
End shear = 3.6/2 = 1.8 kN
Maximum moment = $WL/8 = 3.6 \times 5.0/8 = 2.25$ kN m

With nominal 30% allowance for shear deflection

Anticipated EI required = $1.3 \times 4.34WL^2$
EI required $\cong 5.64WL^2 = 5.64 \times 3.6 \times 5.0^2 = 508$ kN m^2

From Table 8.4, select 150/302 TJI section

Long-term shear capacity = V = 5.70 kN > 1.8 kN, therefore satisfactory
Long-term moment capacity = M = 5.06 kN m > 2.25 kN m, therefore satisfactory
EI = 759 kN m^2 > 508 kN m^2
AG = 1846 kN

$$\text{Bending deflection} = \frac{5 \times W \times L^3}{384 \times EI} = \frac{5 \times 3.6 \times 5.0^3}{384 \times 759} = 0.0077\,\text{m}$$

$$\text{Shear deflection} = \frac{M}{AG} = \frac{2.25}{1846} = 0.0012\,\text{m}$$

Total deflection = bending deflection + shear deflection = 0.0077 + 0.0012 = 0.0089 m
Allowable deflection = 0.003 × span (= 0.015 m) or 0.014 m which ever is the lesser. **O.K.**

Chapter 9
Lateral Stability of Beams

9.1 INTRODUCTION

Solid and built-up beam sections should be checked to ensure that they will not buckle under design load. Stability may be verified by calculation or by adopting deemed-to-satisfy limits as follows:

1. For rectangular solid or glulam sections limit the depth-to-breadth ratio (h/b) as recommended in Table 19 of BS 5268-2 and as presented here as Table 9.1.
2. For built-up thin web beams limit the I_x/I_y ratio as recommended in clause 2.10.10 of BS 5268-2 and as presented here as Table 9.2. It will be noted that there is a similarity between the two tables, there being similar degrees of restraint required for $I_x/I_y = (h/b)^2$.

The simple conservative limits tabulated in Tables 9.1 and 9.2 cover the majority of cases a designer encounters, but there are cases where adequate restraint cannot be provided or where the designer wishes to use a more slender section for economical or architectural reasons. In any of these special cases it is necessary for the designer to see whether the reduced lateral stability leads to a reduction in bending strength capacity.

The lateral buckling of a beam depends not only on the depth-to-breadth ratio (or I_x/I_y ratio) but also on:

- the geometrical and physical properties of the beam section
- the nature of the applied loading
- the position of the applied loading with respect to the neutral axis of the section, and
- the degree of restraint provided at the vertical supports and at points along the span.

Section 9.2 gives a design method which permits a stability assessment for solid or glulam rectangular sections and section 9.4 gives a design method which is applicable to thin web I beams.

9.2 BUCKLING OF RECTANGULAR SOLID AND GLULAM SECTIONS

The anticipated shape of the lateral buckling of the compression flange of a beam which is not fully restrained takes one of the forms sketched in plan in Fig. 9.1 or some intermediate position.

Table 9.1 Maximum depth-to-breadth ratios for solid and glulam beams

Degree of lateral support	Maximum depth-to-breadth ratio h/b
No lateral support	2
Ends held in position	3
Ends held in position and members held in line, as by purlins or tie rods at centres not more than 30 times the breadth of the member	4
Ends held in position and compression edge held in line, as by direct connection of sheathing, deck or joists	5
Ends held in position and compression edge held in line, as by direct connection of sheathing, deck or joists, together with adequate bridging or blocking spaced at intervals not exceeding six times the depth	6
Ends held in position and both edges firmly held in line	7

Table 9.2 Built-up thin web beams

I_x/I_y	Degree of lateral support
Up to 5	No lateral support is required
5–10	The ends of beams should be held in position at the bottom flange at supports
10–20	Beams should be held in line at the ends
20–30	One edge should be held in line
30–40	The beam should be restrained by bridging or other bracing at intervals of not more than 2.4 m
More than 40	The compression flange should be fully restrained

It is difficult to achieve full restraint in direction at the ends of beams as sketched in Fig. 9.1(a) and therefore, in the following design method, the assumption is made that all end joints have no directional restraint laterally, as shown in Figs 9.1(b) and (c). The ends of beams and any point of lateral restraint along the beam are assumed to be held in position laterally but not restrained in direction (i.e. not fixed in direction on plan).

In the following design method L_b is the lateral unrestrained length of beam which is not necessarily the full span L.

For sections which are symmetrical about the x–x and y–y axes, buckling occurs at a critical moment, expressed as:

$$M_{\text{crit}} = \frac{C_1}{L_b}\left[\frac{EI_y GJ}{\alpha}\right]^{\frac{1}{2}}\left[1 - \frac{C_2 h_L}{L_b}\sqrt{\frac{EI_y}{GJ}}\right] \tag{9.1}$$

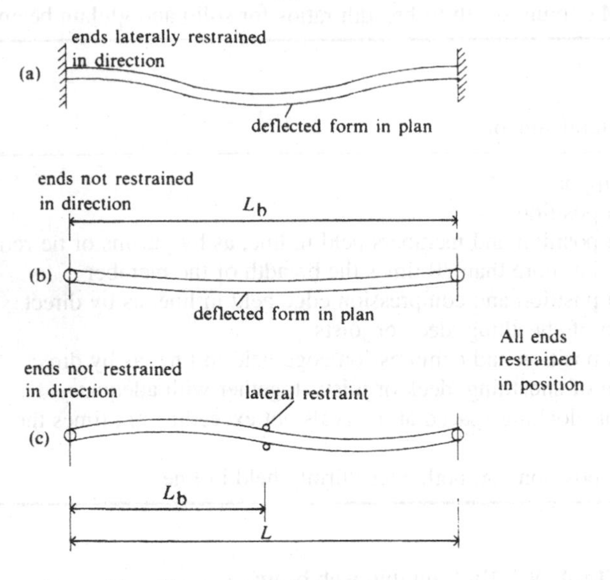

Fig. 9.1 Forms of lateral buckling shown in plan.

where EI_y = effective bending stiffness about the y–y axis

GJ = effective torsional resistance

h_L = height from the x–x axis to the point of application of load, being positive above the neutral axis and negative below the neutral axis when the loading is in the direction shown in Fig. 9.2. If a load is placed on the top flange of a beam, there is a tipping action which increases the instability of the section (i.e. reduces the critical moment), whereas if a load is applied to the bottom flange there is a stabilizing effect

α = $1 - (EI_y/EI_x)$, a factor taking account of the additional stability which occurs when a beam sags below the horizontal. For a rectangular section this may be simplified to $\alpha = 1 - (b/h)^2$. The designer may include α to obtain a slightly higher critical moment, but as most sections requiring a stability check will have $h/b > 3$, for which $\alpha = 0.88$, it will be conservative to adopt $\alpha = 1.0$

L_b = the laterally unrestrained length of beam which is not necessarily the full span between vertical supports.

In eq. (9.1) C_1 and C_2 are constants determined by the nature of the applied loading and the conditions of effective restraint. There are a number of books and publications which derive the values of constants for many typical loading conditions. The cases most likely to be encountered are summarized in Table 9.3. When no external loading is carried between two points of lateral restraint (e.g. cases 6–9), then:

$$C_1 = 5.5 - 3.3\beta + 0.94\beta^2 \quad \text{and} \quad C_2 = 0$$

except that C_1 is not to be taken as more than 7.22 (which occurs at $\beta = -0.46$), where

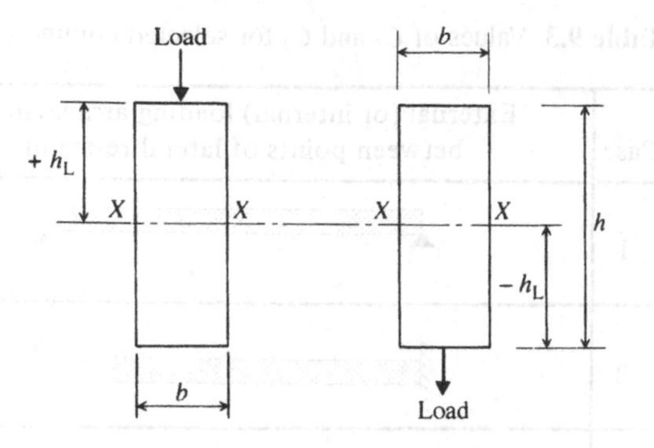

Fig. 9.2 Positive and negative location of the applied load.

$$\beta = \frac{\text{smaller 'end' moment}}{\text{larger 'end' moment}}$$

The sign for β is positive if one moment is anticlockwise and the other clockwise; also if one moment is zero. If both moments are clockwise or anticlockwise, the sign for β is negative.

The function:

$$\left[1 - \frac{C_2 h_L}{L_b}\sqrt{\frac{EI_y}{GJ}}\right]$$

takes account of the point of application of the load above or below the *x–x* axis (Fig. 9.2). If the load is applied at the *x–x* axis, the whole function becomes unity.

The critical moment M_{crit} must be reduced by a suitable factor of safety. A factor of 2.25 is assumed here, therefore the safe buckling moment M_{buc} is found by dividing the critical moment by 2.25. Hence:

$$M_{buc} = \frac{M_{crit}}{2.25}$$

For a beam not to buckle under the design loading, the buckling moment must be equal to or greater than the maximum bending moment produced by the design loading about the *x–x* axis.

To simplify design calculations, it is convenient to introduce coefficients N_1 and N_2 and to further simplify with α = unity, then

$$N_1 = \sqrt{EI_y\, GJ} \tag{9.2}$$

$$N_2 = \sqrt{\frac{EI_y}{GJ}} \tag{9.3}$$

so

$$M_{buc} = \frac{C_1 N_1}{2.25\, L_b}\left(1 - \frac{C_2 h_L N_2}{L_b}\right) \tag{9.4}$$

Table 9.3 Values of C_1 and C_2 for selected common loading conditions

Case	External (or internal) loading arrangement between points of lateral restraint	C_1	C_2
1		+3.55	+1.41
2		+4.08	+4.87
3	= =	+4.24	+1.73
4	= =	+5.34	+4.46
5	¼ ½ ¼	+3.27	+2.64
6	M M	+3.14	+0.0
7	M 0·5M	+4.08	+0.0
8	M	+5.5	+0.0
9	M 0·46 M (β = − 0·46)	+7.22	+0.0
10		+4.1	+1.0
11		+6.42	+1.8

The buckling moment M_{buc} should not be increased to take account of any load–duration factor. If the beam is liable to buckle, it will be an instantaneous action dependent on the magnitude and not the duration of the loading.

For a rectangular solid section the torsional constant J may be taken as $\frac{1}{3}hb^3[1-(0.63\,b/h)]$. If $K_\lambda = \sqrt{1-(0.63\,b/h)}$ and $G = E/16$ then:

$$N_1 = \frac{EI_y\,K_\lambda}{2} \qquad (9.5)$$

$$N_2 = \frac{2}{K_\lambda} \tag{9.6}$$

and

$$M_{buc} = \frac{C_1 E I_y K_\lambda}{4.5\, L_b}\left[1 - \frac{2h_L C_2}{K_\lambda L_b}\right] \tag{9.7}$$

The value of K_λ approaches unity as the slenderness increases, particularly where h/b is more than 4. The relationship between K_λ and h/b for rectangular sections is shown in Fig. 9.3.

In practice the load is most frequently applied at the top edge of the section, and it is unlikely that a stability check will be necessary for any section having a depth-to-breadth ratio less than 3, where from Table 9.1 it can be seen that only the ends would have to be held in position for the full moment capacity to be developed. With these assumptions K_λ may be approximated to 0.9 and h_L may be taken as $+h/2$. This provides a conservative simplification of eq. (9.7) for solid and glulam beams as:

$$M_{buc} = \frac{C_1\, E I_y}{5\, L_b}\left[1 - \frac{C_2\, h}{0.9\, L_b}\right] \tag{9.8}$$

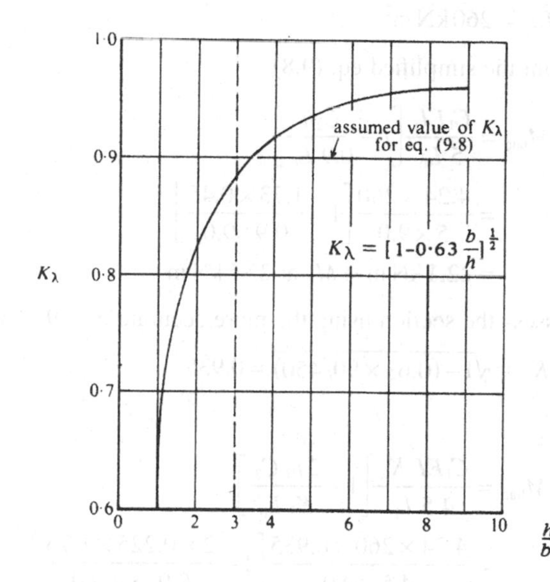

Fig. 9.3 The relationship of K_λ to h/b for solid and glulam beams.

9.3 DESIGN EXAMPLES

9.3.1 Design of a glulam beam with no lateral restraint

A glulam beam spanning 9.0 m supports a medium-term central point load of 10 kN applied by a pin-ended post which gives no lateral restraint to the beam. Design a suitable glulam beam of C24 grade and check lateral stability.

Maximum shear = reaction = 5.0 kN

$$\text{Maximum moment} = M_x = \frac{10.0 \times 9.0}{4} = 22.5\,\text{kN m}$$

The central point load may be converted to an equivalent UDL to determine the *EI* required.

From Table 4.17 at $n = 0.5$, $K_m = 1.60$ and $W_e = 1.6 \times 10 = 16\,\text{kN}$:

$$EI \text{ required} \cong 4.8 W_e L^2 = 4.8 \times 16 \times 9^2 = 6220\,\text{kN m}^2$$

From Table 7.4 select 90 × 450 section.

Shear capacity = 56.1 kN > 5.0 kN
Moment capacity = 37.5 kN m > 22.5 kN m
EI capacity = 7898 kN m^2 > 6220 kN m^2

Lateral stability check
Lateral buckling will take the form of plan (b) in Fig. 9.1; therefore $L_b = 9.0$ m. The loading condition, when related to Table 9.3 is case 3, therefore $C_1 = 4.24$ and $C_2 = 1.73$.
From Table 7.9,

$$EI_y = 260\,\text{kN m}^2$$

From the simplified eq. (9.8)

$$M_{buc} = \frac{C_1 EI_y}{5\,L_b}\left[1 - \frac{C_2 h}{0.9\,L_b}\right]$$

$$= \frac{4.24 \times 260}{5 \times 9.0}\left[1 - \frac{1.73 \times 0.45}{0.9 \times 9.0}\right]$$

$$= 22.1\,\text{kN m} < M_x \text{ at } 22.5\,\text{kN m}$$

Re-assess the section using the more accurate eq. (9.7) with

$$K_\lambda = \sqrt{1 - (0.63 \times 90/450)} = 0.935$$

then

$$M_{buc} = \frac{C_1 EI_y K_\lambda}{4.5\,L_b}\left[1 - \frac{2\,h_L C_2}{K_\lambda L_b}\right]$$

$$= \frac{4.24 \times 260 \times 0.935}{4.5 \times 9.0}\left[1 - \frac{2 \times 0.225 \times 1.73}{0.935 \times 9.0}\right]$$

$$= 23.1\,\text{kN m} > M_x \text{ at } 22.5\,\text{kN m}$$

The section is now satisfactory but being close to the buckling limit it may be better to specify a less slender section.

Try 115 × 405 section: Re-assess using the more accurate eq. (9.7) with

$$K_\lambda = \sqrt{1 - (0.63 \times 115/405)} = 0.906$$

From Table 7.10,

$$EI_y = 488\,\text{kN m}^2$$

From eq. (9.7)

$$M_{buc} = \frac{4.24 \times 488 \times 0.906}{4.5 \times 9.0}\left[1 - \frac{2 \times 0.202 \times 1.73}{0.906 \times 9.0}\right] = 42.3\,\text{kN m}$$

which gives an 83% improvement in buckling moment.

9.3.2 Design of a glulam beam with central lateral restraint

A glulam beam spanning 14.0 m supports a medium-term central point load of 20 kN applied by a secondary beam which also gives lateral restraint to the beam. Design a suitably slender glulam beam of C24 grade and check lateral stability.

Maximum shear = reaction = 10 kN
Maximum moment = $M_x = 20 \times 14/4 = 70$ kN m

The central point load may be converted to an equivalent UDL to determine the EI required.
From Table 4.17 at $n = 0.5$, $K_m = 1.60$ and $W_e = 1.6 \times 20 = 32$ kN

EI required $= 4.8W_e L^2 = 4.8 \times 32 \times 14^2 = 30\,100$ kN m^2

From Table 7.4, try 90 × 720 section.

Shear capacity = 89.7 kN > 10 kN
Moment capacity = 93.3 kN m > 70 kN m
EI capacity = 32 349 kN m^2 > 30 100 kN m^2

Lateral stability check
Lateral buckling will take the form of plan (c) in Fig. 9.1; therefore

$$L_b = \frac{\text{span}}{2} = 7.0\,\text{m}$$

The loading condition, when related to Table 9.3, is case 8, giving $C_1 = 5.5$ and $C_2 = 0$. From Table 7.9

$EI_y = 416$ kN m^2

From the simplified eq. (9.8), with $C_2 = 0$,

$$M_{buc} = \frac{C_1 EI_y}{4.5L_b} = \frac{5.5 \times 416}{5 \times 7.0} = 65.4\,\text{kN m} < 70\,\text{kN m}$$

Re-assess the section using the more accurate eq. (9.7) with

$$K_\lambda = \sqrt{1 - (0.63 \times 90/720)} = 0.96$$

then

$$M_{buc} = \frac{C_1 EI_y K_\lambda}{4.5L_b} = \frac{5.5 \times 416 \times 0.96}{4.5 \times 7.0} = 69.7\,\text{kN m} < 70\,\text{kN m}$$

The section is close to being 'acceptable' but it would be better to improve the buckling moment by adopting a wider section, say 115 × 675. From Table 7.10,

$$EI_y = 813\,\text{kN m}^2 \quad \text{and} \quad M_{buc} = \frac{5.5 \times 813}{4.5 \times 7.0} = 142\,\text{kN m}$$

9.3.3 Comparison design of a glulam beam with restrained/unrestrained conditions

A glulam beam of 12 m span is loaded on the top edge by medium-term concentrated loads of 6 kN in Fig. 9.4. Design a suitable glulam beam of C24 grade for the following options.

Case (1): If the method of applying the point loads restrains the beam.
Case (2): If the method of applying the point loads does not restrain the beam.

Maximum shear = reaction = 6 kN
Maximum moment = $M_x = 6 \times 3 = 18\,\text{kN m}$

The point loads may be converted to an equivalent UDL to determine the EI required. From Table 4.17 at $n = 3/12 = 0.25$, $K_m = 1.10$ and $W_e = 2 \times 6 \times 1.1 = 13.2\,\text{kN}$:

$$EI \text{ required} = 4.8 W_e L^2 = 4.8 \times 13.2 \times 12^2 = 9124\,\text{kN m}^2$$

From Table 7.4 select 90 × 495 section.

Shear capacity = 61.7 kN > 6 kN
Moment capacity = 44.9 kN m > 22.5 kN m
EI capacity = $10512\,\text{kN m}^2 > 9124\,\text{kN m}^2$

From Table 7.9,

$$EI_y = 286\,\text{kN m}^2$$

Case (1)
With restraint at each point load, lateral buckling takes the form of plan (c) in Fig. 9.1, and it is necessary to consider separately the parts of span between lateral restraints.
Consider part AB (part DC being identical).
Load case 8 in Table 9.3 applies with $C_1 = 5.5$ and $C_2 = 0$:

$$L_b = 3.0\,\text{m}$$

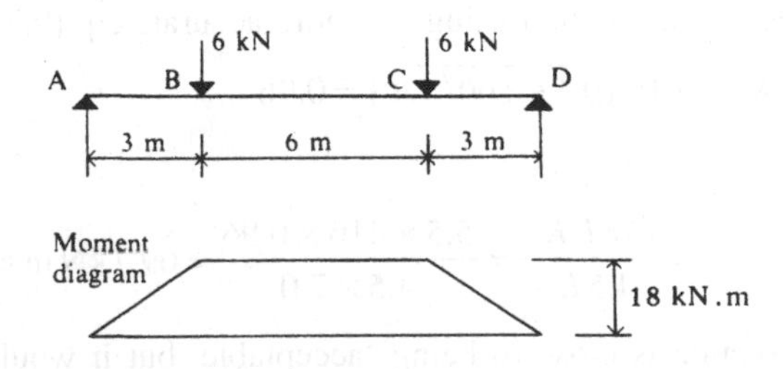

Fig. 9.4

From eq. (9.8),

$$M_{\text{buc}} = \frac{C_1 EI_y}{5L_b} = \frac{5.5 \times 286}{5 \times 3.0} = 105\,\text{kN m} > 18\,\text{kN m, therefore } \textbf{O.K.}$$

Consider part BC
Load case 6 in Table 9.3 applies with $C_1 = 3.14$ and $C_2 = 0$

$$L_b = 6.0\,\text{m}$$

From eq. (9.8),

$$M_{\text{buc}} = \frac{C_1 EI_y}{5L_b} = \frac{3.142 \times 286}{5 \times 6.0} = 30\,\text{kN m} > 18\,\text{kN m, therefore } \textbf{O.K.}$$

Case (2)
If the beam is not restrained at the positions of applied load buckling will be as plan (b) in Fig. 9.1. Loading case 5 in Table 9.3 applies with $C_1 = 3.27$ and $C_2 = 2.64$:

$$L_b = 12.0\,\text{m}$$

From eq. (9.8),

$$M_{\text{buc}} = \frac{3.27 \times 286}{5 \times 12.0}\left[1 - \frac{2.64 \times 0.495}{0.9 \times 12.0}\right] = 13.7\,\text{kN m} < 18\,\text{kN m}$$

This condition would not be stable against lateral buckling and an alternative section should be adopted.

From Table 7.5 try 115 × 450 section.

Shear capacity = 70.6 kN > 6 kN
Moment capacity = 47.9 kN m > 22.5 kN m
EI capacity = 10 092 kN m^2 > 9124 kN m^2

From Table 7.10,

$$EI_y = 542\,\text{kN m}^2$$

From eq. (9.8),

$$M_{\text{buc}} = \frac{3.27 \times 542}{5 \times 12.0}\left[1 - \frac{2.64 \times 0.45}{0.9 \times 12.0}\right] = 26.3\,\text{kN m} > 18\,\text{kN m, therefore } \textbf{O.K.}$$

9.4 PARTIALLY RESTRAINED THIN WEB I BEAMS

With a built-up thin web I beam the full permissible bending stress can be used providing the I_x/I_y ratio does not exceed the values given in Table 9.2 for the stated conditions of lateral stability. When these I_x/I_y values are exceeded the simplest design option, to avoid complex buckling calculations, is to consider the compression flange acting as a column deflecting sideways between points of lateral support (an option recommended in Clause 2.10.10 of BS 5268-2). The assumed

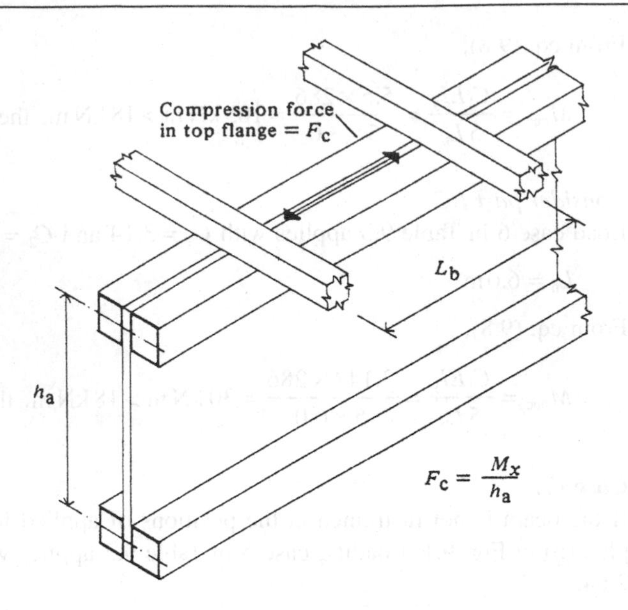

Fig. 9.5

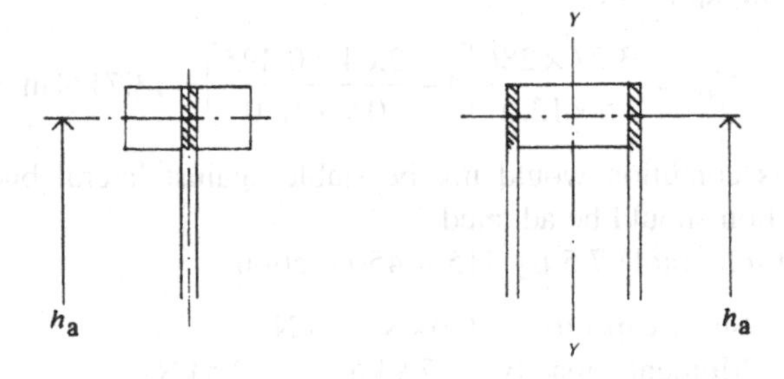

Fig. 9.6

axial load may be taken as the bending moment divided by the effective depth, h_a, as shown in Fig. 9.5.

The effective length of this horizontal compression member for sideways buckling is a function of the distance L_b between centres of lateral restraints. Usually the effective length is taken as $1.0 \times L_b$ although, if the restraints are fixed to the beam with a connection capable of giving some fixity on plan or sheet materials are fixed to the restraints, the effective length may be reduced to $0.85L_b$ or $0.70L_b$.

The flange may be considered to be fully restrained in one direction by the web. The radius of gyration about the other axis may be calculated as

$$r_y = \sqrt{I_{yc}/A_c}$$

where I_{yc} is the inertia of the compression flange about the vertical axis and A_c is the area of the compression flange.

It is quite normal to include the area of the web bounded by the timber flanges when determining I_y and A_c, as shown in Fig. 9.6, assuming for simplicity that this area is of the same material as the flanges.

Chapter 10
Structural Composite Lumber

10.1 INTRODUCTION

Solid timber is limited structurally by the presence of natural characteristics such as knots, slope of grain and fissures which reduce the potential performance of the timber. To improve this performance we have already seen that timber may be peeled into veneers and bonded usually with cross-laminated veneers to form plywood sheathing or small rectangular sections may be bonded together to form laminated members.

An alternative method of improving the structural performance is to break the timber down into much smaller veneers or strands and to rebond with generally all grains parallel to make a structural component in which all the defects are either eliminated or dispersed throughout the product. This is generally known as Structural Composite Lumber (SCL) and is produced in various forms.

In general, SCL is produced by converting logs into rectangular sections by taking veneers either in whole or in strands and feeding them into a press and gluing them together to form large and long billets which are then ripped down into smaller commercial sections.

10.2 KERTO-LVL (LAMINATED VENEER LUMBER)

10.2.1 Description

Kerto-LVL is manufactured by Finnforest Corporation and is specified in two qualities: Kerto-S and Kerto-Q.

Kerto-LVL consists of long layers of wood veneers typically 3.2 mm thick taken from rotary-peeled logs. These veneers are generally end scarf jointed and fed into a press with their end joints staggered by at least 100 mm. In the press they are glued under heat and pressure to produce large billets up to 90 mm wide, 1200 mm deep and up to 26 m long. The billet is then converted into a range of smaller sections.

10.2.2 Structural performance

Design is carried out in accordance with BS 5268-2 using the grade stresses shown in Table 10.1 as taken from Agrément certificate No. 00/3717.

Table 10.1 Grade stresses and moduli for Kerto-LVL

Grade stresses and moduli (N/mm^2)		Kerto-S Service class 1	Kerto-S Service class 2	Kerto-Q Service classes 1 and 2
Bending	edgewise (as a joist)	19.5	17.5	13.2
	flatwise (as a plank)	19.7	17.6	14.8
Tension	parallel to grain	13.5	12.1	8.9
	edgewise perpendicular to grain	0.33	0.30	2.0
Compression	parallel to grain	16.5	14.8	11.1
	perpendicular to grain (as a joist)	3.2	2.9	5.2
	perpendicular to grain (as a plank)	1.6	1.4	1.4
Shear	edgewise (as a joist)	2.2	2.0	2.3
	flatwise (as a plank)	1.6	1.5	0.6
Modulus of elasticity	E_{min}	12000	11500	8360
	E_{mean}	13500	12850	9500
Modulus of rigidity	G_{min}	400	380	380
	G_{mean}	600	570	570

Modification factors K_3, K_4, K_5, K_7, K_{12} and K_{13} may be applied to these stresses but K_8 should have a reduced value of 1.04 to account for the lower coefficient of variation when compared with solid sawn timber.

Deflection is calculated using E_{mean}.

A length factor K_L should be applied to tension members

$$K_L = (2440/L)^{0.125}$$

where L = length (mm) with a minimum value of 2440.

For the purpose of joint design Kerto-LVL should be taken as C27 strength class.

Table 10.2 combines the relevant stresses in Table 10.1 with the properties for a range of Kerto-S sections. The designer should check availability as stockists carry a reduced range usually limited to the smaller sections.

10.3 VERSA-LAM SP LVL (LAMINATED VENEER LUMBER)

10.3.1 Description

Versa-Lam SP LVL is manufactured by Boise Cascade, Boise, Idaho, and marketed in the UK by Boise Cascade Sales Ltd.

Versa-Lam SP LVL consistes of 2.5 m long layers of wood veneers typically 3.2–4.2 mm thick taken from rotary-peeled logs. These veneers are generally end scarf jointed and fed into a press with their end joints staggered by at least 125 mm. In the press they are glued under heat and pressure to produce large billets

Table 10.2 Kerto-S (LVL): Principal members, service class 2

Section				Section properties		Long term		Medium term		
b (mm)	h (mm)	h/b	Depth factor K_7	Z ($\times10^6\,mm^3$)	I ($\times10^6\,mm^4$)	Shear (kN)	Moment (kN m)	Shear (kN)	Moment (kN m)	EI (kN m^2)
45	147	3.3	1.08	0.162	11.9	8.82	3.07	11.03	3.83	153
45	184	4.1	1.06	0.254	23.4	11.04	4.69	13.80	5.86	300
45	195	4.3	1.05	0.285	27.8	11.70	5.23	14.63	6.54	357
45	220	4.9	1.03	0.363	39.9	13.20	6.57	16.50	8.22	513
45	235	5.2	1.03	0.414	48.7	14.10	7.45	17.63	9.31	625
45	245	5.4	1.02	0.450	55.1	14.70	8.06	18.38	10.07	709
45	260	5.8	1.02	0.507	65.9	15.60	9.01	19.50	11.27	847
45	300	6.7	1.01	0.675	101.3	18.00	11.88	22.50	14.85	1301
45	350	6.7	1.01	0.675	101.3	18.00	11.88	22.50	14.85	1301
75	220	2.9	1.03	0.605	66.6	22.00	10.95	27.50	13.69	855
75	245	3.3	1.02	0.750	91.9	24.50	13.43	30.63	16.78	1181
75	300	4.0	1.01	1.125	168.8	30.00	19.80	37.50	24.75	2168
75	400	5.3	0.94	2.000	400.0	40.00	32.99	50.00	41.24	5140
90	184	2.0	1.06	0.508	46.7	22.08	9.38	27.60	11.72	600
90	200	2.2	1.05	0.600	60.0	26.40	12.08	33.00	15.10	879
90	220	2.4	1.03	0.726	79.9	29.04	14.46	36.30	18.08	1242
90	265	2.9	1.01	1.053	139.6	34.98	20.56	43.73	25.70	2224

Coefficients

Duration of load

Long term $K_3 = 1.00$

Medium term $K_3 = 1.25$

Depth factor

$K_7 = (300/h)^{0.11}$ for $72 < h < 300$

$K_7 = 0.81(h^2 + 92300)/(h^2 + 56800)$ for $h > 300$

Derivation of capacities

Shear

Grade stress = 2 N/mm²

Capacity = $(2bh)/3 \times$ grade stress $\times K_3$

Moment

Grade stress = 175 N/mm²

Capacity = $Z \times$ grade stress $\times K_3 \times K_7$

Deflection

$E_{mean} = 12850\,N/mm^2$

Table 10.3 Grade stresses and moduli for Versa-Lam

Grade stresses and moduli (N/mm^2)		Dry exposure condition
Bending	edgewise (as a joist)	17.1
	flatwise (as a plank)	17.1
Tension	parallel to grain	13.5
Compression	parallel to grain	17.5
	perpendicular to grain (as a joist)	5.0
	perpendicular to grain (as a plank)	2.8
Shear	edgewise (as a joist)	1.8
	flatwise (as a plank)	1.2
Modulus of elasticity	E_{mean}	13000
Modulus of rigidity	G_{mean}	812

up to 1200 mm wide, 1200 mm deep and up to 20 m long. The billet is then converted into a range of smaller sections.

10.3.2 Structural performance

Design is carried out in accordance with BS 5268-2 using the grade stresses shown in Table 10.3 as taken from Agrément certificate No. 99/3619.

Modification factors K_3, K_4, K_5, K_7, K_{12} and K_{13} may be applied to these stresses but K_8 should have a reduced value of 1.04 to account for the lower coefficient of variation when compared with solid sawn timber.

Deflection is calculated using E_{mean}.

A length factor K_L should be applied to tension members

$$K_L = (2440/L)^{0.125}$$

where L = length (mm) with a minimum value of 2440.

For the purpose of joint design Versa-Lam SP LVL should be taken as C27 strength class. The maximum diameter of nails inserted parallel to the glue line should be 4 mm.

Table 10.4 combines the relevant stresses in Table 10.3 with the properties for a range of Versa-Lam sections. The designer should check availabilty as stockists carry a reduced range usually limited to the smaller sections.

10.4 PARALLAM PSL (PARALLEL STRAND LUMBER)

10.4.1 Description

Parallam is marketed in the UK by Trus Joist MacMillan Ltd.

Parallam PSL consists of strands of Douglas fir or Southern pine taken from

Table 10.4 Versa-Lam SP LVL: Principal members, dry exposure condition

Section				Section properties		Long term		Medium term		
b (mm)	h (mm)	h/b	Depth factor K_7	Z ($\times 10^6$ mm^3)	I ($\times 10^6$ mm^4)	Shear (kN)	Moment (kN m)	Shear (kN)	Moment (kN m)	EI (kN m^2)
45	140	3.1	1.09	0.147	10.3	7.56	2.73	9.45	3.42	134
45	184	4.1	1.06	0.254	23.4	9.94	4.58	12.42	5.73	304
45	235	5.2	1.03	0.414	48.7	12.69	7.28	15.86	9.09	633
45	241	5.4	1.02	0.436	52.5	13.01	7.63	16.27	9.54	682
45	286	6.4	1.01	0.613	87.7	15.44	10.55	19.31	13.18	1140
45	302	6.7	1.00	0.684	103.3	16.31	11.75	20.39	14.68	1343
45	356	7.9	0.97	0.951	169.2	19.22	15.71	24.03	19.64	2200
45	406	9.0	0.94	1.236	251.0	21.92	19.87	27.41	24.83	3263
45	457	10.2	0.92	1.566	357.9	24.68	24.60	30.85	30.74	4653

Coefficients

Duration of load

Long term $K_3 = 1.00$

Medium term $K_3 = 1.25$

Depth factor

$K_7 = (300/h)^{0.11}$ for $72 < h < 300$

$K_7 = 0.81(h^2 + 92300)/(h^2 + 56800)$ for $h > 300$

Derivation of capacities

Shear

Grade stress = 1.8 N/mm^2

Capacity = $(2bh)/3 \times$ grade stress $\times K_3$

Moment

Grade stress = 17.1 N/mm^2

Capacity = $Z \times$ grade stress $\times K_3 \times K_7$

Deflection

$E_{mean} = 13000$ N/mm^2

Table 10.5 Grade stresses and moduli for Parallam

Grade stresses and moduli (N/mm^2)		Dry exposure condition
Bending	parallel to grain	16.8
Tension	parallel to grain	14.8
Compression	parallel to grain	15.1
	perpendicular to grain (as a joist)	3.6
	perpendicular to grain (as a plank)	2.8
Shear	edgewise (as a joist)	2.2
	flatwise (as a plank)	1.6
Modulus of elasticity	E_{mean}	12750
Modulus of rigidity	G_{mean}	637

wood veneers typically 3.2 mm thick in turn taken from rotary-peeled logs. These veneers are clipped into strands about 20 mm long. The strands are glued and fed into a press with the strands oriented to the length of the member. In the press they are glued under heat and pressure to produce large section billets, typically 280 mm by 480 mm in lengths up to a maximum of 20 m. The billet is then converted into a range of smaller sections.

10.4.2 Structural performance

Design is carried out in accordance with BS 5268-2 using the grade stresses shown in Table 10.5 as taken from Agrément certificate No. 92/2813.

Modification factors K_3, K_4, K_5, K_7, K_{12} and K_{13} may be applied to these stresses but K_8 should have a reduced value of 1.04 to account for the lower coefficient of variation when compared with solid sawn timber.

Deflection is calculated using E_{mean}. No length factor K_L is applicable to Parallam. For the purpose of joint design, Parallam should be taken as C27 strength class.

Table 10.6 combines the relevant stresses in Table 10.5 with the properties for a range of Parallam sections. The designer should check availabilty as stockists carry a reduced range usually limited to the smaller sections.

10.5 TIMBERSTRAND (LAMINATED STRAND LUMBER)

10.5.1 Description

TimberStrand (LSL) is marketed in the UK by Trus Joist MacMillan Ltd and is available in two grades: grade 1.3E and grade 1.5E.

TimberStrand (LSL) is manufactured by stranding all the timber, excluding bark, from a log to create strands which are aproximately 300 mm long and

Table 10.6 Parallam PSL: Principal members, dry exposure condition

Section			Depth	Section properties		Long term		Medium term		
b (mm)	h (mm)	h/b	factor K_7	Z ($\times 10^6$ mm^3)	I ($\times 10^6$ mm^4)	Shear (kN)	Moment (kN m)	Shear (kN)	Moment (kN m)	EI (kN m^2)
89	184	2.1	1.06	0.502	46.2	24.02	8.90	30.02	11.13	589
89	200	2.2	1.05	0.593	59.3	26.11	10.42	32.63	13.03	757
89	235	2.6	1.03	0.819	96.3	30.68	14.14	38.34	17.67	1227
89	241	2.7	1.02	0.862	103.8	31.46	14.83	39.32	18.53	1324
89	286	3.2	1.01	1.213	173.5	37.33	20.49	46.67	25.61	2212
89	302	3.4	1.00	1.353	204.3	39.42	22.83	49.28	28.53	2605
89	356	4.0	0.97	1.880	334.6	46.47	30.53	58.09	38.16	4266
89	406	4.6	0.94	2.445	496.3	53.00	38.60	66.25	48.25	6328
133	184	1.4	1.06	0.750	69.0	35.89	13.30	44.87	16.63	880
133	200	1.5	1.05	0.887	88.7	39.01	15.58	48.77	19.47	1131
133	235	1.8	1.03	1.224	143.8	45.84	21.13	57.30	26.41	1834
133	241	1.8	1.02	1.287	155.1	47.01	22.16	58.76	27.70	1978
133	286	2.2	1.01	1.813	259.3	55.79	30.62	69.74	38.28	3306
133	302	2.3	1.00	2.022	305.3	64.80	37.52	81.00	46.90	4437
133	356	2.7	0.97	2.809	500.1	69.44	45.62	86.80	57.03	6376
133	406	3.1	0.94	3.654	741.7	87.12	63.45	108.90	79.32	10781

Coefficients

Duration of load

Long term $K_3 = 1.00$

Medium term $K_3 = 1.25$

Depth factor

$K_7 = (300/h)^{0.11}$ for $72 < h < 300$

$K_7 = 0.81(h^2 + 92300)/(h^2 + 56800)$ for $h > 300$

Derivation of capacities

Shear

Grade stress = 2.2 N/mm^2

Capacity = $(2bh)/3 \times$ grade stress $\times K_3$

Moment

Grade stress = 16.8 N/mm^2

Capacity = $Z \times$ grade stress $\times K_3 \times K_7$

Deflection

$E_{mean} = 12750$ N/mm^2

Table 10.7 Grade stresses and moduli for TimberStrand

Grade stresses and moduli (N/mm²)		Dry exposure condition Grade 1.3E	Grade 1.5E
Bending	parallel to grain (as a joist)	10.6	13.1
	parallel to grain (as a plank)	11.2	15.1
Tension	parallel to grain	8.0	11.3
Compression	parallel to grain	8.5	11.7
	perpendicular to grain (as a joist)	3.7	5.3
	perpendicular to grain (as a plank)	2.1	2.7
Shear	edgewise (as a joist)	2.8	2.8
	flatwise (as a plank)	1.0	1.0
Modulus of elasticity	E_{mean}	9000	10300
Modulus of rigidity	G_{mean}	450	515

Table 10.8 Special fastener requirements for TimberStrand

Fastener type	Diameter (mm)	Pre-drill	Strength class	Comment/restrictions
Bolt/dowel	M8 to M12	Yes	C24	Fastener parallel to wide face of strands not permitted
	M12 to M24	Yes	C27	Fastener parallel to wide face of strands not permitted
Nail	Up to 3.35	No	C16	Nail and load parallel to wide face of strands; penetration depth of nail not to exceed half member depth
			C27	All load and nail orientations other than parallel to wide face of strands
	3.35 to 5.0	Yes	C16	Nail and load parallel to wide face of strands; penetration depth of nail not to exceed half member depth
		No	C27	All load and nail orientations other than parallel to wide face of strands

0.8–1.3 mm thick. The strands are oriented so that the are essentially parallel to the length of the section and are long layers of wood veneers typically 3.2 mm thick taken from rotary-peeled logs. These veneers are generally end scarf jointed and fed into a press with their end joints staggered by at least 100 mm. In the press they are glued under heat and pressure to produce large billets up to 90 mm wide, 1200 mm deep and up to 10.7 m long. The billet is then converted into a range of smaller sections.

Table 10.9 TimberStrand LSL Grade 1.5E: Principal members, dry exposure condition

Section			Depth	Section properties		Long term		Medium term		
b (mm)	h (mm)	h/b	factor K_7	Z ($\times 10^6$ mm^3)	I ($\times 10^6$ mm^4)	Shear (kN)	Moment (kN m)	Shear (kN)	Moment (kN m)	EI (kN m^2)
45	184	4.1	1.06	0.254	23.4	15.46	3.51	19.32	4.39	241
45	200	4.4	1.05	0.300	30.0	16.80	4.11	21.00	5.14	309
45	235	5.2	1.03	0.414	48.7	19.74	5.57	24.68	6.97	501
45	241	5.4	1.02	0.436	52.5	20.24	5.85	25.31	7.31	541
45	286	6.4	1.01	0.613	87.7	24.02	8.08	30.03	10.10	904
45	302	6.7	1.00	0.684	103.3	25.37	9.00	31.71	11.25	1064
45	356	7.9	0.97	0.951	169.2	29.90	12.04	37.38	15.05	1743
89	184	2.1	1.06	0.502	46.2	30.57	6.94	38.21	8.68	476
89	200	2.2	1.05	0.593	59.3	33.23	8.13	41.53	10.16	611
89	235	2.6	1.03	0.819	96.3	39.04	11.02	48.80	13.78	991
89	241	2.7	1.02	0.862	103.8	40.04	11.56	50.05	14.45	1069
89	286	3.2	1.01	1.213	173.5	47.51	15.98	59.39	19.97	1787
89	302	3.4	1.00	1.353	204.3	55.19	19.58	68.99	24.47	2399
89	356	4.0	0.97	1.880	334.6	59.14	23.81	73.93	29.76	3447
89	406	4.6	0.94	2.445	496.3	74.20	33.11	92.74	41.39	5828
89	457	5.1	0.92	3.098	707.9	83.52	40.99	104.39	51.24	8822

Coefficients

Duration of load

Long term $K_3 = 1.00$

Medium term $K_3 = 1.25$

Depth factor

$K_7 = (300/h)^{0.11}$ for $72 < h < 300$

$K_7 = 0.81(h^2 + 92300)/(h^2 + 56800)$ for $h > 300$

Derivation of capacities

Shear

Grade stress = 2.8 N/mm^2

Capacity = $(2bh)/3 \times$ grade stress $\times K_3$

Moment

Grade stress = 13.1 N/mm^2

Capacity = $Z \times$ grade stress $\times K_3 \times K_7$

Deflection

E_{mean} = 10 300 N/mm^2

10.5.2 Structural performance

Design is carried out in accordance with BS 5268-2 using the grade stresses shown in Table 10.7 as taken from Agrément certificate No. 97/3369.

Modification factors K_3, K_4, K_5, K_7, K_{12} and K_{13} may be applied to these stresses but K_8 should have a reduced value of 1.03 to account for the lower coefficient of variation when compared with solid sawn timber.

Deflection is calculated using E_{mean}.

A length factor K_L should be applied to tension members

$$K_L = (1200/L)^{0.083}$$

where L = length (mm) with a minimum value of 1200.

Joints using bolts/dowels and nails should be designed in accordance with BS 5268-2 incorporating the special requirements of Table 10.8.

Table 10.9 combines the relevant stresses from Table 10.7 with the properties for a range of TimberStrand sections. The designer should check availabilty as stockists carry a reduced range usually limited to the smaller sections.

Chapter 11
Solid Timber Decking

11.1 INTRODUCTION

Solid timber decking is mainly used in roof constructions where the soffit of the decking is exposed to view and the decking spans between glulam beams to give a solid, permanent roof deck which also serves as a ceiling with excellent appearance. Decking is frequently used in swimming pools where the general moisture conditions are a disadvantage for the use of many other structural materials and where ceiling cavities are considered by many to present condensation problems.

Traditionally decking in the UK was undertaken in Western red cedar, usually in what is known as a 'random layup', in which end joints of individual boards can occur in the span rather than only at points of support. With random layups a high degree of site attention is required to ensure that boards are laid less randomly than the term suggests. With the introduction of finger jointing machinery some simplification of layup has been made possible without high length wastage. This, together with the economic trend towards European timbers, has led most companies to finger joint and profile boards in lengths equal to one, two or three bay modules instead of purchasing random-length boards of Canadian origin already profiled at source, and to use European whitewood or redwood.

Basic thicknesses of 38, 50, 63 and 75 mm are used by the larger producers. All decking is tongued and grooved, the thinner boards being tongued only once and the thicker boards being double-tongued. Typical profiles are shown in Fig. 11.1. Decking is usually machined from 150 mm wide sections, although sections less than 150 mm wide can be used at a higher cost (both for machining and laying to a given area).

As there is little possibility of damaging ceiling finishes, the deflection of solid decking is usually relaxed to span/240 under total load.

11.2 SPAN AND END JOINT ARRANGEMENTS

11.2.1 Layup 1. Single span

All boards bear on and are discontinuous at each support. The planks in this arrangement will deflect more than with any other possible arrangement. Depending upon the deck thickness this arrangement may require large widths for the

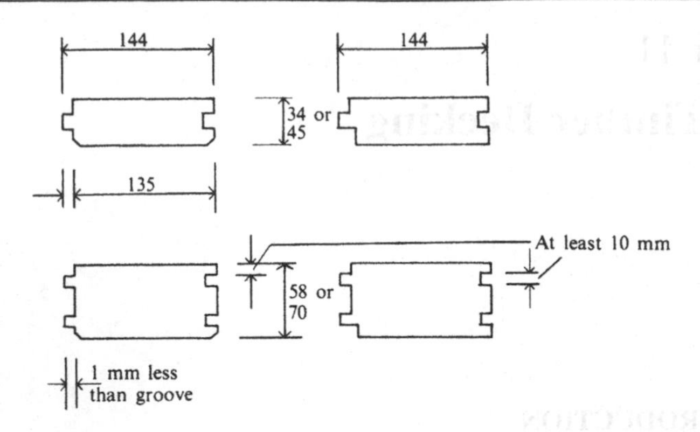

Fig. 11.1 Typical profiles.

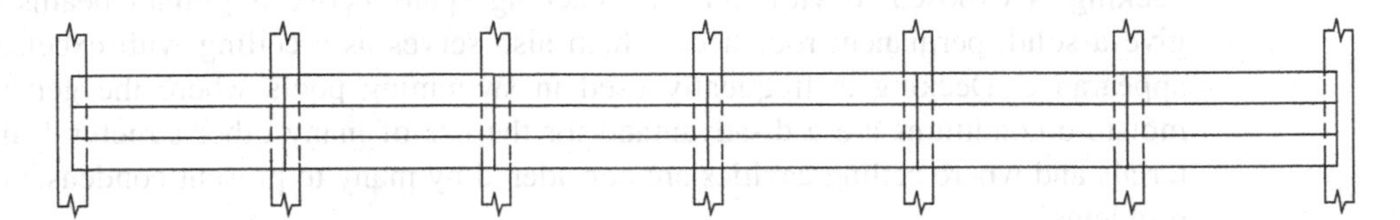

Fig. 11.2 Layup 1.

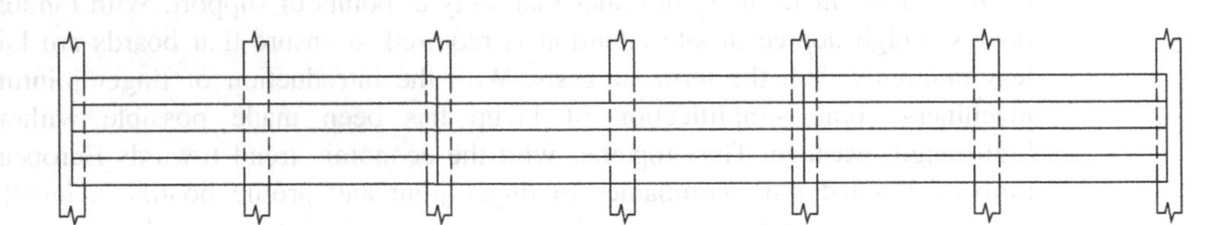

Fig. 11.3 Layup 2.

supporting beams to accommodate nail fixings. End fixing is particularly difficult to resolve when the thicker decks are specified, because the longer nails, available only in the larger gauges, require large end distances and spacings. (The requirements for nail fixings are discussed fully in Chapter 18 and particular fixing details for decking are illustrated in section 11.3.)

11.2.2 Layup 2. Double span

All boards bear on each support and are continuous over two bays. This arrangement has the least deflection of any arrangement, but the loading is not transferred uniformly to supports and beams supporting the decking at mid-length carry more load. For architectural reasons it is usual to require all support beams to be of the same section which, with this particular arrangement, leads to alternate beam sections being over-designed. Similar problems for end fixing as encountered in Layup 1 also apply with this arrangement.

11.2.3 Layup 3. Treble span

All boards bear on supports and are continuous over three spans, breaking regularly at every third beam. This arrangement is usually applied with caution, since there are practical difficulties in manufacturing long straight lengths of boarding and in handling individual boards. It is recommended from practical experience that the length of individual boards should not exceed 9 m wherever possible.

11.2.4 Layup 4. Single and double span

All boards bear on supports. Alternate boards in end bays are single span with the adjacent board being two-span continuous. End joints occur over beam supports, being staggered in adjacent courses (Fig. 11.5). Deflections in the end bay are intermediate between those of Layups 1 and 2, while internal bays deflect approximately the same as for a two-span continuous arrangement. This arrangement is probably the one most usually adopted because it avoids the high deflection characteristic of Layup 1 without overloading alternate supporting beams, and the staggering of end joints permits easier end nailing of boards (Fig. 11.12).

11.2.5 Layup 5. Double and treble span

All boards bear on supports (Fig. 11.6). Alternate boards commence at one end and span three bays continuously with adjacent boards spanning two bays continuously. Internal spans revert to the two-span arrangement of Layup 4. This layup is used only rarely to minimize deflection in the end bays if Layup 4 is not suitable. Care should be taken to avoid board lengths exceeding 9 m, or great care should be taken in selecting the longer boards.

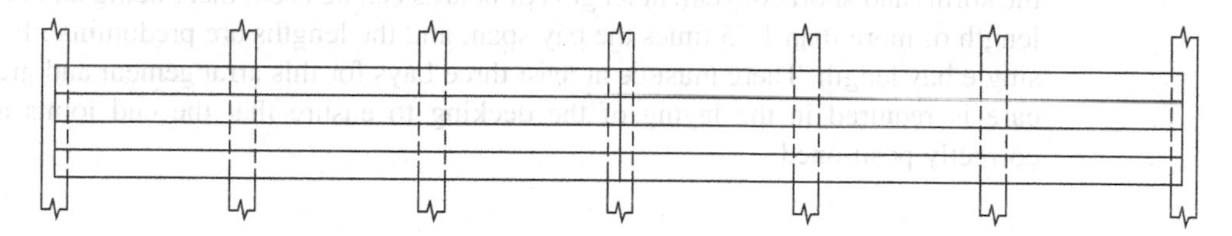

Fig. 11.4 Layup 3.

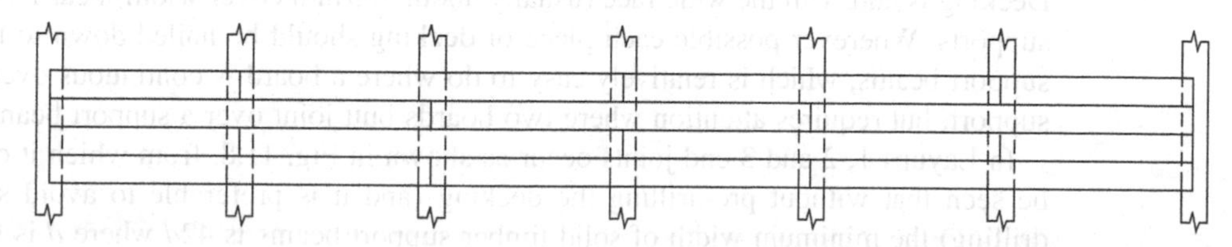

Fig. 11.5 Layup 4.

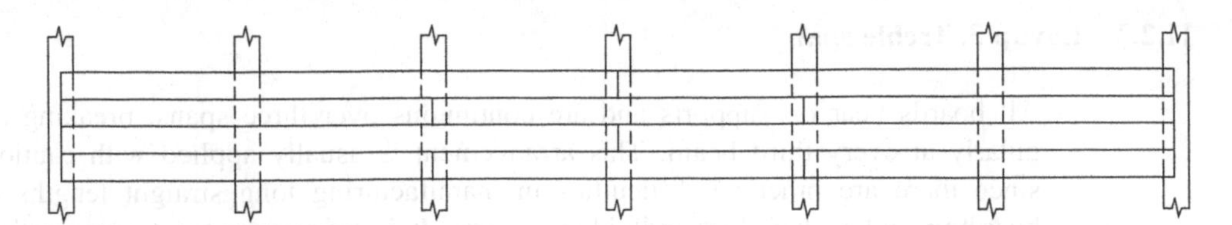

Fig. 11.6 Layup 5.

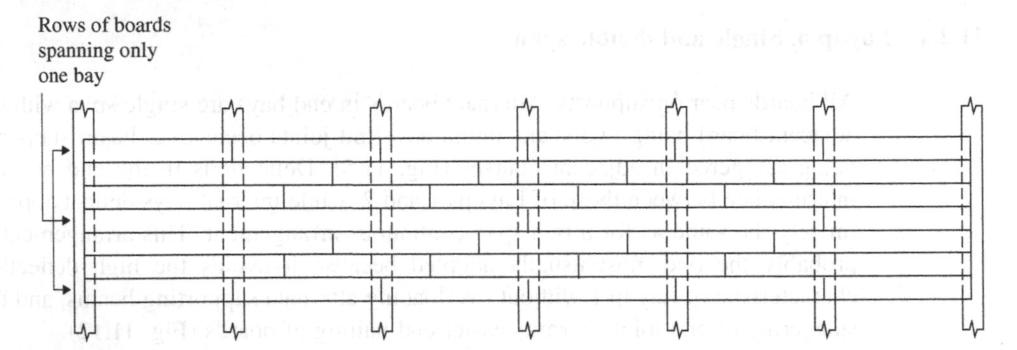

Fig. 11.7 Layup 6.

11.2.6 Layup 6. Cantilevered pieces intermixed

With this arrangement every third course of boarding is simple span (as Layup 1) and intermixed between these courses, is a series of cantilever pieces with butt end joints at between 1/3 and 1/4 points of span. Each piece of the cantilever system rests on at least one support. This arrangement has some advantageous features. For example, glued end jointing is avoided (although end joints will be visible on the soffit) and short convenient lengths of boards can be used, there being no board length of more than 1.75 times the bay span, and the lengths are predominantly of single bay length. There must be at least three bays for this arrangement and great care is required in the laying of the decking to ensure that the end joints are correctly positioned.

11.3 NAILING OF DECKING

Decking is laid with the wide face (usually about 135 mm cover width) bearing on supports. Wherever possible each piece of decking should be nailed down to the support beams, which is relatively easy to do where a board is continuous over a support, but requires attention where two boards butt joint over a support beam.

In Layups 1, 2 and 3 end joints occur as shown in Fig. 11.8, from which it can be seen that without pre-drilling the decking (and it is preferable to avoid site drilling) the minimum width of solid timber support beams is $42d$ where d is the nail diameter. This width may be reduced to $26d$ if the decking is pre-drilled.

It is recommended that two nails per end are used for a width of 135 mm.

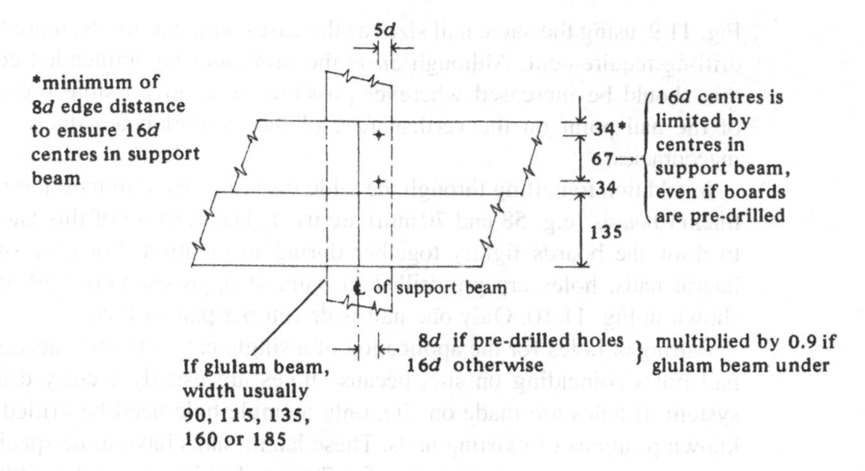

Fig. 11.8 Nail fixing for 135 mm wide decking.

Table 11.1 Nails for fixing 135 mm face width softwood decking (excluding Douglas fir) to timber beams

Decking thickness (mm)	Maximum nail sizes to be used for typical support beam widths (mm) d = diameter (mm)				
	90	115	135	160	185
34	Pre-drill 3.35d × 75	2.65d × 65	3d × 65	3.75d × 75	4d × 100
45	Pre-drill 3.35d × 75	2.65d × 65	3d × 65	3.75d × 75	4d × 100
58	Not recommended with standard nails	Pre-drill 4d × 100	Pre-drill 4d × 100	Pre-drill 4d × 100	4d × 100
70		Pre-drill 4d × 100	Pre-drill 4d × 100	Pre-drill 4d × 100	4d × 100

The centres of nails is obtained from Table 60 of BS 5268-2 multiplied by 0.8 which is a factor that can be applied to all softwoods except Douglas fir. If the support beam is glulam the spacings can be multiplied by a further factor of 0.9.

Table 11.1 indicates the maximum nail sizes which can be taken from normal standard lists to achieve the end distances and spacings shown in Fig. 11.8 to suit 135 mm face width decking of typical thicknesses and support beam widths. Those conditions which require the holes to be pre-drilled are noted. In establishing this table the priorities were:

- to try to avoid pre-drilling
- to use the largest diameter nail in the longest length commercially available.

Where decking is continuous over a support there is no criterion for end distance spacing in the board. In practice the fixing in these cases would be as shown in

Fig. 11.9, using the same nail sizes as the cases with end joints, there being no pre-drilling requirement. Although $5d$ is the minimum recommended edge distance, this should be increased wherever possible to avoid possible breaking through of the nail point on the vertical face of the support beam due to possible site inaccuracies.

In addition to nailing through the wide face, it is usual to nail horizontally when thicker boards (e.g. 58 and 70 mm) are used. The purpose of this lateral nailing is to draw the boards tightly together during installation. For ease of driving the lateral nails, holes are pre-drilled in pairs at approximately 750 mm centres as shown in Fig. 11.10. Only one nail is driven per pair of holes.

Pairing of holes for the application of a single nail is usually necessary to avoid nail holes coinciding on site, because holes are usually factory drilled on a jig system. If holes are made on site, only a single hole need be drilled to avoid the known positions of existing nails. These lateral nails have to be specially ordered, being of non-standard sizes (i.e. for 70 mm decking use 6.7 × 200 mm and for 58 mm decking use 5.6 × 200 mm).

It is not practical to nail laterally through 34 mm and 45 mm thick decking. Instead, it is necessary to slant nails at 0.75 m centres as illustrated in Fig. 11.11. Nails should be driven at an angle of 30° to the horizontal, with an initial edge distance of half the nail length. To meet this recommendation, 2.65 × 50 and 3.35 × 75 mm nails can be used for 34 and 45 mm decking respectively.

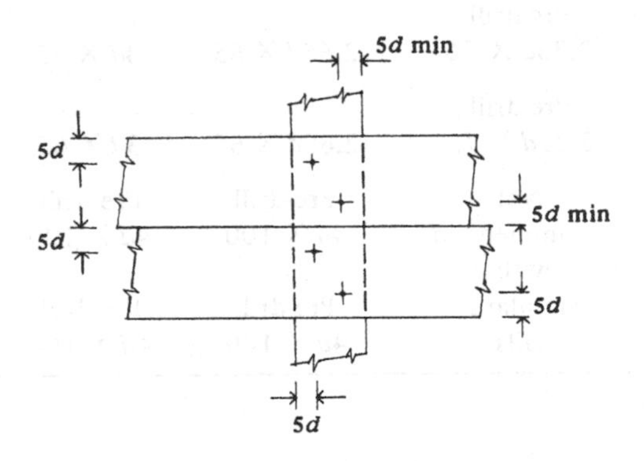

Fig. 11.9

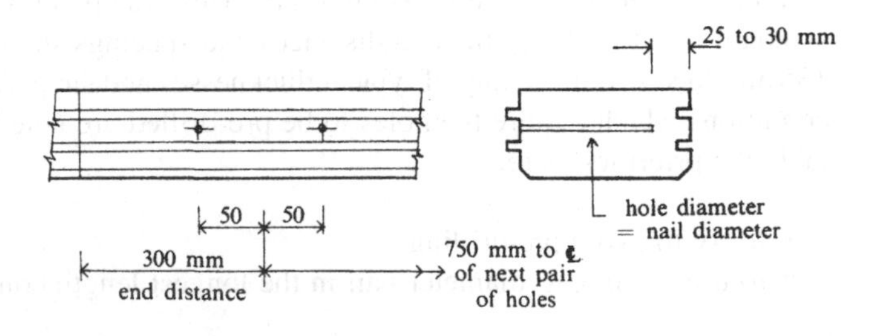

Fig. 11.10

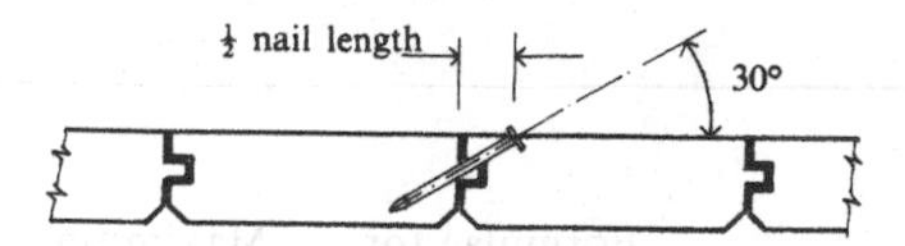

Fig. 11.11

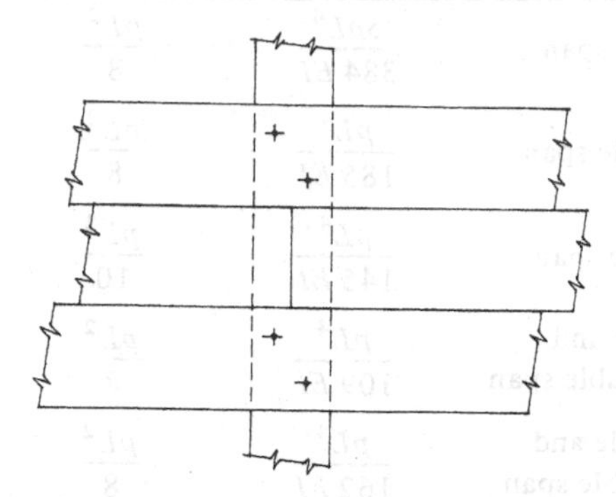

Fig. 11.12

If the preceding guidelines for lateral nailing are implemented, it can usually be assumed that sufficient lateral shear transfer will occur between boards to give full lateral restraint to any supporting beam. In the particular cases of Layups 4, 5 and 6 the end nailing schedule of Table 11.1 can be relaxed and reliance placed upon the fixings in adjacent boards (Fig. 11.12).

11.4 DESIGN PROCEDURE

All boards are tongued and grooved and act together as a unit to resist vertical shears, moments and deflections. However, an exception for which the designer must make allowance is where moments are not always resisted by the full width of the decking system because butt end joints occur over supports. The effect of concentrated loadings may be ignored because of the lateral distribution achieved by the tongues and the fixing of lateral or slant nails.

Shear stresses need not be considered in the design of decking as they can be shown to be of a very low magnitude. It is therefore necessary to check only deflection and bending stresses. With Layup 4 deflection will be a maximum in the end bay (where continuity is a minimum) and bending stresses will be greatest at the first beam support from the end where the net resisting section of deck is least due to end joints. For Layups 2 and 3 the full section resists bending stresses over the first support.

Table 11.2 summarizes the formulae for end bay deflection, maximum moment under balanced loading and the effective width of decking which resists this moment. It is usual to disregard shear deflection which can be shown to be negligible.

Table 11.2

Layup	Description	Formulae for bending deflection in end bay	Maximum moment under balanced load	Ratio of effective design width as laid at position of M_{max}	Required EI (kN.m²) with deflection = $L/240$
1	Single span	$\frac{5pL^4}{384\,EI}$	$\frac{pL^2}{8}$	1.0	$3.12pL^3$
2	Double span	$\frac{pL^4}{185\,EI}$	$\frac{pL^2}{8}$	1.0	$1.30pL^3$
3	Treble span	$\frac{pL^4}{145\,EI}$	$\frac{pL^2}{10}$	1.0	$1.66pL^3$
4	Single and double span	$\frac{pL^4}{109\,EI}$	$\frac{pL^2}{8}$	0.5	$2.20pL^3$
5	Double and treble span	$\frac{pL^4}{162\,EI}$	$\frac{pL^2}{8}$	0.5	$1.48pL^3$
6	Cantilever pieces intermixed	$\frac{pL^4}{105\,EI}$	$\frac{pL^2}{10}$	0.667	$2.29pL^3$

where p = the load per metre run of 1 metre width of decking as laid

11.5 SPECIES OF DECKING, GRADES AND CAPACITIES

Currently the species most frequently used for decking are whitewood and redwood. The choice is usually primarily one of cost, balanced by acceptable architectural appearance. With whitewood and redwood, the grading required for appearance usually means that the decking can be considered to match the visual SS grade. The decking system as a whole may be considered as a load-sharing system permitting the use of E_{mean} at 10 500 N/mm² for deflection calculations. The grade bending stress of 7.5 N/mm² is modified by the depth factor K_7 taken at 1.17 for all thickness less than 72 mm and is further modified by a consideration of load sharing at the locations of high bending moment. For Layups 1, 2 and 3 all boards are available to resist the applied bending moment and the load-sharing factor K_8 = 1.1 is applicable. For Layups 4, 5 and 6 end joints in individual boards frequently occur at locations of maximum bending moment. This is accounted for by reducing the section modulus accordingly and by conservatively assuming no load sharing with respect to bending.

11.6 EXAMPLE OF DESIGN OF DECKING

Calculate the thickness of roof decking of Layup 4, single and double span (Fig. 11.5) required on a span of 3.4 m using redwood of SS grade and permitting a

Table 11.3 Whitewood and redwood decking of SS grade. Stiffness (EI) and moment capacity ($\bar{M}$) for 1 m width as laid

Deck thickness (mm)	I ($10^6 \times mm^4$)	EI ($kN\,m^2$)	Layup	Z_{net} ($10^6 \times mm^3$)	$\sigma_{m,adm,par}$ long term (N/mm)	Moment capacity (kN m)			
						$\bar{M}_L$	$\bar{M}_M$	$\bar{M}_S$	$\bar{M}_{VS}$
34	3.27	34.3	1, 2 and 3	0.193	9.65	1.86	2.33	2.79	3.26
			4 and 5	0.0963	8.77	0.84	1.05	1.26	1.48
			6	0.128	8.77	1.12	1.40	1.68	1.96
45	7.59	79.7	1, 2 and 3	0.337	9.65	3.25	4.07	4.88	5.69
			4 and 5	0.169	8.77	1.48	1.85	2.22	2.59
			6	0.225	8.77	1.97	2.47	2.96	3.45
58	16.3	171	1, 2 and 3	0.561	9.65	5.41	6.76	8.12	9.47
			4 and 5	0.280	8.77	2.45	3.07	3.68	4.30
			6	0.374	8.77	3.28	4.10	4.92	5.74
70	28.6	300	1, 2 and 3	0.817	9.65	7.88	9.85	11.80	13.8
			4 and 5	0.408	8.77	3.57	4.47	5.36	6.26
			6	0.544	8.77	4.77	5.96	7.15	8.35

$\bar{M}_L$ = moment capacity, long term.
$\bar{M}_M$ = moment capacity, medium term.
$\bar{M}_S$ = moment capacity, short term.
$\bar{M}_{VS}$ = moment capacity, very short term.

deflection of $L/240$. The dead loading including self-weight is 0.40 kN/m^2 and the roof imposed load is 0.75 kN/m^2.

Consider 1.0 m width of decking.

Long-term loading = 0.40 kN/m^2
Medium-term loading = 1.15 kN/m^2

From Table 11.2,

$$M_{max} = \frac{pL^2}{8}$$

Maximum long-term moment = $0.4 \times 3.4^2/8 = 0.58$ kNm
Maximum medium-term moment = $(0.4 + 0.75) \times 3.4^2/8 = 1.66$ kNm

From Table 11.2,

$$EI \text{ required} = 2.20pL^3 = 2.20 \times 1.15 \times 3.4^3 = 99.4 \text{ kN m}^2$$

From Table 11.3, select 58 mm decking for which

$\overline{M}_L = 2.45$ kN m > 0.58 kN m
$\overline{M}_M = 3.07$ kN m > 1.66 kN m
$EI = 171$ kN m^2 > 99.4 kN m^2

Check deflection under total load:

$$\text{Deflection} = pL^4/(109\ EI) = 1.15 \times 3.4^4/(109 \times 171) = 0.0082 \text{ m}$$

Allowable deflection at $L/240 = 3400/240 = 0.014$ m, therefore **O.K.**

Chapter 12
Deflection. Practical and Special Considerations

12.1 DEFLECTION LIMITS

Any deflection limit which is set can be for functional reasons or purely for visual reasons. For example, it has been found by experience that deflections below the horizontal not exceeding $0.003L$ are usually visually acceptable to anyone in a room below, unless there is a horizontal feature which makes the deflection much more obvious than it otherwise would be. With beams cambered to off-set dead load deflection, the deflection limit therefore applies to movement under imposed loading only. BS 5268-2 sets a limit 'for most general purposes' of 0.003 of the span. In addition, 'to avoid undue vibration' a further overall limit of 14 mm is set for domestic floor joists.

For beams or decking spanning a short distance (say less than 2 m) it seems rather unnecessary to be too pedantic about observing a deflection limit of $0.003L$. At a span of 1.2 m this would give a limit of 3.6 mm. If the design permitted a deflection of 4.5 mm, this would be quite a large increase in percentage terms but hardly significant in real terms. It is important to realize that an over-deflection can take place without the beam necessarily approaching anywhere near to a stress-failure situation. This is a point often missed by architects, building inspectors, etc. With short spans it is probably more important that the 'feel' of the beam is acceptable to any personnel who walk on it, and a simple test is often more satisfactory to determine suitability on short spans rather than deflection calculations.

In the design of beams for storage racking, it is usually acceptable to permit deflections in the order of 1/180 of the span if there is no access or only occasional access for personnel, and 1/240 of the span if there is general access for staff (not the general public).

If the designer is in the position of having to decide a special deflection limit the following points should be considered:

- the span
- the type of structure and the usage
- the possibility of damage to the ceiling or covering material
- aesthetic requirements, particularly linked to any horizontal feature
- the number of times and length of time when maximum deflection is likely to occur and whether there will be a camber
- roof drainage
- the effect on such items as partitions over or under the position of deflection

- special items (see section 12.6)
- avoidance of secondary stresses at joints due to excessive rotation or distortion.

Only in exceptional circumstances is it normally necessary to tighten the deflection limit of 0.003 × span.

12.2 CAMBER

When a beam is prefabricated from several parts, as with glulam or ply web beams, it may be possible to introduce a camber into simply supported beams (although the more usual practice for proprietary glulam beams and I beams is to manufacture without camber). The usual object of providing a camber is to aim to have the beam deflect to a horizontal position under the action of the dead or permanent loading. In this way the beam has a horizontal soffit for the majority of the time, and deflects below the horizontal only when imposed load is added, and deflects to the maximum permitted position only in the possible extreme cases when the total imposed loading occurs, and then only for short periods of time.

With the variation in the weight of building materials and the *E* value of timber it is impossible for a designer to guarantee that a cambered beam will deflect exactly to the horizontal when the dead loading is in place. It should be possible, however, to design the beam to deflect to this position within the usual degree of building tolerance unless the builder departs from the design loadings to a large extent. It is well established that with a beam which is perfectly horizontal, there is an optical illusion that the beam is deflecting. (This is particularly so with a lattice beam.) Also, it is usually easier for a builder to pack down to obtain a level soffit from the centre of a beam which has a small residual camber, than to pack down from the ends of an under-cambered beam as this may interfere with an edge detail. The designer should usually therefore consider providing a camber slightly larger than the camber theoretically required.

It should be remembered that there is no such thing as a factor of safety on deflection. Double the load and the deflection doubles. If the agreed design loading is changed the beam deflection will be different from the calculated deflection. There is no way in which a beam can be designed to deflect to a required level, then deflect no more if further load is applied. If the designer is informed of an over-deflection having taken place on site and is satisfied that his calculations are correct, experience has shown that the first aspect to check on site is the weight of the applied loading. It may well be that this is excessive.

When a camber has been built-in with the object of bringing the beam to a horizontal position under the action of the permanent loading, the deflection limit (0.003 × span or other agreed limit) applies to deflection under the imposed loading only. If the designer has called for a camber slightly larger than the theoretical movement under permanent loading, it is not permissible to add the over-camber to the permitted deflection under imposed loading in checking the stiffness of the beam. However, if a designer opts for an 'under-camber' there is a case for checking the stiffness of the beam for the total loading minus the loading which would cause the beam to settle to the horizontal.

It is usual for a manufacturer to build in a circular camber even though the beam may deflect in a parabolic manner. This is usually satisfactory for normal spans, but on spans of around 15 m or more, the designer would be wise to calculate several points on the ideal camber curve and instruct the manufacturer accordingly.

12.3 DEFLECTION DUE TO DEAD LOAD ONLY ON UNCAMBERED BEAMS

BS 5268-2 does not lay down any criteria for limiting the deflection under dead loading only on uncambered beams (unless of course there is no imposed loading at all, in which case the limit automatically becomes 0.003 × span). With normal floor or roof beams, the dead loading is rarely more than 60% of the total loading, and therefore the deflected form under dead loading only, which is the position the beam will take up for the majority of its life, is approximately 0.002 × span or less. This is usually perfectly acceptable for the spans encountered with uncambered beams. If, however, for some reason it is impossible to camber a long-span beam, and the percentage of permanent loading to total loading is particularly high, the designer would be wise to discuss the possibility of tightening the deflection limit with the architect and/or the building user.

12.4 DEFLECTION DUE TO WIND UPLIFT ON ROOFS OR WIND ON WALLS

With the large wind uplifts on flat roofs, it is quite possible to find that the residual uplift from the loading condition 'dead load + wind' exceeds the loading from 'dead + snow'. Because there is no load–duration factor for E values, it is possible that the loading case of dead + wind uplift could be critical in setting the size of the beam from the point of view of upward deflection. When one considers that wind loading on beams is based on a five-second gust which occurs very occasionally, if ever, and remembers that any deflection limit is purely an arbitrary limit, it seems rather unnecessary to apply the limit of 0.003 × span to this case, particularly bearing in mind that timber would not deflect to its full calculated amount in five seconds. The purpose of this manual is certainly not to instruct a designer to disregard any part of any code of practice, but this is certainly a case where the designer could exercise some engineering discretion and discuss his opinions with the approving authority before the design is finalized.

Likewise it seems unnecessarily conservative to consider the full calculated deflection as occurring on a column in an external wall, particularly if a stud in a timber-framed house. At the very least it seems fair to claim that, if a beam is supposed to deflect no more than 80% of the calculated amount (see BS 5268-2 on testing) in the first 24 hours, a factor of 0.8 may be applied to the calculated deflection. By using this factor and a small amount of end restraint in the calculation of deflection of a stud, the calculated deflection is about one half of the 'full' simply supported calculated deflection.

12.5 DEFLECTION STAGES DUE TO SEQUENCE OF ERECTION

When an architect or building draughtsman draws a cross-section through a building with a flat roof and/or floor it will show beam soffits and the ceiling as straight horizontal lines, and will not usually consider the stage at which the lines are horizontal. At the time of drawing the cross-section it may not even be known if the beams are to be cambered or not. On certain points, however, it is essential that someone coordinates the sequence of erection of items, either below or above the beams, bearing in mind that initial loading and further imposed loading will be applied to the beams in varying degrees throughout the lifetime of the structure. Some of these points are discussed here in general terms, and in slightly more detail in section 12.6, but the main object here is to make the designer aware of the type of situation that can occur, for which the designer, the architect or the builder must have a solution.

If a non-load-bearing partition is placed under a beam before the total loading (including imposed) is applied to the beam, an adequate gap and vertical sliding connection must be provided at the top of the partition otherwise load will be transmitted to it.

If a builder adjusts the soffit of beams with packings etc. before all the dead loading is applied, the soffit will not be horizontal when the total dead loading has been applied.

If the gap between the top of a bottom run door and the underside of a roof beam is virtually nil once the dead loading is in place on the beam, the door will probably jam when snow falls on the roof. The designer must be alive to such potential troubles, some of which are discussed in section 12.6.

12.6 EXAMPLES OF CASES WHICH REQUIRE SPECIAL CONSIDERATION IN DEFLECTION/CAMBER CALCULATIONS

12.6.1 Beams over room with a change in width

Before the loading is applied to the beams, the undeflected form of cambered beams A and B is as indicated in Fig. 12.1, section XX. The erectors will have difficulty in fitting the secondary members between the beams, particularly if the secondary members are continuous and particularly on the line of wall C. One method of easing the situation is to complete the roof to the left of beam A and to the right of beam B as far as is convenient, then infill between A and B. By the time the ceiling boards are to be fixed the beams will have deflected nearer to the horizontal.

12.6.2 Beams over room with tapered walls

The beams shown in plan in Fig. 12.2 would probably have the same cross section, or at least the same depth, but, if cambered, each beam should be individually cambered to give a uniform soffit under permanent loading.

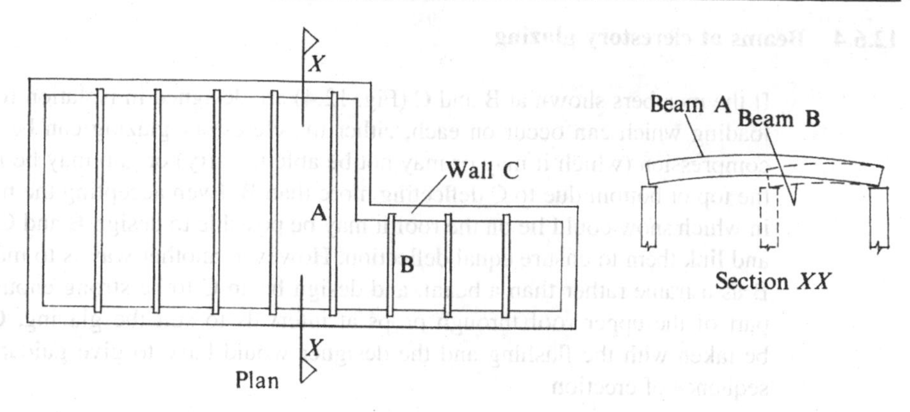

Fig. 12.1

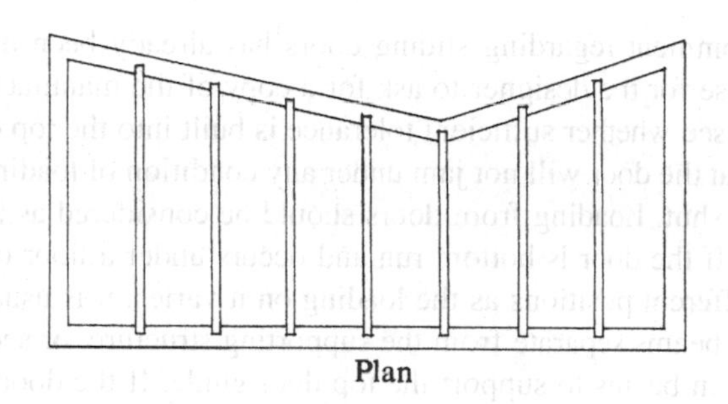

Fig. 12.2

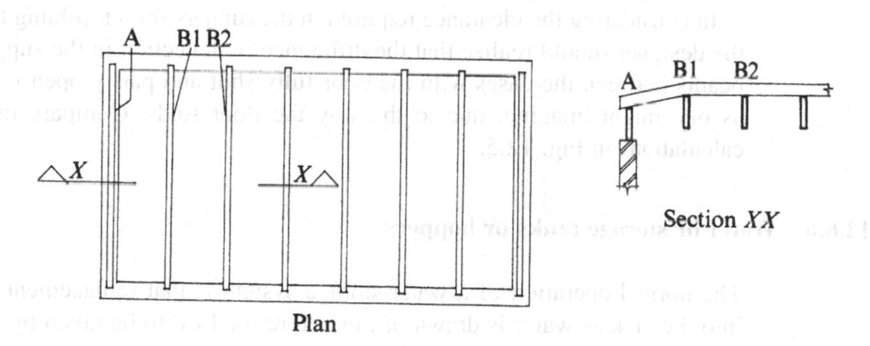

Fig. 12.3

12.6.3 Edge condition with cambered beams

If beams B1, B2, etc., in the plan in Fig. 12.3 are built up, they should preferably be cambered. Beam A would be uncambered or could be a frame just sitting on the wall, rather than a beam. At the time of erecting the decking or secondary members between A and B1, the erectors would have to accept the difference in profile.

12.6.4 Beams at clerestory glazing

If the members shown at B and C (Fig. 12.4) are designed in isolation for the roof loading which can occur on each, either the clerestory glazing can be subject to compression (which it may or may not be able to carry) or gap may be induced at the top or bottom due to C deflecting more than B. Even accepting the many ways in which snow could lie on the roof it may be possible to design B and C as beams and link them to ensure equal deflection. However, another way is to manufacture B as a frame rather than a beam, and design beam C to be strong enough to take part of the upper roof through props at intervals to suit the glazing. Care must be taken with the flashing and the designer would have to give guidance on the sequence of erection.

12.6.5 Sliding doors

Comment regarding sliding doors has already been made in section 3.12.1. It is wise for the designer to ask for a copy of the manufacturing drawing of the doors to see whether sufficient tolerance is built into the top or bottom runners to ensure that the door will not jam under any condition of loading, whether the door is open or shut. Loading from doors should be considered as long term.

If the door is bottom run and occurs under a floor or roof which will deflect to different positions as the loading on it varies, it is usually wise to provide a beam or beams separate from the supporting structure, or secondary beams between the main beams to support the top door guide. If the door is top hung, then it is wise to hang it from one or two beams separate from the main support beams, so that deflection of the main beams will not cause the door to jam (Fig. 3.6), or from a secondary beam (Fig. 3.7).

In considering the clearance required in the runners for a top-hung folding door, the designer should realize that the difference in deflection in the support beam or beams between the cases with the door fully shut and partly open is not as great as one might imagine, due to the way the door folds. Compare the deflection calculations in Fig. 12.5.

12.6.6 Water or storage tanks or hoppers

The normal operation of a water-storage system is that replacement water is fed into the tank as water is drawn off, therefore the load to be taken by any support-

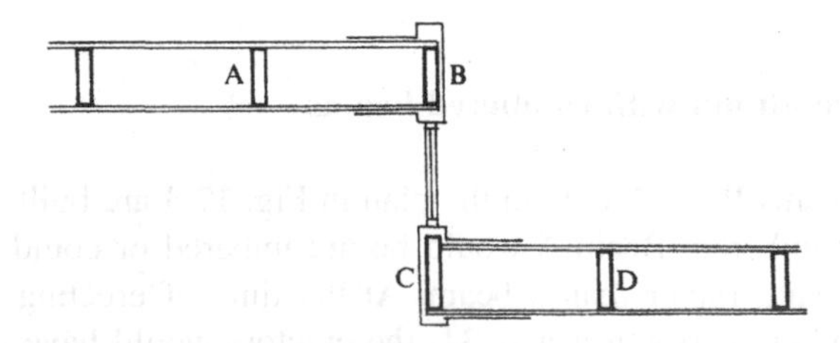

Fig. 12.4

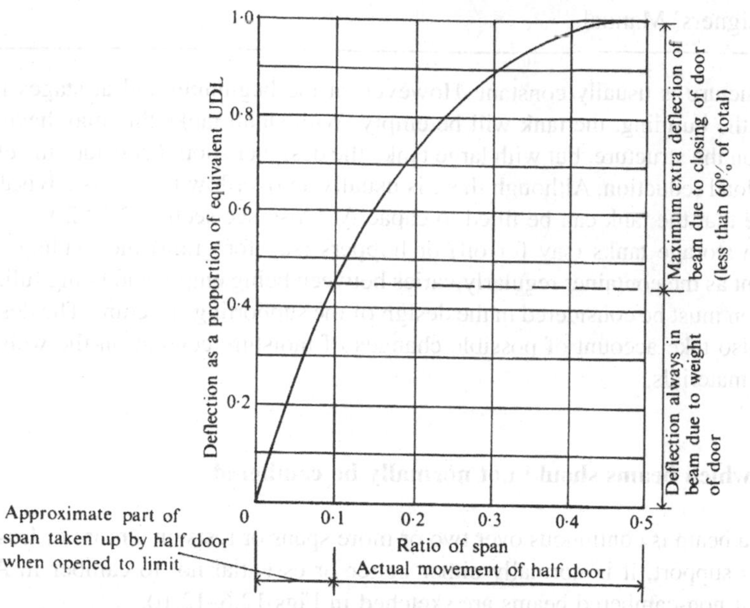

Note that shear deflection is omitted from this comparison exercise.

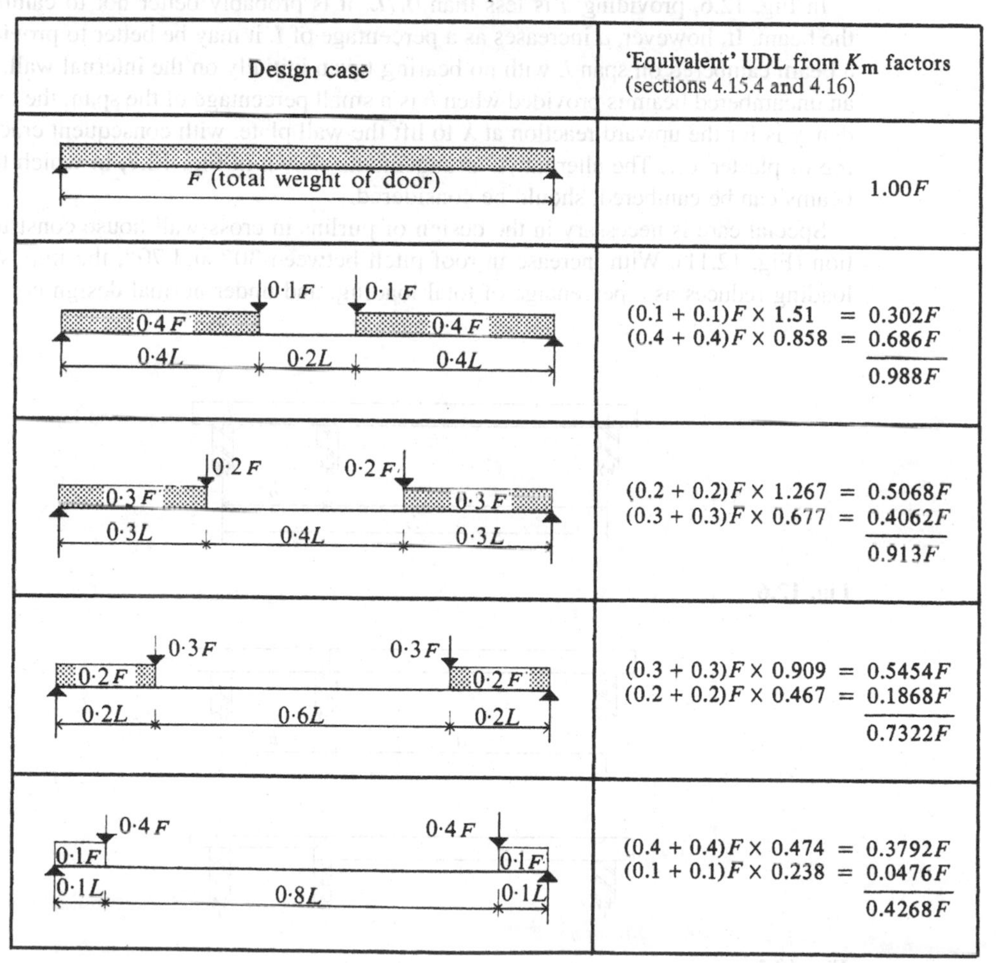

Design case	'Equivalent' UDL from K_m factors (sections 4.15.4 and 4.16)
F (total weight of door)	$1.00F$
$0.1F$, $0.1F$; $0.4F$, $0.4F$; $0.4L$, $0.2L$, $0.4L$	$(0.1 + 0.1)F \times 1.51 = 0.302F$ $(0.4 + 0.4)F \times 0.858 = 0.686F$ $0.988F$
$0.2F$, $0.2F$; $0.3F$, $0.3F$; $0.3L$, $0.4L$, $0.3L$	$(0.2 + 0.2)F \times 1.267 = 0.5068F$ $(0.3 + 0.3)F \times 0.677 = 0.4062F$ $0.913F$
$0.3F$, $0.3F$; $0.2F$, $0.2F$; $0.2L$, $0.6L$, $0.2L$	$(0.3 + 0.3)F \times 0.909 = 0.5454F$ $(0.2 + 0.2)F \times 0.467 = 0.1868F$ $0.7322F$
$0.4F$, $0.4F$; $0.1F$, $0.1F$; $0.1L$, $0.8L$, $0.1L$	$(0.4 + 0.4)F \times 0.474 = 0.3792F$ $(0.1 + 0.1)F \times 0.238 = 0.0476F$ $0.4268F$

Fig. 12.5

ing structure is usually constant. However, at the beginning and at stages in the life of the building, the tank will be empty. With small tanks this may have little effect on the structure, but with large tanks the designer should consider the effects of the load reduction. Although there is usually an overflow tube, it is advisable to assume that the tank can be filled to capacity. (Also see section 3.12.2.)

With storage tanks (say for oil) or hoppers (say for grain) the loading is not constant as the container regularly varies between being empty and being full. This variation must be considered in the design of the supporting structure. The designer must also take account of possible changes of moisture content on the weight of stored materials.

12.6.7 Cases in which beams should not normally be cambered

When a beam is continuous over two or more spans or rests on a more or less continuous support, it is normally either better or essential not to camber it. A few cases of non-cambered beams are sketched in Figs 12.6–12.10.

In Fig. 12.6, providing a is less than $0.7L$, it is probably better not to camber the beam. If, however, a increases as a percentage of L it may be better to provide a beam cambered on span L with no bearing taken initially on the internal wall. If an uncambered beam is provided when b is a small percentage of the span, the tendency is for the upward reaction at X to lift the wall plate, with consequent cracking of plaster, etc. The alternative arrangements shown in Fig. 12.7, in which the beams can be cambered, should be considered.

Special care is necessary in the design of purlins in cross-wall house construction (Fig. 12.11). With increase in roof pitch between 30° and 70°, the imposed loading reduces as a percentage of total loading, and under normal design condi-

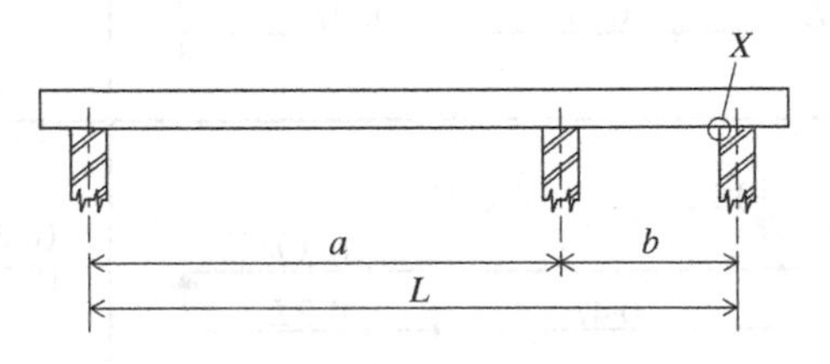

Fig. 12.6

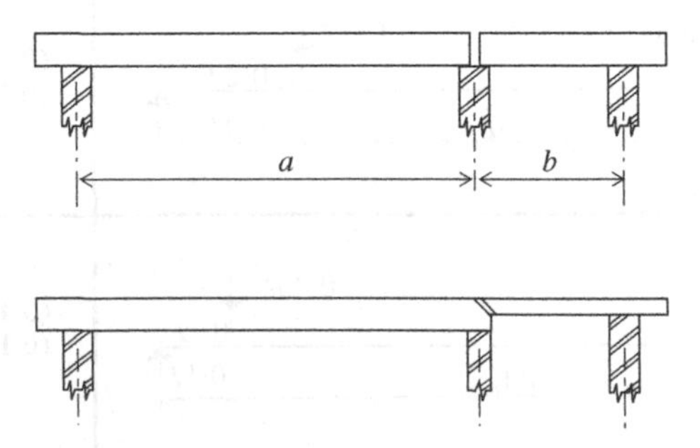

Fig. 12.7

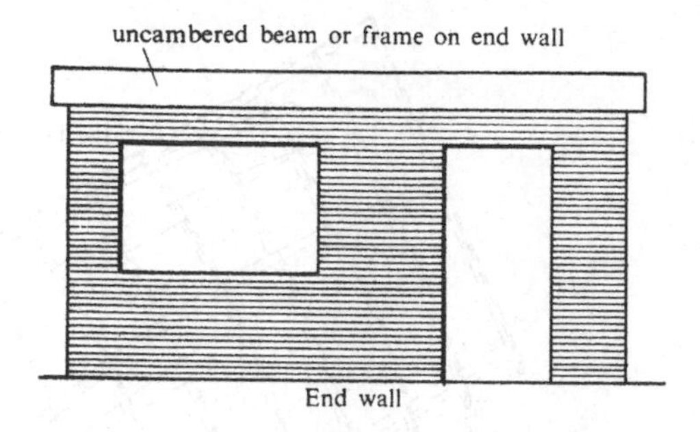

Fig. 12.8

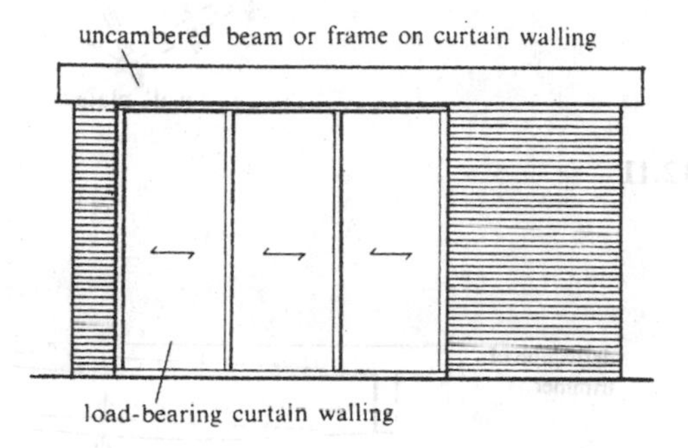

Fig. 12.9

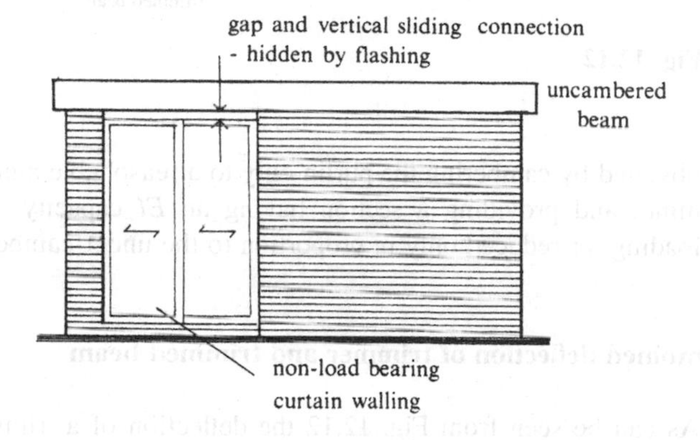

Fig. 12.10

tions, with a beam cambered to off-set dead load deflection, the theoretical *EI* capacity required at 70° roof slope to off-set live load deflection is nil. The effect of this approach may be to introduce excessive camber, because the *EI* capacity obtained with a purlin designed to satisfy only bending and shear is probably exceptionally low. The effect of excessive camber is to change significantly the triangulation of the roof section at midspan of the purlin compared to the gable triangulation. The secondary continuous rafters have to be arched and sprung into place, which presents an unreasonable problem to the builder. This can be

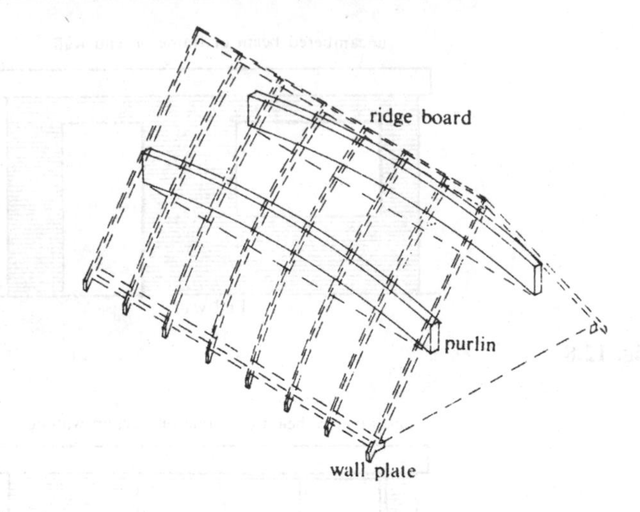

Fig. 12.11

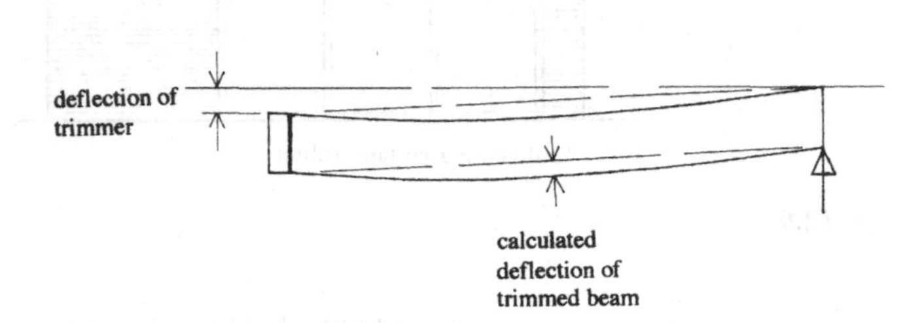

Fig. 12.12

obviated by cambering the purlin only to a reasonable amount bearing erection in mind, and providing a section having an *EI* capacity appropriate to the total loading, or reduced only in proportion to the under-camber.

12.6.8 Combined deflection of trimmer and trimmed beam

As can be seen from Fig. 12.12 the deflection of a trimmer beam can have the visual effect of increasing the deflection of a trimmed beam. In certain cases, particularly if the trimmer is long span, it may be necessary to reduce the deflection of one or both to less than 0.003 times their span.

12.6.9 Deflection of large overhangs

If a beam has a large overhang at one or both ends, it tends to deflect under uniform loading, as sketched in Fig. 12.13.

The designer has two choices for the successful design of a beam with a cantilever:

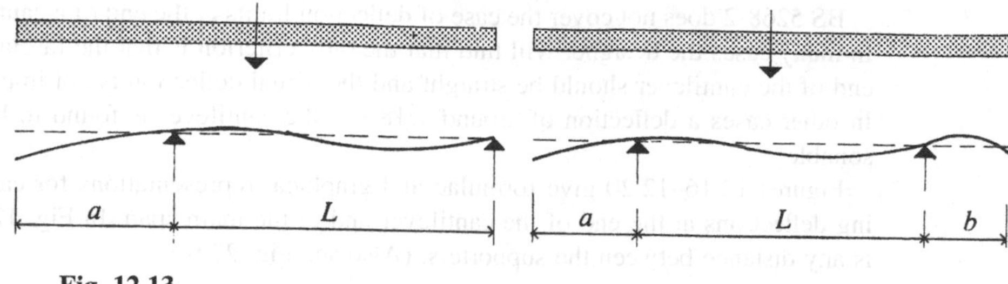

Fig. 12.13

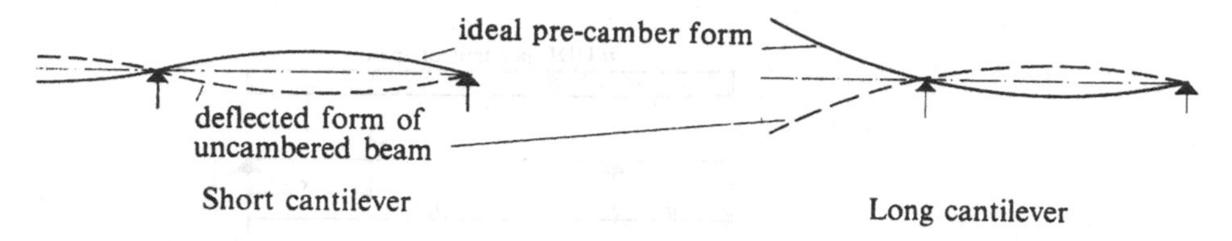

Fig. 12.14

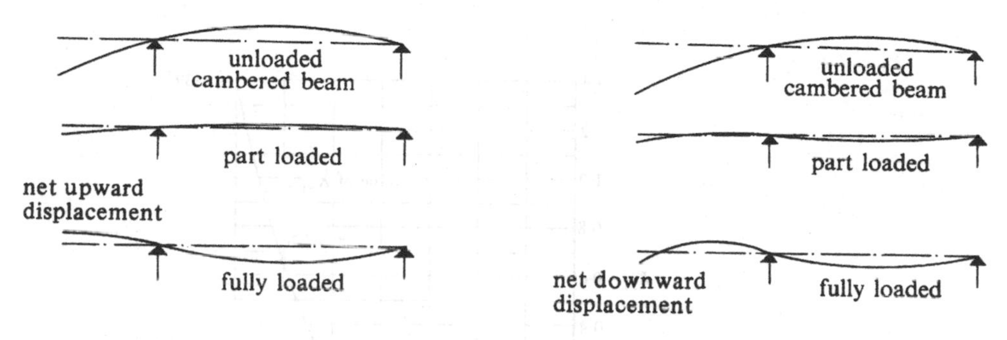

Fig. 12.15

- to decide against camber and limit the deflection all along the length to an acceptable amount
- to camber for economy or functional reasons. In this case, by doing so, care must be taken not to cause an unacceptable situation at the end of the cantilever either before or after loading.

The method of choosing the best camber is to calculate and plot the deflected form under dead loading of an uncambered beam and then provide a camber shape as a 'mirror image' of this (Fig. 12.14). To suit manufacture, a modified camber curve may have to be accepted by the designer.

Figure 12.15 shows typical deflected forms of cambered beams with cantilever, before, during and after loading, which should assist the designer in visualizing the conditions which can be encountered.

When the cantilever is more than about 2 m long it may not be advantageous to use a cambered beam.

BS 5268-2 does not cover the case of deflection limits at the end of a cantilever. In many cases the designer will find that the only criterion is that the fascia at the end of the cantilever should be straight and the actual deflection is not important. In other cases a deflection of around 1/180 of the cantilever is found to be reasonable.

Figures 12.16–12.20 give formulae and graphical representations for calculating deflections at the end of the cantilever and in the main span. In Fig. 12.16, x is any distance between the supporters. (Also see Fig. 27.6.)

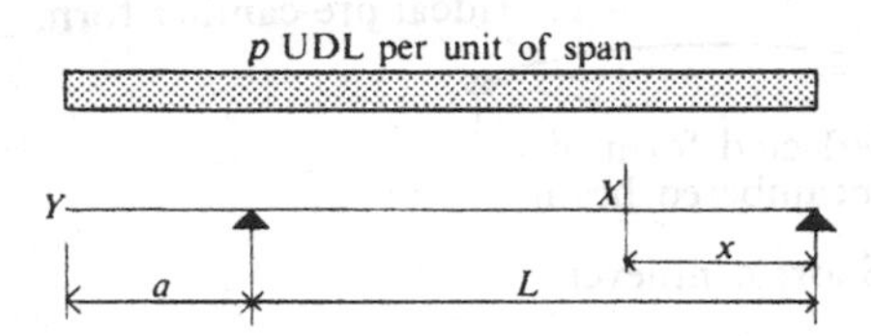

Fig. 12.16

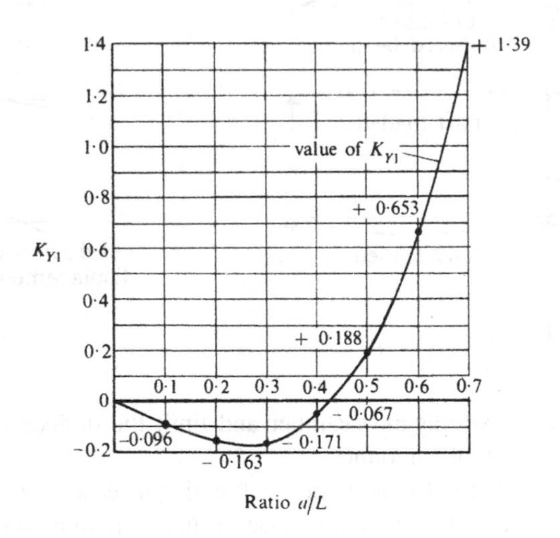

Fig. 12.17

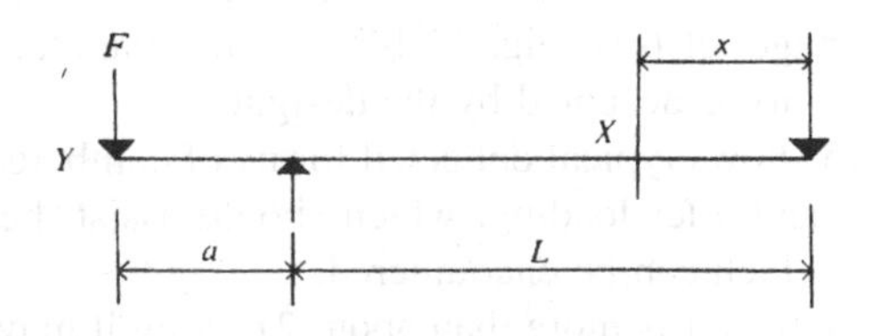

Fig. 12.18

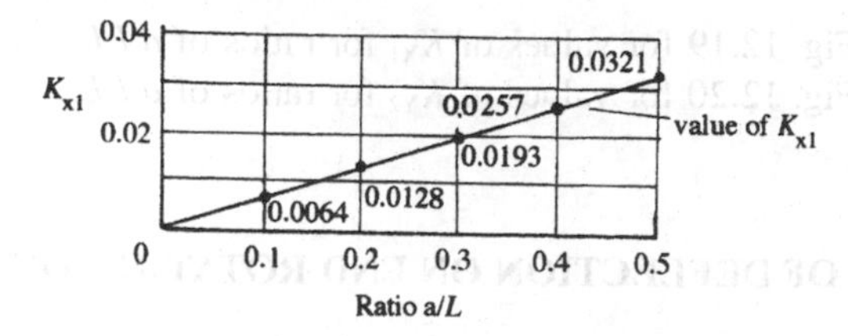

Fig. 12.19

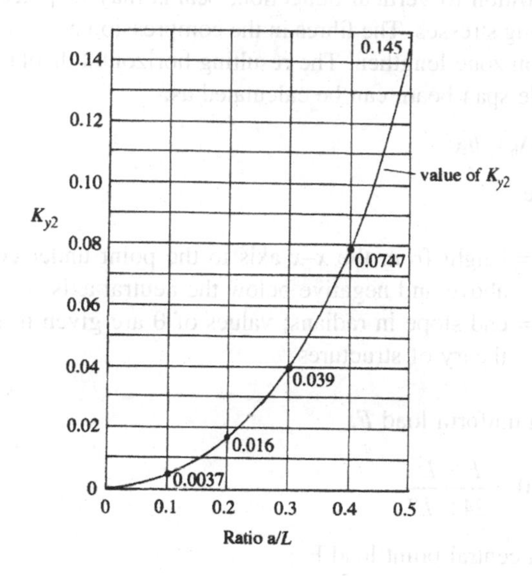

Fig. 12.20

$$\text{Bending deflection at } X = \frac{px}{24EIL}(L^4 - 2L^2x^2 + Lx^3 - 2a^2L^2 + 2a^2x^2)$$

$$\text{Bending deflection at } Y = \frac{pa}{24EI}(4a^2L - L^3 + 3a^3) = \frac{pL^4}{24EI}(K_{Y1})$$

In evaluation of either of these formulae, a positive result indicates downward deflection and a negative value indicates upward deflection. See Fig. 12.17 for values of K_{Y1} for ratios of a / L.

To find the maximum deflection between the reactions, one way is to calculate deflection at two or three positions of x, calculate deflection at Y and plot the deflected form of the beam.

In Fig. 12.18, the distance, x, to the point of maximum bending deflection between the supports is:

$x = L/\sqrt{3}$

$$\text{Bending deflection at } X = \frac{0.0642FaL^2}{EI}\text{(upwards)} = \frac{FL^3}{EI}(K_{X1})$$

$$\text{Bending deflection at } Y = \frac{Fa^2}{3EI}(L + a)\text{(downwards)} = \frac{FL^3}{EI}(K_{Y2})$$

See Fig. 12.19 for values of K_{X1} for ratios of a / L.
See Fig. 12.20 for values of K_{Y2} for ratios of a / L.

12.7 EFFECT OF DEFLECTION ON END ROTATION OF BEAMS

In addition to vertical deflection, beams may displace horizontally as a result of bending stresses. The fibres in the compression zone shorten while the fibres in the tension zone lengthen. The resulting horizontal displacement, Δ_h, at each end of a simple span beam can be calculated as:

$$\Delta_h = h_f\theta$$

where

h_f = height from the x–x axis to the point under consideration, being positive above and negative below the neutral axis
θ = end slope in radians; values of θ are given in most standard textbooks on theory of structures.

For a uniform load F,

$$\theta = \frac{F \times L^2}{24 \times EI}$$

For a central point load F,

$$\theta = \frac{F \times L^2}{16 \times EI}$$

It may he assumed that end rotation is unaffected by shear deflection.

A positive value for Δ_h indicates a shortening of the beam and, conversely, a negative value indicates a lengthening of the beam.

To illustrate the method, calculate the horizontal end displacement of the upper and lower extremes of the 90 × 450 glulam beam in section 7.10.1.

Span = 6.0 m, total uniform load = 38.4 kN and EI = 7898 kNm²

For a 450 mm deep section

$$h_f = \pm\, 0.225\,\text{mm}$$

For a uniform load:

$$\Delta_h = \frac{h_f \times F \times L^2}{24 \times EI} = \pm \frac{0.225 \times 38.4 \times 6.0^2}{24 \times 7898} = \pm\, 0.00164\,\text{m}$$

i.e. 1.64 mm movement horizontally to the left and right of the neutral axis or 3.28 mm total horizontal displacement between top and bottom corners of the section.

The displacement at points between the neutral axis and the extreme fibres may be calculated proportionally.

The detailing of 'pin' joints, particularly between beam and post connections, must allow for this horizontal displacement. For beams of shallow depth the displacement may be sufficiently small to be accommodated within normal manufacturing tolerances and bolt-hole clearances. For deep sections the joint should be designed specifically to permit end rotation without developing secondary stresses. This is achieved by introducing positive clearances between faces of beam and column and by providing slotted holes for bolts, etc.

Chapter 13
Tension Members

13.1 AXIAL TENSILE LOADING

The permissible tensile load in a member carrying only axial loading is the product of the permissible tensile stress and the effective cross-sectional area of the member after deducting the area of any bolt holes, connector dappings or any other notches or cuts. In calculating the net area, it is not necessary to make a reduction for the wane permitted for the grade because wane is taken into account in setting the grade stress. Nor is it usual to consider stress concentrations as occurring at the net area, the adoption of the recommended permissible stresses and net area being considered to make provision for this aspect of design.

When assessing the effective cross section of a member at a multiple joint, all fixings within a given length measured parallel to grain should be considered as occuring at that cross section.

With a member in tension there is no tendency to buckle, therefore the ratio of length to thickness is not critical, although some tension members, particularly in truss frameworks, may be subject to short term or very short term compression due to wind loading, and then the slenderness ratio λ must be limited to 250.

The capacity of a member taking axial tension only may be determined by the strength at end connections or that of any joint along its length.

13.2 WIDTH FACTOR

In general the grade tension stresses tabulated in BS 5268-2 apply to timbers having a width (i.e. the greater transverse dimension), h, of 300 mm to which a width factor K_{14} is applied to other widths. The exception to this requirement is solid timbers graded to North American MSR rules for which no adjustment for width is required.

At a width of 72 mm, $K_{14} = 1.17$ and for greater widths $K_{14} = (300/h)^{0.11}$.

13.3 EFFECTIVE CROSS SECTION

All holes for fixings at a joint will reduce the available area to resist tension in the member so the projected area of connections should be deducted from the gross area to give the effective area. Generally the most unfavourable section is the normal section but it may be neccesary to consider fixings on an oblique section according to the type of fastener.

For a given cross section the projected area of all fasteners that lie within a given distance of the section shall be considered as occurring at that cross section. The specified distances are:

Nails/Screws:	5 diameters for all nails and screws 5 mm diameter or more; no reduction in section area for nails/screws less than 5 mm diameter
Bolts and dowels:	2 bolt diameters
Toothed plate connectors:	0.75 times the nominal connector size
Split ring connectors:	0.75 times the nominal connector size
Shear plate connectors:	0.75 times the nominal connector size.

13.4 COMBINED BENDING AND TENSILE LOADING

With a tension member of uniform cross section also taking lateral loading (Fig. 13.1), the position of maximum stress occurs at the position of maximum bending moment. The sum of the tension and bending stress ratios must not exceed unity.

For bending about one axis:

$$\frac{\sigma_{t,a,par}}{\sigma_{t,adm,par}}+\frac{\sigma_{m,a,par}}{\sigma_{m,adm,par}}\leqslant 1.0$$

For bending about two axes:

$$\frac{\sigma_{t,a,par}}{\sigma_{t,adm,par}}+\frac{\sigma_{mx,a,par}}{\sigma_{mx,adm,par}}+\frac{\sigma_{my,a,par}}{\sigma_{my,adm,par}}\leqslant 1.0$$

These formulae can similarly be expressed in terms of the moment and tensile capacities:

$$\frac{M}{\overline{M}}+\frac{F_t}{\overline{T}}\leqslant 1.0 \quad \text{or} \quad \frac{M_x}{\overline{M}_x}+\frac{M_y}{\overline{M}_y}+\frac{F_t}{\overline{T}}\leqslant 1.0$$

where M = applied moment
T = applied tension
$\overline{M}$ = moment capacity
$\overline{T}$ = tension capacity.

Tension capacities for several solid sections of C16 grade are given in Table 13.1.

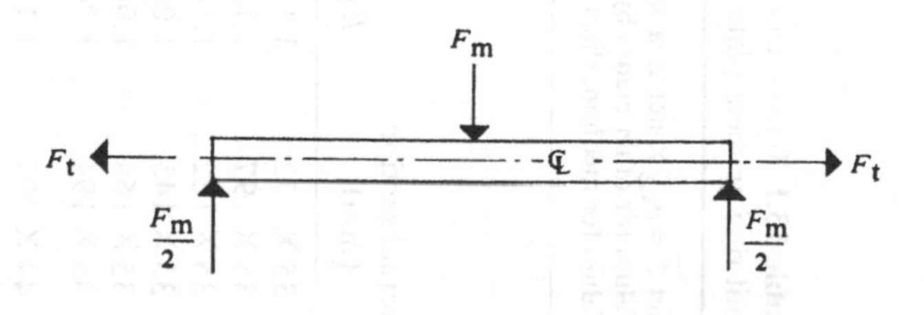

Fig. 13.1

Table 13.1 Tensile capacities for solid sections of C16 strength class, long-term loading for gross and net (effective) sections with deductions for split ring or shear plate connectors and their associated bolts

Net $\overline{T} = \sigma_{t,g} \times$ net area $\times K_{14} = 3.2 \times A_{Ne} \times K_{14}$ (long term × **C16** grade).
Values for other grades by proportion of $\sigma_{t,g}$.
Values for medium, short or very short term loading by proportion of K_3.

				Net tensile capacity $\overline{T}$ (kN)											
		Gross section		Split-rings						Shear-plates					
Actual section (mm)	K_{14}	Area ($mm^2 \times 10^3$)	$\overline{T}$ (kN)	64 mm diameter				102 mm diameter		67 mm diameter				102 mm diameter	
35 × 72	1.170	2.520	9.4	—	—	—	—	—	—	+	+	+	+	+	+
35 × 97	1.132	3.395	12.3	8.4	6.3	—	—	—	—	+	+	+	+	+	+
35 × 122	1.104	4.200	15.1	11.3	9.3	—	—	—	—	+	+	+	+	+	+
35 × 145	1.083	5.075	17.6	13.9	11.9	—	—	10.8	+	+	+	+	+	+	+
35 × 169	1.065	5.915	20.2	16.5	14.5	—	—	13.5	+	+	+	+	+	+	+
35 × 194	1.049	6.790	22.8	19.2	17.3	15.6	11.8	16.2	+	+	+	+	+	+	+
44 × 97	1.132	4.268	15.5	11.1	9.0	—	—	—	—	10.0	8.2	—	—	—	—
44 × 122	1.104	5.368	19.0	14.7	12.7	—	—	—	—	13.7	11.9	—	—	—	—
44 × 145	1.083	6.380	22.1	18.0	16.0	—	—	14.6	10.6	16.9	15.1	—	—	14.1	9.5
44 × 169	1.065	7.436	25.3	21.2	19.3	—	—	18.0	14.0	20.2	18.5	—	—	17.5	13.0
44 × 194	1.049	8.536	28.7	24.6	22.7	20.7	16.8	21.4	17.5	23.6	21.9	18.6	15.2	20.9	16.5
44 × 219	1.035	9.636	31.9	27.9	26.0	24.0	20.2	24.8	20.9	27.0	25.2	22.0	18.6	24.3	19.9

47 X 97	1.132	4.559	16.5	12.0	10.0	—	—	—	—	10.8	9.0	—	—	—	—
47 X 122	1.104	5.640	20.3	15.9	13.8	—	—	—	—	14.7	12.9	—	—	—	—
47 X 145	1.083	6.815	23.6	19.3	17.3	—	—	15.9	11.8	18.2	16.4	—	—	15.4	10.8
47 X 169	1.065	7.943	27.1	22.8	20.9	—	—	19.5	15.5	21.7	20.0	—	—	19.0	14.5
47 X 194	1.049	9.118	30.6	26.4	24.5	22.3	18.5	23.1	19.2	25.4	23.6	20.1	16.7	22.6	18.2
47 X 219	1.035	10.293	34.1	30.0	28.1	25.9	22.1	26.7	22.8	28.9	27.2	23.8	20.3	26.2	21.8
60 X 145	1.083	8.70	30.2	25.2	23.2	—	—	21.5	17.4	23.7	22.0	—	—	21.0	16.3
60 X 169	1.065	10.14	34.6	29.7	27.7	—	—	26.0	22.0	28.3	26.5	—	—	25.5	21.0
60 X 194	1.049	11.64	39.1	34.3	32.4	29.5	25.7	30.6	26.7	32.9	31.1	26.7	23.2	30.1	25.7
60 X 219	1.035	13.14	43.5	38.8	36.9	34.1	30.3	35.2	31.3	37.4	35.7	31.3	27.9	34.7	30.3
72 X 145	1.083	10.440	36.2	30.7	28.7	—	—	26.6	22.5	28.9	27.1	—	—	26.0	21.4
72 X 169	1.065	12.168	41.5	36.0	34.1	—	—	32.0	28.0	34.3	32.5	—	—	31.5	27.0
72 X 194	1.049	13.968	46.9	41.5	39.6	36.2	32.4	37.6	33.6	39.8	38.0	32.7	29.3	37.1	32.6
72 X 219	1.035	15.768	52.2	47.0	45.0	41.7	37.9	43.0	39.1	45.2	43.5	38.3	34.8	42.5	38.1
Connector arrangement		N_b		1	1	2	2	1	1	1	1	2	2	1	1
(see Fig. 13.4)		N_c		1	2	1	2	1	2	1	2	1	2	1	2

— Not permitted owing to insufficient depth of section.
+ Not permitted owing to insufficient breadth of section.

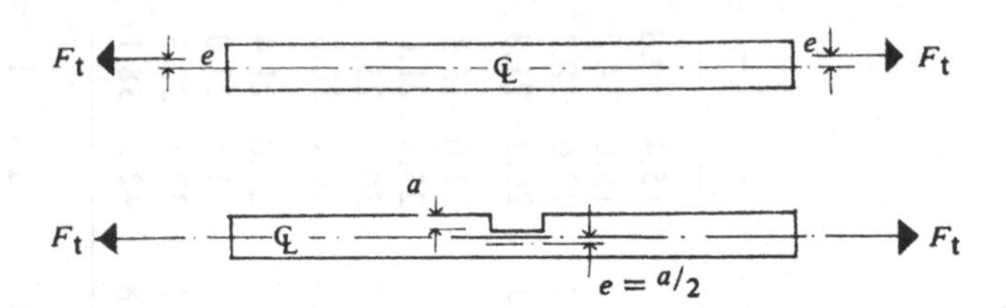

Fig. 13.2

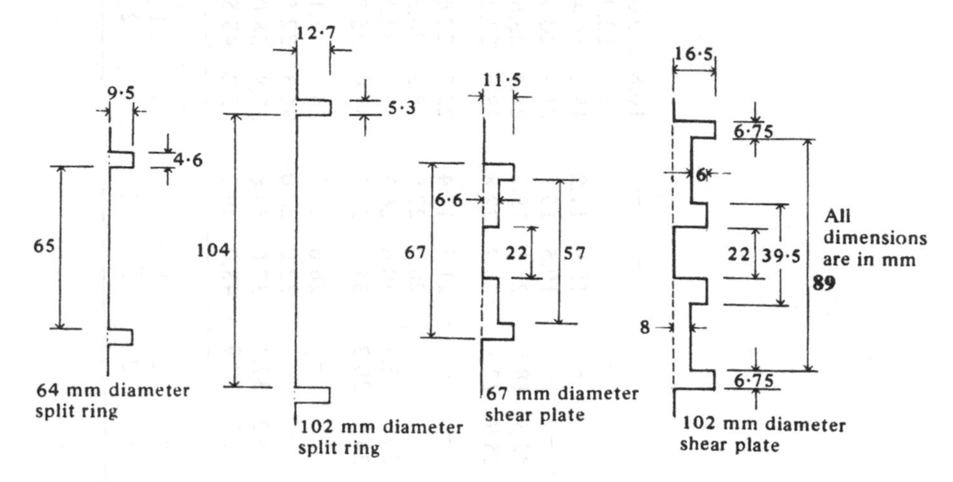

Fig. 13.3 Dimensions of grooves.

The eccentricity of an applied load at an end connection or the provision of a notch along the tie length will create a moment $M = F_t.e$ which may be additive to the external moment at that point (see Fig. 13.2).

13.4 TENSION CAPACITIES OF SOLID TIMBER SECTIONS CONTAINING SPLIT RING OR SHEAR PLATE CONNECTORS

To simplify the design of tension members, particularly in trusses, tensile capacities for solid timbers with gross and net (or effective) areas are given in Table 13.1. The effective area is calculated as the gross area less the projected area of the connector grooves (as required by clauses 6.8.2 and 6.9.2 of BS 5268-2) and the projected area of the bolt hole not contained within the projected area of the connector grooves. Bolt holes are taken as bolt diameter plus 2 mm.

Table 13.1 is prepared for C16 timber with a long-term grade stress of $3.2\,N/mm^2$. Capacities for the same sections in other grades or species can be calculated pro rata to the permissible grade tension stress. (The capacities cannot be pro-rated for timbers graded to North American MSR rules (see section 13.2).) Pro rata modifications for other strengh classes are:

Table 13.2

Type	Connector: Nominal dia. (mm)	Overall dia. (mm)	Depth of groove t_g (mm)	Projected area A_c (mm^2)	Bolt dia. (mm)	Bolt hole dia. d (mm)
Split-ring	64	74.2	9.5	705	12	14
	102	114.6	12.7	1455	20	22
Shear-plate	67	67	11.5	770	20	22
	102	102.5	16.5	1690	20	22

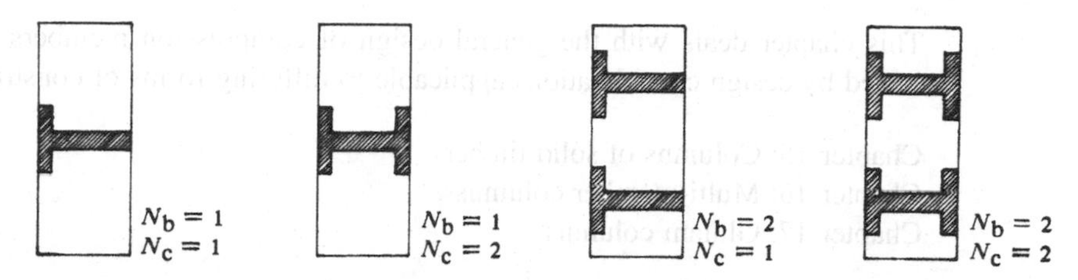

Fig. 13.4

C14	C16	C18	C22	C24	TR26	C27	C30
0.78	1.00	1.09	1.28	1.41	1.87	1.87	2.06

The dimensions of grooves which must be cut for split ring and shear plate connectors are shown in Fig. 13.3, and critical dimensions and areas are given in Table 13.2. From the values of A_c, t_g and d in this table, together with the values of N_b and N_c (Fig. 13.4), a general formula is presented for the effective area of the more commonly occurring connector arrangements on which Table 13.1 is based.

The possible connector arrangements for solid timber of up to 219 mm depth are shown in Fig. 13.4. N_b is the number of bolts and N_c is the number of connectors on each bolt.

For these arrangements the effective area *is:*

$$A_{N_e} = bh - N_b N_c A_c - N_b d(b - N_c t_g)$$

$$\text{Tensile capacity} = \sigma_{t,adm} \times K_3 \times K_{14} \times A_{N_e}$$

Chapter 14
General Design of Compression Members

14.1 RELATED CHAPTERS

This chapter deals with the general design of compression members and is followed by design considerations applicable to differing forms of construction, viz.

Chapter 15: Columns of solid timber
Chapter 16: Multi-member columns
Chapter 17: Glulam columns

Compression members in triangulated frameworks are dealt with in Chapter 21.

14.2 DESIGN CONSIDERATIONS

The principal considerations in the design of compression members are:

- axial stress
- positional restraint at ends
- directional restraint at ends (i.e. fixity)
- lateral restraint along length
- effective length and slenderness ratio
- deflected form
- bearing at each end.

The relevant permissible stresses are computed by modifying the grade stresses by the appropriate modification factors from BS 5268-2.

BS 5268-2 gives grade stresses for services classes 1 and 2 in Tables 8–15 with modification factor K_2 to adjust these stresses for strength class 3.

14.3 EFFECTIVE LENGTH

14.3.1 Introduction

The effective length is determined by:

- positional restraint at each end of the column (i.e. whether or not there is relative sway between the two ends)

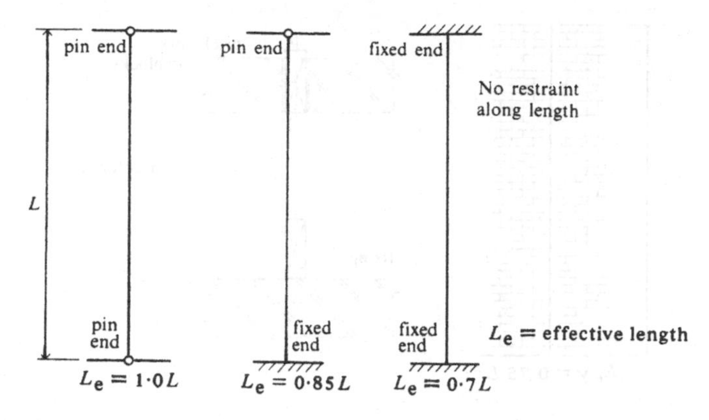

Fig. 14.1 Effective lengths x–x axis. No sway.

- directional restraint at ends (i.e. whether or not there is fixity at one or both ends)
- lateral restraint along the length.

Examples of effective lengths are sketched in Figs 14.1–14.3. It is emphasized, however, that there can be slight differences of opinion between engineers on deciding the effective length for actual cases.

14.3.2 Effective lengths. No sway

Examples of effective lengths about the x–x axis are sketched in Fig. 14.1 for cases in which there is no relative sway between the two ends of the column. Examples about the y–y axis are sketched in Fig. 14.2.

14.3.3 Effcctive lengths. Sway possible

Examples of effective lengths about the x–x axis are skctched in Fig. 14.3 for cases in which sway is possible.

14.3.4 Effective lengths. Points of contraflexure

In certain cases (e.g. in the frame sketched in Fig. 14.4), two points of contraflexure can occur in one member caused by bending about the x–x axis. The effective length, L_{ex}, of the member about the x–x axis can be taken as the distance between the points of contraflexure (BS 5286-2, clause 2.11.3). When considering buckling about the y–y axis due to bending about the x–x axis and axial loading the effective length L_{ey} is determined by the degree of lateral restraint.

In the case of the propped cantilever with sway sketched in Fig. 14.5, L_{ex} can be taken as $2h$, when considering part AC of the column and $2(h - h_c)$, for part CB.

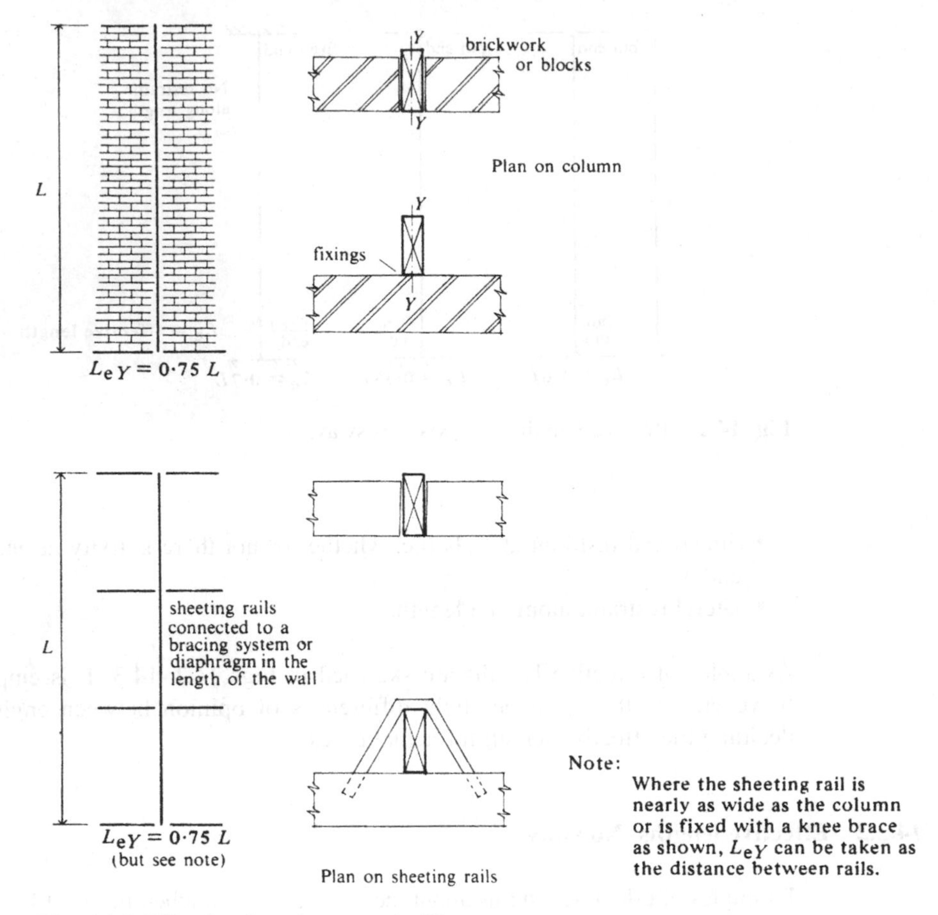

Fig. 14.2 Effective lengths *y–y* axis. No sway.

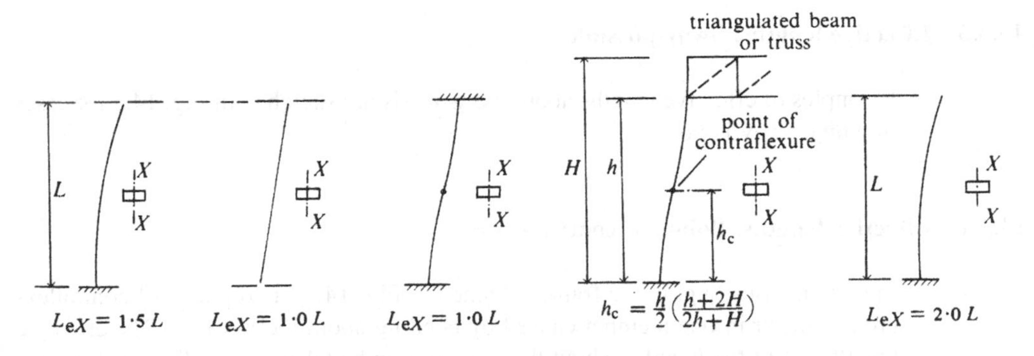

Fig. 14.3 Effective lengths *x–x* axis. Sway possible.

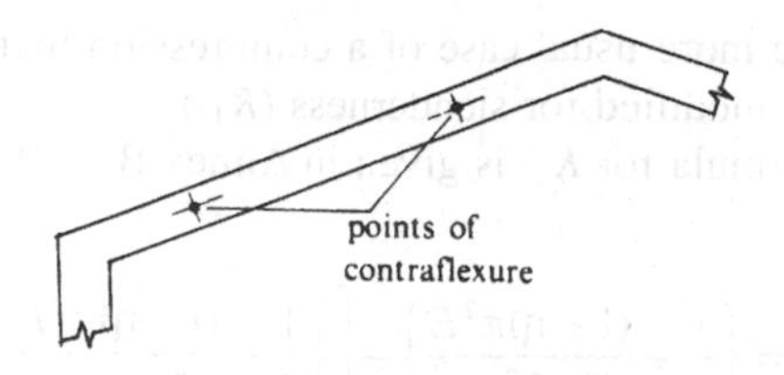

Fig. 14.4

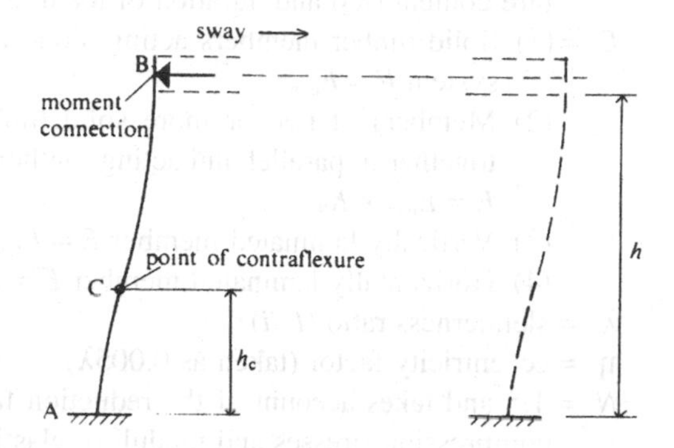

Fig. 14.5

The value of L_{ey} for axial loading and bending about the x–x axis is determined from the degree of lateral restraint.

14.4 PERMISSIBLE COMPRESSIVE STRESS

For any axis through the centre of gravity of a cross section, the radius of gyration may be regarded as a measure of the dispersal of the area about that axis. The radius of gyration, i, is defined as $i = \sqrt{I/A}$, where I is the second moment of area and A is the area.

For a member having the same effective length about both principal axes, buckling under axial compression occurs about the axis with the smaller radius of gyration, i.e. in the direction of the maximum slenderness ratio, where slenderness ratio = effective length/radius of gyration = $L_e/i = \lambda$.

In many practical cases members have differing L_e and i values about each principal axis and hence differing slenderness ratios. If one denotes the principal directions of buckling by the suffixes x and y, then the larger value of $(L_e/i)_x$ and $(L_e/i)_y$ will determine the permissible compression stress.

Timber is used mainly in the form of rectangular sections and as the relationship of radius of gyration (i) to breadth (b) is constant [$i = b/\sqrt{12}$] the slenderness ratio can also be expressed as $\lambda = (L_e\sqrt{12})/b$.

For a compression member with $\lambda \leq 5$ the permissible stress is taken as the grade stress modified as appropriate for moisture content (K_2), duration of loading (K_3) and load sharing (K_8).

For the more usual case of a compression member with $\lambda \geq 5$ the grade stress is further modified for slenderness (K_{12}).

The formula for K_{12} is given in Annex B of BS 5268-2 as:

$$K_{12} = \left\{\frac{1}{2} + \frac{(1+\eta)\pi^2 E}{2N\lambda^2\sigma_c}\right\} - \left[\left\{\frac{1}{2} + \frac{(1+\eta)\pi^2 E}{2N\lambda^2\sigma_c}\right\}^2 - \frac{\pi^2 E}{N\lambda^2\sigma_c}\right]^{\frac{1}{2}}$$

where σ_c = compression parallel to the grain grade stress modified only for moisture content (K_2) and duration of loading (K_3)

E = (1) Solid timber members acting alone or as part of a load-sharing system $E = E_{min}$
(2) Members of two or more solid timber components connected together in parallel and acting togther to support a common load $E = E_{min} \times K_9$
(3) Vertically laminated member $E = E_{min} \times K_{28}$
(4) Horizontally laminated member $E = E_{mean} \times K_{17}$

λ = slenderness ratio (L_e/i)

η = eccentricity factor (taken as 0.005λ)

N = 1.5 and takes account of the reduction factors used to derive grade compression stresses and moduli of elasticity.

Table 22 of BS 5268-2 gives values of K_{12} determined from the above formula.

14.5 MAXIMUM SLENDERNESS RATIO

BS 5268-2 limits slenderness ratio λ for a compression member to 180 unless it is a member normally subject to tension, or combined tension and bending arising from dead and imposed loads, but subject to a reversal of stress solely from the effects of wind, or a compression member such as wind bracing carrying self-weight and wind loads only, in which cases the slenderness ratio λ should not exceed 250.

14.6 COMBINED BENDING AND AXIAL LOADING

14.6.1 Introduction

A column is often subject to bending either about one or about both axes and the combined effect of the bending and axial loading must be considered.

It should be appreciated that, in considering two or more maximum stresses, they coincide at only one plane or one point in the section and the combined stresses do not therefore occur to the same extent over the whole section. This is an important consideration as the plane or point being considered may have different permissible stresses to other parts of the section.

Figure 14.6 illustrates how actual stresses may combine.

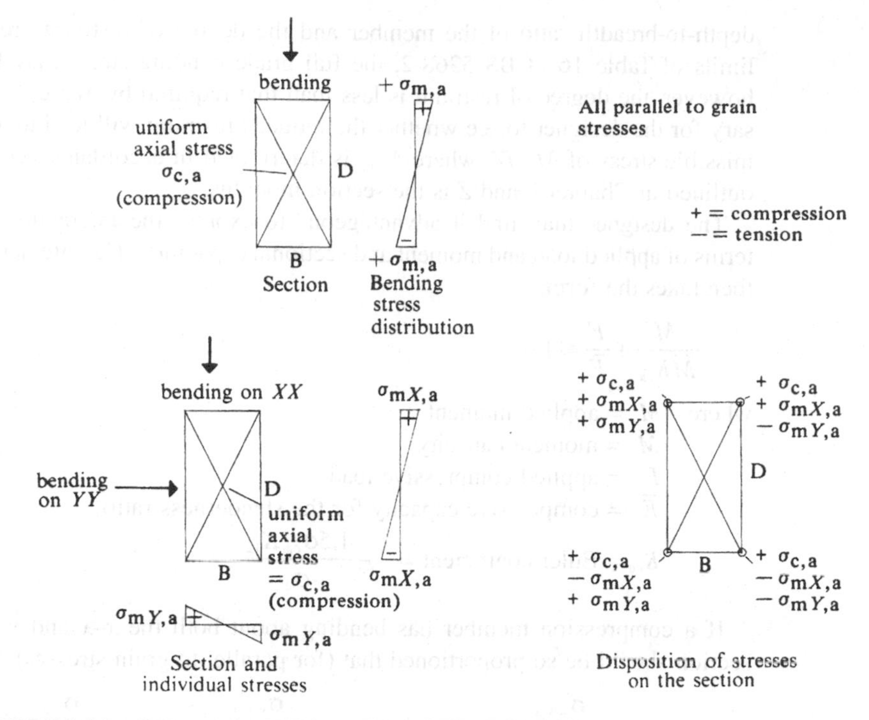

Fig. 14.6

14.6.2 Interaction formula

A compression member subjected also to bending about one axis should be so proportioned that the interaction quantity (for parallel to grain stresses):

$$\frac{\sigma_{m,a}}{\sigma_{m,adm}\left[1-\dfrac{1.5\sigma_{c,a}}{\sigma_e}K_{12}\right]}+\frac{\sigma_{c,a}}{\sigma_{c,adm}}\leqslant 1$$

where $\sigma_{m,a}$ = applied flexural stress parallel to the grain
$\sigma_{m,adm}$ = permissible flexural stress parallel to the grain
$\sigma_{c,a}$ = applied compressive stress
$\sigma_{c,adm}$ = permissible compressive stress
σ_e = Euler critical stress = $\pi^2E/(L_e/i)^2$ (utilizing the appropriate minimum or modified minimum value of E).

Both fractions are influenced by slenderness, the compression part by buckling under axial loading and the bending part by lateral buckling due to flexure. As a column begins to buckle under axial loading a bending moment occurs related to the axial loading and the eccentricity of buckling. This is considered in determining the eccentricity factor η (see formulae in section 14.4). To take account of this, clause 2.11.1 of BS 5268-2 requires a compression member to be straight as erected (e.g. not more than 1/300 of the length out of line).

The permissible bending stress is determined from a consideration of the degree of lateral restraint to the compression edge of the member acting as a beam. If the depth-to-breadth ratio of the member and the degree of restraint are within the limits of Table 16 of BS 5268-2, the full grade bending stress may be used. If, however, the degree of restraint is less than that required by Table 19 it is necessary for the designer to see whether the reduced restraint will lead to a lower permissible stress of M_{buc}/Z, where M_{buc} is determined in accordance with a method outlined in Chapter 9 and Z is the section modulus.

The designer may find it advantageous to express the interaction formula in terms of applied load and moment and sectional capacities. The interaction formula then takes the form:

$$\frac{M}{\overline{M}K_{eu}} + \frac{F}{\overline{F}} \leqslant 1.0$$

where M = applied moment
$\overline{M}$ = moment capacity
F = applied compressive load
$\overline{F}$ = compressive capacity for the slenderness ratio

$$K_{eu} = \text{Euler coefficient} = 1 - \frac{1.5\sigma_{c,a}K_{12}}{\sigma_e}$$

If a compression member has bending about both the x–x and y–y axes, the section should be so proportioned that (for parallel to grain stresses):

$$\frac{\sigma_{mx,a}}{\sigma_{mx,adm}\left(1 - \dfrac{1.5\sigma_{c,a} \times K_{12}}{\sigma_e}\right)} + \frac{\sigma_{my,a}}{\sigma_{my,adm}\left(1 - \dfrac{1.5\sigma_{c,a} \times K_{12}}{\sigma_e}\right)} + \frac{\sigma_{c,a}}{\sigma_{c,adm}} \leqslant 1.0$$

where $\sigma_e = \pi^2 E/\lambda^2$.

The assumption in writing the formula above with plus signs is that there is a point in the section where all three compressive stresses can occur simulaneously. If not, algebraic addition of compression and bending stress ratios applies and the designer is advised to check that the chosen point is where the combination of stress ratios is maximum. This need not necessarily be where the value of one of the applied stresses appears to dominate the case. In the majority of cases with columns of solid timber, the values of $\sigma_{mx,adm}$ and $\sigma_{my,adm}$ will be the same but, in the case of glulam, they are likely to differ (see Chapter 7).

14.7 EFFECTIVE AREA FOR COMPRESSION

For the purpose of calculating the actual compressive stress, open holes and notches must be deducted from the area.

Clause 2.11.7 of BS 5268-2 notes that the effect of circular holes with a diameter not exceeding 25% of the section width need not be calculated providing the hole is positioned on the neutral axis within a distance of 0.25 and 0.4 of the actual length from the end or from a support.

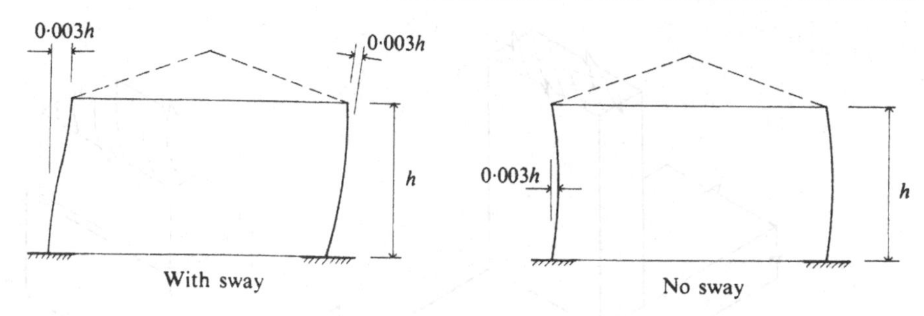

Fig. 14.7

14.8 DEFLECTION AND SWAY OF COLUMNS

Where a column is subject to lateral loading over the whole or part of its height the column deflects laterally. No limit for this deflection is given in BS 5368-2 but it seems logical to adopt a value similar to that for flexural members of 0.003 × height.

Where the lateral deflection is caused by wind forces it should be appreciated that this is a gust force with a probability of occurring only one in say 50 years. Composite action with cladding and partial end fixity will probably reduce the deflection so that a column will not deflect to its full calculated amount during a short gust period. Deflection criteria should be regarded as a guide rather than an absolute limit.

When a building or frame can sway laterally a limit should be set on the sway. No limit for this deflection is given in BS 5368-2 but it seems logical to adopt a value of, say, 0.003 of height (Fig. 14.7). When designing wind girders to restrain the head of a column or stud wall system a reduced deflection must be set for the wind girder deflection compatible with the allowable head delection rather than 0.003 of the wind girder span.

14.9 BEARING AT BASES

The permissible grade bearing stress at the base and top of columns is the value of $\sigma_{c,par}$ for $\lambda = 0$. However, if the column bears on a cross piece of timber (Fig. 14.8) the permissible grade bearing stress on the bearing area is limited to the permissible grade bearing stress perperpendicular to grain. If wane is excluded, the stress relevant to the full area can be taken (section 4.10.1) modified by bearing factor K_4 when appropriate.

The detail at a column base often takes the form of a steel shoe bearing on concrete (Fig. 14.9). This type of base plate can be designed as a hinge or to give fixity.

If, for any reason (e.g. tolerance of manufacture), it is felt that the fit of the timber to the steel plate or shoe will not be sufficiently accurate, then an epoxy resin–sand mix can be used to obtain a tight fit or bearing.

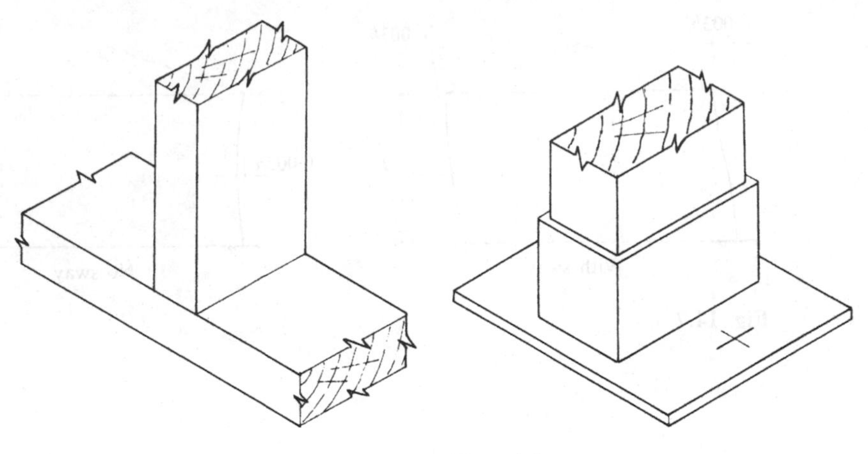

Fig. 14.8

Fig. 14.9

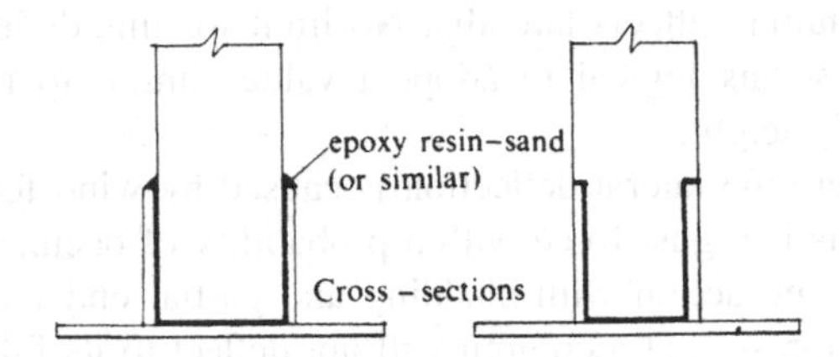

Fig. 14.10

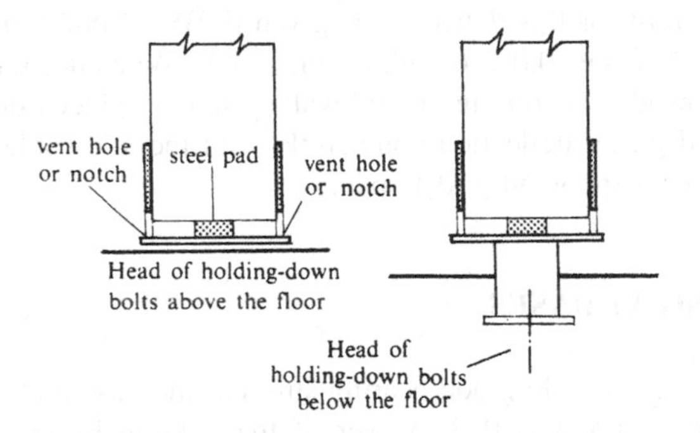

Fig. 14.11

If, for example in a humid service condition, it is felt that a means should be provided to prevent moisture being trapped between the timber and the steel shoe, consider making the inside of the steel shoe slightly oversize and introduce an epoxy resin–sand mix as shown in Fig. 14.10.

Alternatively one can assume that moisture will intrude and allow it to drain or be ventilated away as shown in Fig. 14.11. In doing so the bearing area of timber to steel is considerably reduced but sufficient area can usually be provided.

14.10 BEARING AT AN ANGLE TO GRAIN

With the bearing surface inclined to the grain, the permissible compressive stress for the inclined surface is given by:

$$\sigma_{c,adm,\alpha} = \sigma_{c,adm,0} - (\sigma_{c,adm,0} - \sigma_{c,adm,90}) \sin \alpha$$

where α = the angle between the load and the direction of grain
$\sigma_{c,adm,0}$ = permissible compression stress parallel to grain
$\sigma_{c,adm,90}$ = permissible compressive stress perpendicular to grain.

Load duration factors apply.

For bearing at 90° to grain, load-sharing factor K_9 is permitted when applicable. For bearing at 0° to grain, no load sharing is permitted (see section 4.4.1)

Chapter 15

Columns of Solid Timber

15.1 INTRODUCTION

Chapter 14 details the various general factors which must be taken into account and considers the factors involved in determining the slenderness factor K_{12}.

Values of the grade compressive parallel to grain stress for strength classes are given in BS 5268-2 Tables 8 and 9 and for species/grades in Tables 10–15.

BS 5268-2 Tables 2–5 list the grades/species that satisfy the requirements of each of the strength classes with over 80% of all grades/species being allocated to strength classes C14, C16 or C24.

Calculating K_{12} values from Appendix B of BS 5268-2 or selecting a K_{12} value from Table 19, which requires interpolation in two directions, can be time consuming. Tables 15.1 to 15.4 are therefore provided to simplify the design process for strength classes C14, C16 and C24 for service classes 1, 2 and 3.

15.2 DESIGN EXAMPLE

The column shown in Fig. 15.1 is supporting a medium-term axial compression of 6 kN and a very short duration bending moment of 2.05 kN m caused by a wind load of 4.00 kN on the x–x axis. The wind can either cause pressure or suction on the wall. Check that the chosen section is adequate. The top and bottom of the column are restrained in position but not fixed in direction. The column is restrained on its weak axis by rails at 1.4 m centres. Assume service class 1 and no load sharing. Check suitability of a 60 × 194 mm C16 section.

Medium term loading (no wind)

For axial loading:

$$\lambda_x = \frac{4200\sqrt{12}}{194} = 75$$

$$\lambda_y = \frac{1400\sqrt{12}}{60} = 81$$

From Table 15.3 (with $\lambda = 81$), $K_{12} = 0.454$, so that

Table 15.1 Determination of E/σ_c for service classes 1 and 2 and 3

Service classes 1 and 2

		Long		Medium		Short		Very short	
	E	σ_c	E/σ_c	σ_c	E/σ_c	σ_c	E/σ_c	σ_c	E/σ_c
C14	4600	5.2	885	6.50	708	7.80	590	9.10	505
C16	5800	6.8	853	8.50	682	10.20	569	11.90	487
C18	6000	7.1	845	8.88	676	10.65	563	12.43	483
C22	6500	7.5	867	9.38	693	11.25	578	13.13	495
C24	7200	7.9	911	9.88	729	11.85	608	13.83	521
TR26	7400	8.2	902	10.25	722	12.30	602	14.35	516
C27	8200	8.2	1000	10.25	800	12.30	667	14.35	571

Service class 3

		Long		Medium		Short		Very short	
	E	σ_c	E/σ_c	σ_c	E/σ_c	σ_c	E/σ_c	σ_c	E/σ_c
C14	3680	3.12	1179	3.90	944	4.68	786	5.46	674
C16	4640	4.08	1137	5.10	910	6.12	758	7.14	650
C18	4800	4.26	1127	5.33	901	6.39	751	7.46	644
C22	5200	4.5	1156	5.63	924	6.75	770	7.88	660
C24	5760	4.74	1215	5.93	972	7.11	810	8.30	694
TR26	5920	4.92	1203	6.15	963	7.38	802	8.61	688
C27	6560	4.92	1333	6.15	1067	7.38	889	8.61	762

$$\sigma_{c,adm} = \sigma_{c,g} \times K_3 \times K_{12} = 6.8 \times 1.25 \times 0.454 = 3.86\ \text{N/mm}^2$$

$$\sigma_{c,a} = \frac{6000}{60 \times 194} = 0.52\ \text{N/mm}^2$$

Section satisfactory for medium-term condition.

Very short term loading

(1) Axial loading: From Table 15.3 (with $\lambda = 81$), $K_{12} = 0.372$, so that

$$\sigma_{c,adm} = \sigma_{c,g} \times K_3 \times K_{12} = 6.8 \times 1.75 \times 0.372 = 4.43\ \text{N/mm}^2$$

$$\sigma_{c,a} = \frac{6000}{60 \times 194} = 0.52\ \text{N/mm}^2$$

(2) Bending: Depth-to-breadth ratio = h/b = 194/60 = 3.2. From Table 19 of BS 5268-2 minimum degree of lateral support is *'Ends held in position and member held in line as by purlins or tie rods at centres not more than 30 times breadth of member'*.

With the member held in line by rails at 1.4 m centres (30 × b = 1800 mm) the full grade bending stress is permitted.

Table 15.2 K_{12} values for C14 timber

	Service classes 1 and 2 ($E/\sigma_{c,par}$)				Service class 3 ($E/\sigma_{c,par}$)			
λ	Long 885	Medium 708	Short 590	V. short 505	Long 1179	Medium 944	Short 786	V. short 674
5	0.976	0.975	0.975	0.975	0.976	0.976	0.975	0.975
10	0.952	0.951	0.951	0.951	0.952	0.952	0.952	0.951
15	0.928	0.927	0.927	0.926	0.928	0.928	0.928	0.927
20	0.904	0.902	0.901	0.899	0.905	0.904	0.903	0.902
25	0.879	0.876	0.873	0.870	0.881	0.879	0.877	0.875
30	0.853	0.848	0.843	0.838	0.857	0.854	0.850	0.847
35	0.825	0.818	0.810	0.801	0.832	0.827	0.822	0.816
40	0.796	0.785	0.773	0.760	0.807	0.799	0.791	0.782
45	0.765	0.750	0.733	0.715	0.780	0.769	0.758	0.746
50	0.733	0.712	0.690	0.666	0.752	0.738	0.722	0.706
55	0.699	0.672	0.644	0.615	0.723	0.705	0.685	0.665
60	0.663	0.631	0.598	0.564	0.693	0.671	0.647	0.622
65	0.627	0.589	0.552	0.515	0.663	0.636	0.608	0.580
70	0.590	0.548	0.507	0.469	0.632	0.601	0.569	0.537
75	0.554	0.508	0.465	0.426	0.601	0.565	0.530	0.497
80	0.518	0.470	0.426	0.388	0.570	0.531	0.493	0.458
85	0.484	0.434	0.391	0.353	0.539	0.498	0.458	0.423
90	0.452	0.402	0.359	0.322	0.510	0.466	0.426	0.390
95	0.422	0.371	0.330	0.295	0.481	0.436	0.395	0.360
100	0.393	0.344	0.304	0.271	0.454	0.408	0.367	0.333
105	0.367	0.319	0.280	0.249	0.428	0.381	0.341	0.308
110	0.343	0.296	0.259	0.230	0.403	0.357	0.318	0.286
115	0.321	0.275	0.240	0.212	0.380	0.334	0.296	0.266
120	0.300	0.257	0.223	0.197	0.358	0.313	0.277	0.247
125	0.281	0.239	0.208	0.183	0.338	0.294	0.259	0.231
130	0.264	0.224	0.194	0.170	0.319	0.276	0.242	0.216
135	0.248	0.210	0.181	0.159	0.302	0.260	0.227	0.202
140	0.234	0.197	0.170	0.149	0.285	0.245	0.214	0.189
145	0.220	0.185	0.159	0.139	0.270	0.231	0.201	0.178
150	0.208	0.174	0.150	0.131	0.256	0.218	0.190	0.167
155	0.196	0.164	0.141	0.123	0.243	0.206	0.179	0.158
160	0.186	0.155	0.133	0.116	0.230	0.195	0.169	0.149
165	0.176	0.147	0.126	0.110	0.219	0.185	0.160	0.141
170	0.167	0.139	0.119	0.104	0.208	0.176	0.152	0.133
175	0.159	0.132	0.113	0.098	0.198	0.167	0.144	0.126
180	0.151	0.125	0.107	0.093	0.189	0.159	0.137	0.120
185			0.102	0.088			0.130	0.114
190			0.097	0.084			0.124	0.109
195			0.092	0.080			0.118	0.104
200			0.088	0.076			0.113	0.099
205			0.084	0.073			0.108	0.094
210			0.080	0.069			0.103	0.090
215			0.077	0.066			0.099	0.086
220			0.073	0.064			0.095	0.083
225			0.070	0.061			0.091	0.079
230			0.067	0.058			0.087	0.076
235			0.065	0.056			0.084	0.073
240			0.062	0.054			0.080	0.070
245			0.060	0.052			0.077	0.067
250			0.057	0.050			0.074	0.065

Table 15.3 K_{12} values for C16 timber

	Service classes 1 and 2 (E/σ_c)				Service class 3 (E/σ_c)			
λ	Long 853	Medium 682	Short 569	V. short 487	Long 1137	Medium 910	Short 758	V. short 650
5	0.976	0.975	0.975	0.975	0.976	0.976	0.975	0.975
10	0.952	0.951	0.951	0.951	0.952	0.952	0.951	0.951
15	0.928	0.927	0.926	0.926	0.928	0.928	0.927	0.927
20	0.903	0.902	0.900	0.899	0.905	0.904	0.903	0.902
25	0.878	0.875	0.872	0.869	0.881	0.879	0.877	0.875
30	0.852	0.847	0.842	0.836	0.857	0.853	0.850	0.846
35	0.824	0.816	0.808	0.799	0.832	0.826	0.820	0.814
40	0.795	0.783	0.770	0.757	0.805	0.797	0.789	0.780
45	0.763	0.747	0.729	0.710	0.778	0.767	0.755	0.742
50	0.730	0.708	0.685	0.660	0.750	0.735	0.719	0.702
55	0.695	0.667	0.638	0.608	0.720	0.701	0.681	0.660
60	0.658	0.625	0.590	0.556	0.690	0.666	0.642	0.616
65	0.621	0.582	0.543	0.506	0.659	0.631	0.601	0.572
70	0.584	0.540	0.498	0.459	0.627	0.595	0.562	0.529
75	0.547	0.499	0.456	0.417	0.596	0.559	0.523	0.488
80	0.511	0.461	0.417	0.379	0.564	0.524	0.485	0.450
85	0.476	0.426	0.382	0.344	0.533	0.490	0.450	0.414
90	0.444	0.393	0.350	0.314	0.503	0.458	0.417	0.381
95	0.414	0.363	0.321	0.287	0.474	0.428	0.387	0.352
100	0.385	0.335	0.296	0.263	0.446	0.400	0.359	0.325
105	0.359	0.311	0.273	0.242	0.420	0.373	0.334	0.300
110	0.335	0.288	0.252	0.223	0.396	0.349	0.310	0.278
115	0.313	0.268	0.234	0.206	0.373	0.327	0.289	0.259
120	0.293	0.249	0.217	0.191	0.351	0.306	0.270	0.241
125	0.274	0.233	0.202	0.177	0.331	0.287	0.252	0.224
130	0.257	0.218	0.188	0.165	0.312	0.269	0.236	0.209
135	0.242	0.204	0.176	0.154	0.295	0.253	0.221	0.196
140	0.227	0.191	0.165	0.144	0.279	0.239	0.208	0.184
145	0.214	0.180	0.154	0.135	0.264	0.225	0.195	0.173
150	0.202	0.169	0.145	0.127	0.250	0.212	0.184	0.162
155	0.191	0.159	0.137	0.119	0.237	0.201	0.174	0.153
160	0.181	0.150	0.129	0.112	0.225	0.190	0.164	0.144
165	0.171	0.142	0.122	0.106	0.213	0.180	0.155	0.137
170	0.162	0.135	0.115	0.100	0.203	0.171	0.147	0.129
175	0.154	0.128	0.109	0.095	0.193	0.162	0.140	0.123
180	0.146	0.121	0.104	0.090	0.184	0.154	0.133	0.116
185			0.098	0.086			0.126	0.111
190			0.094	0.081			0.120	0.105
195			0.089	0.077			0.115	0.100
200			0.085	0.074			0.109	0.096
205			0.081	0.070			0.104	0.091
210			0.077	0.067			0.100	0.087
215			0.074	0.064			0.096	0.083
220			0.071	0.061			0.092	0.080
225			0.068	0.059			0.088	0.077
230			0.065	0.056			0.084	0.073
235			0.062	0.054			0.081	0.071
240			0.060	0.052			0.078	0.068
245			0.058	0.050			0.075	0.065
250			0.056	0.048			0.072	0.063

Table 15.4 K_{12} values for C24 timber

	Service classes 1 and 2 (E/σ_c)				Service class 3 (E/σ_c)			
λ	Long 911	Medium 729	Short 608	V. short 521	Long 1215	Medium 972	Short 810	V. short 694
5	0.976	0.975	0.975	0.975	0.976	0.976	0.976	0.975
10	0.952	0.951	0.951	0.951	0.952	0.952	0.952	0.951
15	0.928	0.927	0.927	0.926	0.928	0.928	0.928	0.927
20	0.904	0.902	0.901	0.900	0.905	0.904	0.903	0.902
25	0.879	0.876	0.874	0.871	0.882	0.880	0.878	0.876
30	0.853	0.849	0.844	0.839	0.858	0.854	0.851	0.847
35	0.826	0.819	0.811	0.803	0.833	0.828	0.823	0.817
40	0.797	0.787	0.775	0.763	0.807	0.800	0.792	0.784
45	0.767	0.752	0.736	0.719	0.781	0.771	0.760	0.748
50	0.735	0.715	0.694	0.671	0.754	0.740	0.725	0.710
55	0.701	0.676	0.649	0.622	0.725	0.708	0.689	0.669
60	0.667	0.636	0.604	0.571	0.696	0.674	0.651	0.627
65	0.631	0.594	0.558	0.523	0.666	0.640	0.613	0.585
70	0.595	0.554	0.514	0.477	0.636	0.605	0.574	0.544
75	0.559	0.514	0.472	0.434	0.605	0.571	0.536	0.503
80	0.524	0.477	0.434	0.395	0.575	0.537	0.500	0.465
85	0.490	0.441	0.398	0.361	0.544	0.504	0.465	0.430
90	0.458	0.408	0.366	0.330	0.515	0.472	0.432	0.397
95	0.428	0.378	0.336	0.302	0.487	0.442	0.402	0.367
100	0.400	0.350	0.310	0.277	0.460	0.414	0.374	0.339
105	0.374	0.325	0.286	0.255	0.434	0.387	0.348	0.314
110	0.349	0.302	0.265	0.235	0.409	0.363	0.324	0.292
115	0.327	0.281	0.246	0.218	0.386	0.340	0.303	0.271
120	0.306	0.262	0.228	0.202	0.364	0.319	0.283	0.253
125	0.287	0.245	0.213	0.188	0.344	0.300	0.264	0.236
130	0.270	0.229	0.199	0.175	0.325	0.282	0.248	0.221
135	0.254	0.215	0.186	0.163	0.307	0.265	0.233	0.207
140	0.239	0.202	0.174	0.153	0.291	0.250	0.219	0.194
145	0.225	0.189	0.163	0.143	0.275	0.236	0.206	0.182
150	0.212	0.178	0.154	0.135	0.261	0.223	0.194	0.171
155	0.201	0.168	0.145	0.127	0.248	0.211	0.183	0.162
160	0.190	0.159	0.137	0.119	0.235	0.200	0.173	0.153
165	0.180	0.150	0.129	0.113	0.224	0.190	0.164	0.144
170	0.171	0.143	0.122	0.107	0.213	0.180	0.156	0.137
175	0.163	0.135	0.116	0.101	0.203	0.171	0.148	0.130
180	0.155	0.128	0.110	0.096	0.193	0.163	0.140	0.123
185			0.104	0.091			0.134	0.117
190			0.099	0.086			0.127	0.111
195			0.095	0.082			0.121	0.106
200			0.090	0.078			0.116	0.101
205			0.086	0.075			0.111	0.097
210			0.082	0.071			0.106	0.092
215			0.079	0.068			0.101	0.088
220			0.075	0.065			0.097	0.085
225			0.072	0.063			0.093	0.081
230			0.069	0.060			0.089	0.078
235			0.066	0.058			0.086	0.075
240			0.064	0.055			0.083	0.072
245			0.061	0.053			0.079	0.069
250			0.059	0.051			0.076	0.067

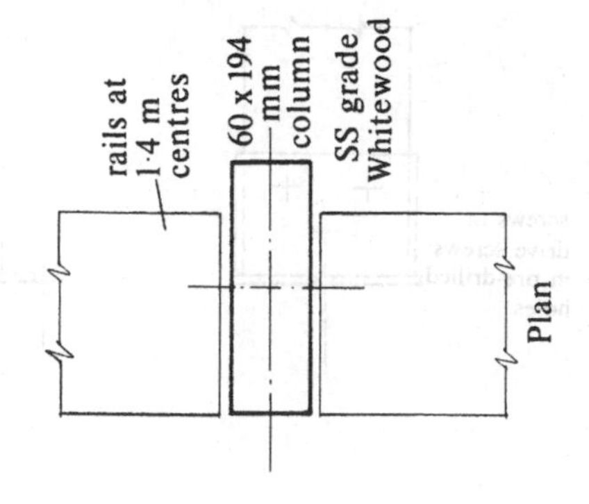
rails at
1·4 m
centres
60 x 194
mm
column
SS grade
Whitewood
Plan

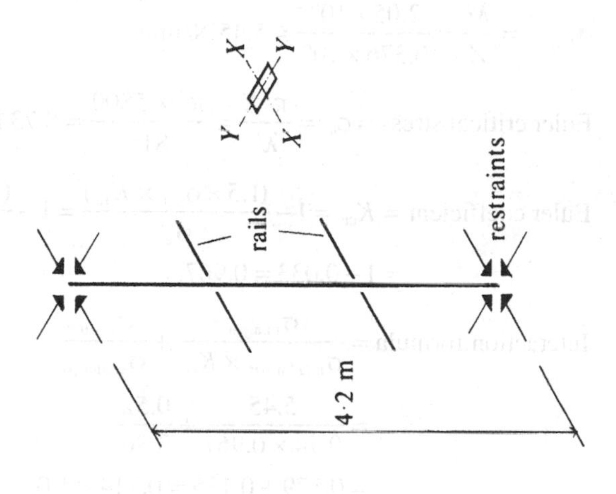
X
Y
Y
X
restraints
rails
4·2 m

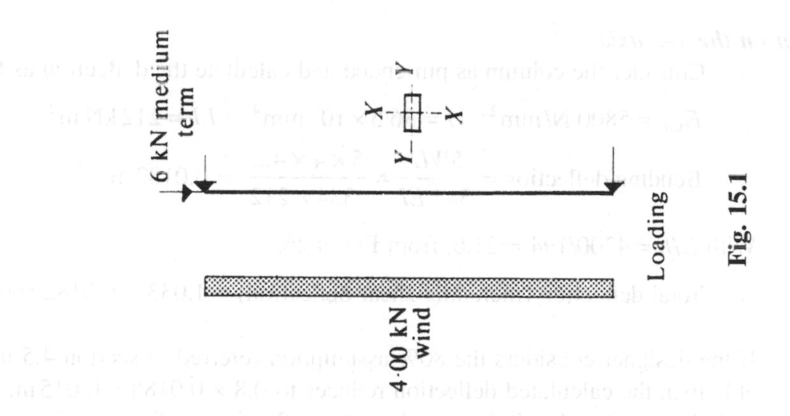
6 kN medium
term
Y
X
Y
X
4·00 kN
wind
Loading

Fig. 15.1

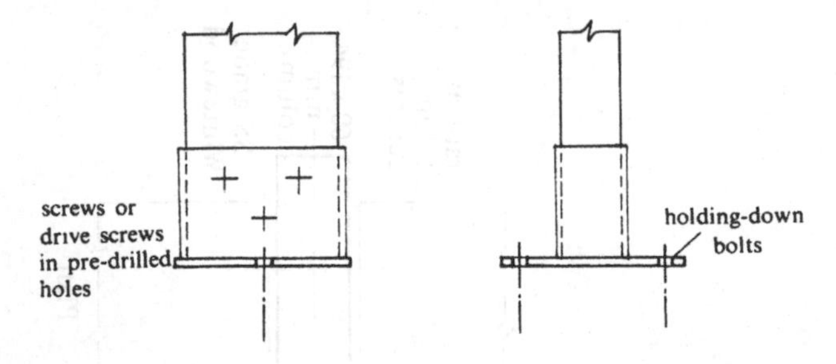

Fig. 15.2

$$\sigma_{m,adm} = \sigma_{m,g} \times K_3 \times K_7 = 5.3 \times 1.75 \times 1.05 = 9.74 \text{ N/mm}^2$$

$$\sigma_{m,a} = \frac{M}{Z} = \frac{2.05 \times 10^6}{0.376 \times 10^6} = 5.45 \text{ N/mm}^2$$

$$\text{Euler critical stress} = \sigma_e = \frac{\pi^2 E}{\lambda^2} = \frac{\pi^2 \times 5800}{81^2} = 8.73 \text{ N/mm}^2$$

$$\text{Euler coefficient} = K_{eu} = 1 - \frac{(1.5 \times \sigma_{c,a} \times K_{12})}{\sigma_e} = 1 - \frac{(1.5 \times 0.52 \times 0.372)}{8.73}$$

$$= 1 - 0.033 = 0.967$$

$$\text{Interaction formula} = \frac{\sigma_{m,a,par}}{\sigma_{m,adm,par} \times K_{eu}} + \frac{\sigma_{c,a,par}}{\sigma_{c,adm,par}}$$

$$= \frac{5.45}{9.74 \times 0.967} + \frac{0.52}{3.86}$$

$$= 0.579 + 0.135 = 0.714 < 1.0, \text{ satisfactory.}$$

Deflection on the x–x axis

Consider the column as pin-ended and calculate the deflection as for a beam.

$$E_{min} = 5800 \text{ N/mm}^2, \quad I = 36.5 \times 10^6 \text{ mm}^4, \quad EI = 212 \text{ kN m}^2$$

$$\text{Bending deflection} = \frac{5WL^3}{384\,EI} = \frac{5 \times 4 \times 4.2^3}{384 \times 212} = 0.0182 \text{ m}$$

With $L/h = 4200/194 = 21.6$, from Fig. 4.26,

Total deflection (including shear deflection) = 1.033 × 0.0182 = 0.0188 m

If the designer considers the 80% assumption referred to section 4.5 to be justifiable then the calculated deflection reduces to 0.8 × 0.0188 = 0.015 m.

A degree of end fixity may reduce the deflection as discussed in section 15.3.

End bearing

Actual bearing stress at the base (Fig. 15.2) = 6000/(60 × 194) = 0.515 N/mm^2. Permissible medium-term bearing stress (parallel to grain) = 6.8 × 1.25 = 8.5 N/mm^2 from which it can be seen that bearing with this type of detail is not

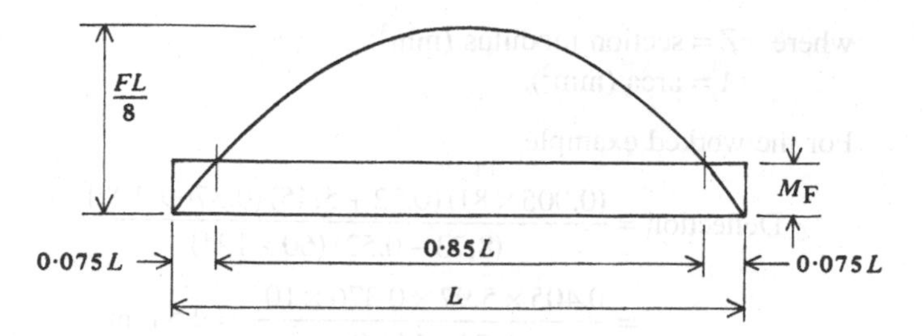

Fig. 15.3

likely to be critical on any column, particularly as the slenderness ratio increases, unless the end is severely reduced to suit a fixing detail.

15.3 DEFLECTION OF COMPRESSION MEMBERS

Although in calculating λ_x the effective length was taken as 1.0 × actual length, partial end fixing of connections can usually justify the effective length to be taken as 0.85 × actual length. BS 5268-6.1 recommends an effective length of 0.85*L* for wall studs restrained by sheathing and plasterboard.

This partial fixity has an effect on the deflection of the column. By setting the distance between points of contraflexure as 0.85*L* the effect on bending deflection can be calculated (see Fig. 15.3) by working back from the BM diagram.

From the properties of a parabola, the fixing moment is

$$M_F = (FL/8) \times \frac{0.075L \times 0.925L}{0.5L \times 0.5L} = 0.0347FL$$

where F = total UDL and L = length.

$$\text{Deflection at mid-length} = \frac{5FL^3}{384EI} - \frac{M_F L^2}{8EI}$$

$$= \frac{(0.01302 - 0.00434)FL^3}{EI}$$

$$= \frac{0.00868FL^3}{EI}$$

which is two-thirds of the deflection for the simply supported case.

Therefore, if one takes account of this small amount of partial directional restraint and the 0.8 times factor, the actual bending deflection can be shown to be 0.8 × 66.6% = 53% of the 'full' calculated value. In the worked example the deflection would then be 0.0188 × 0.53 = 0.010 m.

As a comparison, TRADA Publication TBL52 gives the following formula for the approximate deflection of studs when subject to both compression and bending forces:

$$\text{Deflection} = \frac{0.005\lambda(\sigma_{c,a} + \sigma_{m,a})}{\sigma_e - \sigma_{c,a}} \times \frac{Z}{A}$$

where Z = section modulus (mm^3)
A = area (mm^2).

For the worked example,

$$\text{Deflection} = \frac{(0.005 \times 81)(0.52 + 5.45)(0.376 \times 10^6)}{(8.73 - 0.52)(60 \times 194)}$$

$$= \frac{0.405 \times 5.97 \times 0.376 \times 10^6}{8.21 \times 11640} = 9.5 \text{ mm}$$

Chapter 16
Multi-member Columns

16.1 INTRODUCTION

Multi-member columns are defined as columns consisting of more than one member which do not classify as glulam. The examples in this chapter are limited to components of two or three members.

Obviously it is possible to nail, screw or bolt two or more pieces of timber together to form a column and such a column will have considerably more strength than the sum of the strength of the two or more pieces acting alone. However, there is some doubt as to whether or not such mechanically jointed columns provide full composite action. This depends on the strength, spacing and characteristics of the fastenings and the method of applying the loading; therefore this chapter deals only with glued composites other than those covered in the section on spaced columns. If a designer wishes to use a mechanically jointed member (other than a spaced column for which design rules are detailed in BS 5268-2 and section 16.4) it is suggested that testing is undertaken or the design errs on the side of safety in calculating the permissible axial stresses. (See notes on composite action in section 6.1.) Typical glued columns of two or three members are sketched in Fig. 16.1.

When designing the columns shown in Fig. 16.1 see the notes in section 14.4 for guidance on the E value to use in calculations.

16.2 COMBINED BENDING AND AXIAL LOADING FOR TEE SECTIONS

The method of combining bending and axial compression stresses is described in section 14.6 but there is one additional point to note with tee sections. The section is not symmetrical about the x–x axis, and therefore bending about the x–x axis leads to a different stress in each of the extreme fibres, the maxinium stress being at the end of the stalk. This maximum may be a tensile bending stress, in which case the critical compression in the section will be the combination of axial compressive stress plus the bending compressive stress on the extreme fibre of the table. The axial compressive stress is uniform throughout the section.

The shape of a tee section is such that with compressive bending stress on the table, the full permissible bending stress will be realized before any lateral instability takes place. For cases with the stalk of the tee in compressive bending, a conservative estimate for lateral stability is to ensure that the proportion of the stalk member on its own complies with the requirements of section 4.6.1. Properties of a range of tee sections are given in Table 16.1.

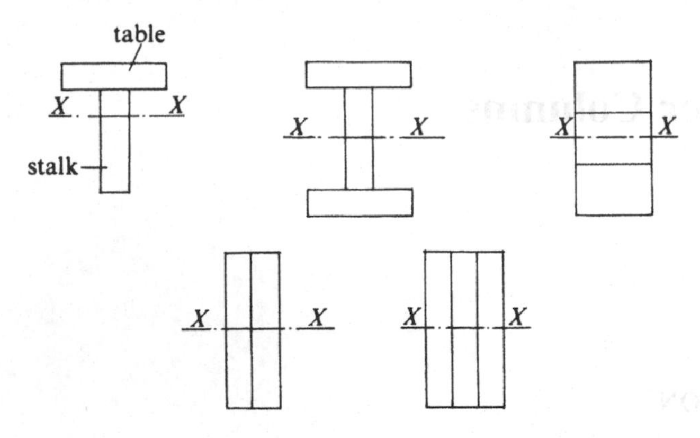

Fig. 16.1

16.3 TEE SECTION: DESIGN EXAMPLE

A composite 47T97 section (see Table 16.1) supports a long-term axial compression of 12 kN and a long-term bending moment of 0.90 kN m caused by a uniform load of 2.4 kN over a span of 3 m, as shown in Fig. 16.2.

Assume the effective length about the *x–x* axis is 1.0*L* and that there is a degree of end fixity about the *y–y* axis from end connections such that the effective length about the *y–y* axis is 0.85*L*. Check the suitability of the section with C24 timber.

Data for the section from Table 16.1 is as follows:

$A = 9.12 \times 10^3\,\text{mm}^2$
$I_x = 16.2 \times 10^6\,\text{mm}^4$
$Z_x\,(\text{stalk}) = 0.192 \times 10^6\,\text{mm}^3$
$Z_x\,(\text{table}) = 0.273 \times 10^6\,\text{mm}^3$
$i_x = 42.2\,\text{mm}$
$i_y = 22.0\,\text{mm}$

For axial loading

$$\lambda_x = \frac{3000}{42.2} = 71$$

$$\lambda_y = \frac{3000 \times 0.85}{22.0} = 116$$

For E/σ_c adopt

$$E = E_{\text{min}} \times K_9(\text{for 2 pieces}) = 7200 \times 1.14 = 8208\ \text{N/mm}^2$$

$$\sigma_c = \sigma_{c,g} \times K_3 = 7.9 \times 1.0 = 7.9\ \text{N/mm}^2$$

then

$$\frac{E}{\sigma_c} = \frac{8208}{7.9} = 1039$$

Table 16.1 Geometrical properties of tee sections

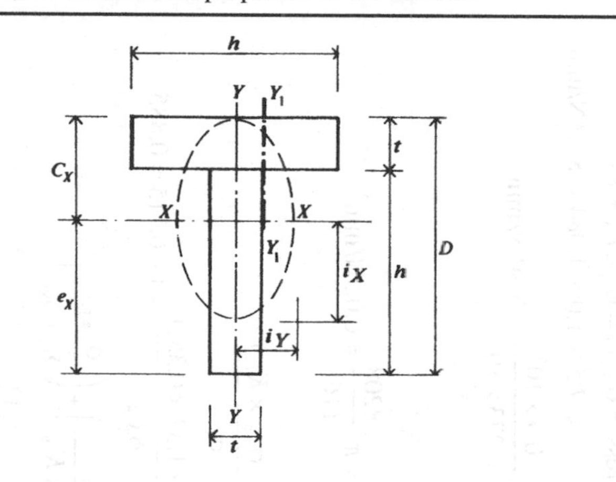

$$A = 2ht$$
$$C_X = 0.75t + 0.25h$$
$$e_X = 0.25t + 0.75h$$
$$I_Y = \frac{ht^3}{12} + \frac{th^3}{12}$$
$$I_X = \frac{ht(t+h)^2}{8} + I_Y$$
$$i_X = \sqrt{(I_X/A)}$$
$$i_Y = \sqrt{(I_Y/A)}$$
$$S_X = 0.5\,e_X^{\,2}\,t$$
$$S_{Y1} = \frac{t(h-t)^2}{8} \text{ at axis } Y_1Y_1$$

				Centre of gravity		Area	First moment of area		Second moment of area		Section modulus			Radius of gyration	
											Z_X				
Section	D (mm)	h (mm)	t (mm)	C_X (mm)	e_X (mm)	A (mm² × 10³)	S_X (mm² × 10⁶)	S_{Y1} (mm² × 10⁶)	I_X (mm⁴ × 10⁶)	I_Y (mm⁴ × 10⁶)	max (mm³ × 10⁶)	min (mm³ × 10⁶)	Z_Y (mm³ × 10⁶)	i_X (mm)	i_Y (mm)
$_{35}T_{97}$	132	97	35	50.50	81.50	6.79	0.116	0.0168	10.4	3.00	0.206	0.128	0.062	39.1	21.0
$_{35}T_{122}$	157	122	35	56.75	100.25	8.54	0.176	0.0331	18.9	5.73	0.333	0.188	0.094	47.0	25.9
$_{35}T_{147}$	182	147	35	63.00	119.00	10.3	0.248	0.0549	31.1	9.79	0.494	0.261	0.133	55.0	30.8
$_{35}T_{170}$	205	170	35	68.75	136.00	11.9	0.325	0.0797	46.2	14.9	0.672	0.339	0.176	62.3	35.4
$_{35}T_{195}$	230	195	35	75.00	155.00	13.7	0.420	0.112	67.5	22.3	0.899	0.435	0.229	70.3	40.4
$_{35}T_{220}$	255	220	35	81.30	174.00	15.4	0.528	0.150	94.4	31.8	1.16	0.543	0.289	78.3	45.5
$_{47}T_{97}$	144	97	47	59.50	84.50	9.12	0.168	0.0147	16.2	4.41	0.273	0.192	0.091	42.2	22.0
$_{47}T_{122}$	169	122	47	65.75	103.25	11.5	0.250	0.0330	28.6	8.17	0.436	0.277	0.134	50.0	26.7
$_{47}T_{147}$	194	147	47	72.00	122.00	13.8	0.350	0.0587	46.6	14.1	0.648	0.382	0.192	58.1	32.0
$_{47}T_{170}$	217	170	47	77.80	139.00	16.0	0.456	0.0888	67.7	20.7	0.871	0.486	0.244	65.1	36.0
$_{47}T_{195}$	242	195	47	84.00	58.00	18.3	0.587	0.129	97.8	30.7	1.16	0.619	0.315	73.1	40.9
$_{47}T_{220}$	267	220	47	90.30	177.00	20.7	0.734	0.176	136.0	43.6	1.50	0.768	0.396	81.0	45.9

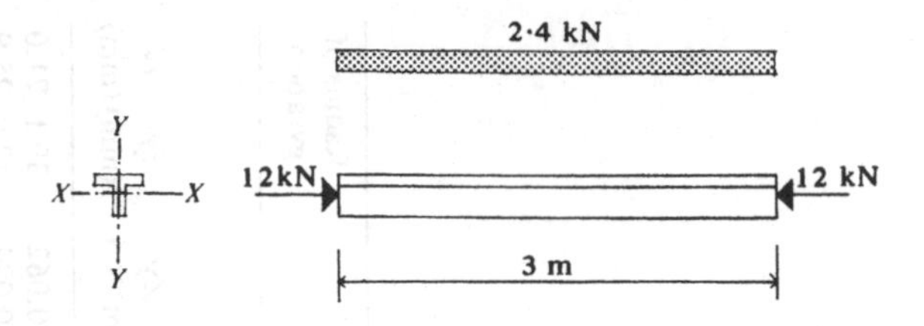

Fig. 16.2

From Table 19 of BS 5268-2, at $\lambda_{max} = 136$, $K_{12} = 0.351$. (A conservative value of K_{12} could be taken as 0.323 from Table 15.4.) Then

$$\sigma_{c,adm} = \sigma_{c,g} \times K_3 \times K_{12} = 7.9 \times 1.0 \times 0.351 = 2.77 \text{ N/mm}^2$$

$$\sigma_{c,a} = \frac{12000}{9120} = 1.32 \text{ N/mm}^2$$

For bending

The shape of the section permits it to be considered as fully laterally restrained against bending on the x–x axis.

There is no specific guidance in BS 5268-2 for the depth factor K_7 for multi-member sections but as the K_7 factor is applicable to the full depth of glued laminated beams (BS clause 3.2) it appears appropriate to apply the factor to the full section depth of 144 mm. Therefore, adopt $K_7 = (300/144)^{0.11} = 1.084$.

$$\text{Long-term permissible bending stress} = \sigma_{m,adm} = \sigma_g \times K_3 \times K_7$$
$$= 7.5 \times 1.0 \times 1.084 = 8.13 \text{ N/mm}^2$$

$$\sigma_{m,a} \text{(compression in table)} = \frac{M}{Z} = \frac{0.9 \times 10^6}{0.273 \times 10^6} = 3.30 \text{ N/mm}^2$$

For combined axial and compression

$$\text{Euler critical stress} = \sigma_e = \pi^2 \frac{E}{\lambda^2} = \pi^2 \frac{8208}{116^2} = 6.02 \text{ N/mm}^2$$

$$\text{Euler coefficient} = K_{eu} = 1 - \frac{(1.5 \times \sigma_{c,a} \times K_{12})}{\sigma_e}$$
$$= 1 - \frac{(1.5 \times 1.32 \times 0.351)}{6.02} = 1 - 0.115 = 0.885$$

$$\text{Interaction formula} = \left(\frac{\sigma_{m,a,par}}{\sigma_{m,adm,par} \times K_{eu}}\right) + \left(\frac{\sigma_{c,a,par}}{\sigma_{c,adm,par}}\right)$$
$$= \frac{3.30}{8.13 \times 0.885} + \frac{1.32}{2.77}$$
$$= 0.458 + 0.476 = 0.93 < 1.0 \quad \text{Satisfactory}$$

Check deflection about the x–x axis neglecting shear deflection.

$$EI_x = 8208 \times 16.2 \times 10^6 = 133 \times 10^9 = 133\,\text{k Nm}^2$$

$$\text{Deflection} = \frac{5WL^3}{384EI} = \frac{5 \times 2.4 \times 3^3}{384 \times 129} = 0.0065\,\text{m}$$

Permissible deflection at $0.003L = 0.009\,\text{m}$ Satisfactory

16.4 SPACED COLUMNS

Spaced columns are defined as two or more equal rectangular shafts spaced apart by end and intermediate blocking pieces suitably glued, bolted, screwed or otherwise adequately connected together. Figure 16.3 shows a typical assembly of a spaced column. In the figure

L = overall length of the composite unit
L_1 = distance between the centroids of end and closest intermediate spacer
L_2 = distance between the centroids of intermediate spacers.

Under an axial compressive load it is possible for the unit to buckle about three axes, these being the w–w, x–x and y–y axes shown in Fig. 16.3. The x–x axis and

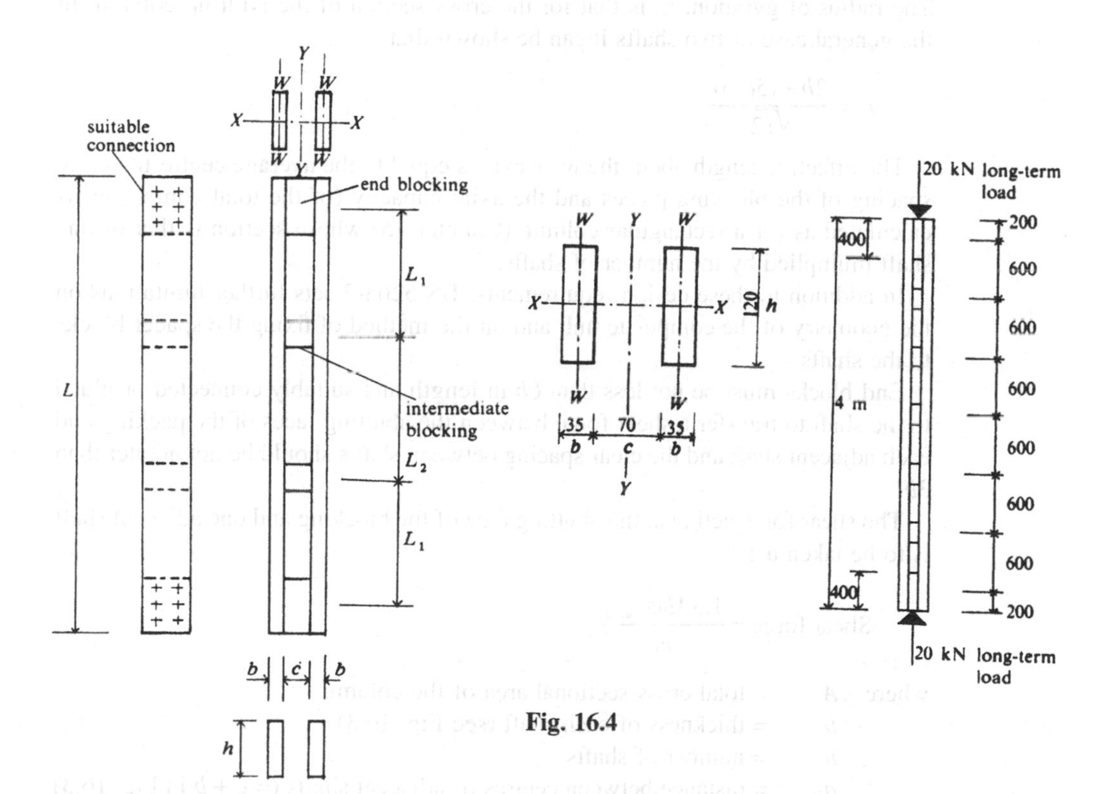

Fig. 16.4

Fig. 16.3

Table 16.2 Modification factor K_{13}

Method of connection	Ratio c/b 0	1	2	3
Nailed	1.8	2.6	3.1	3.5
Screwed or bolted	1.7	2.4	2.8	3.1
Attached by connector	1.4	1.8	2.2	2.4
Glued	1.1	1.1	1.3	1.4

the y–y axis relate to the composite section and the w–w axis relates to the individual sections. The lowest axial capacity of the section determined by consideration of these three axes will govern the design.

The effective length about the x–x axis is assessed in accordance with Chapter 15. The radius of gyration, i_x, being the same as that for the individual members. The axial capacity is that for a solid column with an area equal to the total area of the shaft.

The effective length about the y–y axis is first assessed in accordance with Chapter 15, but is then multiplied by a modification factor K_{13} which provides for the method of connecting the blocking pieces to the shafts (see Table 16.2) and the ratio c/b, i.e. the ratio of space between shafts to the shaft thickness (see Fig. 16.3). The radius of gyration, i_y, is that for the cross section of the built-up column. In the general case of two shafts it can be shown that

$$i_y = \frac{2b + (5c/3)}{\sqrt{12}}$$

The effective length about the w–w axis is equal to the average centre-to-centre spacing of the blocking pieces and the axial capacity (of the total composite) is calculated as for a rectangular column (Chapter 15) whose section is that of one shaft multiplied by the number of shafts.

In addition to these design requirements, BS 5268-2 sets further limitations on the geometry of the composite unit and on the method of fixing the spacer blocks to the shafts.

End blocks must be not less than $6b$ in length and suitably connected or glued to the shaft to transfer a shear force between the abutting faces of the packing and each adjacent shaft and the clear spacing between shafts should be not greater than $3b$.

The shear force acting at the abutting face of the blocking and one adjacent shaft is to be taken as:

$$\text{Shear force} = \frac{1.3Ab\sigma_{c,a,par}}{na}$$

where A = total cross-sectional area of the column
b = thickness of each shaft (see Fig. 16.3)
n = number of shafts
a = distance between centres of adjacent shafts (= $c + b$ in Fig. 16.3)
$\sigma_{c,a,par}$ = applied compression stress.

But, as $A\sigma_{c,a,par}$ is the total applied axial load on the column, the required shear force to be developed between the abutting face and one shaft may be expressed as

$$\text{shear force} = \frac{1.3Pb}{na}$$

where P = total applied axial load on column.

Intermediate blocking pieces should be at least 230 mm long and should be designed to transmit, between the abutting face of the blocking and one adjacent shaft, a shear force of half the corresponding shear force for the end packing, i.e. a shear force of $0.65Pb/na$.

If using glued packings, screws or bolts may be relied on to give adequate glue line pressure. In this case there should be at least four screws or bolts per packing, spaced to provide uniform pressure over the area of the packing.

BS 5268-2 sets slenderness limitations on the individual shaft members. Where the length of the column does not exceed 30 times the thickness of the shaft, only one intermediate blocking needs be provided. Sufficient blockings should be provided to ensure that the greater slenderness ratio of the local portion of an individual shaft between packings is limited to 70, or 0.7 times the slenderness ratio of the whole column, whichever is the lesser.

For the purpose of calculating the slenderness ratio of the local portion of an individual shaft, the effective length should be taken as the length between centroids of the groups of mechanical connectors or glue areas in adjacent packings.

With so many conditions to fulfil simultaneously, spaced column design is one of trial and error.

16.5 EXAMPLE OF SPACED COLUMN DESIGN

Check that the long-term load of 20 kN applied to the spaced column shown in Fig. 16.4 complies with the design requirements of BS 5268-2. The column consists of two 35 × 120 C24 timbers 70 mm apart. All joints are glued and intermediate packs are 250 mm long.

Establish capacity about x–x axis

$$\lambda_x = \frac{4000\sqrt{12}}{120} = 115$$

Although the two members are connected together to support the applied load, an unmodified value of E_{min} will be adopted for the determination of E/σ_c:

$$E = 7200 \text{ N/mm}^2$$

$$\sigma_c = \sigma_{c,g} \times K_3 = 7.9 \times 1.0 = 7.9 \text{ N/mm}^2$$

then

$$\frac{E}{\sigma_c} = \frac{7200}{7.9} = 911$$

From Table 15.4, $K_{12} = 0.327$

$$\sigma_{c,adm,par} = 7.9 \times K_3 \times K_{12} = 7.9 \times 1.0 \times 0.327 = 2.58\ \text{N/mm}^2$$

$$\text{Axial capacity } (x\text{–}x) = \frac{2 \times 35 \times 120 \times 2.58}{1000} = 21.7\ \text{kN}$$

Establish capacity about y–y axis

Radius of gyration about y–y axis is

$$i_y = \frac{2b + (5c/3)}{\sqrt{12}} = \frac{186.7}{3.464}$$
$$= 53.9\ \text{mm}$$

$$\frac{c}{b} = \frac{70}{35} = 2 < 3 \quad \text{staisfactory}$$

For $c/b = 2$, $K_{13} = 1.3$ (Table 16.2)

$$\lambda_y = \frac{L_e}{i_y} \times K_{13} = \frac{4000}{53.9} \times 1.3 = 96.5$$

From Table 15.4, $K_{12} = 0.42$

$$\sigma_{c,adm,par} = 7.9 \times K_3 \times K_{12} = 7.9 \times 1.0 \times 0.42 = 3.32\ \text{N/mm}^2$$

$$\text{Axial capacity } (y\text{–}y) = \frac{2 \times 35 \times 120 \times 3.32}{1000} = 27.9\ \text{kN}$$

Establish capacity about w–w axis

$$\lambda_w = \frac{600\sqrt{12}}{35} = 59.4$$

$$\frac{E}{\sigma_c} = \frac{7200}{7.9} = 911$$

From Table 15.4, $K_{12} = 0.67$

$$\sigma_{c,adm,par} = 7.9 \times K_3 \times K_{12} = 7.9 \times 1.0 \times 0.67 = 5.29\ \text{N/mm}^2$$

$$\text{Axial capacity } (w\text{–}w) = \frac{2 \times 35 \times 120 \times 5.29}{1000} = 44.4\ \text{kN}$$

Check end blocks for compliance with geometrical limits and spacer requirements. Minimum length required for end pack = $6b = 6 \times 35 = 210\,\text{mm} < 400\,\text{mm}$ satisfactory

Shear force between end blocking and one shaft is

$$\frac{1.3Pb}{na} = \frac{1.3 \times 20\,000 \times 35}{2 \times 105}$$
$$= 4333\ \text{N}$$

With 400 mm long end packing,

$$\text{glue line stress on end pack parallel to grain} = \frac{4333}{400 \times 120}$$
$$= 0.09\ \text{N/mm}^2$$

$$\text{permissible long-term glue line stress} = \text{permissible long-term shear stress parallel to grain}$$
$$= 0.70\ \text{N/mm}^2 \quad \text{which is acceptable}$$

Check intermediate blocks for compliance with geometrical limits and spacer requirements.

Minimum length required for intermediate pack = 230 mm < 250 mm Satisfactory

Shear force between intermediate blocking and one shaft is

$$\frac{0.65Pb}{na} = \frac{0.65 \times 20\,000 \times 35}{2 \times 105}$$
$$= 2166\ \text{N}$$

With 250 mm long end packing,

glue line stress on end pack parallel to grain

$$= \frac{2166}{250 \times 120} = 0.072\ \text{N/mm}^2 \quad \text{which is acceptable}$$

16.6 COMPRESSION MEMBERS IN TRIANGULATED FRAMEWORKS

16.6.1 Single compression members

Single compression members in triangulated frameworks should be designed using the same principles outlined in Chapter 15 for columns of solid timber.

For a continuous compression member the effective length may be taken as between 0.85 and 1.0 (according to the degree of end fixity) times the distance between node points for buckling in the plane of the framework, and times the actual distance between effective restraints for buckling perpendicular to the plane of the framework. Clause 15.10 of BS 5268: Part 2 defines some of the more usual types of restraint which may be considered as effective.

For a non-continuous compression member such as the internal web members of a truss, the effective length will depend upon the degree of end fixity. Most commonly the web members are assumed to be pin jointed at each end as they are frequently jointed at each end by a single connector unit. The effective length in this case is taken as the actual distance between the connector units at each end of the member. In the special case of web members held at each end by glued gusset plates (usually plywood) it may be assumed that a degree of fixity is developed and the effective length both in and out of the plane of the truss may be taken as 0.9 times the distance between the centre lines of the members connected.

16.6.2 Spaced compression members

Compression members in trusses frequently take the form of a twin member and the question arises as to when such an arrangement should be designed as a spaced column.

Consider a twin compression member which is restrained at the node points only. This would be designed as a spaced column adopting the principles given in section 16.4 except that the recommendations regarding the design of end packs will not apply (BS 5268-2, clause 2.11.11). Here it is assumed that the node point connections to transfer web/chord forces achieves the interface shear force.

The effective length for the calculation of λ_x may be taken between 0.85 and 1.0 (according to the degree of end fixity) times the distance between node points.

The effective length for the calculation of λ_y should be taken as the actual length between node points times K_{13} (see Table 16.2) depending on the method of connection of intermediate packs. At least one intermediate pack must be provided if advantage is to be taken of the improved strength offered by spaced column action, otherwise the twin compression member should be designed as two individual components.

The effective length for the calculation of λ_w is the distance between centres of packs or between centres of packs and node points. λ_w should not exceed 70, or $0.7\lambda_x$ or $0.7\lambda_y$, whichever is the lesser.

Although BS 5268 clearly indicates that end packings and adjacent shafts need not be designed for a transfer of shear force it appears that intermediate packs must be designed to transmit not less than half of the shear force calculated for the end packings. From section 16.4 the shear force would be

$$\text{shear force} = \frac{0.65Pb}{na}$$

where P = total axial compression load in shafts
b = width of shaft
n = number of shafts
a = distance between centres of adjacent shafts (= $b + c$ in Fig. 16.3).

This will frequently be found to be a large force not easily developed by nails or screws, so that packings will usually require to be bolted, connected or glued.

If, instead of being restrained only at node points, there are further lateral restraints to the y–y and w–w axes from joisting, etc., at intermediate points, then

1. The effective length for λ_x is calculated as above.
2. The effective length for λ_w is taken as the lesser of (a) the distance between effective lateral restraints, (b) the centres of intermediate packings, or (c) the centres between a packing and node point.
3. The effective length for λ_y is taken as the distance between effective restraints modified by K_{13}. Consequently, with close spacings of lateral restraints (for example, such as joisting at 600 mm centres) λ_y will be less than λ_w and intermediate packings are unnecessary. In such a situation it would not be necessary to design the member as a spaced column.

Chapter 17
Glulam Columns

17.1 INTRODUCTION

A glulam section is one manufactured by gluing together at least four laminations with their grain essentially parallel. The notes in the sections listed in Chapter 7, 'Glulam beams', apply equally to columns.

17.2 TIMBER STRESS GRADES FOR GLULAM COLUMNS

Chapter 14 details the various general factors which must be taken into account and considers factors involved in determining the slenderness factor K_{12}. Values of the grade compressive parallel to grain stresses for strength classes are given in BS 5268-2, Table 8, and for species/grades in Tables 9–15. BS 5268-2, Tables 2–7, list the grades/species that satisfy the requirements of each strength class with over 80% of all species/grades being allocated to strength classes C14, C16 or C24.

As mentioned in section 7.9.1, the most popular grade used by fabricators is C24, which will be adopted in this chapter.

Calculating K_{12} values from Appendix B of BS 5268-2 or selecting K_{12} values from Table 19, which requires interpolation in two directions, can be time consuming. Tables 17.1 and 17.2 are provided to simplify the design precedure for C24 strength glulam.

The following stress values have been used in preparing the K_{12} values given in Table 17.1.

For service classes 1 and 2:

$$E = E_{\text{mean}} \times K_{20}\ (\text{number of laminations}) = 10\,800 \times 1.07 = 11\,556\ \text{N/mm}^2$$

$$\begin{aligned}\sigma_{\text{c,par}} &= \sigma_{\text{grade}} \times K_3\ (\text{duration of load}) \times K_{17}\ (\text{number of laminations})\\ &= 7.9 \times 1.0 \times 1.04 = 8.21\ \text{N/mm}^2\ (\text{long term})\\ &= 7.9 \times 1.25 \times 1.04 = 10.27\ \text{N/mm}^2\ (\text{medium term})\\ &= 7.9 \times 1.50 \times 1.04 = 12.32\ \text{N/mm}^2\ (\text{short term})\\ &= 7.9 \times 1.75 \times 1.04 = 14.37\ \text{N/mm}^2\ (\text{very short term})\end{aligned}$$

$$\begin{aligned}E/\sigma_{\text{c,par}} &= 11556/8.21 = 1407\ (\text{long term})\\ &= 11556/10.27 = 1125\ (\text{medium term})\\ &= 11556/12.32 = 938\ (\text{short term})\\ &= 11556/14.37 = 804\ (\text{very short term})\end{aligned}$$

Table 17.1 K_{12} and Euler stress values for C24 glulam

	Service classes 1 and 2				Euler stress	Service class 3				Euler stress
	Long	Medium	Short	Very short	(N/mm^2)	Long	Medium	Short	Very short	(N/mm^2)
λ \ $\frac{E}{\sigma_{c,par}}$	1407	1125	938	804		1875	1500	1251	1071	
5	0.976	0.976	0.976	0.975	4563	0.976	0.976	0.976	0.976	3651
10	0.952	0.952	0.952	0.952	1141	0.952	0.952	0.952	0.952	913
15	0.929	0.928	0.928	0.928	507	0.929	0.929	0.929	0.928	406
20	0.906	0.905	0.904	0.903	285	0.907	0.906	0.905	0.905	228
25	0.883	0.881	0.879	0.878	183	0.884	0.883	0.882	0.881	146
30	0.859	0.857	0.854	0.851	127	0.862	0.860	0.858	0.856	101
35	0.836	0.831	0.827	0.822	93	0.840	0.837	0.833	0.830	75
40	0.811	0.805	0.799	0.792	71	0.817	0.813	0.808	0.803	57
45	0.786	0.778	0.769	0.759	56	0.794	0.788	0.782	0.775	45
50	0.761	0.749	0.737	0.725	46	0.771	0.763	0.755	0.746	37
55	0.734	0.720	0.704	0.688	38	0.748	0.738	0.727	0.716	30
60	0.707	0.689	0.670	0.650	32	0.724	0.712	0.698	0.684	25
65	0.680	0.658	0.635	0.612	27	0.700	0.685	0.669	0.652	22
70	0.652	0.626	0.600	0.573	23	0.676	0.658	0.639	0.619	19
75	0.624	0.594	0.564	0.535	20	0.652	0.631	0.609	0.586	16
80	0.595	0.562	0.530	0.498	18	0.628	0.603	0.579	0.554	14
85	0.567	0.531	0.496	0.463	16	0.603	0.576	0.549	0.522	13
90	0.540	0.501	0.465	0.431	14	0.579	0.550	0.520	0.492	11
95	0.513	0.472	0.434	0.400	13	0.556	0.523	0.492	0.462	10
100	0.487	0.444	0.406	0.372	11	0.532	0.498	0.465	0.434	9
105	0.462	0.418	0.380	0.346	10	0.509	0.473	0.439	0.408	8
110	0.438	0.393	0.355	0.323	9	0.487	0.449	0.415	0.383	8
115	0.415	0.370	0.333	0.301	9	0.466	0.427	0.392	0.360	7
120	0.393	0.349	0.312	0.281	8	0.445	0.405	0.370	0.339	6
125	0.373	0.329	0.293	0.263	7	0.425	0.385	0.350	0.319	6
130	0.353	0.310	0.275	0.246	7	0.406	0.365	0.331	0.301	5
135	0.335	0.293	0.259	0.231	6	0.388	0.347	0.313	0.283	5
140	0.318	0.277	0.244	0.217	6	0.370	0.330	0.296	0.268	5
145	0.302	0.262	0.230	0.205	5	0.354	0.314	0.281	0.253	4
150	0.287	0.248	0.217	0.193	5	0.338	0.299	0.266	0.239	4
155	0.273	0.235	0.205	0.182	5	0.323	0.284	0.253	0.227	4
160	0.260	0.223	0.195	0.172	4	0.309	0.271	0.240	0.215	4
165	0.248	0.212	0.184	0.163	4	0.296	0.258	0.228	0.204	3
170	0.236	0.201	0.175	0.155	4	0.283	0.247	0.218	0.194	3
175	0.225	0.192	0.166	0.147	4	0.271	0.235	0.207	0.185	3
180	0.215	0.183	0.158	0.139	4	0.260	0.225	0.198	0.176	3
185			0.151	0.133	3			0.189	0.168	3
190			0.144	0.126	3			0.180	0.160	3
195			0.137	0.121	3			0.172	0.153	2
200			0.131	0.115	3			0.165	0.146	2
205			0.125	0.110	3			0.158	0.140	2
210			0.120	0.105	3			0.152	0.134	2
215			0.115	0.101	2			0.145	0.128	2
220			0.110	0.096	2			0.140	0.123	2
225			0.106	0.093	2			0.134	0.118	2
230			0.102	0.089	2			0.129	0.114	2
235			0.098	0.085	2			0.124	0.109	2
240			0.094	0.082	2			0.120	0.105	2
245			0.090	0.079	2			0.115	0.101	2
250			0.087	0.076	2			0.111	0.098	1

For service class 3:

$$E = E_{\text{mean}} \times K_2 \text{ (service class)} \times K_{20} \text{ (number of laminations)}$$
$$= 10\,800 \times 0.8 \times 1.07 = 9245 \text{ N/mm}^2$$

$$\sigma_{\text{c,par}} = \sigma_{\text{grade}} \times K_2 \text{ (service class)} \times K_3 \text{ (duration of load)}$$
$$\times K_{17} \text{ (number of laminations)}$$
$$= 7.9 \times 0.6 \times 1.0 \times 1.04 = 4.93 \text{ N/mm}^2 \text{ (long term)}$$
$$= 7.9 \times 0.6 \times 1.25 \times 1.04 = 6.16 \text{ N/mm}^2 \text{ (medium term)}$$
$$= 7.9 \times 0.6 \times 1.50 \times 1.04 = 7.39 \text{ N/mm}^2 \text{ (short term)}$$
$$= 7.9 \times 0.6 \times 1.75 \times 1.04 = 8.63 \text{ N/mm}^2 \text{ (very short term)}$$

$$E/\sigma_{\text{c,par}} = 9245/4.93 = 1875 \text{ (long term)}$$
$$= 9245/6.16 = 1500 \text{ (medium term)}$$
$$= 9245/7.39 = 1251 \text{ (short term)}$$
$$= 9245/8.63 = 1071 \text{ (very short term)}$$

17.3 JOINTS IN LAMINATIONS

With structural glulam the end jointing of individual laminates is carried out almost certainly by finger jointing or scarf joints. If a butt joint is used, then the laminate in which it occurs has to be disregarded in the stress calculations, and this will usually make the member uneconomical.

In a column taking pure compression or a combination of compression and bending, either the joint must be strong enough to develop the full design strength of the laminations or the permissible strength must be reduced accordingly. Joint efficiencies and design are covered in sections 19.7 and 7.5. Rather than repeat the notes here, the designer is referred to these sections. Also, the worked example in section 17.5 includes a check on a finger joint in a member taking combined compression and bending.

17.4 EXAMPLE OF COMBINED BENDING AND COMPRESSION IN A GLULAM SECTION

Determine a suitable rafter section for the three-pin frame illustrated in Fig. 17.1, using C24 glulam. The frames are at 4 m centres, supporting purlins, which may be assumed to restrain the frames laterally at 1.5 m centres on true plan. Location is service class 2.

F_L (long-term point load) = 8 kN (each)

F_M (medium-term point load) = 12.5 kN total (each)

$$\theta = 21.8°$$

Purlin spacing up the slope = 1.62 m

Table 17.2 Axial capacity (kN) C24 glulam at zero slenderness ratio

Service classes 1 and 2

	Long					Medium				
h (mm) \ b (mm)	90	115	135	160	185	90	115	135	160	185
180	133	170	200	236	273	166	212	249	296	342
225	166	212	249	296	342	208	266	312	369	427
270	200	255	299	355	410	249	319	374	443	513
315	233	297	349	414	478	291	372	436	517	598
360	266	340	399	473	547	333	425	499	591	683
405	299	382	449	532	615	374	478	561	665	769
450	333	425	499	591	683	416	531	623	739	854
495	366	467	549	650	752	457	584	686	813	940
540	399	510	599	709	820	499	637	748	887	1025
585	432	552	648	768	889	540	690	810	961	1111
630	466	595	698	828	957	582	744	873	1034	1196

Service class 3

	Long					Medium				
h (mm) \ b (mm)	90	115	135	160	185	90	115	135	160	185
180	80	102	120	142	164	100	128	150	177	205
225	100	128	150	177	205	125	159	187	222	257
270	120	153	180	213	246	150	191	225	266	308
315	140	179	210	248	287	175	223	262	311	359
360	160	204	240	284	328	200	255	299	355	410
405	180	230	270	319	369	225	287	337	399	462
450	200	255	299	355	410	250	319	374	444	513
495	220	281	329	390	451	275	351	412	488	564
540	240	306	359	426	493	299	383	449	532	616
585	260	332	389	461	534	324	415	487	577	667
630	280	357	419	497	575	349	446	524	621	718

To determine the moments and forces, analyse the loaded frame in two parts, firstly the local bending on the rafter, and secondly the axial loading in the rafter (Fig. 17.2). Consider the medium-term loading condition:

The maximum rafter moment is

$$M_{\mathrm{M}} = (2F_{\mathrm{M}} \times 3) - (F_{\mathrm{M}} \times 1.5) = 56.25\ \mathrm{kNm}$$

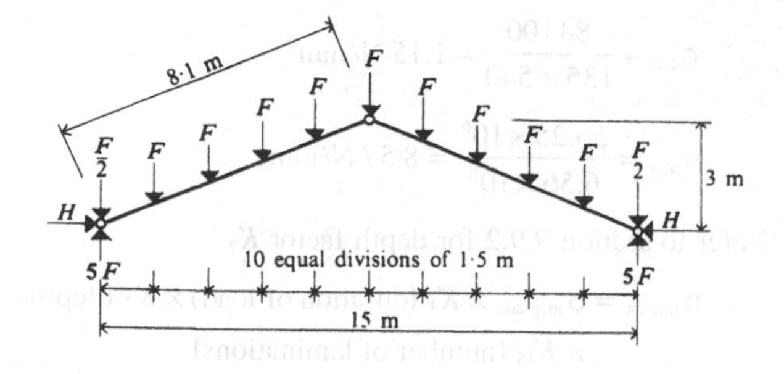

Fig. 17.1

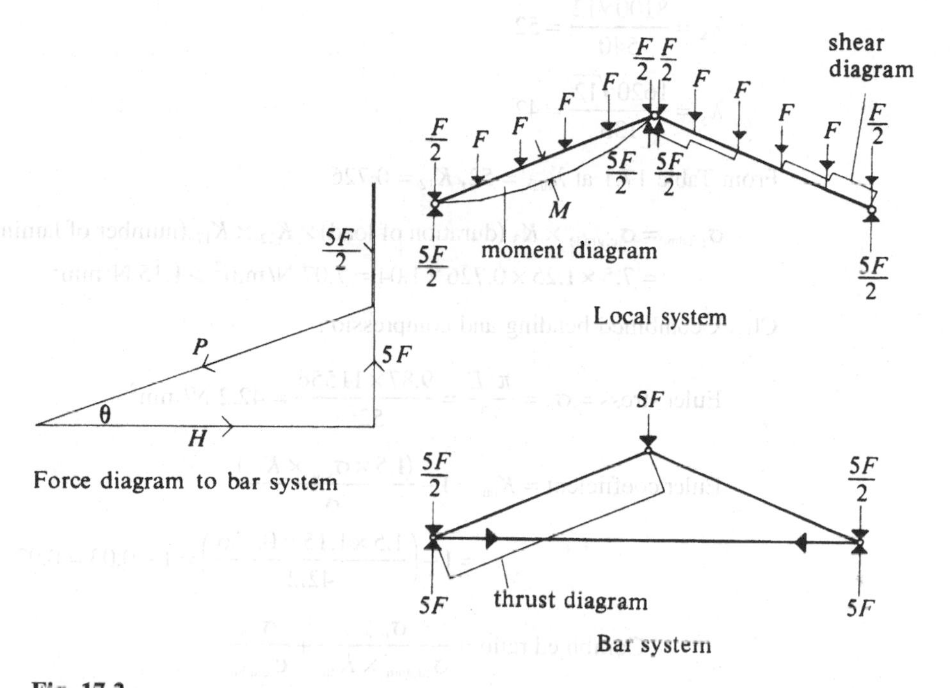

Fig. 17.2

The axial force in the rafter is

$$P = \frac{2.5 \times F_M}{\sin\theta} = \frac{2.5 \times 12.5}{0.3714} = 84.1\,\text{kN}$$

The horizontal force is

$$H = \frac{2.5 \times F_M}{\tan\theta} = \frac{2.5 \times 12.5}{0.4} = 78.1\,\text{kN}$$

Effective lengths for the rafter members are 8.1 m about the x–x axis and 1.62 m about the y–y axis. A trial section must be assumed. Experience shows that a fair first trial section is one having a moment capacity about 1.4 times the actual bending moment. Try a 135 × 540 mm C24 with 12 laminations at 45 mm:

$$\sigma_{c,a} = \frac{84100}{135 \times 540} = 1.15 \text{ N/mm}^2$$

$$\sigma_{m,a} = \frac{56.25 \times 10^6}{6.56 \times 10^6} = 8.57 \text{ N/mm}$$

Refer to section 7.9.2 for depth factor K_7

$$\sigma_{m,adm} = \sigma_{m,grade} \times K_3 \text{ (duration of load)} \times K_7 \text{ (depth)}$$
$$\times K_{15} \text{ (number of laminations)}$$
$$= 7.5 \times 1.25 \times 0.89 \times 1.45 = 12.1 \text{ N/mm}^2 > 8.57 \text{ N/mm}^2$$

$$\lambda_x = \frac{8100\sqrt{12}}{540} = 52$$

$$\lambda_y = \frac{1620\sqrt{12}}{135} = 42$$

From Table 17.1 at $\lambda_{max} = 52$, $K_{12} = 0.726$

$$\sigma_{c,adm} = \sigma_{c,grade} \times K_3 \text{ (duration of load)} \times K_{12} \times K_{17} \text{ (number of laminations)}$$
$$= 7.5 \times 1.25 \times 0.726 \times 1.04 = 7.07 \text{ N/mm}^2 > 1.15 \text{ N/mm}^2$$

Check combined bending and compression

$$\text{Euler stress} = \sigma_e = \frac{\pi^2 E}{\lambda^2} = \frac{9.87 \times 11556}{52^2} = 42.2 \text{ N/mm}^2$$

$$\text{Euler coefficient} = K_{eu} = 1 - \frac{(1.5 \times \sigma_{c,a} \times K_{12})}{\sigma_e}$$
$$= 1 - \left(\frac{1.5 \times 1.15 \times 0.726}{42.2}\right) = 1 - 0.03 = 0.97$$

$$\text{Combined ratio} = \frac{\sigma_{m,a}}{\sigma_{m,adm} \times K_{eu}} + \frac{\sigma_{c,a}}{\sigma_{c,adm}}$$
$$= \frac{8.57}{12.1 \times 0.97} + \frac{1.15}{7.07}$$
$$= 0.73 + 0.16 = 0.89 < 1.0 \quad \text{which is satisfactory}$$

The combined ratio value of 0.89 is based on the assumption that the axial compression occurs at the neutral axis of the section. If the joints are carefully detailed so as to provide an off-centre application of the compressive force (i.e. below the neutral axis as in Fig. 17.3), then a moment equal to P_e is artificially introduced into the system to relieve the moment due to lateral forces on the member.

A calculation for the required bearing area both vertically and horizontally and the consequent inclusion of adequate bearing plates determines the location of the intersection of the horizontal and vertical forces at the point of support. This leads to the determination of the eccentricity e of the thrust P. The apex joint must then be accurately notched at the upper edge to ensure that the same relieving moment is introduced into the member.

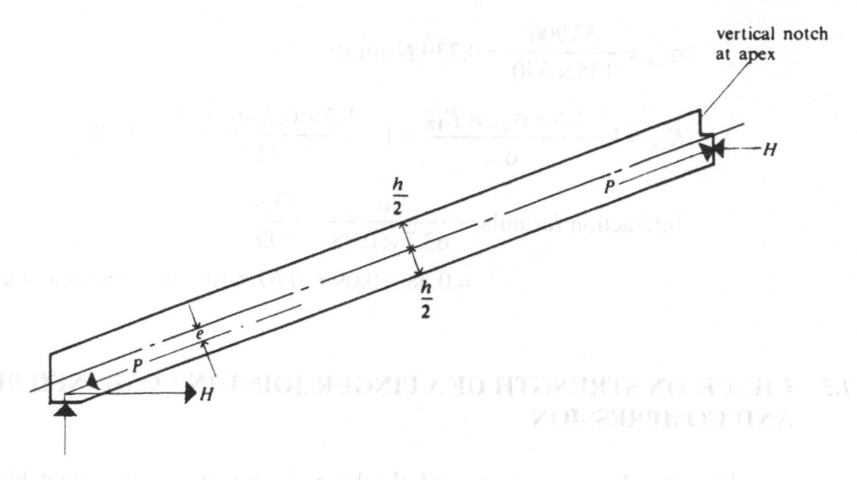

Fig. 17.3

In the example quoted, an eccentricity of approximately 135 mm could be accommodated, giving a relieving moment of 84.1 × 0.135 = 11.3 kN m. The applied moment would reduce to 56.25 − 11.3 = 44.95 kN m. This will reduce the sum of the bending and compression ratios to 0.67 and in turn may justify checking whether a smaller section could be used.

In the design so far only medium-term loading has been considered. A consideration of long-term loading is also necessary. This will be undertaken taking advantage of the available capacity tables.

$$\frac{F_L}{F_M} = \frac{8}{12.5} = 0.64$$

Referring to section 14.6.2, the interaction formula takes the form

$$\frac{M}{\overline{M}K_{eu}} + \frac{F}{\overline{F}} < 1.0$$

The long-term applied moment is

$$M = 0.64 \times 56.25 = 36\,\text{kN m}$$

The long-term applied axial load is

$$F = 0.64 \times 84.1 = 53.9\,\text{kN m}$$

From Table 7.6,

$$\overline{M}_L = 63.7\,\text{kN m}$$

From Table 17.1, at $\lambda_{max} = 52$,

$$K_{12} = 0.75 \quad \text{and} \quad \text{Euler stress} = \sigma_{eu} = 42.2\ \text{N/mm}^2$$

From Table 17.2,

$$\overline{F}_L = 599\,\text{kN}$$

$$\sigma_{c,a} = \frac{53\,900}{135 \times 540} = 0.739 \text{ N/mm}^2$$

$$K_{eu} = 1 - \frac{1.5 \times \sigma_{c,a} \times K_{12}}{\sigma_e} = 1 - \frac{1.5 \times 0.739 \times 0.75}{42.2} = 0.98$$

$$\text{Interaction formula} = \frac{36}{63.7 \times 0.98} + \frac{53.9}{599}$$

$$= 0.58 + 0.09 = 0.67 < 1.0 \quad \text{which is satisfactory}$$

17.5 CHECK ON STRENGTH OF A FINGER JOINT IN COMBINED BENDING AND COMPRESSION

BS 5268-2 gives two ways of checking if a finger joint is acceptable to end joint individual laminates. One is to provide a joint of a stated efficiency in bending, no matter how under-stressed the laminate may be and even if it is stressed largely in compression, while the other is to design for the actual strength required for the actual loading.

Assuming C24 laminates are used in the member being checked, reference to BS 5268-2 and section 7.5.1 shows that a joint having 70% efficiency in bending may be used without any further design check. If, however, only the 12.5/3.0/0.5 joint (see Table 7.3) is available, which has an efficiency in bending of 65% and an efficiency in compression of 83%, it will be necessary to carry out a design check. One method is detailed below.

Check for the medium-term loading.

The permissible *bending stress* in the member as limited by the joint being used is calculated as:

C24 grade stress for the species (7.5 N/mm^2)
× relevant load–duration factor K_3 (1.25 for medium term)
× relevant moisture content factor K_2 (1.0 for service classes 1 and 2)
× modification factor for depth of member, K_7 (0.893 for 540 mm)
× ratio for efficiency of the joint in bending (0.65 in this case)
× factor K_{30} (1.85)

$$\text{Permissible bending stress} = 7.5 \times 1.25 \times 1.0 \times 0.893 \times 0.65 \times 1.85$$
$$= 10.0 \text{ N/mm}^2$$

The permissible *compression stress* in the member as limited by the joint being used is calculated as:

C24 grade stress for the species (7.9 N/mm^2)
× relevant load–duration factor K_3 (1.25 for medium term)
× relevant moisture content factor K_2 (1.0 for service classes 1 and 2)
× ratio for efficiency of the joint in compression (0.83 in this case)
× factor K_{32} (1.15)

$$\text{Permissible bending stress} = 7.9 \times 1.25 \times 1.0 \times 0.83 \times 1.15 = 9.43 \text{ N/mm}^2$$

With medium-term loading in the member being jointed:

$$\sigma_{c,a} = 1.15\ \text{N/mm}^2 \quad \text{and} \quad \sigma_{m,a} = 8.58\ \text{N/mm}^2$$

Checking the combined effect of compression and bending on the joint:

$$\text{Combined ratio} = \frac{1.15}{9.43} + \frac{8.58}{10.0} = 0.122 + 0.858 = 0.98 < 1.0$$

Therefore the 12.5/3.0/0.5 finger-jointed laminations with 65% efficiency in bending and 83% efficiency in compression are acceptable in this case.

Chapter 18
Mechanical Joints

18.1 GENERAL

BS 5268-2 recommends basic loading for fasteners which fall into four categories related to the strength classes into which the species and stress grade combinations of the timber being used can be allocated. The level of these categories is indicated in Table 18.1.

The reader is referred to Table 2–15 of BS 5268-2 for the complete schedule of species/stress grade combinations and the strength classes they satisfy. The species most frequently used in timber engineering are softwoods of grades C16 and C24.

BS 5268-2 gives recommended loadings for nails, screws, steel bolts, steel dowels, toothed plate connectors, split ring connectors and shear plate connectors.

For nails, screws, bolts and dowels the basic loading is related to a standard minimum spacings, edge distances and end distances with no increase in load allowed for spacings, etc., greater than the minimum specified. The permissible loading is then modified for duration of loading, moisture content and number of fixing in line in order to establish the permissible load.

For toothed plate, split ring and shear plate connectors the basic loading is given for a standard spacing, edge distance and end distance with no increase in load allowed for spacings, etc., greater than the standard specified but requiring a reduction in loading for sub-standard spacing, etc. The permissible loading is then modified for duration of loading, moisture content, the number of fixing in line in order and any sub-standard spacing, edge distance or end distance in order to establish the permissible load.

18.1.1 Duration of loading factor

The modification factors for duration of load are:

K_{48} for nails
K_{52} for screws
K_{58} for toothed plates
K_{62} for split rings
K_{66} for shear plates.

Table 18.2 summarizes all these modification factors, and a value of 1.0 applies to all conditions other than those listed. The duration of load factors for bolts and dowels is allowed for in the basic loading values.

Table 18.1

Strength class	General description
C14	Low stress grade softwoods
C16/18/22/24	The majority of softwoods
C27/30	High stress grade softwoods
D30/40	Low stress grade hardwoods
D60/70	High stress grade hardwoods

Table 18.2 Modification factors for duration of load

Fastener	Modification factor	Duration of load		
		Long	Medium	Short/ Very short
Nails (partical board to timber)	K_{48}	1.00	1.40	2.10
Nails (tempered hardboard to timber)	K_{48}	1.00	1.25	1.62
Nails (other than above)	K_{48}	1.00	1.12	1.25
Screws	K_{52}	1.00	1.12	1.25
Toothed plate connectors	K_{58}	1.00	1.12	1.25
Split ring connectors	K_{62}	1.00	1.25	1.50
Shear plate connectors	K_{66}	1.00	1.25	1.50

18.1.2 Modification factors for moisture content

The permissible load on a fastener is further influenced by moisture content. The tabulated basic values are service classes 1 and 2, for other cases, the basic load is multiplied by the 'moisture content' modification factor.

The modifications for moisture content are

K_{49} for nails
K_{53} for screws
K_{56} for bolts
K_{59} for toothed plates
K_{63} for split rings
K_{67} for shear plates.

Table 18.3 summarizes all these modification factors and a value of 1.0 applies to all conditions other than those listed.

18.1.3 Modification factors for number of fixings in line

Potter, reporting on work on nailed joints undertaken at Imperial College in 1969, showed that there is a distinct regression of apparent load per nail as the number

Table 18.3 Modification factors for moisture content

	Service class	Nails* K_{49}	Screws K_{53}	Bolts K_{56}	Toothed plates K_{59}	Split rings K_{63}	Shear plates K_{67}
Lateral	1 and 2	1.00	1.00	1.00	1.00	1.00	1.00
	3	0.70	0.70	0.70	0.70	0.70	0.70
	3 to 2† or 3 to 1†			0.40			
Withdrawal	1 and 2	1.00	1.00				
	3	1.00	0.70				
	cyclic‡	0.25					

* K_{49} = 1.0 for lateral loads using annular ringed shank nails and helical threaded shank nails in all service class conditions.
† Bolted joint made in timber of service class 3 and used in service classes 1 and 2.
‡ Cyclic changes in moisture content after nailing.

of nails increases in line with the load. Similarly, from other sources there is evidence of a reduction in the apparent permissible load per bolt with an increasing number of bolts in line with the load.

The modification factors for the number of fixings in line are:

K_{50} for nails
K_{54} for screws
K_{57} for bolts
K_{61} for toothed plate connectors
K_{65} for split ring connectors
K_{69} for shear plate connectors.

The 'in-line' modification factor applies where a number of fasteners of the same size are arranged symmetrically in one or more lines parallel to the line of action of the load in an axially loaded member.

Nails	$K_{50} = 1.0$ for $n < 10$
	$K_{50} = 0.9$ for $n \geq 10$
Screws	$K_{54} = 1.0$ for $n < 10$
	$K_{54} = 0.9$ for $n \geq 10$
Bolts	$K_{57} = 1 - 0.003(n - 1)$ for $n \leq 10$
	$K_{57} = 0.7$ for $n > 10$
Toothed plates	$K_{61} = 1 - 0.003(n - 1)$ for $n \leq 10$
	$K_{61} = 0.7$ for $n > 10$
Split rings	$K_{65} = 1 - 0.003(n - 1)$ for $n \leq 10$
	$K_{65} = 0.7$ for $n > 10$
Shear plates	$K_{69} = 1 - 0.003(n - 1)$ for $n \leq 10$
	$K_{69} = 0.7$ for $n > 10$

where n = the number of fasteners in each line.

For all other cases where more than one fastener is used the factors above should be taken as 1.0.

As far as practicable, fasteners should be arranged so that the line of force in a member passes through the centroid of the group. When this is not practicable, account should be taken of the secondary stresses induced through the full or partial rigidity of the joint and of the effect of rotation imposing higher loads on the fasteners furthest from the centroid of the group.

The loads specified for nails, screws and bolts apply to those which are not treated against corrosion. The loads specified for toothed plate, split ring and shear plate connectors apply to those which are treated against corrosion. Some forms of anti-corrosion treatment may affect fastener performance particularly when preservative or fire-retardant timber impregnation treatments are specified. When the designer is in doubt the manufacturer of the fastener should be consulted.

Because of the anisotropic nature of timber, the bearing stresses permitted in a direction parallel to the grain are higher than those permitted in a direction perpendicular to the grain. Similarly, with the exception of lateral loads on nails and screws, connector units carry maximum loading when loaded parallel to the grain.

Permissible loads for intermediate angles of load to the grain from 0° (parallel) to 90° (perpendicular) are calculated using the Hankinson formula

$$F_\alpha = \frac{F_0 F_{90}}{F_0 \sin^2\alpha + F_{90}\cos^2\alpha}$$

where F_α = value of the load at angle α to grain
F_0 = value of the load parallel to the grain
F_{90} = value of the load perpendicular to the grain.

Using the function $\cos^2\alpha + \sin^2\alpha = 1$ simplifies the Hankinson formula to

$$F_\alpha = \frac{F_0}{1+[(F_0/F_{90})-1]\sin^2\alpha}$$

Several connector units remove part of the cross-sectional area of the timber and the designer must consider the effect of loading on the net section. The net area at a section is the full cross-sectional area of the timber less the total projected area of that portion of the connector within the member at the cross section (e.g. the projected area of a split ring and the projected area of the associated bolt not within the projected area of the split ring). It is not usual to deduct the projected area of nails/screws (<5 mm diameter), or the teeth of toothed plate connectors in calculating the strength of a joint.

The correct location of a mechanical fastener with respect to the boundaries of a timber component is of utmost importance to the satisfactory performance of a joint and to the development of the design load. It cannot be emphasized too strongly that care in detailing, particularly of tension end distances, is all important to the performance of a joint.

Table 18.4 Stock sizes for ordinary round wire nails

Length (mm)	Diameter (mm) 6.0	5.6	5.0	4.5	4.0	3.75	3.35	3.0	2.65
150	☐ ☒								
125		☐ ☒	☐						
100			☐	☐ ☒	☐ ☒				
90					☐ ☒	☒	☒		
75					☐ ☒	☐ ☒	☐ ☒	☐	
65							☐ ☒		☐ ☒
50							☒	☐ ☒	☐ ☒

☐ = ordinary ☒ = galvanized

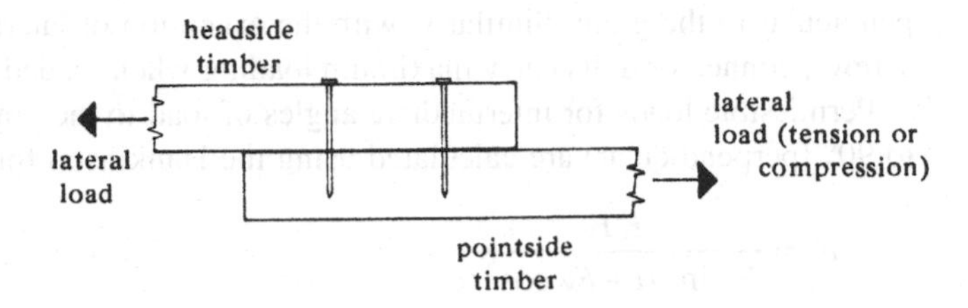

Fig. 18.1

18.2 NAILED JOINTS

18.2.1 Stock sizes for ordinary round wire nails

A typical range of stock sizes for ordinary round wire nails is given in Table 18.4.

Special nails are subject to a cost premium and longer delivery period. Manufacturers will generally make nails of almost any diameter and length providing that the quantities ordered will present a commercial proposition. To ensure that nails will be readily available only stock sizes should generally be specified.

18.2.2 Lateral loads in single shear timber-to-timber joints

The basic lateral load for a nail in single shear (Fig. 18.1) when inserted at right angles to the side grain of timber is given in Table 61 of BS 5268-2 for the full range of softwood and hardwood strength classes.

Nails in hardwoods should always be pre-drilled as should nails of 6 mm diameter or more in softwoods of strength classes C27–40. Although pre-drilling is not required for the other softwoods it should be noted that the lateral load on all nails into pre-drilled softwood may be multiplied by 1.15 (clause 6.4.4.1 of BS 5268-2).

Values are extracted for C16 and C24 strength classes as shown in Table 18.5. The tabulated values are applicable to loads at any angle to the side grain. Nailed joints should normally contain at least two nails and the lateral load on a nail driven

Table 18.5 Basic single shear lateral loads for round wire nails in timber–timber joints

Nail diameter (mm)	Standard penetration* (mm)	Basic single shear lateral load (N) Strength class	
		C16	C24
2.65	32	258	274
3.00	36	306	326
3.35	41	377	400
3.75	46	453	481
4.20	50	534	567
4.50	55	620	659
5.00	60	712	756
5.60	66	833	885
6.00	72	962	1022
7.00	84	1240	1318
8.00	96	1546	1643

*These values apply to headside thickness and pointside penetration.

into end grain should not exceed the recommended load for a similar nail driven into the side grain multiplied by the end grain modification factor K_{43}, which has a value of 0.7.

For the basic load to apply the standard headside thickness and the standard pointside penetration, which are equal in value, must be provided. It can be seen that the standard penetration is equal to 12 nail diameters. Greater thicknesses of timber and larger pointside penetration do not permit loadings higher than the basic.

When the headside thickness or pointside penetration is less than the standard value the basic load should be multiplied by the smaller of the two ratios K_{pen} (defined for later reference).

$$K_{pen} = \frac{\text{actual headside penetration}}{\text{standard headside penetration}} \quad \text{or} \quad \frac{\text{actual pointside penetration}}{\text{standard pointside penetration}}$$

For softwoods K_{pen} must not be less than 0.66 (and for hardwoods not less than 1.0). It follows from this limit for K_{pen} that as the minimum standard penetration give in Table 54, BS 5268-2, is 32 mm no load-bearing capacity can be assumed for nails driven into timber less than 21 mm thick.

The basic single shear lateral loads are determined from Annex G.2 of BS 5268-2 calculated from equations given in Eurocode 5 which considers six possible modes of failure from which the lowest value is tabulated.

Example

Calculate the permissible medium-term lateral load for two nails in single shear in C16 timber as shown in Fig. 18.2 using 3.35 mm diameter × 75 mm ordinary round nails, service class 1.

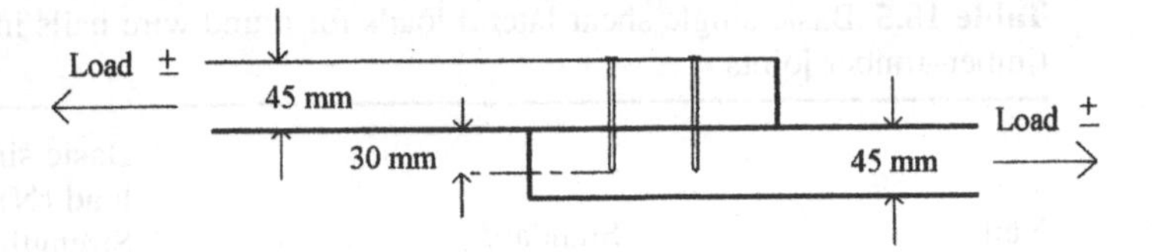

Fig. 18.2

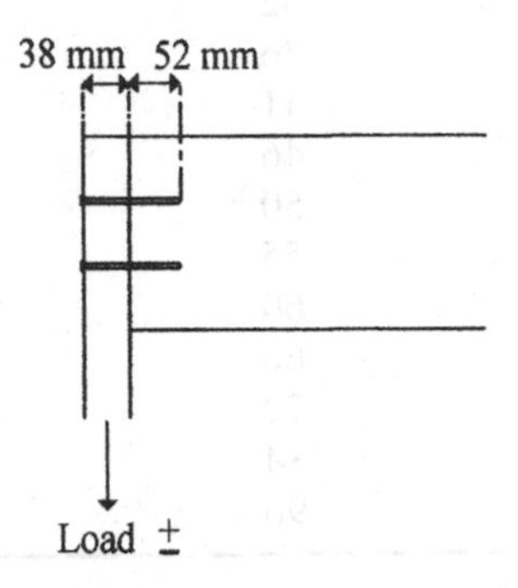

Fig. 18.3

Basic single shear capacity (from Table 18.5) = 377 N
Basic load for two nails = 2 × 377 = 754 N
For duration of load, $K_{48} = 1.12$
For service class, $K_{49} = 1.0$
For 'in-line fixing', $K_{50} = 1.0$

The actual headside thickness is 45 mm and the standard headside thickness required is 41 mm, therefore the headside thickness permits the full basic load of 377 N.

The pointside penetration is 30 mm whereas the standard pointside penetration required is also 41 mm, therefore the basic load must be reduced by $K_{pen} = 30/41 = 0.73$.

Note that K_{pen} must not be less than 0.66 (clause 6.4.4.1, BS 5268-2).

$$\text{Permissible load} = \text{basic load} \times K_{pen} \times K_{48} \times K_{49} \times K_{50}$$
$$= 754 \times 0.73 \times 1.12 \times 1.0 \times 1.0 = 616\,\text{N}$$

Example

Calculate the permissible medium-term lateral load for two nails in single shear in C16 timber as shown in Fig. 18.3 using 3.75 mm diameter. × 90 mm ordinary round nails, service class 1.

Basic single shear capacity (from Table 18.5) = 453 N
Basic load for two nails = 2 × 453 = 906 N
For K_{48}, K_{49} and K_{50} see sections 18.1.1, 18.1.2 and 18.1.3 respectively

For duration of load, $K_{48} = 1.12$
For service class, $K_{49} = 1.0$
For 'in-line fixing', $K_{50} = 1.0$

The actual headside thickness is 38 mm and the standard headside thickness required is 46 mm, therefore the basic load must be reduced by $K_{pen} = 38/46 = 0.826$.

Note that K_{pen} must not be less than 0.66 (clause 4.4.1, BS 5268-2).

Permissible load for headside = basic load $\times K_{pen} \times K_{48} \times K_{49} \times K_{50}$
$= 906 \times 0.826 \times 1.12 \times 1.0 \times 1.0 = 838\,\text{N}$

The actual pointside penetration is 52 mm and the standard pointside penetration required is 46 mm, therefore $K_{pen} = 1.0$.

As nail is into end grain, the load capacity must be modified by $K_{43} = 0.7$.

Permissible load for pointside = basic load $\times K_{pen} \times K_{43} \times K_{48} \times K_{49} \times K_{50}$
$= 906 \times 1.0 \times 0.7 \times 1.12 \times 1.0 \times 1.0 = 710\,\text{N}$

The joint capacity is limited by pointside penetration to 710 N.

18.2.3 Lateral loads in double shear timber-to-timber joints

The basic multiple shear lateral load for a nail is determined by multiplying the basic single lateral load by the number of shear planes, provided that the thickness of the inner member is not less than 0.85 times the standard tabulated thickness. Where the headside thickness and/or pointside penetration are less than the standard penetration or where the thickness of the inner member is less than 0.85 of the standard thickness the nail value shall be reduced pro rata.

Example

Calculate the permissible medium-term lateral load for three nails in double shear in C24 timber as shown in Fig. 18.4 using 3.35 mm diameter × 90 mm ordinary round nails, service class 1.

Basic single shear capacity (from Table 18.5) = 400 N
Number of shear planes = 2
Basic load for three nails = 3 × 400 × 2 = 2400 N
For K_{48}, K_{49} and K_{50} see sections 18.1.1, 18.1.2 and 18.1.3 respectively
For duration of load, $K_{48} = 1.12$

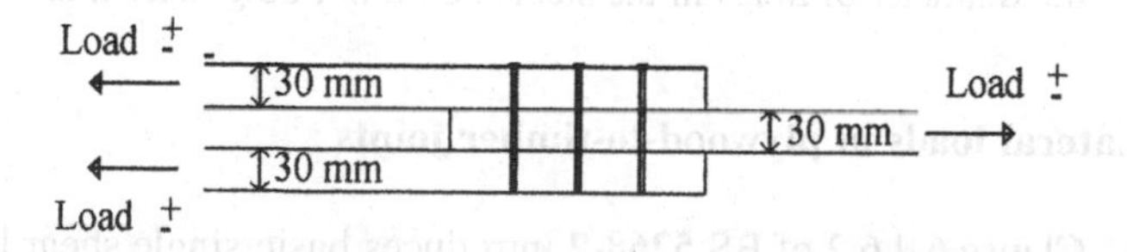

Fig. 18.4

For service class, $K_{49} = 1.0$
For 'in-line fixing', $K_{50} = 1.0$

The headside penetration is 30 mm and the standard headside penetration required is 41 mm, therefore the basic load must be reduced by

$$K_{pen} = 30/41 = 0.73\ (>0.66)$$

The inner member is 30 mm thick whereas the minimum thickness required to adopt the tabulated basic nail value is 0.85 × 41 mm = 35 mm. Therefore the basic load must be reduced by

$$K_{pen} = 30/35 = 0.86\ (>0.66)$$

The pointside penetration is 30 mm whereas the standard pointside penetration required is 41 mm, therefore the basic load must be reduced by

$$K_{pen} = 30/41 = 0.73\ (>0.66)$$

Note that K_{pen} must not be less than 0.66 (clause 6.4.4.1, BS 5268-2).
Adopt lowest value of $K_{pen} = 0.73$.

$$\text{Permissible load} = \text{basic load} \times K_{pen} \times K_{48} \times K_{49} \times K_{50}$$
$$= 2400 \times 0.73 \times 1.12 \times 1.0 \times 1.0 = 1962\,\text{N}$$

The designer may experience some difficulties in designing a multiple shear joint owing to the incompatibility of available nail diameter/lengths, required timber thicknesses and the limitations set by K_{pen} not having to exceed 0.66. The pointside penetration into the outer member may be improved by adopting longer nails penetrating the full assembly and clenched on the outside face.

18.2.4 Lateral loads for improved nails

The basic lateral load may be multiplied by modification factor K_{44}, which has a value of 1.20, if square grooved or square twisted shank nails of steel with a yield stress of not less than 375 N/mm^2 are used. The side width of the nail should be adopted as the nominal diameter. These nails have a limited stock range and availability should be checked before specifying.

18.2.5 Lateral loads for steel plate-to-timber joints

If the headside timber is replaced by a metal plate the basic lateral load may be increased by the modification factor K_{46} which has the value of 1.25. For this modification factor to apply the steel must be at least 1.2 mm or 0.3 × nail diameter thick and the diameter of holes in the steel should not be greater than the diameter of the nail.

18.2.6 Lateral loads in plywood-to-timber joints

Clause 6.4.6.2 of BS 5268-2 introduces basic single shear lateral loads for nails in plywood-to-timber joints. The basic loads are given in Table 63 of BS 5268-2 according to timber strength class, nail diameter, plywood species and thickness, together with minimum lengths of nails.

Nails shall be fully embedded and have a length not less than that given in Table 63 of BS 5268-2.

For K_{48}, K_{49} and K_{50} see sections 18.1.1, 18.1.2 and 18.1.3 respectively
Permissible load per nail = tabulated basic load × K_{48} × K_{49} × K_{50}

18.2.7 Withdrawal loads

Ordinary round wire nails are relatively weak when loaded in withdrawal and should be used only for relatively light loadings. Basic withdrawal loads per mm of pointside penetration are given in Table 62 of BS 5268-2 according to strength class and nail diameter.

The basic values in withdrawal apply to all service classes. When cyclic conditions of moisture content can occur after nailing the basic values should be modified by the moisture content factor $K_{49} = 0.25$, unless annular-ringed shank or helical-threaded shank nails are used in which case $K_{49} = 1.0$. No load in withdrawal should be carried by a nail driven into the end grain of timber.

Annular-ringed shank and helical-threaded shank nails have a greater resistance to withdrawal than ordinary round nails due to the effective surface roughening of the nail profile. Their resistance to withdrawal is not affected by changes in moisture content. BS 5268 consequently permits the basic withdrawal load for the 'threaded' part to be multiplied by factor $K_{45} = 1.5$ (clause 6.4.4.4) for these particular forms of improved nail and does not require a reduction in the basic withdrawal resistance even if there are likely to be subsequent changes in moisture content.

The range of available sizes for annular-ringed shank nails is given in Table 18.6 and the basic withdrawal values for both ordinary round, annular-ringed and helical-threaded shank nails are given in Table 18.7 for timber in strength classes C16 and C24.

18.2.8 Spacing of nails

To avoid splitting of timber, nails should be positioned to give spacings, end distances and edge distances not less than those recommended in Table 53 of BS 5268-2. All softwoods, except Douglas fir, may have values of spacing (but not edge distance) multiplied by 0.8. If a nail is driven into a glue-laminated section at right angles to the glue surface, then spacings (but not edge distances) may be multiplied by a further factor of 0.9 because the glued surface provides a restraining action against cleavage of the face lamination. Although glue-laminated components by definition have four or more laminations, there appears to be no reason why the reduced spacing should not also apply to composites of two or three laminations suitably glued. The spacing, end distance and edge distance are functions of the diameter of the nail and are related to the direction of grain (but not to the direction of load) and are summarized in Fig. 18.5.

Table 18.6 Stock sizes for annular-ringed shank nails

Length (mm)	Diameter (mm) 5	3.75	3.35	3	2.65
100	●				
75		●			
65			●		
60			●		
50			●	●	●
45					●
40					●

BS 1202: Part 1 gives sizes up to 200 × 8 mm.

Table 18.7 Basic resistance to withdrawal of ordinary round wire nails and annular-ringed shank and helical-threaded shank nails at right angles to grain

Nail diameter (mm)	Ordinary round nails (N/mm)		Improved nails (N/mm of threaded length)	
	Strength class C16	C24	Strength class C16	C24
2.7	1.53	2.08	2.29	3.12
3	1.71	2.31	2.56	3.46
3.4	1.93	2.62	2.89	3.93
3.8	2.16	2.93	3.45	4.39
4.2	2.39	3.23		
4.6	2.62	3.54		
5	2.84	3.85	4.26	5.77
5.5	3.13	4.24		
6	3.41	4.62		
7	3.98	5.39		
8	4.55	6.16		

18.3 SCREW JOINTS

18.3.1 Stock and special sizes

Screws are used principally for fixings which require a resistance to withdrawal greater than that provided by either ordinary or improved nails. Although BS 5268-2 gives permissible lateral loadings for screws it should be recognized that nails of a similar diameter offer better lateral capacities and can be driven more economically. If, however, the depth of penetration is limited, some advantage may be gained by using screws.

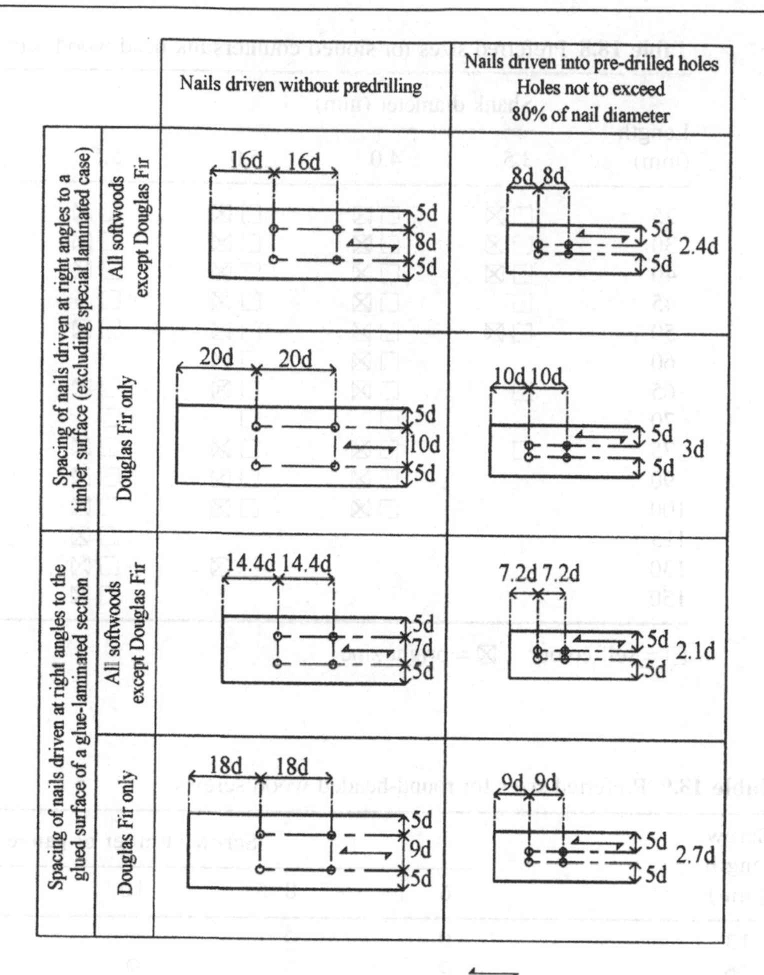

Fig. 18.5

Tables 66 and 67 of BS 5268-2 give lateral and withdrawal values respectively for screws made from steel complying with BS 1210 with a minimum tensile strength of 550 N/mm^2. Screws are referred to by their shank diameter and length.

The most common form of wood screw is the slotted countersunk head wood screw available in the preferred sizes given in Table 18.8 which is threaded for approximately 67% of the length under the head.

Round-headed screws may be used to advantage with metalwork as countersinking holes for the screw head is avoided. The range of round-headed wood screws is, however, less comprehensive than that for countersunk wood screws. Preferred sizes are given in Table 18.9.

Steel Twinfast® Pozidriv® woodscrews have twin threads and a plain shank diameter less than the diameter of the thread. This eliminates the wedge action of the ordinary screw and reduces the danger of splitting. Pre-drilling can also be

Table 18.8 Preferred sizes for slotted countersunk head wood screws

Length (mm)	Shank diameter (mm) 3.5	4.0	5.0	5.5	6.0	7.0
25	□ ☒	□ ☒	□ ☒	□ ☒		
30	□ ☒	□ ☒	□ ☒	□ ☒	□	
40	□ ☒	□ ☒	□ ☒	□ ☒	□	
45	□	□ ☒	□ ☒	□	□	
50	□ ☒	□ ☒	□ ☒	□ ☒	□ ☒	
60		□ ☒	□ ☒	□		
65	□	□ ☒	□ ☒	□ ☒	□	□
70		□	□	□		
75	□	□ ☒	□ ☒	□ ☒	□ ☒	□
90		□ ☒	□ ☒	□ ☒		
100		□ ☒	□ ☒	□ ☒	□ ☒	□
115				□ ☒		
130			□ ☒	□ ☒	□	□
150				□ ☒	□	□

□ = self-colour ☒ = bright zinc

Table 18.9 Preferred sizes for round-headed wood screws

Screw length (mm)	Screw number or gauge 6	8	10	12	14
13	○	○			
16	○	○	○		
19	●	●	○		
25	●	●	●	○	
32	○	●	●		
38	○	○	●	○	○
45		○	○		
50		○	○	○	○
63			○		
Diameter (mm)	3.4	4.2	4.9	5.6	6.3

2/3 length (approx)
length

● = Also available Sherardised.

simplified to the drilling of a single hole. (See Fig. 18.7 for pre-drilling for ordinary screws.) Twinfast screws are threaded for their full length which may give additional depth of penetration than normal screws. Preferred sizes are given in Table 18.10 from which it can be noted that the maximum length is less than that available with ordinary countersunk head wood screws.

Table 18.10 Preferred sizes for Twinfast® Pozidriv® screws

Screw length (mm)	Screw number or gauge			
	6	8	10	12
13	●	○		
16	●	●		
19	●	●	○	
25	●	●	●	○
32	●	●	●	○
38	●	●	●	○
45		●	●	○
50		●	○	○
57			○	
63			○	○
Diameter (mm)	3.4	4.2	4.9	5.6

○ = Also available bright zinc plated.

Table 18.11 Basic lateral loads for wood screws inserted at right angles to the grain for timber of strength classes C16 and C24

Screw shank diameter (mm)	Standard penetration		Basic single shear lateral load (N)	
	Headside (mm)	Pointside (mm)	C16	C24
3	11	21	205	205
3.5	12	25	278	310
4	14	28	361	395
4.5	16	32	454	490
5	18	35	550	593
5.5	19	39	654	705
6	21	42	765	826
7	25	49	1011	1093
8	28	56	1286	1391
10	35	70	1608	1741

18.3.2 Lateral loads

Basic lateral loads for screws are given in Table 66 of BS 5268-2. Values for strength classes C16 and C24 are extracted and given in Table 18.11.

The basic single shear values given in Table 18.11 are applicable to service classes 1 and 2. For screws in service class 3 these values must be modified by the moisture content factor $K_{53} = 0.7$.

If the actual headside penetration is less than the standard headside penetration the basic load should be multiplied by the ratio of actual to standard headside penetration and the pointside screw penetration should then be at least equal to twice the actual (reduced) headside thickness (as distinct from the standard pointside penetration).

The lateral load on a screw driven into end grain should be reduced by the end grain modification factor $K_{43} = 0.7$.

When a steel plate is screwed to a timber member, lateral loading values may be multiplied by the steel-to-timber modification factor $K_{46} = 1.25$. For this factor to apply, the steel must be at least 1.2 mm thick or 0.3 times screw shank diameter and the holes pre-drilled in the steel plate should be no greater than the screw shank diameter.

Table 68 of BS 5268-2 provides basic single shear lateral loads for screws in plywood-to-timber joints which depend upon screw diameter, plywood species and thickness, together with recommended minimum lengths of screws.

$$\text{Permissible lateral load} = \text{basic load} \times K_{52} \times K_{53} \times K_{54}$$

where K_{52}, K_{53} and K_{54} are given in sections 18.1.1, 18.1.2 and 18.1.3 respectively.

18.3.3 Withdrawal loads

Basic withdrawal loads for screws are given in Table 67 of BS 5268-2. Values for strength classes C16 and C24 are extracted and given in Table 18.12.

These tabulated values apply to the actual pointside penetration of the threaded portion of the screw. The threaded portion of a screw may be taken as 67% of the actual screw length. The penetration of the screw point should not be less than 15 mm.

No withdrawal load should be carried by a screw driven into end grain of timber.

Table 18.12 Basic withdrawal loads for wood screws inserted at right angles to the grain in timber of strength classes C16 and C24

Screw shank diameter (mm)	Basic withdrawal load per mm of pointside penetration (N)	
	C16	C24
3	8.65	11.02
3.5	10.09	12.86
4	11.53	14.70
4.5	12.97	16.54
5	14.41	18.37
5.5	15.86	20.21
6	17.30	22.05
7	20.18	25.72
8	23.06	29.40
10	28.83	36.75

Permissible withdrawal = basic load × K_{52} × K_{53} × K_{54}

where K_{52}, K_{53} and K_{54} are given in sections 18.1.1, 18.1.2 and 18.1.3 respectively.

18.3.4 Spacing of screws

Clause 6.5.1 of BS 5268-2 requires that screws should be turned, not hammered, into pre-drilled holes. The required spacing and pilot hole dimensions for countersunk wood screws in pre-drilled holes are shown in Figs 18.6 and 18.7 respectively.

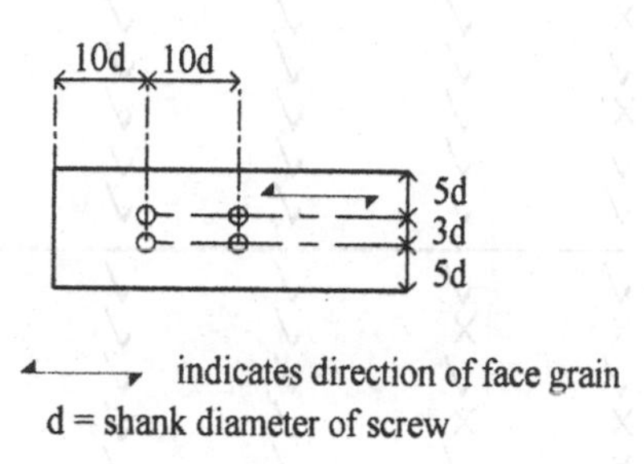

Fig. 18.6

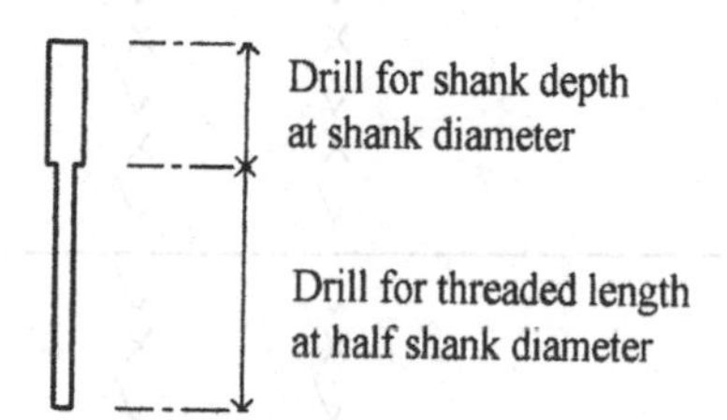

Fig. 18.7

18.4 BOLTED JOINTS

18.4.1 Stock and standard non-stock sizes

Bolts for timber engineering tend to have large *L*/*d* ratios (where *L* is the bolt length and *d* the bolt diameter) compared to those used for structural steelwork connections, resulting from the need to join thick sections of timber. Table 18.13 tabulates stock and standard non-stock sizes.

For joints which will be permanently exposed to the weather or sited in a condition of high corrosion risk, the use of rust-proofing is essential. Also, when joints are in a non-hazard service condition but are to be of architectural merit (i.e. exposed to view) rust-proofing should be specified in order to avoid rust stains appearing on the timber surfaces.

Table 18.13 Stock and standard non-stock ISO metric bolts

Bolt length (mm)	ISO metric hex bolts grade 4.6 BS 4190 metric coarse thread							
	Bolt size							
	M8	M10	M12	M16	M20	M24	M30	M36
35	✓	✓	✓					
40	✓	✓	✓					
45	X	✓	✓					
50	✓	✓	✓	✓	✓			
55	X	X	X	✓	✓			
60	✓	✓	✓	✓	✓			
65	X	✓	✓	✓	✓			
70	✓	✓	✓	✓	✓	✓		
75	X	✓	✓	✓	✓	✓		
80	✓	✓	✓	✓	✓	✓		
90	✓	✓	✓	✓	✓	✓		
100	✓	✓	✓	✓	✓	✓	X	
110	X	X	✓	✓	✓	✓	X	X
120	✓	✓	✓	✓	✓	✓	X	X
130	X	X	✓	✓	✓	X	X	X
140	X	✓	✓	✓	✓	✓	X	X
150	X	X	✓	✓	✓	X	X	X
160			✓	✓	✓	✓	X	X
170			X	X	X	X	X	X
180			✓	✓	✓	✓	X	X
190			X	X	X	X	X	X
200			✓	✓	✓	✓	X	X
220			✓	✓	X	X	X	X
240			X	X	X	X	X	X
260			✓	✓	✓	X	X	X
280			X	X	X	X	X	X
300			✓	✓	✓	X	X	X

✓ = preferred X = non-preferred

If bolts are Sherardized or hot-dip galvanized, either the bolt threads must be run down or the nut tapped out before being treated, or the nut will not fit the thread. These disadvantages in the use of Sherardizing and hot-dip galvanizing favour the use of electro-galvanizing, which is readily undertaken on the bolt sizes specified in Table 18.13, although the thickness of the coating is not as much as obtained by the two other methods.

Bolts are probably the most frequently used fixing for medium/heavy timber engineering components. They are readily available and provide a medium/high loading capacity especially in typical diameters of 12 mm to 20 mm and for timber 45 mm thick or more. They are easier to align with open holes and to insert insitu and any nominal misalignment of holes can usually be easily correct by comparison to shear plates and split rings.

18.4.2 Loading capacities

Unlike nails and screws, bolts have loading capacities which vary according to the direction of the applied load with respect to the direction of grain, as shown in Fig. 18.8.

If a force F acts at an angle θ to the axis of the bolt (Fig. 18.9) the component of load perpendicular to the load, $F \sin\theta$, must not be greater than the allowable lateral load. The component $F \cos\theta$ acting axially on the bolt creates either tension or compression in the bolt and must be suitably resisted.

Annex G of BS 5268-2 sets out six equations (G_1–G_6) representing the six possible failure modes of timber or bolts in a two-member joint and four equations (G_7–G_{10}) representing the four possible failure modes of timber or bolts in a three-member joint. These equations come from the ENV version of EC5 and are based on the theory first developed by Johansen. The allowable lateral load is taken as

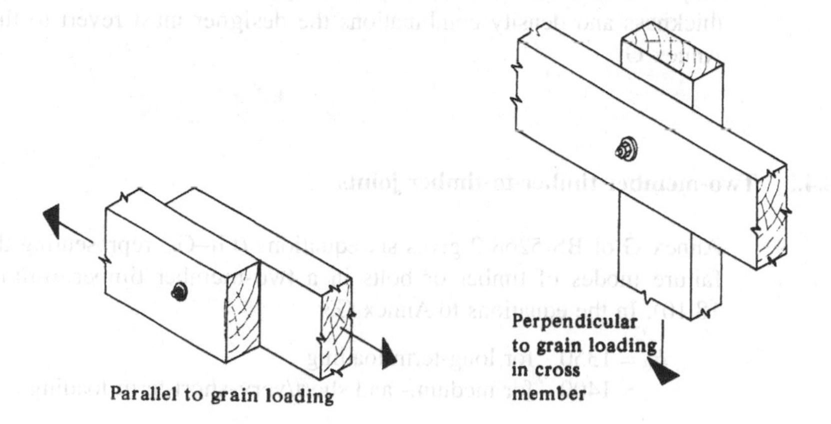

Fig. 18.8

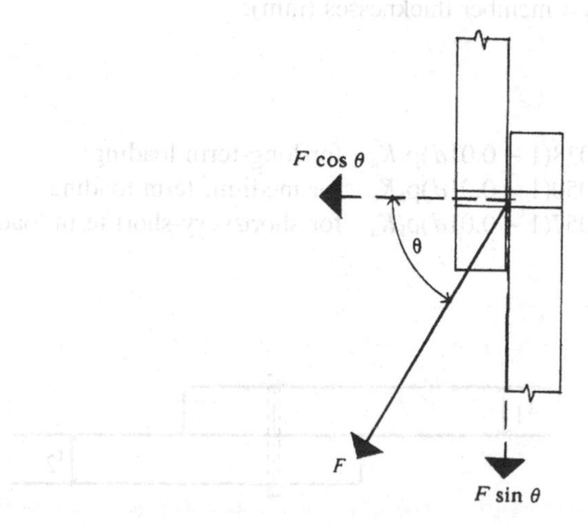

Fig. 18.9

the minimum calculated from the 'six/four failure mode equations' and permits the derivation of loading for all softwoods/hardwood scheduled in BS 5268-2 for any timber thicknesses and for any bolt diameters.

The formulae are somewhat complex and, at first glance, would appear to require the designer to go through the daunting task of calculating six/four possible loadings so as to select the lowest value as the allowable lateral load for any one angle of load to grain.

To assist the designer BS 5268-2 provides a series of basic load tables, (a) Tables 69–74 for basic single shear loads bolts in a two-member joint with members of equal thickness and density, and (b) Tables 75–80 for basic single shear loads bolts in a three-member joint where the central member is twice the thickness of the outer members and members are of equal density.

Only parallel-to-grain and perpendicular-to-grain loads are listed from which intermediate angles of load to grain may be calculated from Hankinson's formula (see BS 5268-2, clause 6.6.4.1 and section 18.4.3 below). Alternatively the designer may calculate the bolt capacity directly from Annex G. For any other timber thickness and density combinations the designer must revert to the equations in Annex G.

18.4.3 Two-member timber-to-timber joints

Annex G of BS 5268-2 gives six equations (G_1–G_6) representing the six possible failure modes of timber or bolts in a two-member timber-to-timber joint (Fig. 18.10). In the equations to Annex G3:

$F_d = 1350$ for long-term loading
$= 1400$ for medium- and short/very-short-term loading

$f_{h,1,d}$ = embedding strength in t_1
$f_{h,2,d}$ = embedding strength in t_2

where t_1 and t_2 = member thicknesses (mm).

$$\beta = \frac{f_{h,2,d}}{f_{h,1,d}}$$

$f_{h,0,d} = 0.038(1 - 0.01d)\rho_k K_a$ for long-term loading
$= 0.050(1 - 0.01d)\rho_k K_a$ for medium-term loading
$= 0.057(1 - 0.01d)\rho_k K_a$ for short/very-short term loading

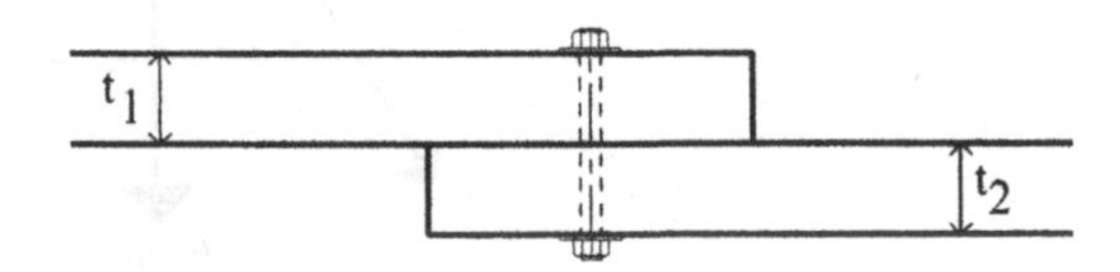

Fig. 18.10

where d = bolt diameter (mm)
ρ_k = characteristic density of timber (kg/m³)
= 290 for strength class C14
= 310 for strength classes C16/18/22
= 350 for strength class C24
= 370 for strength classes TR26/C27/30/35/40.

$$K_a = \sqrt{\frac{a_{par}}{(4+3\cos\alpha)d}}$$

where a_{par} = spacing of bolts parallel to grain (mm)
α = angle of load to grain
K_a = the 'bolt spacing effect'.

Bolt capacity is optimized when the bolt spacing parallel to grain is at least 7 × bolt diameter. The spacing parallel to grain may be reduced to a minimum of 4 × bolt diameter and in this case the load capacity is reduced due to the danger of possible splitting of grain. European practice is to determine bolt capacity at 7 × diameter spacing and apply a reduction factor to account for the spacing reducing to 4 × diameter. BS 5268 has its origins in American/Canadian practice where is was normal to require 4 × diameter parallel to grain spacing with no permitted increase for spacings in excess of 4 × diameter. Table 18.14 shows values for K_a according to angle of load to grain and bolt spacing.

For bolts at any angle of load to grain the embedding strength is

$$f_{h,a,d} = \frac{f_{h,0,d}}{K_{90}\sin^2\alpha + \cos^2\alpha}$$

where K_{90} = 1.35 + 0.015d for softwoods
= 0.90 + 0.015d for hardwoods.

Table 18.14 Values of K_a for bolt spacing between 4d and 7d parallel to grain

Angle of load to grain	Bolt spacing a_{par} parallel to grain							Bolt spacing limit
	4d	4.5d	5d	5.5d	6d	6.5d	7d	
0	0.756	0.802	0.845	0.886	0.926	0.964	1.00	7.00d
10	0.758	0.804	0.848	0.889	0.929	0.967		6.95d
20	0.766	0.812	0.856	0.898	0.938	0.976		6.82d
30	0.779	0.826	0.871	0.913	0.954	0.993		6.60d
40	0.797	0.845	0.891	0.935	0.976			6.30d
50	0.821	0.871	0.918	0.963				5.93d
60	0.853	0.905	0.953	1.00				5.50d
70	0.892	0.946	0.997					5.00d
80	0.941	0.998						4.52d
90	1.0							4.00d

18.4.4 Derivation of bolt capacities in BS5268-2 for two-member timber-to-timber joints

Tables 69–74 of BS 5268-2 are given for members of equal thickness ($t_1 = t_2$), equal density and a bolt spacing of 4 × bolt diameter. Therefore

$$\frac{t_1}{t_2} = 1.0 \quad \text{and} \quad \beta = \frac{f_{h,2,d}}{f_{h,1,d}} = 1.0$$

Equations G_1 to G_6 of Annex G3 then simplify to:

$$G_1 = \frac{f_{h,a,d} t_1 d}{F_d}$$

$$G_2 = \frac{f_{h,a,d} t_2 d}{F_d} = G_1$$

$$G_3 = \frac{1.33}{F_d}\frac{f_{h,a,d} t_1 d}{2}\left\{(1+2[1+1+1^2]+1^2)^{0.5} - (1+1)\right\} = \frac{1.33 G_1\{0.828\}}{2}$$
$$= 0.55\,G_1$$

$$G_4 = \frac{1.1 f_{h,a,d} t_1 d}{F_d(2+1)}\left\{\left(2[1+1] + \frac{4(2+1)M_{y,d}}{f_{h,a,d} d t_1^2}\right)^{0.5} - 1\right\}$$
$$= \frac{1.1\,G_1}{3}\left\{\left(4 + \frac{12 M_{y,d}}{f_{h,a,d} d t_1^2}\right)^{0.5} - 1\right\}$$

$$G_5 = \frac{1.1 f_{h,a,d} t_2 d}{F_d(1+2)}\left\{\left(2[1^2][1+1] + \frac{4(1+2)M_{y,d}}{f_{h,a,d} d t_2^2}\right)^{0.5} - 1\right\}$$
$$= \frac{1.1\,G_1}{3}\left\{\left(4 + \frac{12 M_{y,d}}{f_{h,a,d} d t_2^2}\right)^{0.5} - 1\right\} = G_4$$

$$G_6 = \frac{1.1}{F_d}\sqrt{\frac{2\times 1}{1+1}}\sqrt{2 M_{y,d} f_{h,a,d}}\, d = \frac{1.1}{F_d}\sqrt{2 M_{y,d} f_{h,a,d}}\, d$$

where $M_{y,d} = 0.12 f_y d^3$ ($f_y = 400\,\text{N/mm}^2$ for grade 4.6 steel).

Tables 69–74 of BS 5268-2 are determined as the minimum value calculated from the above simplified equations for G_3, G_4 and G_6 and adopting K_a according to direction of load to grain with bolt spacing of 4 × bolt diameter.

For bolt values parallel to grain

$$f_{h,a,d} = f_{h,0,d} \quad \text{with } K_a = 0.756 \text{ (Table 18.14)}$$

For bolt values perpendicular to grain

$$f_{h,a,d} = \frac{f_{h,0,d}}{K_{90}} \quad \text{with } K_a = 1.0 \text{ (Table 18.14)}$$

Equations G_4 and G_6 include the term $M_{y,d}$ which depends upon the bolt material.

Clause 6.6.1 of BS 5268-2 states that bolts made of higher strength steel may give higher load-carrying capacities than those given in Tables 69–80. For grade 4.6 bolts Annex G3 gives f_y = 400 N/mm^2 based on a minimum ultimate stress of 40 kgf/mm^2. As the ratio of timber thickness to bolt diameter (t/d) increases the critical equation progresses from equation G_3 to G_4 to G_6 with equation G_3 being generally critical for t/d < 5.0 (approximately). The bolt capacities tabulated in BS 5268-2 can therefore be improved upon by using grade 8.8 bolts for higher values of t/d adopting a value of f_y = 800 N/mm^2 based on a minimum ultimate stress of 80 kgf/mm^2. Figure 18.11 compares the capacity of M12 grade 4.6 and 8.8 bolts loaded parallel to grain in strength class C24 timber.

The bolt capacities as determined from Annex G allow for duration of load. The capacities must be modified by K_{56} for moisture content (see section 18.1.2) and K_{57} (see section 18.1.3) if appropriate. For service class 3 use K_{56} = 0.7 and take K_{57} from Table 18.15 according to number of bolts (n) in a line parallel to the line of action of the bolt in a primary axially loaded member.

Bolted members in trusses and laminated construction generally adopt C24 grade timber. Values for grade 4.6 bolts for a two-member timber-to-timber joint in this situation are listed in Table 18.16 and Table 18.17 for loads parallel and perpendicular to load respectively. Member thicknesses are given in 20 mm increments to assist interpolation. Where two load capacities are indicated, the higher capacity may be adopted if grade 8.8 bolts are specified.

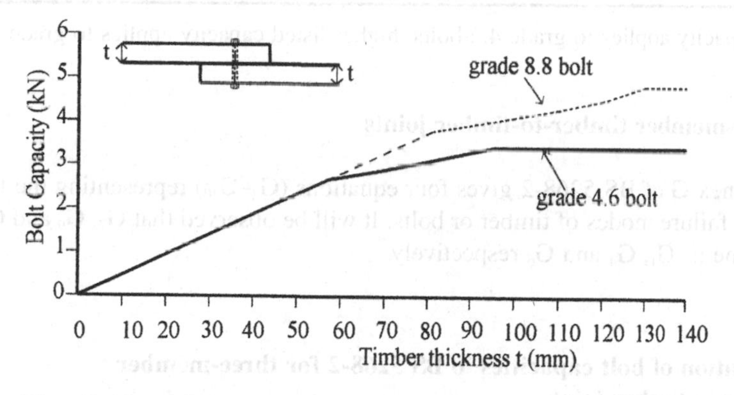

Fig. 18.11 Single shear load for M12 bolt in two-member joint parallel to grain long-term C24 timber.

Table 18.15 K_{57} values

n	1	2	3	4	5	6	7	8	9	10	11 or more
K_{57}	1.00	0.97	0.94	0.91	0.88	0.85	0.82	0.79	0.76	0.73	0.7

Table 18.16 Single shear bolt capacity (kN): two-member joint: C24 timber: load parallel to grain

Load duration	Member thickness (mm)	Bolt diameter (mm) M8	M10	M12	M16	M20	M24
Long term	20	0.60	0.74	0.87	1.10	1.31	1.50
	40	1.21	1.48	1.73	2.11	2.63	2.99
	60	1.52/1.80	2.07/2.22	2.60	3.31	3.94	4.49
	80	1.55/2.07	2.40/2.85	3.09/3.47	4.41	5.25	5.99
	100	1.55/2.20	2.40/3.19	3.42/4.11	5.13/5.51	6.56	7.48
	120	1.55/2.20	2.40/3.40	3.42/4.51	5.65/6.62	7.59/7.88	8.98
	140	1.55/2.20	2.40/3.40	3.42/4.84	5.94/7.22	8.18/9.19	10.4/10.5
Medium term	20	0.77	0.98	1.10	1.40	1.67	1.90
	40	1.46/1.53	1.87	2.20	2.80	3.33	3.80
	60	1.72/2.12	2.45/2.81	3.17/3.30	4.20	5.00	5.70
	80	1.72/2.43	2.66/3.36	3.69/4.38	5.46/5.60	6.66	7.60
	100	1.72/2.43	2.66/3.76	3.78/4.86	6.09/7.00	8.23/8.33	9.49
	120	1.72/2.43	2.66/3.76	3.78/5.35	6.57/7.92	8.97/9.99	11.4
	140	1.72/2.43	2.66/3.76	3.78/5.35	6.57/8.57	9.78/11.6	12.2/13.3
Short term and very short term	20	0.87	1.07	1.25	1.60	1.90	2.16
	40	1.60/1.75	2.14	2.51	3.19	3.80	4.33
	60	1.84/2.34	2.72/3.20	3.49/3.76	4.79	5.70	6.49
	80	1.84/2.60	2.84/3.72	4.04/4.80	6.00/7.49	7.60	8.66
	100	1.84/2.60	2.84/4.01	4.04/5.39	6.75/7.98	9.04/9.49	10.80
	120	1.84/2.60	2.84/4.01	4.04/5.71	7.02/8.73	9.92/11.4	12.5/13.0
	140	1.84/2.60	2.84/4.01	4.04/5.71	7.02/9.50	10.7/12.7	13.5/15.1

Lower listed capacity applies to grade 4.6 bolts; higher listed capacity applies to grade 8.8 bolts.

18.4.5 Three-member timber-to-timber joints

Annex G of BS 5268-2 gives four equations (G_7–G_{10}) representing the four possible failure modes of timber or bolts. It will be observed that G_7, G_9 and G_{10} are the same as G_1, G_4 and G_6 respectively.

18.4.6 Derivation of bolt capacities in BS 5268-2 for three-member timber-to-timber joints

Tables 75–80 of BS 5268-2 assume that

- the inner member is at least twice the thickness of the outer member ($t_2 = 2t_1$)
- members are of equal density
- bolt spacing = 4 × bolt diameter.

Equations G_7–G_{10} of Annex G3 then simplify to:

Table 18.17 Single shear bolt capacity (kN): two-member joint: C24 timber: load perpendicular to grain

Load duration	Member thickness (mm)	Bolt diameter (mm) M8	M10	M12	M16	M20	M24
Long term	20	0.54	0.65	0.75	0.92	1.05	1.16
	40	1.09	1.30	1.50	1.84	2.11	2.32
	60	1.40/1.63	1.89/1.95	2.25	2.75	3.16	3.47
	80	1.47/1.91	2.19/2.60	2.77/3.00	3.67	4.21	4.63
	100	1.47/2.09	2.26/2.89	3.13/3.70	4.49/4.59	5.26	5.79
	120	1.47/2.09	2.26/3.19	3.18/4.03	4.90/5.51	6.32	6.95
	140	1.47/2.09	2.26/3.19	3.18/4.40	5.35/6.32	6.93/7.37	8.11
Medium term	20	0.69	0.83	0.95	1.16	1.34	1.47
	40	1.35/1.38	1.65	1.90	2.33	2.67	2.94
	60	1.63/1.96	2.23/2.48	2.85	3.49	4.01	4.41
	80	1.63/2.29	2.50/3.06	3.29/3.80	4.66	5.34	5.88
	100	1.63/2.31	2.50/3.46	3.52/4.35	5.29/5.82	6.68	7.35
	120	1.63/2.31	2.50/3.53	3.52/4.81	5.86/6.94	7.61/8.01	8.81
	140	1.63/2.31	2.50/3.53	3.52/4.98	5.99/7.44	8.22/9.35	10.2/10.3
Short term and very short term	20	0.79	0.94	1.08	1.33	1.52	1.67
	40	1.48/1.57	1.88	2.17	2.65	3.05	3.35
	60	1.74/2.16	2.46/2.83	3.14/3.25	3.98	4.57	5.02
	80	1.74/2.46	2.66/3.38	3.65/4.33	5.25/5.31	6.09	6.70
	100	1.74/2.46	2.66/3.77	3.76/4.81	5.85/6.64	7.61	8.37
	120	1.74/2.46	2.66/3.77	3.76/5.31	6.40/7.62	8.38/9.14	10.0
	140	1.74/2.46	2.66/3.77	3.76/5.31	6.40/8.22	9.11/10.7	11.2/11.7

Lower listed capacity applies to grade 4.6 bolts; higher listed capacity applies to grade 8.8 bolts.

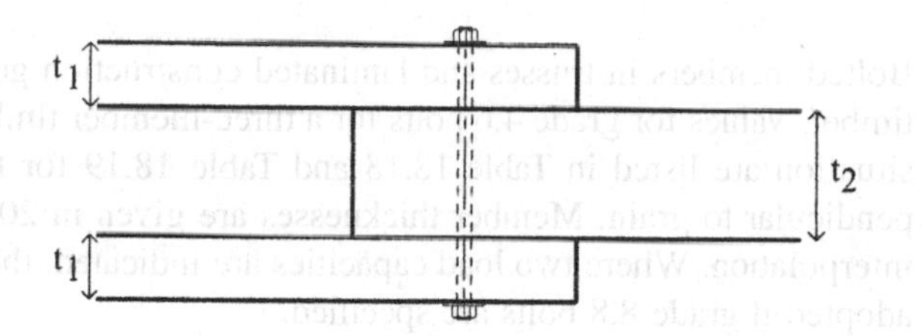

Fig. 18.12

$$G_7 = G_1 = \frac{f_{h,a,d} t_1 d}{F_d}$$

$$G_8 = \frac{0.5 f_{h,a,d} t_2 d}{F_d} = G_7$$

$$G_9 = G_4 = \frac{1.1}{3} G_1 \left\{ \left(4 + \frac{12 M_{y,d}}{f_{h,a,d} d t_1^2} \right)^{0.5} - 1 \right\}$$

$$G_{10} = G_6 = \frac{1.1}{F_d} \sqrt{2 M_{y,d} f_{h,a,d} d}$$

Table 18.18 Single shear bolt capacity: three member joint: C24 timber: load parallel to grain

Load duration	Member thickness (mm)	Bolt diameter (mm)					
		M8	M10	M12	M16	M20	M24
Long term	20	1.10	1.34	1.57	2.00	2.38	2.72
	40	1.24/1.61	1.79/2.42	2.47/3.15	4.00	4.77	5.43
	60	1.52/1.80	2.07/2.57	2.72/3.52	4.35/5.95	6.44/7.15	8.15
	80	1.55/2.07	2.40/2.85	3.09/3.77	4.68/6.09	6.68/9.08	9.10/10.9
	100	1.55/2.20	2.40/3.19	3.42/4.11	5.13/6.38	7.08/9.26	9.40/12.8
	120	1.55/2.20	2.40/3.40	3.42/4.51	5.65/6.76	7.59/9.57	9.85/12.9
	140	1.55/2.20	2.40/3.40	3.42/4.84	5.94/7.22	8.18/9.98	10.4/13.3
Medium term	20	1.22/1.39	1.70	2.00	2.54	3.02	3.45
	40	1.46/2.12	2.06/2.71	2.79/3.79	4.67/5.08	6.05	6.89
	60	1.72/2.12	2.45/2.97	3.17/4.00	4.95/6.63	7.22/9.07	9.97/10.3
	80	1.72/2.43	2.66/3.36	3.69/4.38	5.46/6.91	7.63/10.1	10.2/13.8
	100	1.72/2.43	2.66/3.76	3.78/4.86	6.09/7.36	8.23/10.5	10.7/14.2
	120	1.72/2.43	2.66/3.76	3.78/5.35	6.57/7.92	8.97/11.0	11.4/14.6
	140	1.72/2.43	2.66/3.76	3.78/5.35	6.57/8.57	9.78/11.6	12.2/15.2
Short term and very short term	20	1.32/1.59	1.94	2.28	2.90	3.45	3.93
	40	1.60/1.98	2.25/2.92	3.02/4.07	5.01/5.79	6.89	7.86
	60	1.84/2.34	2.72/3.25	3.49/4.34	5.38/7.12	7.77/10.3	10.7/11.8
	80	1.84/2.60	2.84/3.72	4.04/4.80	6.00/7.49	8.30/10.9	11.1/15.0
	100	1.84/2.60	2.84/4.01	4.04/5.39	6.75/8.05	9.04/11.3	11.7/15.3
	120	1.84/2.60	2.84/4.01	4.04/5.71	7.02/8.73	9.92/12.0	12.5/15.8
	140	1.84/2.60	2.84/4.01	4.04/5.71	7.02/9.50	10.7/12.7	13.5/16.5

Lower listed capacity applies to grade 4.6 bolts; higher listed capacity applies to grade 8.8 bolts.

Bolted members in trusses and laminated construction generally adopt C24 grade timber. Values for grade 4.6 bolts for a three-member timber-to-timber joint in this situation are listed in Table 18.18 and Table 18.19 for loading parallel and perpendicular to grain. Member thicknesses are given in 20 mm increments to assist interpolation. Where two load capacities are indicated, the higher capacity may be adopted if grade 8.8 bolts are specified.

Figure 18.13 compares the capacity of M12 grade 4.6 and 8.8 bolts loaded parallel to grain in strength class C24 timber.

18.4.7 Loads between parallel and perpendicular to grain

The basic load F_α acting at an angle α degrees to grain should be determined by the Hankinson formula:

$$F_\alpha = \frac{F_0 F_{90}}{F_0 \sin^2 \alpha + F_{90} \cos^2 \alpha}$$

which can be simplified using the function $\cos^2 \alpha + \sin^2 \alpha = 1$ to:

Table 18.19 Single shear bolt capacity: three-member joint: C24 timber: load perpendicular to grain

Load duration	Member thickness (mm)	Bolt diameter (mm)					
		M8	M10	M12	M16	M20	M24
Long term	20	0.99	1.18	1.36	1.67	1.91	2.10
	40	1.16/1.51	1.66/2.26	2.28/2.72	3.33	3.82	4.20
	60	1.40/1.67	1.89/2.38	2.47/3.24	3.91/5.00	5.73	6.31
	80	1.47/1.91	2.19/2.60	2.77/3.43	4.15/5.49	5.87/7.64	7.93/8.41
	100	1.47/2.09	2.26/2.89	3.13/3.70	4.49/5.69	6.14/8.19	8.09/10.5
	120	1.47/2.09	2.26/3.19	3.18/4.03	4.90/5.97	6.50/8.38	8.37/11.3
	140	1.47/2.09	2.26/3.19	3.18/4.40	5.35/6.32	6.93/8.66	8.74/11.4
Medium term	20	1.16/1.25	1.50	1.73	2.11	2.42	2.67
	40	1.35/1.71	1.90/2.52	2.56/3.45	4.23	4.85	5.33
	60	1.63/1.96	2.23/2.73	2.86/3.66	4.42/6.01	6.39/7.27	8.00
	80	1.63/2.29	2.50/3.06	3.29/3.96	4.80/6.19	6.65/9.01	8.86/10.7
	100	1.63/2.31	2.50/3.46	3.52/4.35	5.29/6.51	7.07/9.20	9.17/12.4
	120	1.63/2.31	2.50/3.53	3.52/4.81	5.86/6.94	7.61/9.54	9.61/12.6
	140	1.63/2.31	2.50/3.53	3.52/4.98	5.99/7.44	8.22/9.97	10.2/12.9
Short term and very short term	20	1.24/1.43	1.71	1.97	2.41	2.76	3.04
	40	1.48/1.86	2.07/2.71	2.77/3.77	4.54/4.82	5.53	6.08
	60	1.74/2.16	2.46/2.98	3.14/3.96	4.79/6.44	6.87/8.29	9.12
	80	1.74/2.46	2.66/3.38	3.65/4.33	5.25/6.69	7.21/9.66	9.53/12.2
	100	1.74/2.46	2.66/3.77	3.76/4.81	5.85/7.10	7.73/9.93	9.93/13.3
	120	1.74/2.46	2.66/3.77	3.76/5.31	6.40/7.62	8.38/10.4	10.5/13.6
	140	1.74/2.46	2.66/3.77	3.76/5.31	6.40/8.22	9.11/10.9	11.2/14.0

Lower listed capacity applies to grade 4.6 bolts; higher listed capacity applies to grade 8.8 bolts.

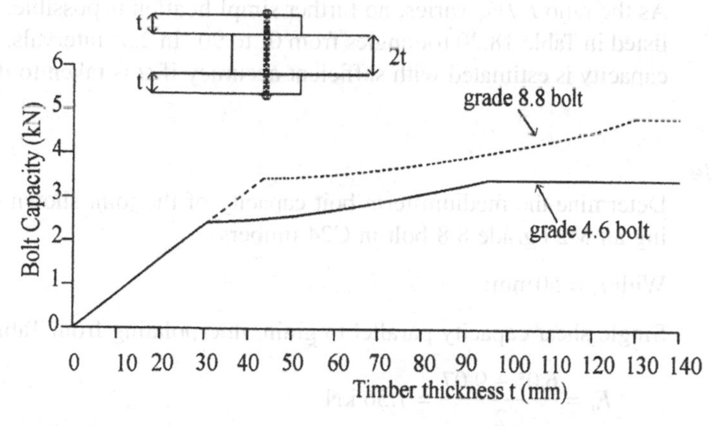

Fig. 18.13 Single shear load for M12 bolt in three-member joint parallel to grain long-term C24 timber.

Table 18.20 Values of $\sin^2 \alpha$

$\alpha°$	0	2.5	5	7.5
0	0	0.001 90	0.007 60	0.0170
10	0.0302	0.046 8	0.067 0	0.0904
20	0.117	0.146	0.179	0.213
30	0.250	0.289	0.329	0.371
40	0.413	0.456	0.500	0.544
50	0.587	0.629	0.671	0.711
60	0.750	0.787	0.821	0.854
70	0.883	0.910	0.933	0.953
80	0.970	0.983	0.992	0.998
90	1.0			

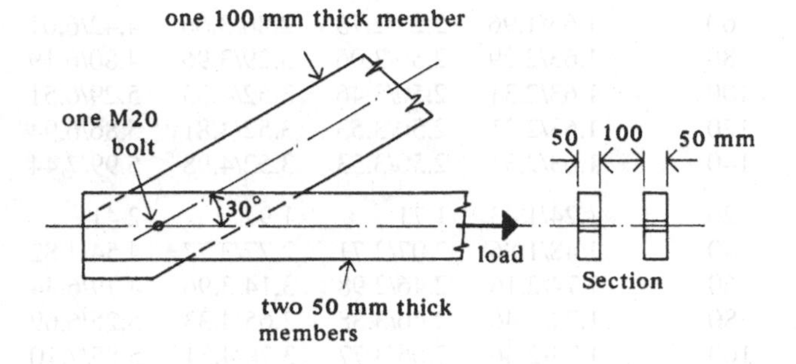

Fig. 18.14

$$F_\alpha = \frac{F_0}{1+[(F_0/F_{90})-1]\sin^2 \alpha}$$

As the ratio F_0/F_{90} varies, no further simplification is possible. Values of $\sin^2 \alpha$ are listed in Table 18.20 for angles from 0° to 90° in 2.5° intervals. In practice the joint capacity is estimated with sufficient accuracy if α is taken to the nearest 5° value.

Example

Determine the medium-term bolt capacity of the joint shown in Fig 18.14 adopting an M20 grade 8.8 bolt in C24 timbers.

With t_1 = 50 mm:

Single shear capacity parallel to grain, interpolating from Table 18.18,

$$F_0 = \frac{6.05+9.07}{2} = 7.56 \text{ kN}$$

Single shear capacity perpendicular to grain, interpolating from Table 18.19,

$$F_{90} = \frac{4.85+7.27}{2} = 6.06 \text{ kN}$$

$$\frac{F_0}{F_{90}} - 1 = (7.56/6.06) - 1 = 0.248$$

At $\alpha = 30°$, $\sin^2\alpha = 0.25$.

$$\text{Single shear capacity at } 30° \text{ to grain} = \frac{7.56}{1+(0.248 \times 0.25)} = \frac{7.56}{1.062} = 7.12 \text{ kN}$$

Bolt is in double shear, therefore

$$\text{total joint bolt capacity} = 2 \times 7.12 = 14.24\,\text{kN}$$

18.4.8 Spacing of bolts

In general, bolt holes should be drilled as closely as possible to the diameter of the specified bolt. Too great a hole tolerance will produce excessive joint slip under load which in turn will lead to additional deflections of a complete assembly, i.e. a truss or similar framed structure. In practice, holes are usually drilled no more than 2 mm larger in diameter than the bolt.

A mild steel washer should be fitted under any head or nut which would otherwise be in dircct contact with a timber surface. Minimum washer sizes (larger than those associated with structural steelwork) are required having a diameter at least 3 times the bolt diameter and a thickness at least 0.25 times the bolt diameter. When appearance is important, round washers are preferred. They are easier to install than square washers, the latter requiring care in alignment if they are not to appear unsightly. The required washer diameters are large to avoid crushing of the fibres of the timber, and are consequently relatively thick in order to avoid cupping as the bolted joint is tightened. To avoid fibre crushing, care should be taken not to overtighten. If square washers are used then the side length and thickness must be not less than the diameter and thickness of the equivalent round washer.

Spacing of bolts should be in accordance with Table 18.21, having due regard to the direction of load to grain. When a load is applied at an angle to the grain, the recommended spacings, both parallel and perpendicular, whichever are the larger, are applicable. No reductions in spacings are permitted even when the applied load is less than the bolt capacity for the joint.

18.5 TOOTHED PLATE CONNECTOR UNITS

18.5.1 General

Toothed plate connectors (section 1.6.7) are frequently used for small to medium span domestic trusses and for other assemblies, when loadings at joints are too high for simple bolts but too low to require the use of split ring or shear plate connectors. Special equipment is needed to form a toothed plate joint as described in 6.7.4.1 of BS 5268-2. Toothed plates may be used in all softwood species and low strength hardwoods of strength classes D30 to D40. Toothed plates are not suitable for denser hardwood. Basic loads in Table 86 of BS 5268-2 are limited to softwoods.

Table 18.21 Minimum spacing, edge and end distances for bolts

Load parallel to grain	Load perpendicular to grain
7d 4d 1.5d 4d 1.5d Loaded End	4d 4d 1.5d 4d 4d
4d 4d 1.5d 4d 1.5d Unloaded End	

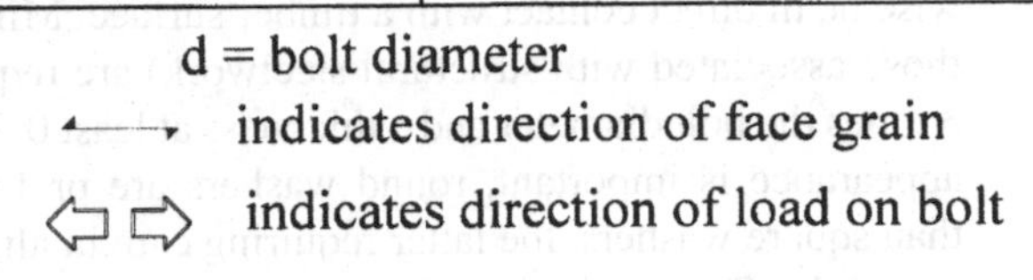

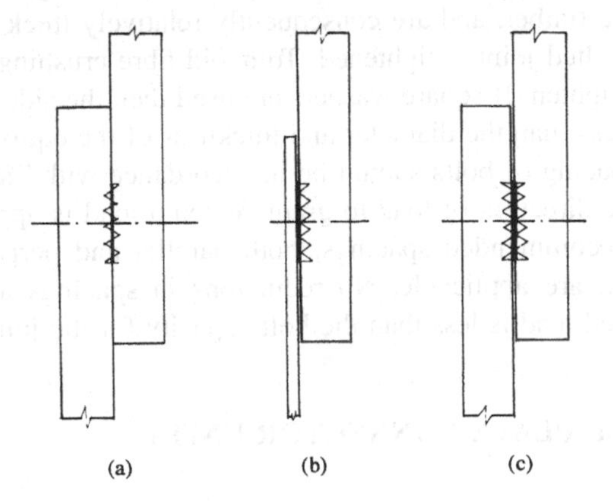

Fig. 18.15

A toothed plate connector unit consists of one or other of the following:

- one double-sided connector with bolt in single shear (Fig. 18.15(a)).
- one single-sided connector with bolt in single shear, used with a steel plate in a timber-to-metal joint (Fig. 18.15(b)).
- two single-sided connectors used back-to-back in the contact faces of a timber-to-timber joint with bolt in single shear (Fig. 18.15(c)).

BS 5268-2 limits designers to single- and double-sided, round and square toothed plates of 38, 51, 64 and 76 mm diameter (or nominal side of square).

Where the nut or head of the bolt occurs on a timber surface a washer must be provided. These washers are larger in diameter than those recommended for bolted joints.

18.5.2 Permissible loadings

Table 86 of BS 5268-2 gives the basic loads permitted for one toothed plate connector unit for loads parallel to grain and perpendicular to grain. Note that hardwoods are not tabulated because the timbers are too dense for the connectors to be successfully embedded, there being a risk that the teeth will deform.

A connector on both sides of a member and on the same bolt requires twice the thickness of timber required for a connector on one side only, if required to carry the same load.

The improved loading capacity over that for bolts is achieved by a reduction in stress concentrations and an enlargement of the timber area made available to resist shear, without simultaneously removing too much timber from the joint. Load is transferred between members by bearing stresses developed between the teeth of the connector and the timber, and by bearing stresses developed between the bolt and the timber. Some 40% of the ultimate load is carried by the bolt, consequently it is important always to include a bolt, and to use the correct bolt diameter associated with the toothed plate. The triangular teeth of the plates are shaped in such a way that they tend to lock into the timber as the joint is formed and therefore tend to give a relatively stiff joint.

The basic load at an angle to the grain is calculated using the Hankinson formula as described for bolts in section 18.4.3.

The permissible load for a toothed plate connector is given by the formula:

$$\text{Permissible load} = \text{basic load} \times K_{58} \times K_{59} \times K_{60} \times K_{61}$$

where K_{58} is the modification factor for duration of loading:

K_{58} = 1.0 for long-term loading
= 1.12 for medium-term loading
= 1.25 for short-term and very short-term loading

K_{59} is the modification factor for moisture content:

K_{59} = 1.0 for a joint in service classes 1 or 2
= 0.7 for a joint in service class 3

K_{60} is the modification factor for connector spacing:

K_{60} = 1.0 for a joint with standard end distance, edge distance and spacing
= the lesser value of K_C, K_D or K_S which are the modification factors for sub-standard end distance, edge distance and spacing respectively

K_{61} is the modification factor for 'in-line' fixings (see section 18.1.3)

$K_{61} = 1 - 0.003(n - 1)$ for $n \leq 10$
= 0.7 for $n > 10$, where n = number of connectors in line parallel to load.

18.5.3 Standard and sub-standard placement of connectors

To develop the full basic load (i.e. for $K_{60} = 1.0$) toothed plate connectors must have the required standard end distance C_{st}, edge distance D_{st} and spacing S_{st}, which can be expressed in terms of the toothed plate diameter d. (In the case of a square connector d is taken as the nominal side dimension plus 6 mm.)

End distance

Referring to Fig. 18.16, end distance C is the distance measured parallel to the grain from the centre of a connector to the square-cut end of the member. If the end is splay cut, the end distance should be taken as the distance measured parallel to the grain taken from a point $0.25d$ from the centreline of the connector.

The end distance is said to be loaded when the force on the connector has a component acting towards the end of the member, and is applicable to forces acting from $\alpha = 0°$ to $90°$ (Fig. 18.17).

For a loaded end, $K_C = 1.0$ at $C_{st} = 0.5d + 64$ mm and K_C reduces for sub-standard end distances to a variable degree, depending on the type and size of connector. The minimum value of K_C for sub-standard end distances varies between 0.85 and 0.5 and always occurs at $C_{min} = 0.5d + 13$ mm for any given case.

There is no change in the values of C_{st} and C_{min} as the angle α varies from 0° to 90° and for intermediate values of loaded end distance linear interpolation is permitted.

The end distance is said to be unloaded when the force on the connector has no component acting towards the end, and applies to forces acting from $\alpha = 90°$ to 0°, as shown in Fig. 18.17.

For an unloaded end, if the load is at $\alpha = 90°$ to grain

$$C_{st} = 0.5d + 64\,\text{mm} \quad \text{and} \quad C_{min} = 0.5d + 13\,\text{mm}$$

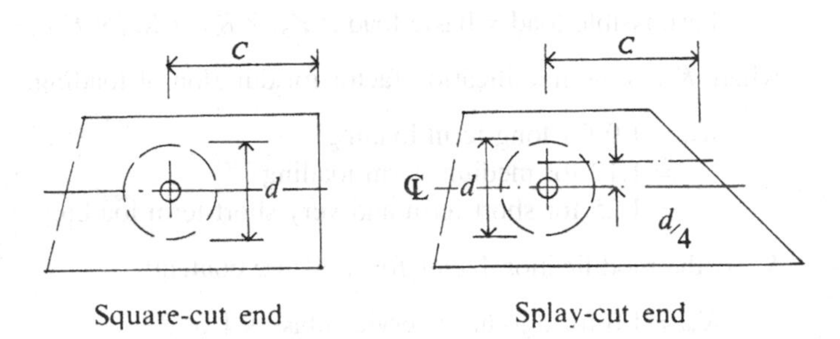

Fig. 18.16

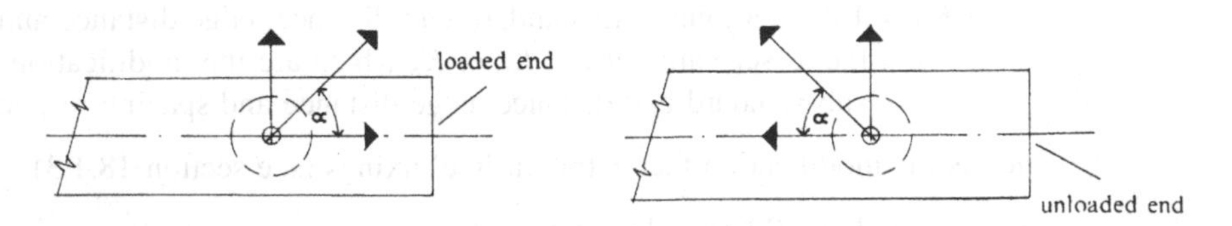

Fig. 18.17

For an unloaded end, if the load is at $\alpha = 0°$ to grain

$$C_{st} = C_{min} = 0.5d + 13\,\text{mm} \quad \text{and} \quad K_C = 1.0$$

For values between $\alpha = 0°$ and $90°$ the value of C_{st} is determined by linear interpolation.

Edge distance

Referring to Fig. 18.18, edge distance D is the distance from the edge of the member to the centre of the connector measured perpendicular to the edge. If the edge is splay cut, the perpendicular distance from the centre of the connector to the splay edge should not be less than the edge distance D.

The edge distance is said to be loaded or unloaded respectively according to whether or not the force on the connector has a component acting towards the edge of the member (Fig. 18.19).

In the case of toothed plate connectors, the standard edge distance D_{st} is also the minimum edge distance and applies to both the loaded and the unloaded edge, and $D_{st} = D_{min} = 0.5d + 6\,\text{mm}$ for all directions of load to grain. Therefore there are no sub-standard edge distances for toothed plates and K_{60} is unaffected by edge distance.

Spacing

Spacing S is the distance between centres of adjacent connectors measured along a line joining their centres and known as the connector axis. Spacing can be parallel, perpendicular or at an angle to the grain, as shown in Fig. 18.20.

The spacing of connectors is determined by the intersection of a diameter of an ellipse with its perimeter. The coordinates A (parallel to grain) and B (perpendicular to grain), as shown in Fig. 18.22, depend on the angle α of load to grain and

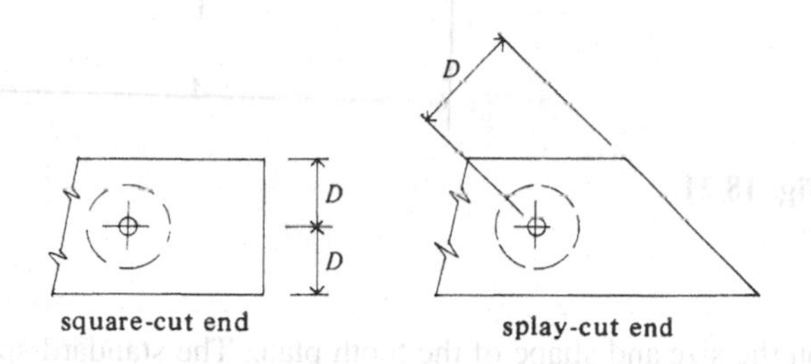

Fig. 18.18

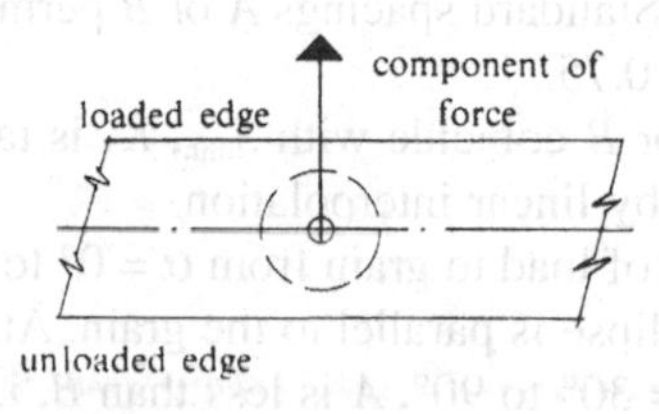

Fig. 18.19

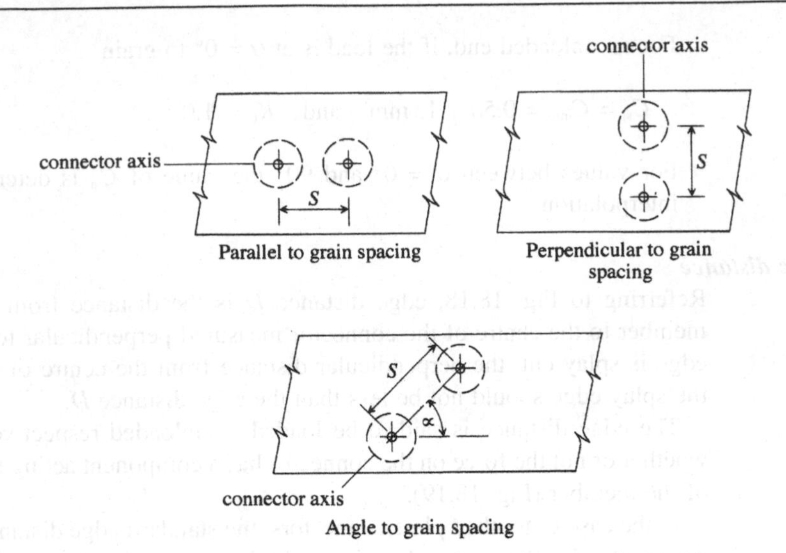

Fig. 18.20

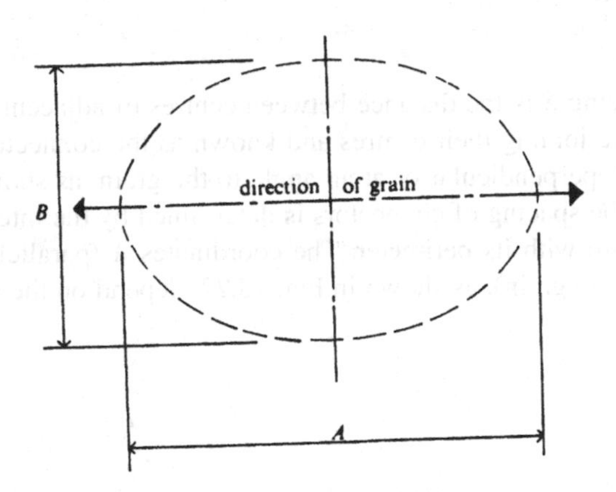

Fig. 18.21

on the size and shape of the tooth plate. The standard spacing A (parallel to grain) and B (perpendicular to grain) and the minimum spacing S_{min} for loads parallel and perpendicular to the grain are expressed in terms of the connector diameter d in Table 18.22. Standard spacings A or B permit $K_S = 1.0$ and minimum spacing S_{min} permits $K_S = 0.75$.

Where A or B coincide with S_{min}, K_S is taken as 1.0. Intermediate values of K_S are obtained by linear interpolation.

For angles of load to grain from $\alpha = 0°$ to $30°$, A is greater than B, i.e. the major axis of the ellipse is parallel to the grain. At $\alpha = 30°$, the ellipse becomes a circle, and from $\alpha = 30°$ to $90°$, A is less than B, i.e. the minor axis of the ellipse is parallel to the grain. These cases are illustrated in Fig. 18.22.

Table 18.22 Spacing of toothed plate connectors

Angle of load to grain α	A	B	S_{min}
$0°$	$1.5d$	$d + 13$	$d + 13$
$90°$	$d + 13$	$1.5d$	$d + 13$

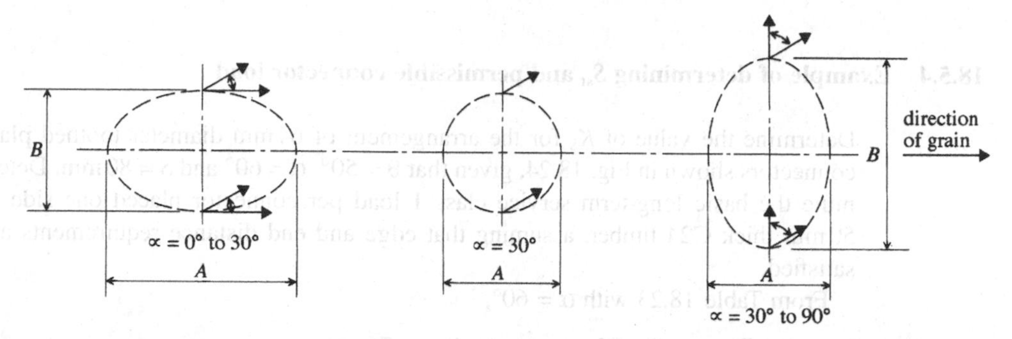

Fig. 18.22

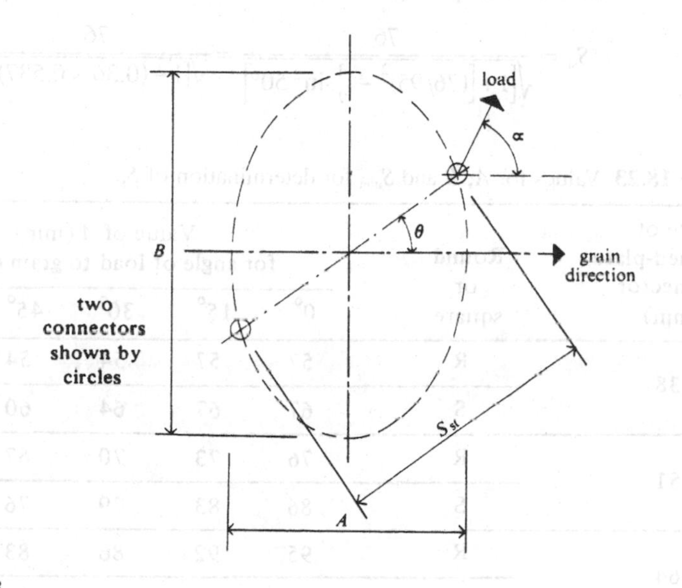

Fig. 18.23

When the load on a connector acts in a direction of α to the grain and the connector axis is at an angle θ to grain (Fig. 18.23), the standard spacing of connectors giving K_S may be determined from the equation:

$$S_{st} = \frac{AB}{\sqrt{(A^2 \sin^2 \theta + B^2 \cos^2 \theta)}} = \frac{A}{\sqrt{\{1 + [(A/B)^2 - 1]\sin^2 \theta\}}}$$

$K_S = 0.75$ at S_{min} and intermediate values of K_S can be obtained by linear interpolation between S_{st} and S_{min}:

$$K_S = 0.75 + 0.25\left[\frac{S - S_{min}}{S_{st} - S_{min}}\right]$$

where S is the actual spacing.

Values of A, B and S_{min} for angle of load to grain are given for toothed plates in Table 18.23.

18.5.4 Example of determining S_{st} and permissible connector load

Determine the value of K_S for the arrangement of 64 mm diameter toothed plate connectors shown in Fig. 18.24, given that $\theta = 50°$, $\alpha = 60°$ and $S = 80$ mm. Determine the basic long-term service class 1 load per connector placed one side in 50 mm thick C24 timber, assuming that edge and end distance requirements are satisfied.

From Table 18.23 with $\alpha = 60°$,

$$A = 76\,\text{mm},\ B = 95\,\text{mm} \quad \text{and} \quad S_{min} = 76\,\text{mm}$$

The standard spacing at $\theta = 50°$ is:

$$S_{st} = \frac{76}{\sqrt{\left[1 + \left[(76/95)^2 - 1\right]\sin^2 50°\right]}} = \frac{76}{\sqrt{[1 - (0.36 \times 0.587)]}} = 86\,\text{mm}$$

Table 18.23 Values for A, B and S_{min} for determination of S_{st}

Size of toothed-plate connector (mm)	Round or square	Value of A (mm) for angle of load to grain α (deg)					
		0°	15°	30°	45°	60–90°	S_{min}
38	R	57	57	54	54	51	51
	S	67	67	64	60	57	57
51	R	76	73	70	67	64	64
	S	86	83	79	76	70	70
64	R	95	92	86	83	76	76
	S	105	102	95	89	83	83
76	R	114	108	102	95	89	89
	S	124	117	111	102	95	95
		90–60°	45°	30°	15°	0°	
		Value of B (mm) for angle of load to grain α (deg)					

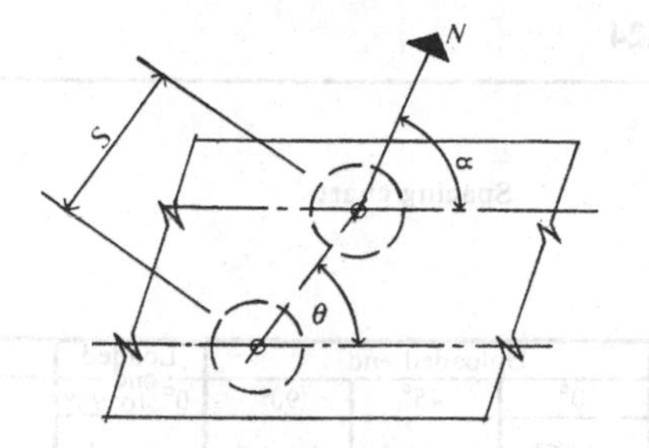

Fig. 18.24

$$K_S = 0.75 + 0.25\left(\frac{80-76}{86-76}\right) = 0.85$$

From Table 80 of BS 5268-2,

$$F_0 = 4.54 \text{ kN}, \quad F_{90} = 3.75 \text{ kN} \quad \text{and} \quad \frac{F_0}{F_{90}} = 1.21$$

$$F_{60} = \frac{F_0}{1+[(F_0/F_{90})-1]\sin^2 60°}$$

$$= \frac{4.54}{1+(0.21\times 0.75)} = 3.92 \text{ kN}$$

The basic load per connector = $F_{60} \times K_S = 3.92 \times 0.85 = 3.33$ kN

18.5.5 Charts for the standard placement of toothed plate connectors

Sections 18.5.2 and 18.5.3 detail the background study into the determination of permissible loads on toothed plates and the formulae for establishing suitable placement of connectors with respect to one another and with respect to ends and edges of timber. To simplify the determination of loads and spacings for office use, Tables 18.24–18.31 give a graphical solution.

Table 18.24

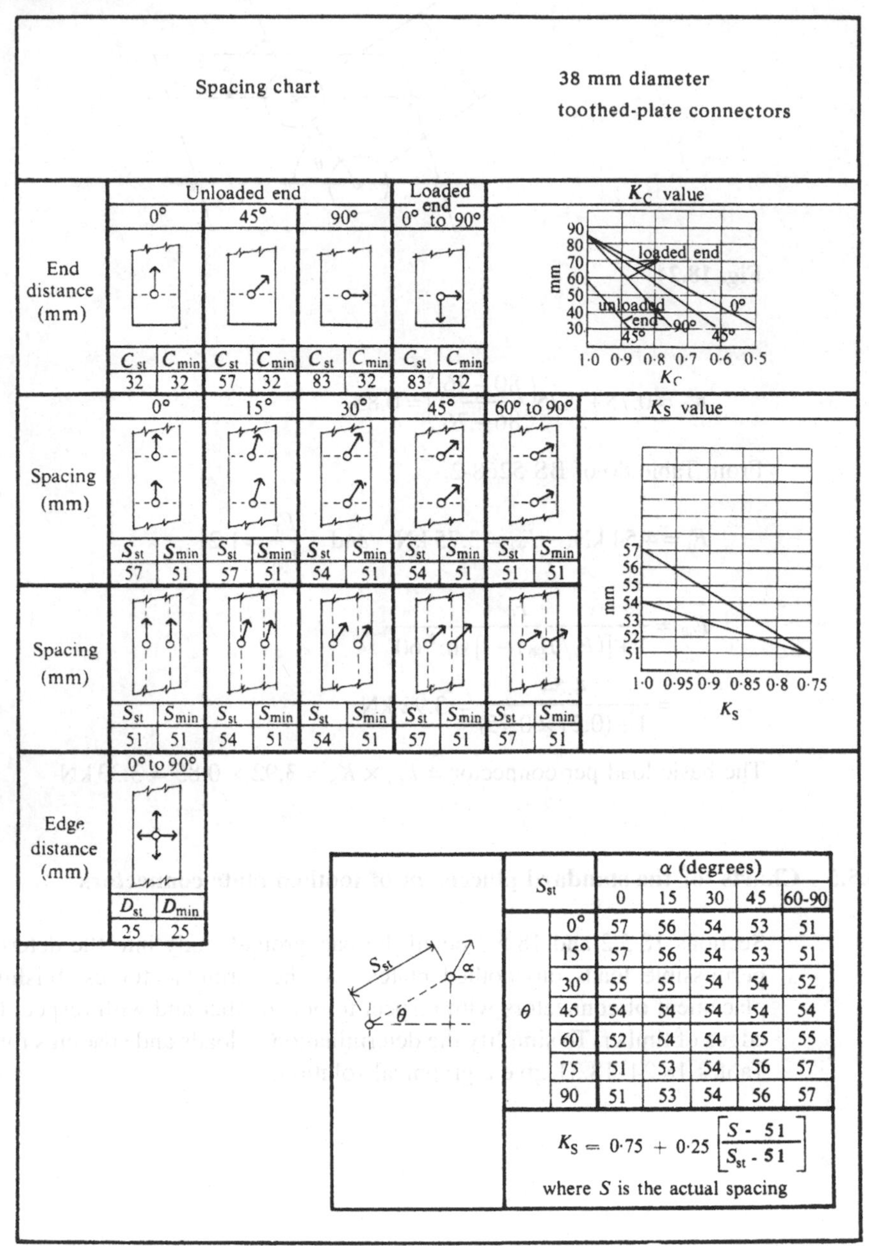

Spacing chart — 38 mm diameter toothed-plate connectors

End distance (mm)

Unloaded end 0°		Unloaded end 45°		Unloaded end 90°		Loaded end 0° to 90°	
C_{st}	C_{min}	C_{st}	C_{min}	C_{st}	C_{min}	C_{st}	C_{min}
32	32	57	32	83	32	83	32

Spacing (mm)

0°		15°		30°		45°		60° to 90°	
S_{st}	S_{min}	S_{st}	S_{min}	S_{st}	S_{min}	S_{st}	S_{min}	S_{st}	S_{min}
57	51	57	51	54	51	54	51	51	51

Spacing (mm)

S_{st}	S_{min}	S_{st}	S_{min}	S_{st}	S_{min}	S_{st}	S_{min}	S_{st}	S_{min}
51	51	54	51	54	51	57	51	57	51

Edge distance (mm)

0° to 90°	
D_{st}	D_{min}
25	25

S_{st}	θ \ α (degrees)	0	15	30	45	60-90
	0°	57	56	54	53	51
	15°	57	56	54	53	51
	30°	55	55	54	54	52
	45	54	54	54	54	54
	60	52	54	54	55	55
	75	51	53	54	56	57
	90	51	53	54	56	57

$$K_S = 0{\cdot}75 + 0{\cdot}25\left[\frac{S - 51}{S_{st} - 51}\right]$$

where S is the actual spacing

Table 18.25

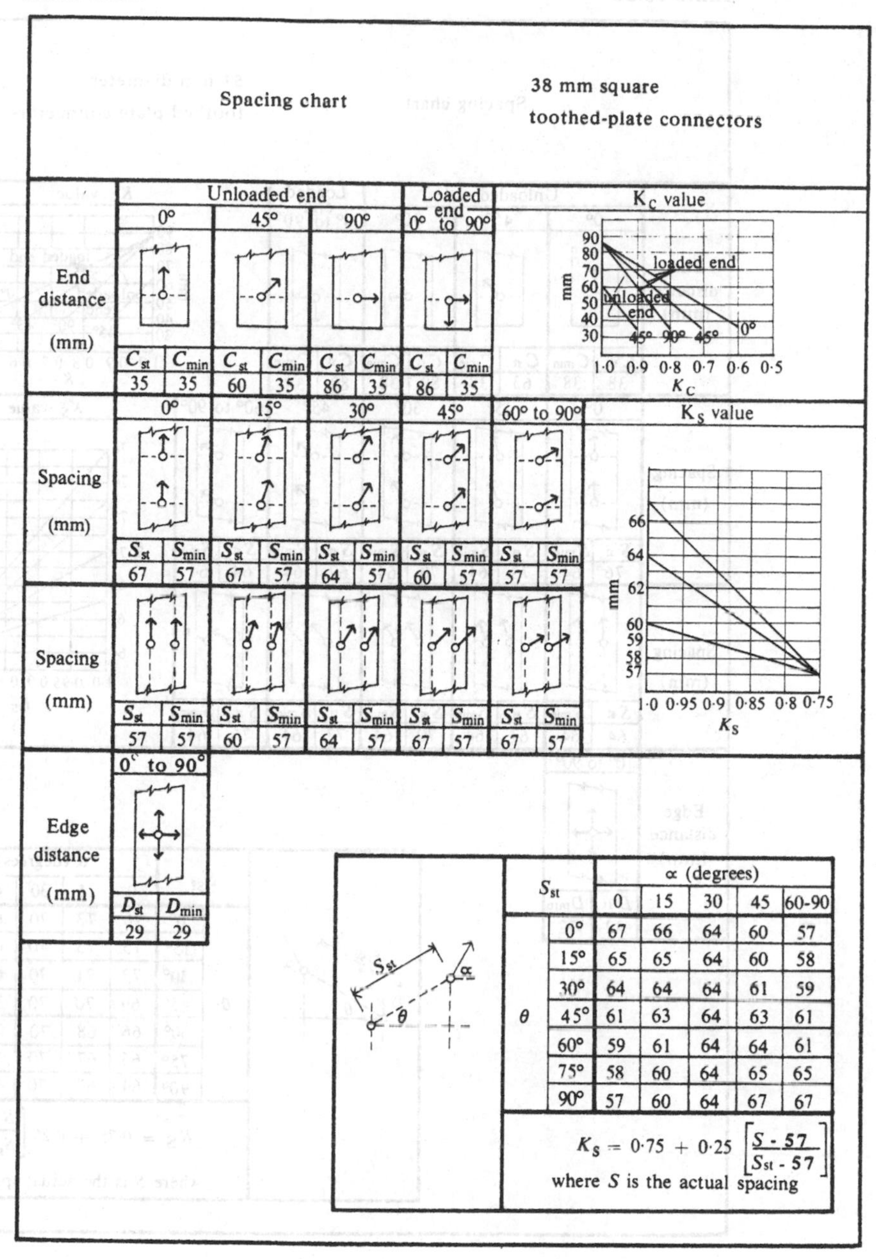

Spacing chart — 38 mm square toothed-plate connectors

End distance (mm)

	Unloaded end 0°		Unloaded end 45°		Unloaded end 90°		Loaded end 0° to 90°	
	C_{st}	C_{min}	C_{st}	C_{min}	C_{st}	C_{min}	C_{st}	C_{min}
End distance (mm)	35	35	60	35	86	35	86	35

	0°		15°		30°		45°		60° to 90°	
	S_{st}	S_{min}	S_{st}	S_{min}	S_{st}	S_{min}	S_{st}	S_{min}	S_{st}	S_{min}
Spacing (mm)	67	57	67	57	64	57	60	57	57	57
Spacing (mm)	57	57	60	57	64	57	67	57	67	57

	0° to 90°	
	D_{st}	D_{min}
Edge distance (mm)	29	29

S_{st}		α (degrees) 0	15	30	45	60-90
θ	0°	67	66	64	60	57
	15°	65	65	64	60	58
	30°	64	64	64	61	59
	45°	61	63	64	63	61
	60°	59	61	64	64	61
	75°	58	60	64	65	65
	90°	57	60	64	67	67

$$K_S = 0{\cdot}75 + 0{\cdot}25\left[\frac{S - 57}{S_{st} - 57}\right]$$

where S is the actual spacing

Table 18.26

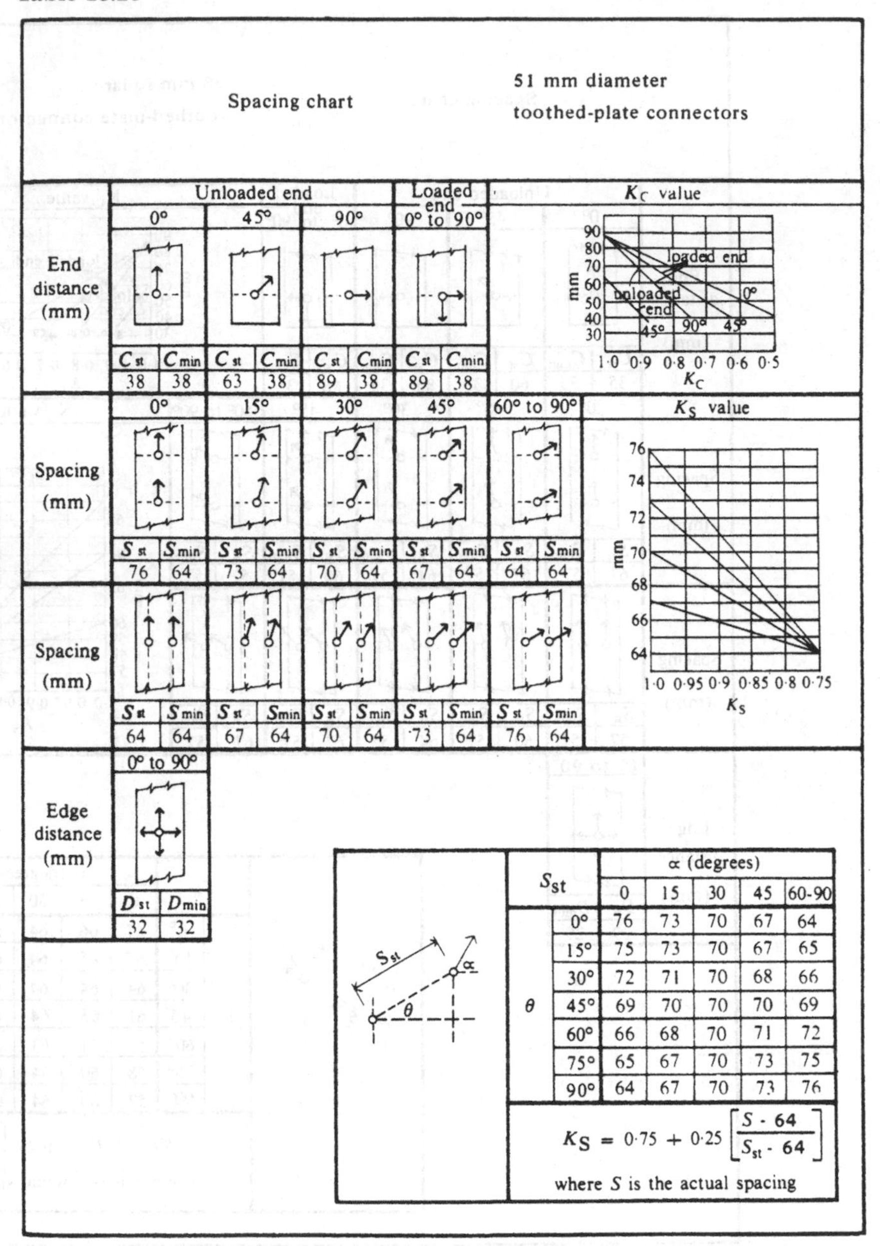

Spacing chart — 51 mm diameter toothed-plate connectors

End distance (mm)

	Unloaded end 0°		Unloaded end 45°		Unloaded end 90°		Loaded end 0° to 90°	
	C_{st}	C_{min}	C_{st}	C_{min}	C_{st}	C_{min}	C_{st}	C_{min}
End distance (mm)	38	38	63	38	89	38	89	38

Spacing (mm)

	0°		15°		30°		45°		60° to 90°	
	S_{st}	S_{min}	S_{st}	S_{min}	S_{st}	S_{min}	S_{st}	S_{min}	S_{st}	S_{min}
Spacing (mm)	76	64	73	64	70	64	67	64	64	64
Spacing (mm)	64	64	67	64	70	64	73	64	76	64

Edge distance (mm)

	0° to 90°	
	D_{st}	D_{min}
Edge distance (mm)	32	32

S_{st}		α (degrees) 0	15	30	45	60-90
θ	0°	76	73	70	67	64
	15°	75	73	70	67	65
	30°	72	71	70	68	66
	45°	69	70	70	70	69
	60°	66	68	70	71	72
	75°	65	67	70	73	75
	90°	64	67	70	73	76

$$K_S = 0{\cdot}75 + 0{\cdot}25\left[\frac{S - 64}{S_{st} - 64}\right]$$

where S is the actual spacing

Table 18.27

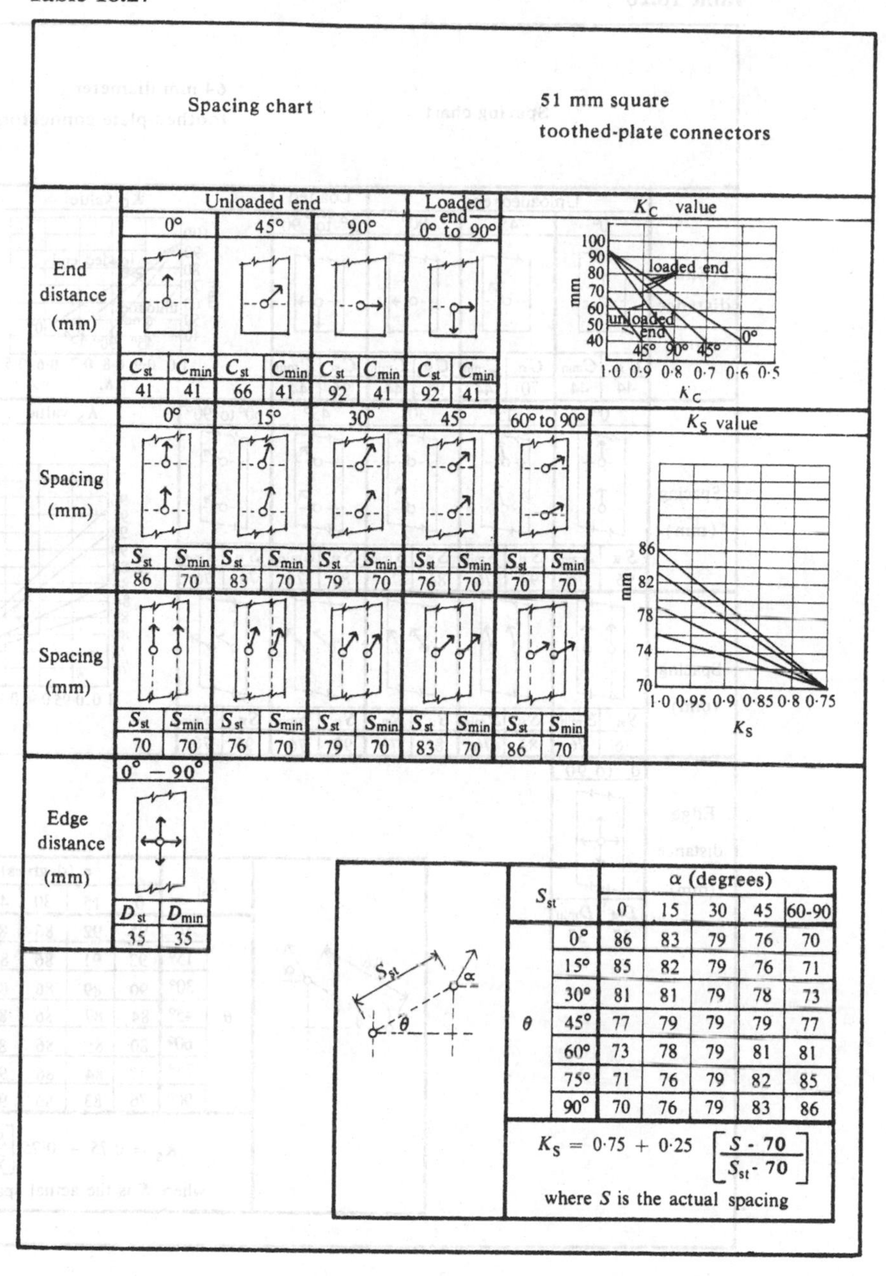
Spacing chart

51 mm square toothed-plate connectors

End distance (mm)

	Unloaded end 0°		Unloaded end 45°		Unloaded end 90°		Loaded end 0° to 90°	
	C_{st}	C_{min}	C_{st}	C_{min}	C_{st}	C_{min}	C_{st}	C_{min}
End distance (mm)	41	41	66	41	92	41	92	41

Spacing (mm)

	0°		15°		30°		45°		60° to 90°	
	S_{st}	S_{min}	S_{st}	S_{min}	S_{st}	S_{min}	S_{st}	S_{min}	S_{st}	S_{min}
Spacing (mm)	86	70	83	70	79	70	76	70	70	70
Spacing (mm)	70	70	76	70	79	70	83	70	86	70

Edge distance (mm)

	0° – 90°	
	D_{st}	D_{min}
Edge distance (mm)	35	35

S_{st}		α (degrees) 0	15	30	45	60-90
θ	0°	86	83	79	76	70
	15°	85	82	79	76	71
	30°	81	81	79	78	73
	45°	77	79	79	79	77
	60°	73	78	79	81	81
	75°	71	76	79	82	85
	90°	70	76	79	83	86

$$K_S = 0{\cdot}75 + 0{\cdot}25 \left[\frac{S - 70}{S_{st} - 70}\right]$$

where S is the actual spacing

Table 18.28

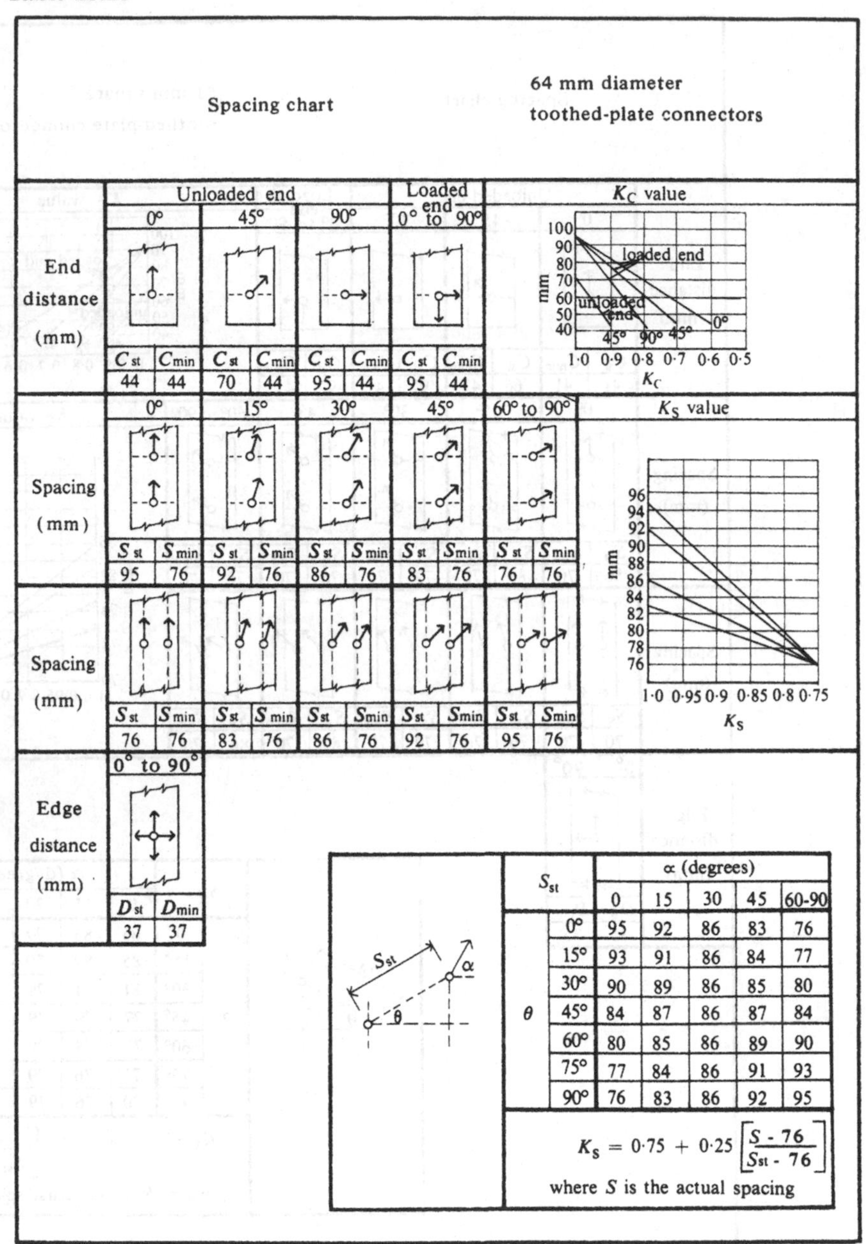

Spacing chart	64 mm diameter toothed-plate connectors

End distance (mm)

	Unloaded end 0°		Unloaded end 45°		Unloaded end 90°		Loaded end 0° to 90°	
	C_{st}	C_{min}	C_{st}	C_{min}	C_{st}	C_{min}	C_{st}	C_{min}
End distance (mm)	44	44	70	44	95	44	95	44

Spacing (mm)

	0°		15°		30°		45°		60° to 90°	
	S_{st}	S_{min}	S_{st}	S_{min}	S_{st}	S_{min}	S_{st}	S_{min}	S_{st}	S_{min}
Spacing (mm)	95	76	92	76	86	76	83	76	76	76
Spacing (mm)	76	76	83	76	86	76	92	76	95	76

Edge distance (mm)

	0° to 90°	
	D_{st}	D_{min}
Edge distance (mm)	37	37

S_{st}		α (degrees) 0	15	30	45	60-90
θ	0°	95	92	86	83	76
	15°	93	91	86	84	77
	30°	90	89	86	85	80
	45°	84	87	86	87	84
	60°	80	85	86	89	90
	75°	77	84	86	91	93
	90°	76	83	86	92	95

$$K_S = 0{\cdot}75 + 0{\cdot}25\left[\frac{S - 76}{S_{st} - 76}\right]$$

where S is the actual spacing

Table 18.29

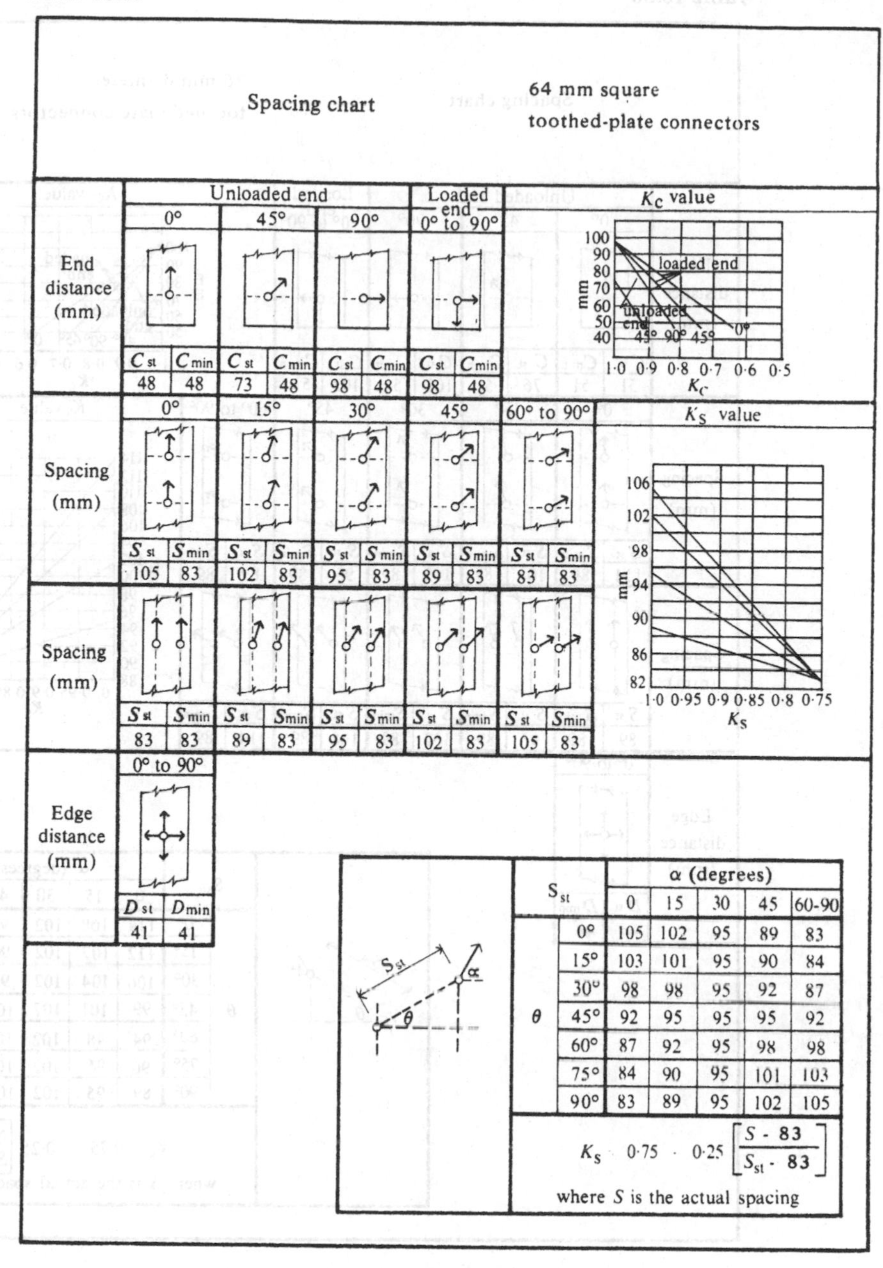

Spacing chart — 64 mm square toothed-plate connectors

End distance (mm)

Unloaded end 0°		Unloaded end 45°		Unloaded end 90°		Loaded end 0° to 90°	
C_{st}	C_{min}	C_{st}	C_{min}	C_{st}	C_{min}	C_{st}	C_{min}
48	48	73	48	98	48	98	48

Spacing (mm) (connectors in line with grain)

0°		15°		30°		45°		60° to 90°	
S_{st}	S_{min}	S_{st}	S_{min}	S_{st}	S_{min}	S_{st}	S_{min}	S_{st}	S_{min}
105	83	102	83	95	83	89	83	83	83

Spacing (mm) (connectors across grain)

0°		15°		30°		45°		60° to 90°	
S_{st}	S_{min}	S_{st}	S_{min}	S_{st}	S_{min}	S_{st}	S_{min}	S_{st}	S_{min}
83	83	89	83	95	83	102	83	105	83

Edge distance (mm)

0° to 90°	
D_{st}	D_{min}
41	41

S_{st}		α (degrees) 0	15	30	45	60-90
θ	0°	105	102	95	89	83
	15°	103	101	95	90	84
	30°	98	98	95	92	87
	45°	92	95	95	95	92
	60°	87	92	95	98	98
	75°	84	90	95	101	103
	90°	83	89	95	102	105

$$K_S = 0{\cdot}75 + 0{\cdot}25\left[\frac{S - 83}{S_{st} - 83}\right]$$

where S is the actual spacing

Table 18.30

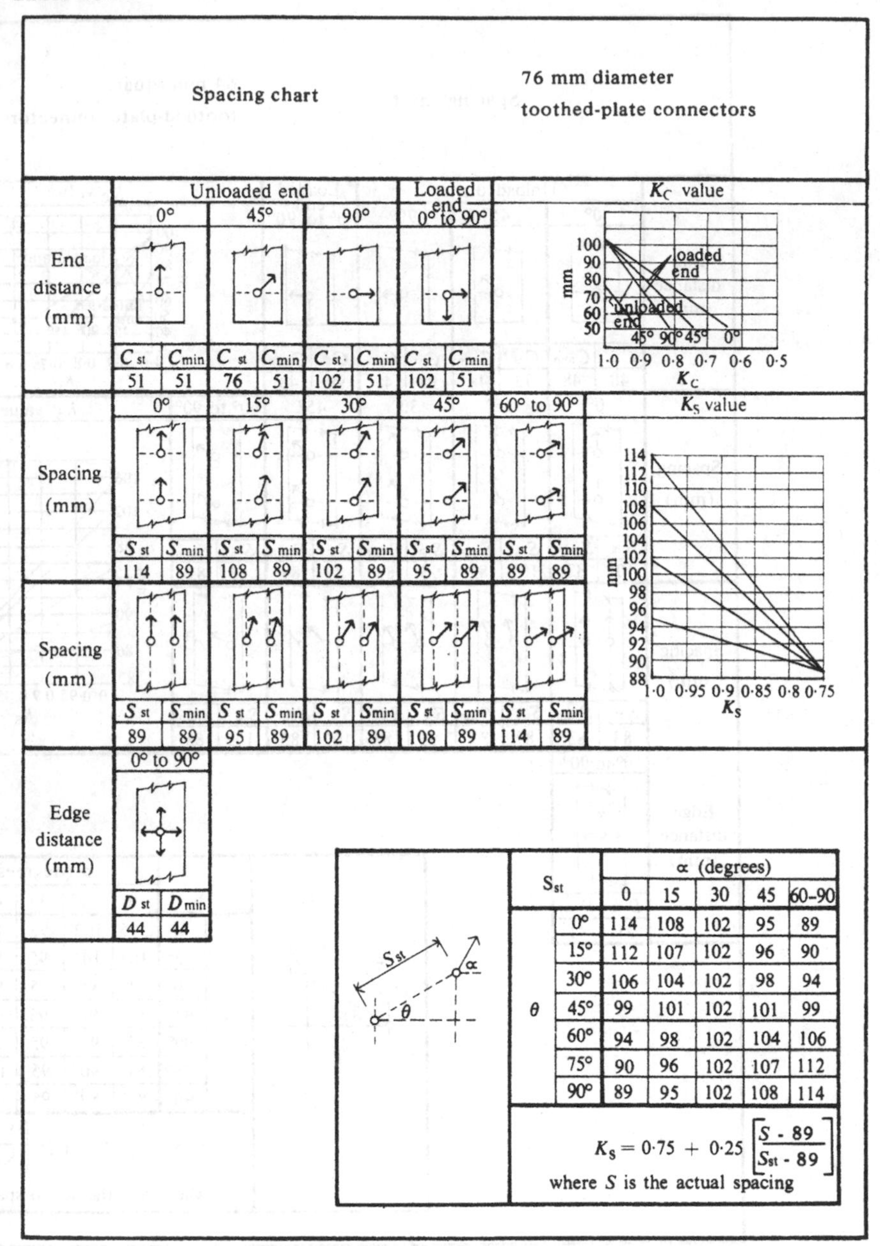

Spacing chart — **76 mm diameter toothed-plate connectors**

End distance (mm)

	Unloaded end 0°		Unloaded end 45°		Unloaded end 90°		Loaded end 0° to 90°	
	C_{st}	C_{min}	C_{st}	C_{min}	C_{st}	C_{min}	C_{st}	C_{min}
End distance (mm)	51	51	76	51	102	51	102	51

Spacing (mm)

	0°		15°		30°		45°		60° to 90°	
	S_{st}	S_{min}	S_{st}	S_{min}	S_{st}	S_{min}	S_{st}	S_{min}	S_{st}	S_{min}
Spacing (mm)	114	89	108	89	102	89	95	89	89	89
Spacing (mm)	89	89	95	89	102	89	108	89	114	89

Edge distance (mm)

	0° to 90°	
	D_{st}	D_{min}
Edge distance (mm)	44	44

S_{st}		α (degrees) 0	15	30	45	60–90
θ	0°	114	108	102	95	89
	15°	112	107	102	96	90
	30°	106	104	102	98	94
	45°	99	101	102	101	99
	60°	94	98	102	104	106
	75°	90	96	102	107	112
	90°	89	95	102	108	114

$$K_S = 0{\cdot}75 + 0{\cdot}25\left[\frac{S - 89}{S_{st} - 89}\right]$$

where S is the actual spacing

Table 18.31

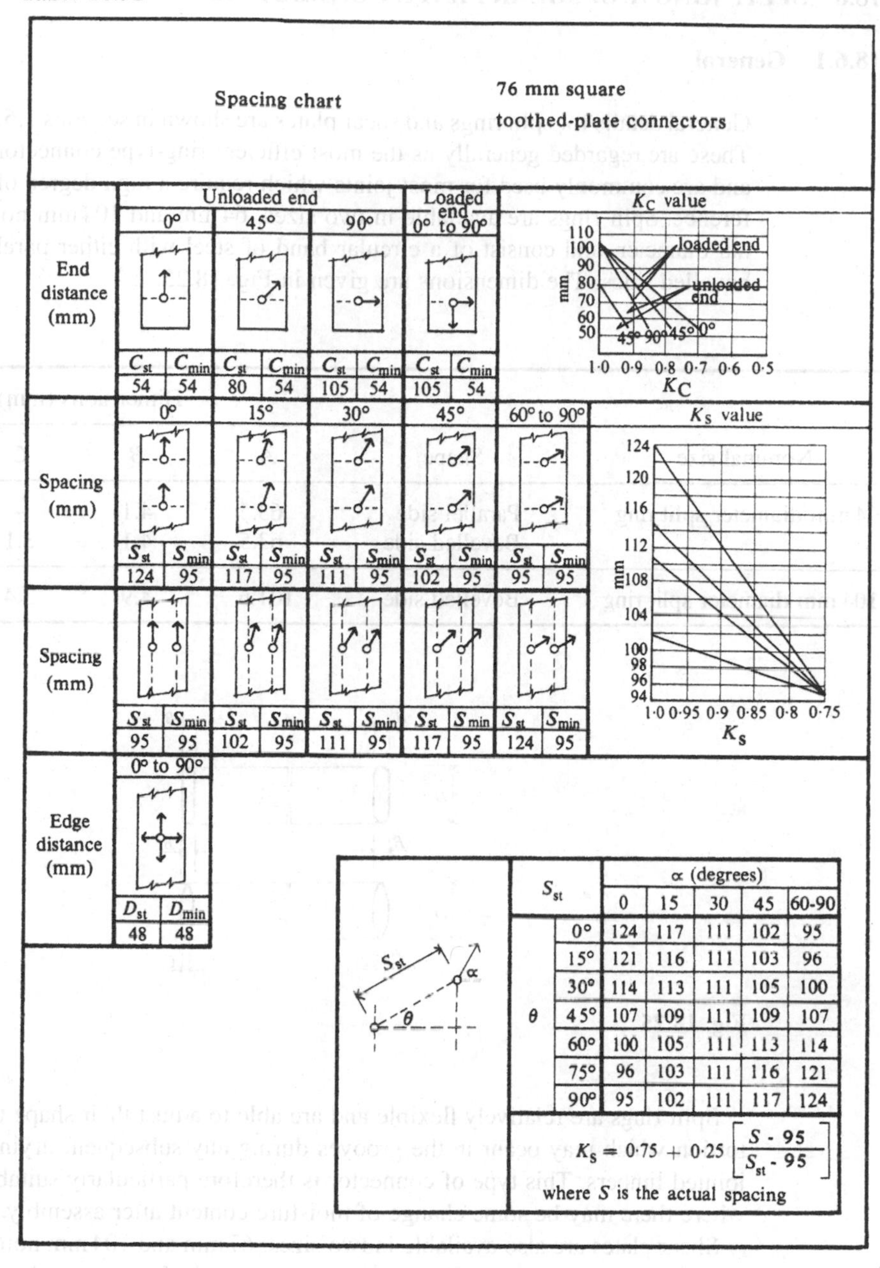

Spacing chart — 76 mm square toothed-plate connectors

End distance (mm)

	Unloaded end 0°		Unloaded end 45°		Unloaded end 90°		Loaded end 0° to 90°	
	C_{st}	C_{min}	C_{st}	C_{min}	C_{st}	C_{min}	C_{st}	C_{min}
End distance (mm)	54	54	80	54	105	54	105	54

Spacing (mm)

	0°		15°		30°		45°		60° to 90°	
	S_{st}	S_{min}	S_{st}	S_{min}	S_{st}	S_{min}	S_{st}	S_{min}	S_{st}	S_{min}
Spacing (mm)	124	95	117	95	111	95	102	95	95	95
Spacing (mm)	95	95	102	95	111	95	117	95	124	95

Edge distance (mm)

	0° to 90°	
	D_{st}	D_{min}
Edge distance (mm)	48	48

S_{st}		α (degrees) 0	15	30	45	60-90
θ	0°	124	117	111	102	95
	15°	121	116	111	103	96
	30°	114	113	111	105	100
	45°	107	109	111	109	107
	60°	100	105	111	113	114
	75°	96	103	111	116	121
	90°	95	102	111	117	124

$$K_s = 0{\cdot}75 + 0{\cdot}25\left[\frac{S - 95}{S_{st} - 95}\right]$$

where S is the actual spacing

18.6 SPLIT RING AND SHEAR PLATE CONNECTORS

18.6.1 General

General details for split rings and shear plates are shown in sections 1.6.8 and 1.6.9. These are regarded generally as the most efficient ring-type connectors available, and are commonly used for most joints which require a high degree of load transference. Split rings are available in two sizes, 64 mm and 104 mm nominal internal diameter, and consist of a circular band of steel with either parallel sides or bevelled sides. The dimensions are given in Fig. 18.25.

Nominal size	Shape	Dimensions (mm) A	B	C	D
64 mm diameter split ring	Parallel side	63.5	4.1	–	19
	Bevelled side	63.5	4.1	3.1	19
104 mm diameter split ring	Bevelled side	101.6	4.9	3.4	25.4

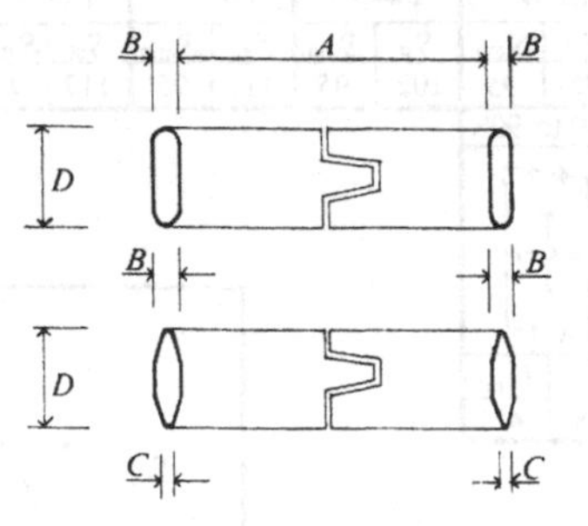

Fig. 18.25

Split rings are relatively flexible and are able to adjust their shape to offset distortion which may occur in the grooves during any subsequent drying out of the jointed timbers. This type of connector is therefore particularly suitable for joints where there may be some change of moisture content after assembly.

Shear plates are also available in two sizes: 67 mm and 104 mm nominal outside diameter. Shear plates of 67 mm diameter are made from pressed steel, whereas 104 mm diameter shear plates are made from malleable cast iron. Each varies significantly in detail, as illustrated in Fig. 18.26.

A split ring unit consists of one split ring with its bolt in single shear (Fig. 18.27). The 64 mm diameter split ring requires a 12 mm diameter bolt and the 104 mm diameter split ring requires a 20 mm diameter bolt. The 20 mm diameter bolt should also be used if 64 mm and 104 mm diameter split rings occur simultaneously at a joint concentric about a single bolt.

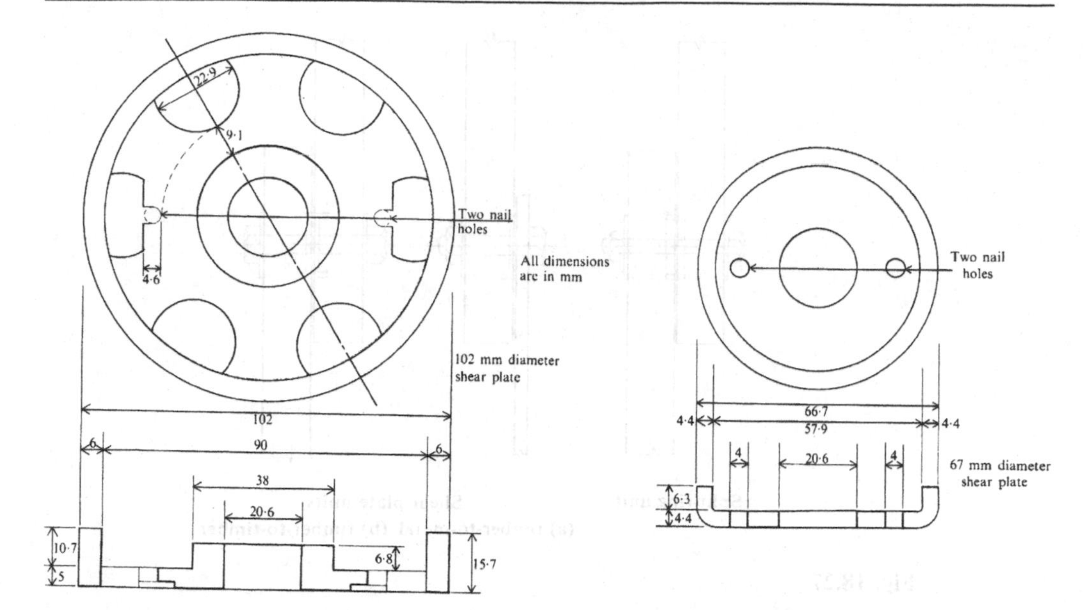

Fig. 18.26 Details of shear plates.

A shear plate unit consists of either (a) one shear plate connector with its bolt in single shear used with a steel side plate in a timber-to-metal joint, or (b) two shear plates used back-to-back in a timber-to-timber joint. Although this latter unit requires two shear plates to develop a load only slightly in excess of that provided by a single split ring, it can be justified where demountable joints are required or where components are site fixed. Split rings must be drawn into a jointing groove under pressure, whereas shear plates can be a sliding fit followed by a bolt insertion.

18.6.2 Permissible loadings

The load-carrying actions of split rings and shear plates differ considerably. A split ring connector is embedded in a pre-drilled groove, the groove being slightly greater in diameter than the ring. The tongue-and-groove slot in the connector (see Fig. 18.25) permits the split ring to expand into the pre-formed groove. The depth of the split ring is shared equally between two mating faces of timber at a joint. The capacity of the unit is the combined values of the connector taking approximately 75% of the total load and the associated bolt taking approximately 25% of the total load. To function to its optimum value, therefore, the bolt (of the correct size) should not be omitted.

If a split ring is subjected to load, having been sprung into position in the pre-formed groove (as shown in Fig. 18.28), compression stresses occur where the

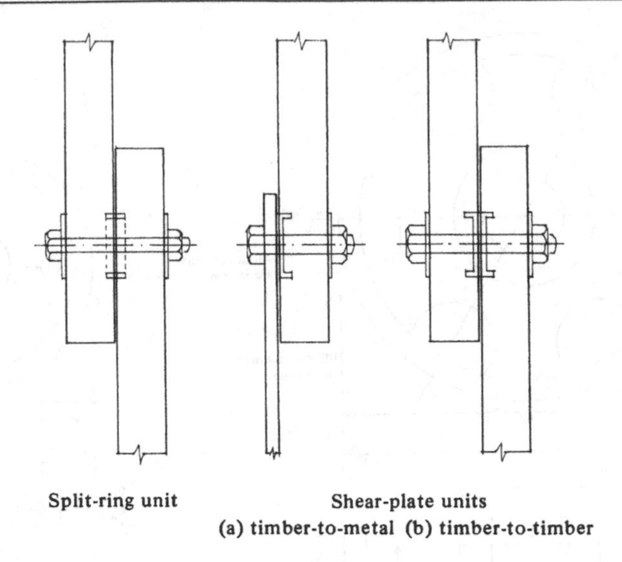

Fig. 18.27

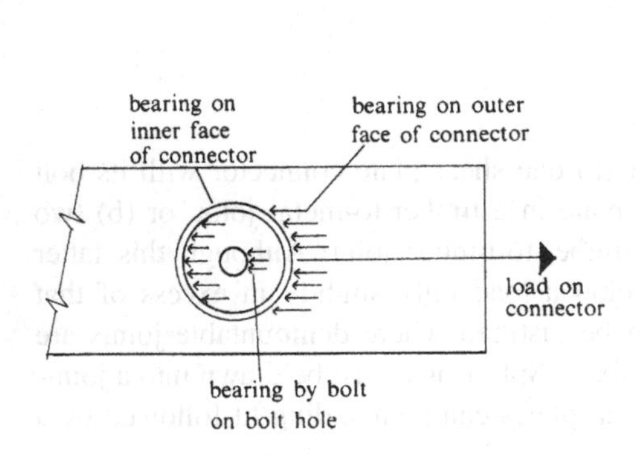

Fig. 18.28

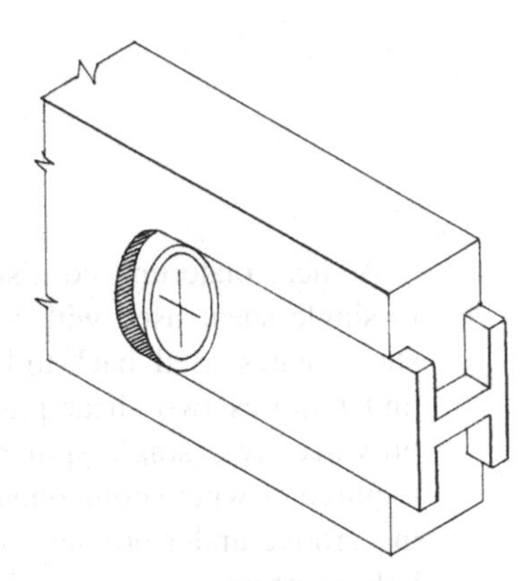

Fig. 18.29 Typical split ring mode of failure.

outside of the connector bears against the side of the groove and where the inside of the connector bears against the internal annulus of the joint. The bolt also bears against the side of the central hole. These compression forces under the rings must be resisted by shear forces in the timber at the end of the member, and consequently, because of timber's low resistance to horizontal shear stresses, the end distance greatly influences the force which can be transferred. Failure of a joint is usually accompanied by a shearing action and the displacement of an 'I section' piece of timber from under the bolt and ring, as illustrated in Fig. 18.29.

The action of a shear plate is to transfer peripheral bearing stresses from the timber via the disc of the shear plate to the central hole in the plate. Load transfer is completed by bearing stress from the plate onto the central bolt and shear resistance in the bolt. The 67 mm diameter pressed steel shear plate is limited by bearing at the central hole, whereas the 102 mm diameter malleable cast iron shear plate with an increased centre hub thickness has a permissible load limited by the shear capacity of the 20 mm diameter bolt.

Table 94 of BS 5268-2 gives the basic loads for one split ring connector unit. The values are listed in Table 18.32 for strength classes C14 to C30.

Table 98 of BS 5268-2 gives the basic loads for one shear plate connector unit. The values are listed in Table 18.33 for strength classes C14 to C30. Under no circumstances should the permissible load exceed the limiting values given in Table 18.34.

In addition to the design check on the adequacy of the connectors, the designer must also ensure that applied stresses on the net timber area of timber at the connection are within permitted limits. The projected area of split ring or shear plate connectors plus the area of the bolt hole which falls outside the projected area of the connector must be deducted from the gross area to give the net area. To aid the designer, particularly in the design of truss joints, values for loading parallel to the grain in C24 timbers surfaced in accordance with the recommendations for constructional timber (section 1.2.4.4) are given in Table 18.35. Strength class C24 is recommended for truss design especially when exposed to view as the higher grade is accompanied by improved appearance.

The basic load at an angle to the grain is calculated using the Hankinson formula described in section 18.4.7.

It has been shown that:

$$F_\alpha = \frac{F_0}{1+[(F_0/F_{90})-1]\sin^2\alpha}$$

A consideration of the ratio F_0/F_{90} for split rings and shear plates gives a standard value of 1.428 for all strength classes. Therefore the the above formula may be simplified to:

$$F_\alpha = K_\alpha F_0$$

where $K_\alpha = 1/(1 + 0.428\sin^2\alpha)$. (Values of K_α are given in Table 18.36.)

For a split ring connector

$$\text{Permissible load} = \text{basic load} \times K_{62} \times K_{63} \times K_{64} \times K_{65}$$

For a shear plate connector

$$\text{Permissible load} = \text{basic load} \times K_{66} \times K_{67} \times K_{68} \times K_{69}$$

where $K_{62} = K_{66}$ = the modification factor for duration of load, being
= 1.0 for permanent loading
= 1.25 for medium-term loading
= 1.5 for short-term or very short-term loading

$K_{63} = K_{67}$ = the modification factor for moisture content as discussed in section 18.1.2

$K_{64} = K_{68}$ = the lesser value of K_C, K_D and K_S which are the respective modification factors for sub-standard end distance, edge distance and spacing

$K_{65} = K_{69}$ = the modification factor for the number of connectors in each line as discussed in section 18.1.3.

Table 18.32 Basic loads for one split ring connector unit

Split ring diameter (mm)	Bolt size	Thickness of members*		Basic load, F_0, parallel to grain (kN)				Basic load, F_{90}, perpendicular to grain (kN)			
		Connector on one side only (mm)	Connectors on both sides and on same bolt (mm)	C14	C16/ 18/22	C24	C27/30	C14	C16/ 18/22	C24	C27/30
64	M12	22	32	5.23	5.38	5.85	6.28	3.66	3.77	4.09	4.39
		25	40	6.32	6.51	7.06	7.58	4.42	4.55	4.95	5.31
		29 or over	50 or over	7.68	7.91	8.58	9.21	5.38	5.54	6.01	6.45
102	M20	29	41	10.10	10.38	11.22	12.04	7.04	7.25	7.87	8.45
		32	50	11.50	11.82	12.78	13.72	8.01	8.25	8.95	9.61
		36	63	13.50	13.88	15.02	16.12	9.42	9.70	10.52	11.30
		40	72	14.30	14.72	15.98	17.14	9.98	10.26	11.12	11.94
		41 or over	75 or over	14.40	14.82	16.08	17.26	10.10	10.38	11.22	12.04

* Actual thickness. Intermediate thicknesses may be obtained by linear interpolation.

Table 18.33 Basic loads for one shear plate connector unit

Shear plate diameter (mm)	Bolt size	Thickness of members*		Basic load, F_0, parallel to grain (kN)				Basic load, F_0, perpendicular to grain (kN)			
		Connector on one side only (mm)	Connectors on both sides and on same bolt (mm)	C14	C16/ 18/22	C24	C27/30	C14	C16/ 18/22	C24	C27/30
67	M20	–	41	6.45	6.64	7.20	7.73	4.51	4.64	5.05	5.42
		–	50	7.72	7.95	8.62	9.26	5.40	5.56	6.04	6.49
		–	63	8.21	8.45	9.18	9.86	5.75	5.92	6.42	6.89
		41 and over	67 and over	8.32	8.56	9.30	9.97	5.82	5.99	6.51	6.99
102	M20	–	44	8.32	8.57	9.30	9.98	5.83	6.00	6.51	6.99
		–	50	9.18	9.44	10.24	10.98	6.42	6.61	7.18	7.71
		–	63	10.50	10.82	11.78	12.66	7.38	7.60	8.25	8.86
		41	–	11.70	12.04	13.06	14.02	8.17	8.41	9.13	9.80
		–	75	11.90	12.24	13.26	14.24	8.33	8.57	9.31	10.00
		44 and over	92 and over	12.50	12.88	14.02	15.06	8.79	9.05	9.84	10.56

* Actual thickness. Intermediate thicknesses may be obtained by linear interpolation.

Table 18.34 Limiting values for permissible loads on one shear plate connector unit*

Shear-plate diameter (mm)	Bolt size	All loading except wind (kN)	All loading including wind (kN)
67	M20	12.9	17.2
102	M20	22.1	29.5

*These values may cause a reduction in permissible loads obtained from Table 18.33.

Table 18.35 Basic connector capacities, F_0, parallel to grain in surfaced timber of strength class C24

			Connector capacity F_0 (kN)					
			Connector on one side of member			Connector on each side of member		
Type of connector	Diameter (mm)	Thickness of member (mm)	Long-term load	Medium-term load	Short- or very short-term load	Long-term load	Medium-term load	Short- or very short-term load
Split ring	64	35	8.6	10.7	12.9	6.3	7.9	9.4
		47	8.6	10.7	12.9	8.12	10.1	12.2
		60	8.6	10.7	12.9	8.58	10.7	12.9
		72	8.6	10.7	12.9	8.58	10.7	12.9
	104	35	14.5	18.1	21.7	×	×	×
		47	16.1	20.1	24.1	12.3	15.3	18.4
		60	16.1	20.1	24.1	14.5	18.1	21.7
		72	16.1	20.1	24.1	16.0	20.0	24.0
Shear plate	67	35	×	×	×	×	×	×
		47	9.3	11.6	13.9	8.2	10.3	12.3
		60	9.3	11.6	13.9	9.0	11.3	13.6
		72	9.3	11.6	13.9	9.3	11.6	13.9
	104	35	×	×	×	×	×	×
		47	14.0	17.5	21.0	9.8	12.2	14.6
		60	14.0	17.5	21.0	11.4	14.3	17.1
		72	14.0	17.5	21.0	12.9	16.1	19.3

× = not permitted.

18.6.3 Standard and sub-standard placement of connectors

To develop the full basic load split rings and shear plates must have the required standard end distance C_{st}, edge distance D_{st} and spacing S_{st}, which can be expressed in terms of the nominal connector diameter d.

In calculating end distance, edge distance and spacing (all as defined in section 18.5.3), d for a 67 mm diameter shear plate should be taken as 64 mm.

Table 18.36

α	K_α
0	1.0
5	0.997
10	0.987
15	0.972
20	0.952
25	0.929
30	0.903
35	0.877
40	0.850
45	0.824
50	0.799
55	0.777
60	0.757
65	0.740
70	0.726
75	0.715
80	0.707
85	0.702
90	0.700

End distance

With loaded end distance (Fig. 18.17), $K_C = 1.0$ at $C_{st} = d + 76$ mm, and the minimum value of K_C for sub-standard end distance is $K_C = 0.625$ at $C_{min} = 0.5C_{st} = 0.5d + 38$ mm. There is no change in the values of C_{st} and C_{min} as the angle of load to grain α varies from 0° to 90°. For intermediate values of loaded end distance linear interpolation of K_C is permitted.

When the end distance is unloaded at $\alpha = 0°$ (see Fig. 18.17), $K_C = 1.0$ at $C_{st} = d + 38$ mm and $K_C = 0.625$ at $C_{min} = 0.5d + 32$ mm with intermediate values of K_C determined by linear interpolation.

When the load acts perpendicular to grain ($\alpha = 90°$), the end distance is to be regarded as loaded; hence $K_C = 1.0$ at $C_{st} = d + 76$ mm and $K_C = 0.625$ at $C_{min} = 0.5d + 38$ mm. When the load acts at an intermediate angle to grain, C_{st} and C_{min} are reduced linearly with $C_{st} = d + 0.422\alpha + 38$ mm and $C_{min} = 0.5d + \alpha/15 + 32$ mm.

Edge distance

The minimum edge distance is $D_{min} = 44$ mm for 64 mm diameter split rings and 67 mm diameter shear plates, and $D_{min} = 70$ mm for 102 mm diameter split rings and shear plates.

For an unloaded edge distance, $D_{st} = D_{min}$ and $K_D = 1.0$ (see Figs 18.18 and 18.19).

For a loaded edge distance, $D_{st} = D_{min}$ at $\alpha = 0°$, whereas at $\alpha = 45°$ to 90°, $D_{st} = 70$ mm for 64 mm diameter split rings and 67 mm diameter shear plates and $D_{st} = 95$ mm for 102 mm diameter split rings and shear plates.

For intermediate angles of load to grain between $\alpha = 0°$ and 45° there is a linear variation in D_{st}.

Modification factor K_D is applied when the edge distance is sub-standard. For 64 mm diameter split rings and 67 mm diameter shear plates:

$$K_D = 1 - 0.17\left[\frac{\alpha}{45} - \frac{(D-44)}{26}\right] \not> \text{unity}$$

where α = angle of load to grain between 0° and 45°
D = actual edge distance (mm) in the range 44 mm to 70 mm.

For 102 mm diameter split rings and shear plates:

$$K_D = 1 - 0.17\left[\frac{\alpha}{45} - \frac{(D-70)}{25}\right] \not> \text{unity}$$

where α = angle of load to grain between 0° and 45°
D = actual edge distance (mm) in the range 70 mm to 95 mm.

If $\alpha > 45°$, $\alpha/45°$ is taken as unity.

Spacing

The spacing for split ring and shear plate connectors is determined in a similar way to that for toothed plate connectors. Referring to Fig. 18.22, the appropriate standard spacings A and B parallel and perpendicular to the grain respectively, and the minimum spacing S_{min} for loads parallel and perpendicular to the grain, can be expressed in terms of the nominal connector diameter as given in Table 18.37.

Values of A, B and S_{min} for various angles of load to grain are given in Table

Table 18.37 Spacing for split rings and shear plates

Angle of load to grain α (degrees)	A	B	S_{min}
0	$1.5d + 76$	$d + 25$	$d + 25$
90	$d + 25$	$1.5d + 12$	$d + 25$

Table 18.38 Values of A, B and S_{min} for the determination of S_{st}

Type and size of connector	Angle of load to grain (degrees)	A (mm)	B (mm)	S_{min} (mm)
64 mm split ring 67 mm shear plate	0	171	89	89
	15	152	95	
	30	130	98	
	45	108	105	
	60–90	89	108	
102 mm split ring 102 mm shear plate	0	229	127	127
	15	203	137	
	30	178	146	
	45	152	156	
	60–90	127	165	

18.38. When the load on a connector acts at an angle α to grain (see Fig. 18.30) and the connector axis is at an angle θ to grain, the standard spacing of connectors, giving $K_S = 1.0$, is determined as for tooth plates:

$$S_{st} = \frac{A}{\sqrt{[1 + [(A/B)^2 - 1]\sin^2\theta]}}$$

similarly $K_S = 0.75$ at S_{min} and intermediate values of K_S are given as:

$$K_S = 0.75 + 0.25\left[\frac{S - S_{min}}{S_{st} - S_{min}}\right]$$

where S is the actual spacing.

When K_S is known, this formula may be transposed to:

$$S = \left[\frac{(K_S - 0.75)(S_{st} - S_{min})}{0.25}\right] + S_{min}$$

18.6.4 Example of determining permissible connector load

Determine the permissible long-term load for each connector as in Fig. 18.30. Connectors are 64 mm diameter split rings placed in two faces of a 47 mm thick C24 timber. θ = 30°, α = 15° and S = 100 mm. Assume that edge and end distance requirements are satisfied.

From Table 18.38 (with α = 15°), A = 152 mm, B = 95 mm and S_{min} = 89 mm:

$$S_{st} = \frac{152}{\sqrt{[1 + [(152/95)^2 - 1]\sin^2 30^\circ]}} = 129 \text{ mm}$$

$$K_S = 0.75 + 0.25\left[\frac{100 - 89}{129 - 89}\right] = 0.819$$

From Table 18.35, F_0 = 8.12 kN. From Table 18.36, K_α = 0.972

$$\text{Permissible load (per connector)} = F_0 \times K_\alpha \times K_S = 8.12 \times 0.972 \times 0.819 = 6.46 \text{ kN}$$

18.6.5 Charts for the standard placement of split ring and shear plate connectors

Sections 18.6.2 and 18.6.3 detail the background study into the determination of permissible loads for shear plates and split rings, giving somewhat complex for-

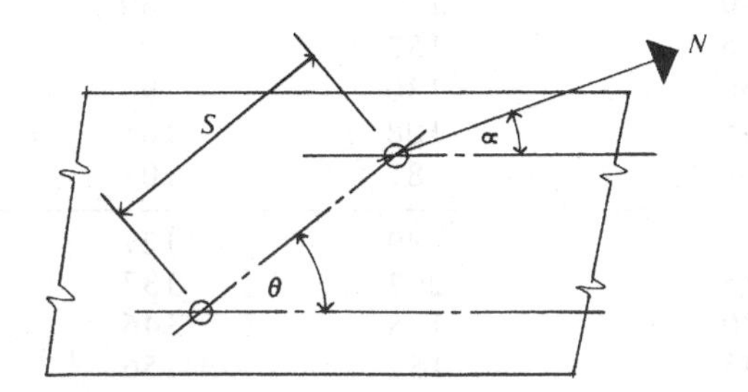

Fig. 18.30

mulae for establishing suitable placements of connectors with respect to one another and with respect to ends and edges of timber. To simplify the determination of loads and spacings for general design use, a graphical solution is given in Tables 18.39 and 18.40 and Figs 18.31 and 18.32.

Table 18.39

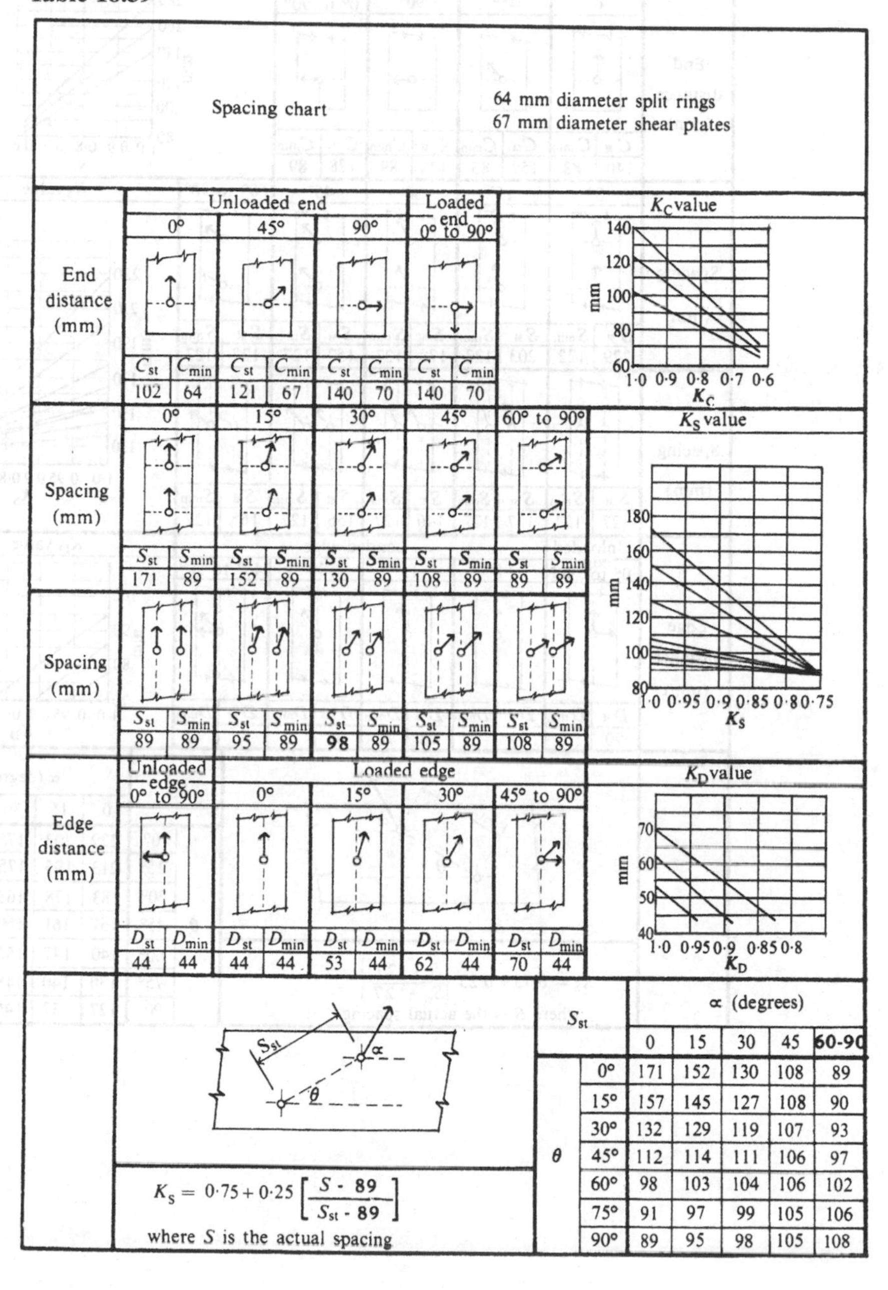

Spacing chart — 64 mm diameter split rings, 67 mm diameter shear plates

End distance (mm)

	Unloaded end 0°		Unloaded end 45°		Unloaded end 90°		Loaded end 0° to 90°	
	C_{st}	C_{min}	C_{st}	C_{min}	C_{st}	C_{min}	C_{st}	C_{min}
End distance (mm)	102	64	121	67	140	70	140	70

Spacing (mm)

	0°		15°		30°		45°		60° to 90°	
	S_{st}	S_{min}	S_{st}	S_{min}	S_{st}	S_{min}	S_{st}	S_{min}	S_{st}	S_{min}
Spacing (mm)	171	89	152	89	130	89	108	89	89	89
Spacing (mm)	89	89	95	89	98	89	105	89	108	89

Edge distance (mm)

	Unloaded edge 0° to 90°		Loaded edge 0°		Loaded edge 15°		Loaded edge 30°		Loaded edge 45° to 90°	
	D_{st}	D_{min}	D_{st}	D_{min}	D_{st}	D_{min}	D_{st}	D_{min}	D_{st}	D_{min}
Edge distance (mm)	44	44	44	44	53	44	62	44	70	44

$$K_S = 0{\cdot}75 + 0{\cdot}25\left[\frac{S - 89}{S_{st} - 89}\right]$$

where S is the actual spacing

S_{st}

θ	α (degrees) 0	15	30	45	60-90
0°	171	152	130	108	89
15°	157	145	127	108	90
30°	132	129	119	107	93
45°	112	114	111	106	97
60°	98	103	104	106	102
75°	91	97	99	105	106
90°	89	95	98	105	108

Table 18.40

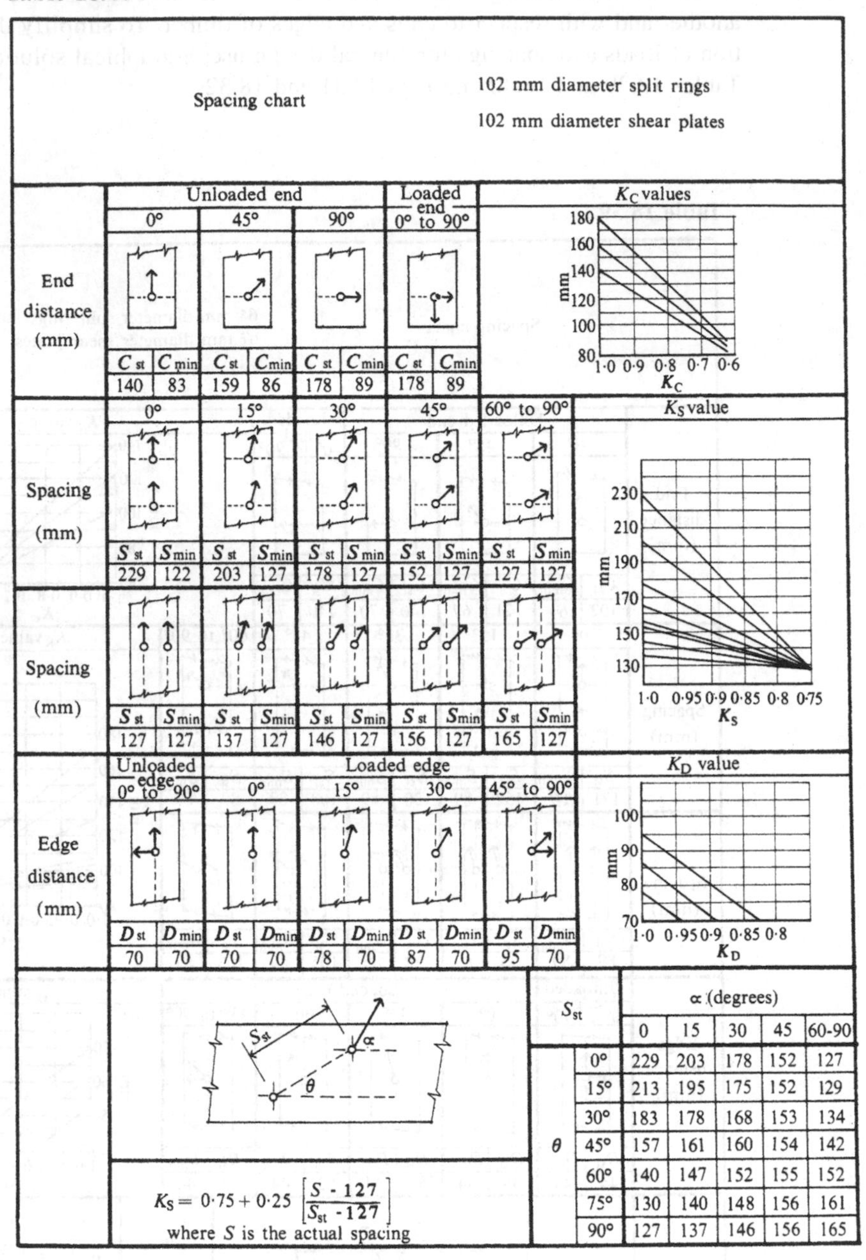

End distance (mm)

Unloaded end 0°		Unloaded end 45°		Unloaded end 90°		Loaded end 0° to 90°	
C_{st}	C_{min}	C_{st}	C_{min}	C_{st}	C_{min}	C_{st}	C_{min}
140	83	159	86	178	89	178	89

Spacing (mm)

0°		15°		30°		45°		60° to 90°	
S_{st}	S_{min}	S_{st}	S_{min}	S_{st}	S_{min}	S_{st}	S_{min}	S_{st}	S_{min}
229	122	203	127	178	127	152	127	127	127

Spacing (mm)

S_{st}	S_{min}	S_{st}	S_{min}	S_{st}	S_{min}	S_{st}	S_{min}	S_{st}	S_{min}
127	127	137	127	146	127	156	127	165	127

Edge distance (mm)

Unloaded edge 0° to 90°		Loaded edge 0°		Loaded edge 15°		Loaded edge 30°		Loaded edge 45° to 90°	
D_{st}	D_{min}	D_{st}	D_{min}	D_{st}	D_{min}	D_{st}	D_{min}	D_{st}	D_{min}
70	70	70	70	78	70	87	70	95	70

$$K_S = 0{\cdot}75 + 0{\cdot}25\left[\frac{S - 127}{S_{st} - 127}\right]$$

where S is the actual spacing

S_{st}		α (degrees)				
		0	15	30	45	60-90
θ	0°	229	203	178	152	127
	15°	213	195	175	152	129
	30°	183	178	168	153	134
	45°	157	161	160	154	142
	60°	140	147	152	155	152
	75°	130	140	148	156	161
	90°	127	137	146	156	165

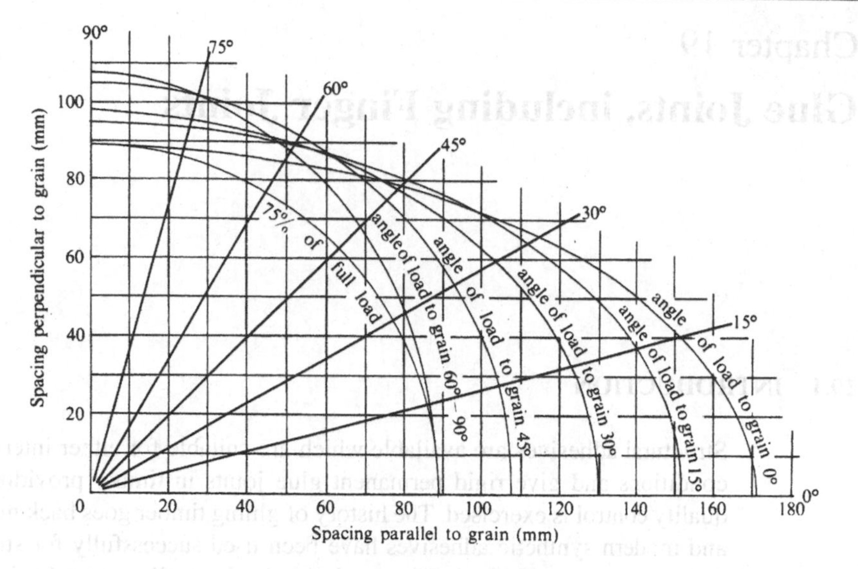

Fig. 18.31 Spacing for 64 mm diameter split rings and 67 mm diameter shear plates.

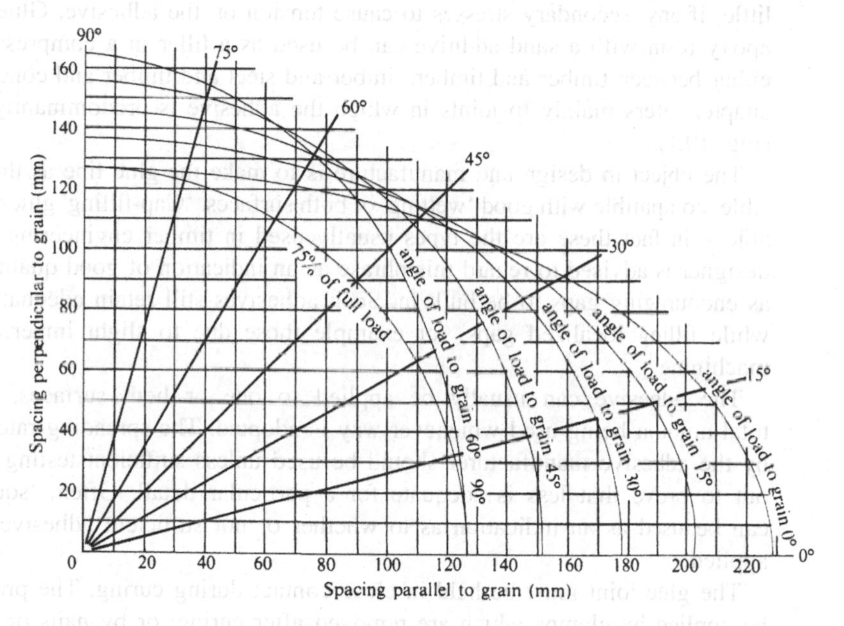

Fig. 18.32 Spacing for 102 mm diameter split rings and shear plates.

Chapter 19
Glue Joints, including Finger Joints

19.1 INTRODUCTION

Structural adhesives are available which are suitable for either interior or exterior conditions and give rigid permanent glue joints in timber providing the correct quality control is exercised. The history of gluing timber goes back many centuries, and modern synthetic adhesives have been used successfully for structural applications for over 100 years. Finger jointing is also well proven, having been developed and used successfully since the 1950s.

Glue joints should be designed so that the adhesive is stressed in shear with little, if any, secondary stresses to cause tension on the adhesive. Glues such as epoxy resin with a sand additive can be used as a filler in a compression joint, either between timber and timber, timber and steel and timber and concrete. This chapter refers mainly to joints in which the adhesive is predominantly in shear (Fig. 19.1).

The object in design and manufacture is to make the glue line as thin as possible, compatible with good 'wetting' of both surfaces. 'Gap-filling' glues are available – in fact these are the types usually used in timber engineering – but the designer is advised to regard this phrase as an indication of good quality and not as encouraging gaps to be built in. Such adhesives still retain adequate strength while filling localized gaps, for example those due to slight imperfections in machining.

The adhesive can usually be applied to one or both surfaces, the same total amount being used whichever way is adopted. The spreading rate specified by the adhesive manufacturer should be used unless sufficient testing is carried out to prove that less is adequate for a particular detail. Often, 'squeeze-out' can be used as an indication as to whether or not sufficient adhesive has been applied.

The glue joint must be held in close contact during curing. The pressure can be applied by clamps which are removed after curing, or by nails or other fasteners which are left in the final assembly. Even if using nails for bonding pressure there are advantages in holding the assembly with clamps until the nailing is carried out.

If a glue joint is tested to destruction, failure should normally take place in the timber or plywood close to the glue face, not in the thickness of the adhesive, and the permissible design stresses used in the design of glue joints are those appropriate to the timber, not the adhesive. A correctly designed structural glue joint may be expected not to slip.

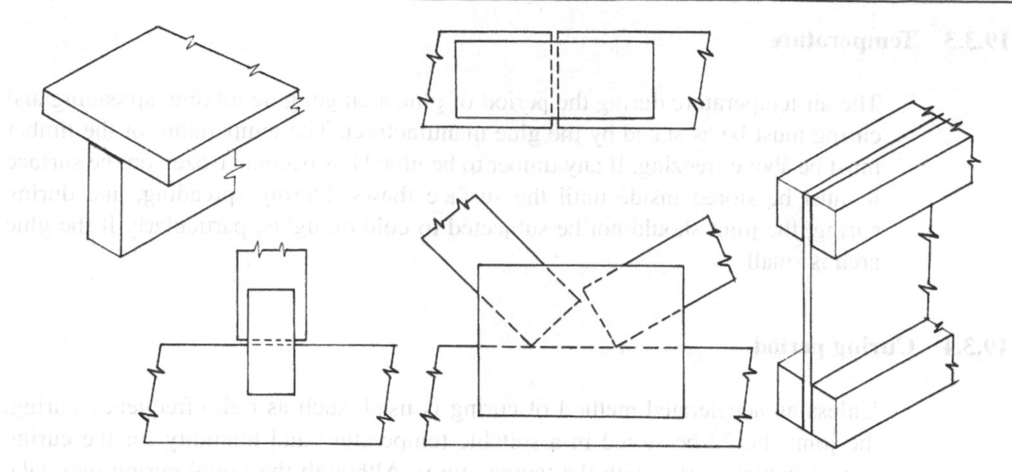

Fig. 19.1 Typical glue joints.

19.2 TYPES OF ADHESIVE USED IN TIMBER ENGINEERING

The choice of adhesive is determined by the conditions the component will encounter in service, the method of manufacture and perhaps the conditions during delivery and construction. Table 100 of BS 5268-2 details the appropriate choice of adhesive according to intended use and exposure category.

19.3 QUALITY CONTROL REQUIREMENTS. GENERAL GLUE JOINTS

19.3.1 Glue mixing and spreading

The weights or volumes of the parts of the mix must be accurately measured and mixed in accordance with the adhesive manufacturer's instructions. The mixing must be carried out in clean containers. With factory gluing there should usually be a separate room or area set aside for mixing. The separate parts should be stored in the correct temperature conditions, and mixing should be carried out with the air above the minimum stated temperature. Normal gel tests and tests for rate of spread of adhesive must be carried out and records kept. The rate of spread must be in accordance with the adhesive manufacturer's instructions unless testing is carried out to determine that less is adequate for a particular joint.

19.3.2 'Open' and 'closed' storage times

Once the adhesive is mixed, spreading must be carried out within the stated 'open storage' time, and once the two mating surfaces are brought together any adjustment in alignment or clamping must take place within the stated 'closed storage' time.

19.3.3 Temperature

The air temperature during the period of glue storage, glue mixing, spreading and curing must be as stated by the glue manufacturer. The temperature of the timber must be above freezing. If any timber to be glued has become frozen on the surface it must be stored inside until the surface thaws. During spreading, and during curing, the joint should not be subjected to cold draughts, particularly if the glue area is small.

19.3.4 Curing period

Unless an accelerated method of curing is used, such as radio-frequency curing, the joint should be stored in a suitable temperature and humidity for the curing period (which varies with the temperature). Although the initial curing may take place within a matter of hours, there are certain glues, even WBP glues, which for seven days must not be placed where rain could affect them. Until the end of that period the glue strength could be reduced by chemicals near the edges being 'washed-out'.

Accelerated curing can be completed in seconds. The methods of achieving this are expensive in themselves and may require the timber to be kilned to around 12% moisture content, but may give overall savings by suiting a particular flow-line technique. Some manufacturing processes utilize a part-accelerated curing technique followed by natural curing. Even with accelerated curing, full cure may still take several days. During the curing period care will be required in handling glued members. If radio-frequency (RF) curing is used it is usually necessary to exclude any metal from the area of the glue joint. Metal can cause serious 'shorting'.

19.3.5 Moisture content

At the time of gluing solid timber to solid timber, the moisture content of the timber must not exceed the limit stated either by the glue manufacturer or in BS 5268: Parts 2 and 3, BS 6446 on glued structural components, BS 5291 on structural finger joints, or BS 4169 on glued-laminated members. This figure is usually around 20% or less.

The two pieces or surfaces being glued must be at approximately the same moisture content, usually within 3–5% and preferably within a few percent of the equilibrium moisture content of the component in service. See section 24.8.1 and the relevant standard for the actual levels laid down for various situations.

If accelerated methods of curing are to be used it may be necessary to dry the timber to around 12% before gluing takes place.

19.3.6 Machining surfaces to be glued. Site gluing

The surfaces of solid timber to be glued must be machined and the gluing must be carried out within a prescribed period of machining unless special precautions are taken to ensure that the surface stays suitable for gluing. Various time limits are given in standards. These vary from 24 to 72 hours for unpreserved timber, down to 12 hours maximum for finger-jointing preserved timber. However, factory conditions, etc., can have a significant influence and generally the time should be kept to a minimum for solid timber. If the timber is left too long, the cut-through cells tend to close over, a condition sometimes called 'case hardening', and if the glue joint is then produced it will have less strength than otherwise. This is one reason why site gluing is not to be encouraged unless carried out under strict supervision with surfaces perhaps dressed just prior to gluing, or if the adhesive is mainly to provide added stiffness rather than to take stress.

The machined surfaces must be kept clean and free from any dust, etc. It is essential to carry out the gluing operations in a clean part of the factory.

When gluing plywood, experience has shown that the time limit does not apply, but the surface of the plywood must be clean. Any paper stickers which coincide with the glue area must be removed. Any previously dried adhesive must be cleaned off before being reglued or the piece must be rejected.

19.3.7 Pressure during curing

The requirement for most adhesives used in timber engineering is that the mating surfaces must be held together in close contact during curing rather than that there must be pressure. In practice this means that pressure is applied but (except for finger joints) the pressure which actually occurs on the glue line is rarely measured. The pressure is applied by clamps or pads which are removed after the curing period, or by nails or staples (occasionally screws), which are left in place after curing although not considered in the design as adding to the shear strength of the glue joint. Except for finger jointing (see section 19.6.6), pressure is usually considered to be adequate when 'squeeze-out' of the glue occurs. If clamps are used to hold several components together, it is necessary to tighten them occasionally during the closed assembly period.

Even when the bonding/cramping pressure is achieved by nails there are advantages in assembling units with clamps before and during nailing. Guidance on the spacing of nails is given in BS 5268-2 and BS 6446 on structural glued assemblies.

When curing at high temperatures the viscosity of the adhesive decreases and the glue line pressure must be sufficient to prevent adhesive 'run-out'.

19.3.8 Quality control tests to destruction

The manufacturer must have a sampling system for testing joints made from standard production glue mixes, usually at least two or more from any shift or major glue mix, and several more if large productions are involved. The tests should be to destruction and records should be kept. It is useful for the manufacturer to plot these results on graphs with degrees of strength on the horizontal axis and number of tests on the vertical axis, allocating one square to each test. If the correct quality control is being exercised, the shape of the plot on the graph will gradually build

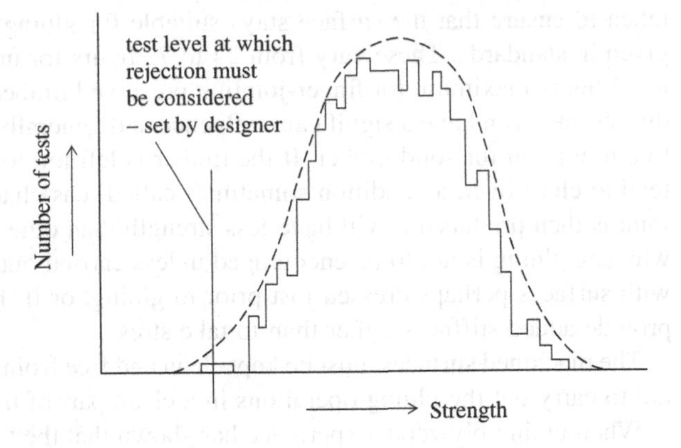

Fig. 19.2

up to the normal Gaussian shape typical of timber strength (Fig. 19.2). If any result falls below a level set by the designer or by quality control in standards (such as BS 5291), the components involved should be inspected and perhaps rejected, the cause ascertained and corrective action taken.

19.3.9 Appearance

Particularly if using a WBP adhesive, the 'squeeze-out' which occurs with most glue joints will affect the appearance of the component. When appearance is important, as with some glulam for example, machining after gluing is desirable or essential. Wiping will not usually remove WBP, BR or MR adhesive and may even spread it around. After machining glulam or finger-jointed timber, the thin appearance of the edge of the glue lines is generally considered neat and acceptable visually, particularly with finger joints if a type with no gap at the tips is being used.

19.3.10 Compatibility with preservatives or fire retardants

When gluing preserved timber or preserving glued timber, care must be taken to ensure that there is compatibility. This is particularly the case with preservatives containing water-repellent waxes or additives such as resins. It is normally possible to glue preserved timber, but extra care has to be taken. There are cases, however, in which gluing is not possible. For example, it is not considered possible to glue timber which has already been treated with a fire retardant containing ammonia or inorganic salts, although it is possible to treat with such a formulation providing a suitable period (usually seven days) has elapsed after gluing and the adhesive has achieved full cure.

When encountering for the first time a combination of glue and preservative, the designer should check the compatibility of each with the manufacturer. When gluing preservative-treated plywood, one would not normally wish to machine or sand the surfaces before gluing, but brushing to remove surplus salts is desirable.

19.3.11 Temperature and moisture content in service

If the temperature in service is likely to be high (particularly if over 50 °C) and other than a WBP adhesive is to be used, the designer should check the suitability of the adhesive with the manufacturer. Likewise, if the moisture content in service is likely to exceed 18% for other than short periods, the designer should check with the adhesive manufacturer.

19.4 THE STRENGTH OF A GLUE JOINT

19.4.1 Permissible stresses

The permissible stresses for a glue joint are determined by the strength of the timber or plywood face to which the glue is adhering and not to the strength within the glue line itself.

The permissible shearing stress parallel to the grain of solid timber is the appropriate stress of the timber for shear parallel to the grain with load–duration and load-sharing factors being applicable. The permissible shearing stress perpendicular to the grain of solid timber is one-third the value parallel to the grain and is referred to as 'rolling shear' (see section 4.14 and 8.2.6).

In the event of a glued joint being designed so that one face of solid timber is loaded at an angle to the direction of the grain, the permissible shear stress is determined by applying the following formula from clause 6.10.1.3 of BS 5268-2:

$$\tau_\alpha = \tau_{adm,\,par}(1 - 0.67\sin\alpha)$$

where $\tau_{adm,\,par}$ = permissible shear stress parallel to the grain for the timber
α = angle between the direction of load and the longitudinal axis of the timber.

When considering the face of plywood, the permissible shear stress is given the name permissible 'rolling shear' stress, described in section 8.2.6 for ply web beams. Even when the shear stress on the face of the plywood is parallel to the face grain, the permissible rolling shear stress should be taken in calculations, because the perpendicular veneer next to the face is so close to the surface that one cannot be certain that full dispersal of a face stress could occur before rolling shear starts in the perpendicular veneer. Therefore the formula given above does not apply to glue lines on plywood faces.

19.4.2 Reduction in permissible stresses for stress concentrations

In the special case of the flange-to-web connection of a ply web beam and the connection of plywood (or other board) to the outermost joist of a glued stress skin panel, clause 4.7 of BS 5268-2 requires that the permissible shear stress at the glue line be multiplied by the K_{37} factor of 0.5. This is an arbitrary factor to take account of likely stress concentrations.

19.4.3 Glued/nailed joints

BS 5268-2 gives certain clauses relating to glued/nailed joints. When gluing plywood to timber the permissible glue line stresses should be multiplied by 0.90 (K_{70}) if assembly is by nailing (see section 6.10.1.3). The maximum spacing of nails required to give bonding pressure is detailed in BS 6446 on glued structural components. See Table 4.5 in section 4.14.

In nail-pressure gluing, the nails are not considered to add to the strength or stiffness of the glue line. Screws, improved wire nails and power-driven fastenings such as staples may be used if proved to be capable of applying pressure to the glue line at least equal to the nailing procedures described in BS 5268 and BS 6446.

19.5 STRUCTURAL FINGER JOINTS

19.5.1 Types

Structural finger joints are generally considered to be of two basic types which are the longer finger joints which are deliberately made with a small gap at the tips to ensure contact on the sloping sides, and the more recent but well-established short joints which have no measurable gap at the tips. Both types are sketched in Fig. 19.3.

For maximum strength a finger joint should have its sloping surfaces as close as possible to the longitudinal direction of the timber, and have as small a tip width as possible commensurate with it being possible to cut the joint. The length of the glue line per unit width of member also affects the strength. Tests show that a correctly made short joint can have a strength as high as most of the longer joints. This is partly due to the increased pressure at which the short joint can be assembled without causing the timber to split. The short joint is suitable for joinery as well as structural use, whereas the gap at the tips of the longer structural joints makes them unsuitable for most joinery uses.

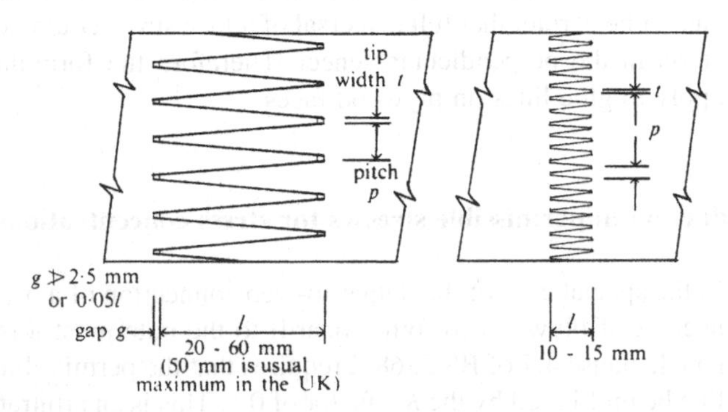

Fig. 19.3 Types of structural finger joint.

19.5.2 Maximum size and length of finger-jointed timber

There is no theoretical limit to the size of timber which can be finger jointed. Size is normally limited only by the capacity of the machine which has been installed. Finger jointing in softwood sizes of 75×200 mm or 50 × 300 mm is quite common. Finger jointing of glulam sections for portal frame haunches is now possible and the reader is referred to section 19.7.4 for design procedures.

The only limit to the length being finger jointed is that of handling in the factory, in transport and on site. Lengths of 12 m are very common and do not represent the maximum by any means.

19.5.3 Appearance and wane

It is possible to leave a piece of finger-jointed timber in the 'as sawn' condition without planing. In this case the adhesive which has squeezed out is very obvious, particularly if WBP, and it is possible for the two pieces at a joint to be off-set due to tolerances and the lining-up of the fingers (Fig. 19.4).

This has little effect on strength but, when appearance is important, the timber should be surfaced after finger jointing. The finger joint then usually has a very neat appearance, particularly if a short joint is being used. Wiping off the glue squeezed out instead of machining is unlikely to improve appearance and may make it worse by spreading the glue over a larger area.

Wane in the length of a finger joint and within a short distance of a finger joint acts as a stress raiser and must be limited. BS 5291 permits wane to occur on one or two corners but gives limits. Within the finger length and within 75 mm of the roots of the fingers, if the efficiency rating of the joint is equal to or less than 60%, the sum of the dimensions of the wane should not exceed 10% of the width plus thickness of the piece (which represents a maximum of about 0.6% of the area). If the efficiency is in excess of 60%, the sum of the dimensions of the wane should not exceed 5% of the width plus thickness of the piece.

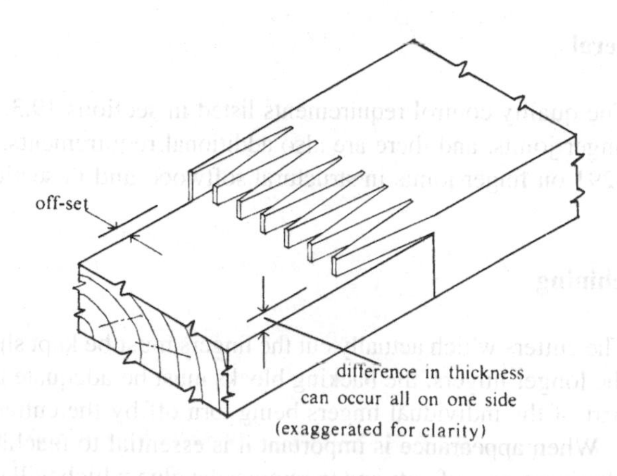

Fig. 19.4

19.5.4 Types of adhesive

Although it is possible to finger joint with an adhesive having only interior classification, it is normal for a manufacturer to use a WBP adhesive or perhaps a BR or MR type, even if using a different type elsewhere in the assembly of the component. The adhesive must be adequate for the service conditions (see Table 100 of BS 5268-2) and BS 5291 emphasizes the use of WBP adhesives for structural finger joints.

19.5.5 Location of finger joints

BS 4169, clause 7.1.2 requires the distance between finger joints in adjacent laminations to be not less than the width of the lamination, i.e. 140 mm apart in a 140 mm wide section.

During fabrication the finger joints will become randomly spaced in a laminated member and there is no reason to suppose that the existence of several finger joints at close centres reduces the strength of a component. However, even if only to avoid the possibility of adverse reaction from site staff or occupants, it is wise to place some limit on the number of joints in a structural member, therefore the distance between the centres of any two finger joints in a piece of timber or single lamination should not usually be less than 1 metre.

19.5.6 Species mix

Most of the experience in the UK of structural finger jointing has been with European whitewood or European redwood. Even though there are known cases where these two species have been finger jointed together successfully, BS 5291 is quite clear in stating that species should not be mixed at a joint.

19.6 QUALITY CONTROL REQUIREMENTS FOR STRUCTURAL FINGER JOINTS

19.6.1 General

The quality control requirements listed in sections 19.3.1–19.3.7 apply equally to finger joints, and there are also additional requirements. These are detailed in BS 5291 on finger joints in structural softwood and in sections 19.6.2–19.6.7.

19.6.2 Machining

The cutters which actually cut the fingers must be kept sharp, and, particularly with the longer fingers, the backing blocks must be adequate to prevent 'spelching' (i.e. part of the individual fingers being torn off by the cutters).

When appearance is important it is essential to machine the completed piece to eliminate any off-sets and to remove the glue which will squeeze out from the joint.

The amount of cup must be limited on pieces to be finger jointed, or splitting is likely to generate from the tips of the fingers, resulting from face pressure applied during assembly of the joint.

19.6.3 Moisture content

The shape and method of assembling finger joints is such an excellent example of a glue joint that it is usually quite acceptable to work to the upper limit of moisture content in the timber relevant to the glue being used. The moisture content of the timber at assembly should not exceed 20% and should be within a few percent of the average equilibrium moisture content expected in service. The moisture content of the pieces to be jointed should not differ by more than 6%.

19.6.4 Knots

Knots must be limited both in and close to the fingers. BS 5291 limits the dimension of knots within the finger length to half the pitch of the fingers or 5 mm whichever is the lesser. Outside the length of the fingers no knot shall be closer to the root of the fingers than three times its maximum dimension d measured parallel to the grain (see Fig. 19.5), although knots with a dimension d of 5 mm or less can be disregarded. In trimming the end of a piece to be finger jointed, disturbed grain should be removed as well as over-size knots.

19.6.5 Wane

Wane within the length of fingers and close to fingers should be limited. The limits of BS 5291 are detailed in section 19.5.3.

19.6.6 End pressure and fissures

The end pressure during assembly of the finger joints must be sufficient for maximum strength to be developed, but not too great or splitting will occur from

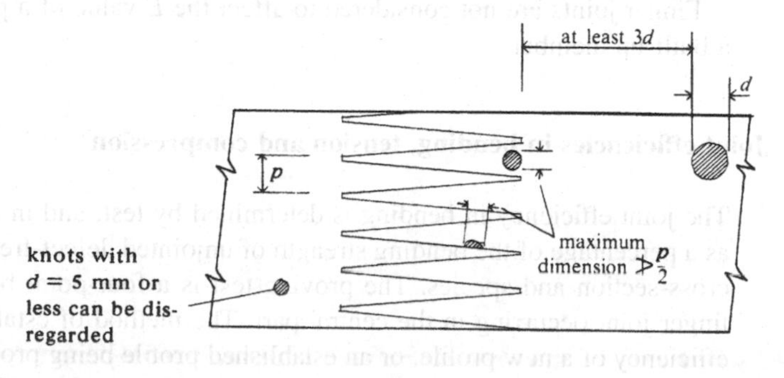

Fig. 19.5 Maximum permitted size of knots.

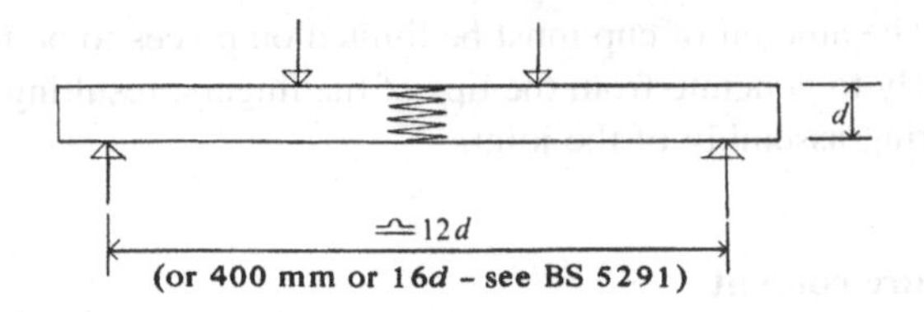

Fig. 19.6

the fingers. With short joints the applied pressure is in the order of 5–15 N/mm^2 reducing to 1.5–5 N/mm^2 for the longest joints.

BS 5291 details the limits on fissures which may occur at a finger joint.

19.6.7 Quality control tests to destruction

It is necessary for a manufacturer of structural finger joints to have access to a test machine to test specimens of standard production joints. The test is by four-point loading (Fig. 19.6), and the manufacturer can consider plotting the results as indicated in Fig. 19.2. BS 5291 details the test and result requirements. If a correctly made joint is found to have low strength, one of the first checks to make is on the density of the timber.

19.7 THE STRENGTH AND DESIGN OF FINGER JOINTS

19.7.1 General

The general philosophy in BS 5268: Part 2 on the use of finger joints in a structural situation is either to require a finger joint to have a certain efficiency in bending when using timber of a certain stress grade without carrying out a design check on the actual stress combinations, or to design for the actual stress combinations and provide a joint accordingly. Design examples are given in section 7.5 for combined bending and tension and in section 17.5 for combined bending and compression, using both methods.

Also see section 19.7.3 on the use of finger joints in a load-sharing system and a non-load-sharing system.

Finger joints are not considered to affect the *E* value of a piece of timber or of a built-up member.

19.7.2 Joint efficiencies in bending, tension and compression

The joint efficiency in bending is determined by test, and in the UK is expressed as a percentage of the bending strength of unjointed defect-free timber of the same cross-section and species. The proving test is a four-point bending test with the finger joint occurring in the central part. The method of establishing the bending efficiency of a new profile, or an established profile being produced on a new production line, is detailed in BS 5291.

Table 19.1

Finger profiles			Efficiency rating in bending and tension (per cent)	Efficiency rating in compression (per cent)
Length l(mm)	Pitch p(mm)	Tip width t(mm)		
55	12.5	1.5	75	88
50*	12.0	2.0	75	83
40	9.0	1.0	65	89
32*	6.2	0.5	75	92
30	6.5	1.5	55	77
30	11.0	2.7	50	75
20*	6.2	1.0	65	84
15*	3.8	0.5	75	87
12.5*	4.0	0.7	65	82
12.5*	3.0	0.5	65	83
10.0	3.7	0.6	65	84
10.0	3.8	0.6	65	84
7.5	2.5	0.2	65	92

*** Profiles more likely to be available.**

The joint efficiency in tension can be established by test but is usually taken in the UK as having the same efficiency value as in bending (even though the grade stress of timber in tension is considerably less than the grade stress in bending).

Tests tend to show that the joint efficiency in compression is 100% or close to 100%. However, in the UK, the efficiency is taken as:

$$\frac{p-t}{t} \times 100\%$$

where p = pitch of the fingers
t = width of the tip.

Normally no finger joint with an efficiency in bending of less than 50% should be used structurally. This is a requirement of BS 5268-2.

Guide efficiency values for well-established joint profiles are given in Table 19.1 with the more common profiles indicated by an asterisk. See the various design examples in this manual on how to apply efficiency ratings.

As stated above, when using a finger joint, one can either use one of a stated efficiency in bending related to various stress grades, or design for the actual stresses or stress combinations encountered in a particular design. Table 19.2 gives joint efficiency ratings in bending which, if matched for a particular strength class, may be used without any further design check.

When it is necessary to carry out a design check on glulam for the actual stresses rather than simply use the efficiency figures given in Table 19.2, see modification factors K_3, K_3, and $K_{3.2}$ of BS 5268: Part 2 and section 7.5 for combined bending and tension, and section 17.5 for combined bending and compression.

Table 19.2

Strength class	Minimum finger jointing efficiency in bending
C14–C16	55%
C18–C24	70%
TR26–C27	75%

19.7.3 Load-sharing/non-load-sharing systems

When finger jointing is used in a load-sharing system such as four or more members acting together and spaced at not more than 600 mm centres, such as rafters, joists, trusses or wall studs, with adequate provision for lateral distribution of loads, there is no restriction on the use of finger jointing.

Clause 6.10.2 of BS 5268-2 states that: '*Finger joints should not be used in principal members, or other members acting alone, where failure of a single joint could lead to collapse, except where the joints have been manufactured under a third part control scheme.*'

This restriction is quite clear when one considers a member such as a trimmer beam consisting of one piece of solid timber and supporting a system of floor or roof joists. In this case, if a finger joint in the trimmer beam fails, the load cannot normally be transmitted laterally and therefore it would not be permissible to use finger jointing in the trimmer beam.

The wording of this clause implies that finger jointing is acceptable provided that a stuctural 'member' consisting of at least two members supports a commom load so as to be equally strained. Typical examples would be twin members forming the flanges of non-load-sharing I beams or box beams and twin chords in trusses.

19.7.4 Finger-jointed eaves haunch to laminated portal frame

A special case of finger-jointing in principal members is that of the finger-jointed eaves haunch construction marketed by Moelven Laminated Timber Structures Ltd and certified by Agrement certificate No. 89/2326.

This is a special haunch construction which permits straight/tapered glulam members to be end jointed to form a portal profile with consequential savings over the more costly alternative of a curved laminated section that requires special jigging of thin laminations.

Such a design approach is conditional upon the eaves haunch moment being in hogging at all times (see Fig. 19.7) and this should be verified by the designer before any commitment is made to this form of construction. It is unlikely that portals supporting a light dead load and subject to high wind loading would be suitable for this method of construction.

Referring to Fig. 19.8, the adequacy of the section at the location of the finger joint is to be checked using the following empirical formula:

$$0.87\sigma_{\mathrm{m,a,par}} + \frac{\sigma_{\mathrm{c,a,par}}}{K_{12}} \leq \sigma_{\mathrm{c,adm,\alpha}}$$

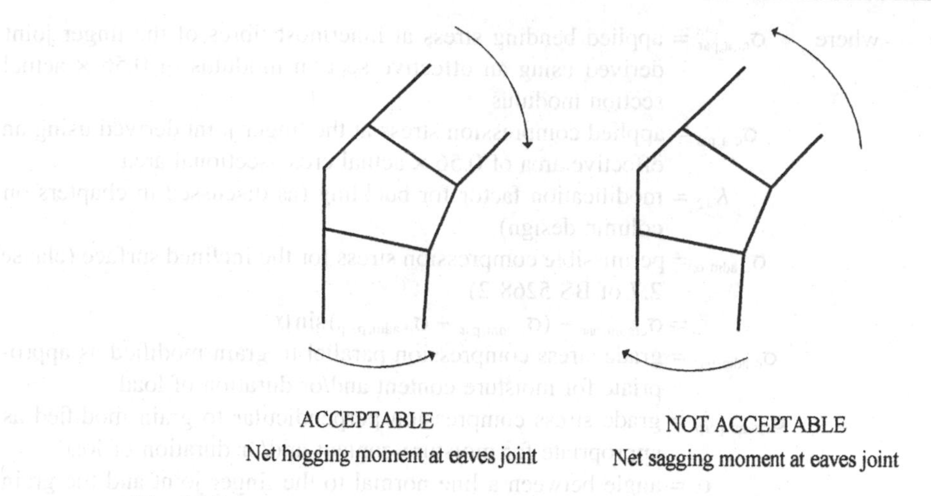

Fig. 19.7

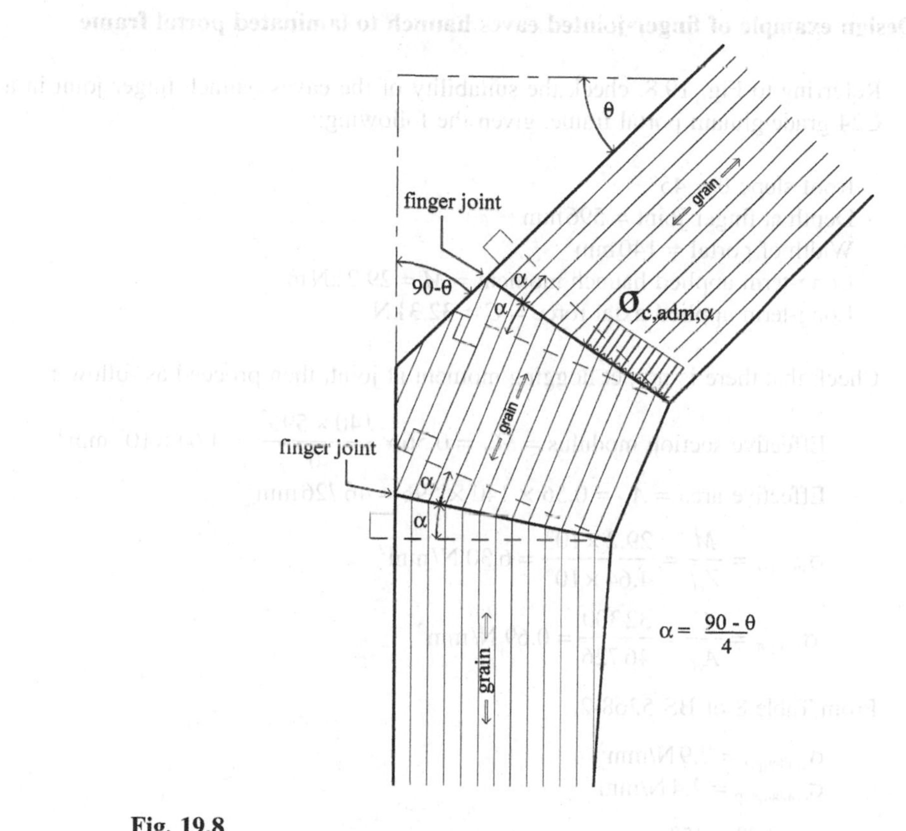

Fig. 19.8

where $\sigma_{m,a,par}$ = applied bending stress at innermost fibres of the finger joint derived using an effective section modulus of 0.56 × actual section modulus

$\sigma_{c,a,par}$ = applied compression stress at the finger joint derived using an effective area of 0.56 × actual cross-sectional area

K_{12} = modification factor for buckling (as discussed in chapters on column design)

$\sigma_{c,adm,\alpha}$ = permissible compression stress for the inclined surface (clause 2.7 of BS 5268-2)

$= \sigma_{c,adm,par} - (\sigma_{c,adm,par} - \sigma_{c,adm,perp}) \sin\alpha$

$\sigma_{c,adm,par}$ = grade stress compression parallel to grain modified as appropriate for moisture content and/or duration of load

$\sigma_{c,adm,perp}$ = grade stress compression perpendicular to grain modified as appropriate for moisture content and/or duration of load

α = angle between a line normal to the finger joint and the grain (Fig. 19.8) = $(90° - \theta)/4$, where θ denotes the slope of top edge of rafter to horizontal (Fig. 19.8).

19.7.5 Design example of finger-jointed eaves haunch to laminated portal frame

Referring to Fig. 19.8, check the suitability of the eaves haunch finger joint in a C24 grade glulam portal frame, given the following:

Roof slope $\theta = 45°$
Depth at finger joint = 596 mm
Width of portal = 140 mm
Long-term applied haunch moment = M = 29.2 kN m
Long-term applied axial force = C = 32.3 kN

Check that there is no net hogging moment at joint, then proceed as follows:

$$\text{Effective section modulus} = M_{ef} = 0.56 \times \frac{140 \times 596^2}{6} = 4.64 \times 10^6 \text{ mm}^3$$

$$\text{Effective area} = A_{ef} = 0.56 \times 140 \times 596 = 46\,726 \text{ mm}^2$$

$$\sigma_{m,a,par} = \frac{M}{Z_{ef}} = \frac{29.2 \times 10^6}{4.64 \times 10^6} = 6.30 \text{ N/mm}^2$$

$$\sigma_{c,a,par} = \frac{C}{A_{ef}} = \frac{32\,300}{46\,726} = 0.69 \text{ N/mm}^2$$

From Table 8 of BS 5268-2,

$$\sigma_{c,adm,par} = 7.9 \text{ N/mm}^2$$
$$\sigma_{c,adm,perp} = 2.4 \text{ N/mm}^2$$

$$\alpha = \frac{90° - 45°}{4} = 11.25°$$

$$\sigma_{c,adm,\alpha} = 7.9 - (7.9 - 2.4)\sin 11.25° = 7.9 - 5.5 \sin 11.25° = 7.9 - 1.07 = 6.8 \text{ N/mm}^2$$

For brevity, assume that slenderness ratios about x–x and y–y have been established and that K_{12} has been determined as 0.81.

$$\text{Interaction formula} = 0.87\sigma_{m,a,par} + \frac{\sigma_{c,a,par}}{K_{12}}$$

$$= (0.87 \times 6.3) + (0.69/0.81) = 6.33\ \text{N/mm}^2 < 6.8\ \text{N/mm}^2$$

Therefore finger joint is adequate.

Chapter 20
Stress Skin Panels

20.1 INTRODUCTION

Plywood stress skin panels consist of plywood sheets attached to longitudinal timber members either by glue (usually glued/nailed joints) or mechanical means (usually nails or staples) to give a composite unit. With this construction it is possible to use smaller longitudinal members than those which would be required in a conventional joisting system or to extend the span of standard joist sizes. In addition, prefabricated panels can be used to reduce site work and speed erection. Most panels are sufficiently light to be erected by hand or with simple lifting gear. They have been used on floors, roofs and walls.

The maximum span of panels with simple joist longitudinal members, finger jointed if necessary, is in the order of 6.0 m. The span of panels can be extended if glulam/composite beams are used for the web members.

Stress skin panels can be either of double- or single-skin construction, the latter being sometimes referred to as 'stiffened panels'. The top flange is usually Canadian softwood plywood or similar 'low' cost plywood whereas the bottom flange may be Finnish birch-faced plywood or similar if appearance is a consideration. The web members are usually C16 or C24 strength class timber.

If the full stress skin effect is required then it is essential to glue the plywood to the timber throughout the length of the member. Glue bonding is usually achieved with nails or staples. If nails or staples are used without adhesive, only a part of the full stress skin effect will be achieved no matter how close the spacing of the nails or staples. It is normal to surface the timber members on all four sides although it could be possible to regularize only the depth of the joists. The timber must be dried to the appropriate moisture content, particularly if in a double-skin construction.

In the designs in this chapter glued/nailed joints are assumed between the plywood and the timber joists, and the joists are surfaced on all four faces.

20.2 FORMS OF CONSTRUCTION

A basic form of double-skin panel is sketched in Fig. 20.1. Architectural details, insulation, ventilation, falls, etc., can be added, and even plumbing or electrics can be fitted in the factory. The edge joists are shown as the same thickness as the inner joists, but usually take half loading and could be thinner. Because of the size of available plywood sheets, panels are usually 1.2 m or 1.22 m wide, and usually have

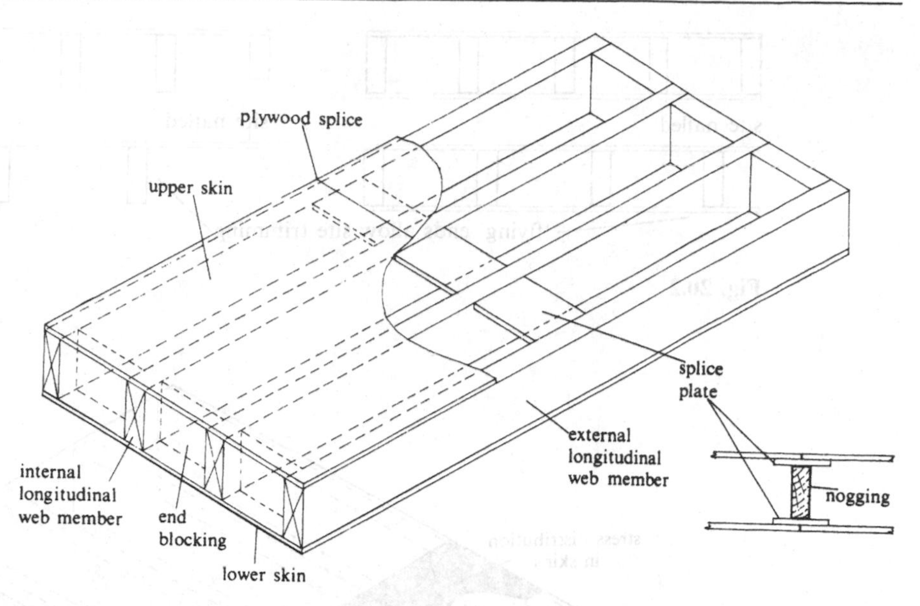

Fig. 20.1

three or four joists in this width, designed as part of a load-sharing system. Maximum economy will be achieved only if the panel width is matched to the available sizes of plywood sheets.

With Canadian plywood skins, the face grain usually runs parallel to the span of the panel whereas with Finnish plywoods the face grain would normally run at right angles to the direction of panel span.

The plywood splice plate shown in Fig. 20.1 can sometimes be replaced by a solid timber part-depth or full-depth nogging, providing a design check is carried out. Alternatively, it is occasionally possible to buy large plywood sheets suitably end scarf jointed, or for the panel manufacturer to scarf standard sheets. If the joists are finger jointed, the joint must have adequate stength.

It is not usual to camber stress skin panels. End blocking is often provided at points of bearing.

Four of the many alternative forms are sketched in Fig. 20.2 and these can have modified edge details to prevent differential deflection.

20.3 SPECIAL DESIGN CONSIDERATIONS

20.3.1 Effect of shear lag on bending stresses and deflection

Tests on double-skin stress skin panels seem to have been carried out first in the US in the early 1930s, and the results showed that the interaction between the plywood skins and the timber webs requires a special design consideration. However, a design method was not developed until around 1940 when, because of the shortage of materials during the Second World War, an added interest was created in the high strength-to-weight ratio of plywood and timber, particularly in

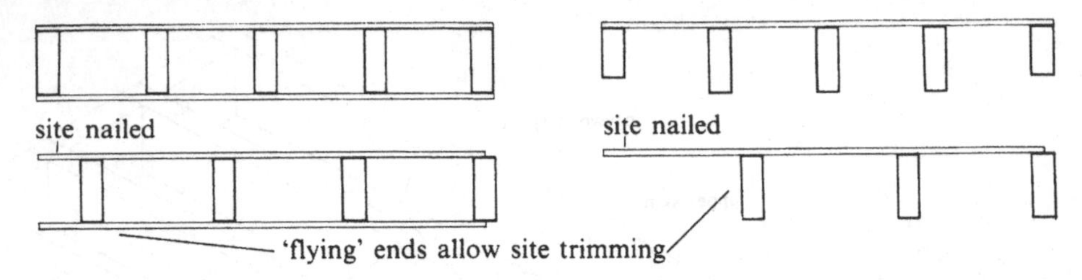

Fig. 20.2

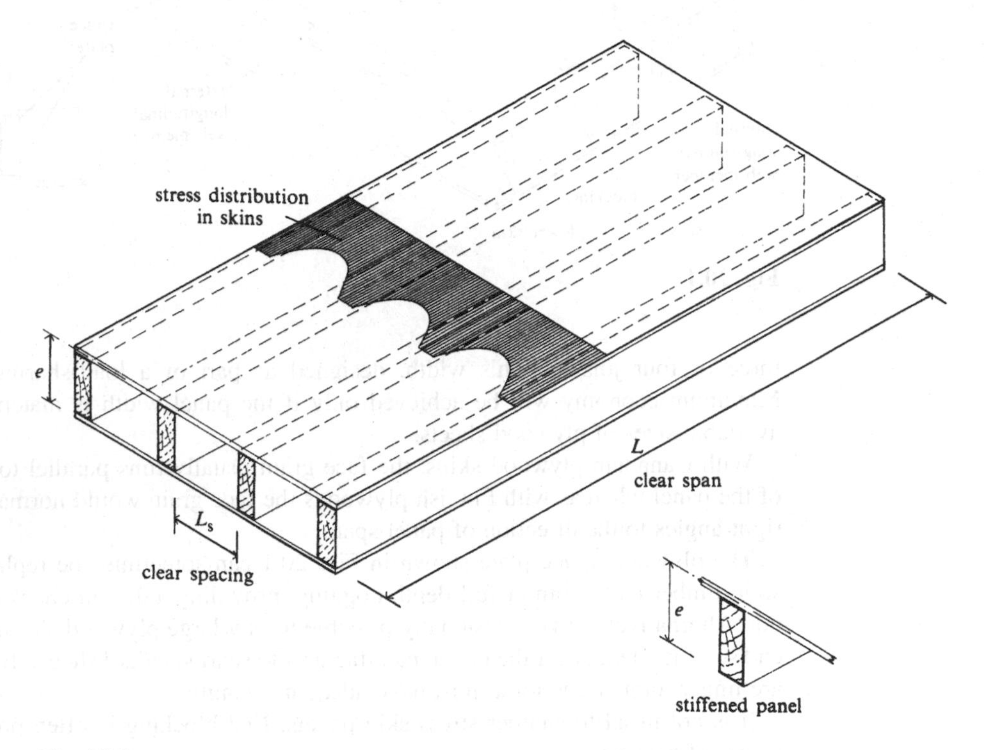

Fig. 20.3

the aircraft industry. The method used a factor known as the 'basic spacing' which was based on the buckling characteristics of plywood loaded uniformly in compression. This method was used until reappraised by the Council of the Forest Industries of British Columbia from which a simplified design method was developed for use with Canadian plywood.

The design philosophy takes account of the fact that the tension or compression stresses in the plywood skins under longitudinal bending result from a shear transfer from the web members into the plywood flanges, these stresses being a maximum at the junctions between web members and plywood and a minimum equidistant between webs. The stress distribution is sketched in Fig. 20.3. The variation is caused by shear deflection, and the variation from the elementary theory (i.e. that tension or compression stresses in the plywood are uniform across the panel) is usually called 'shear lag'.

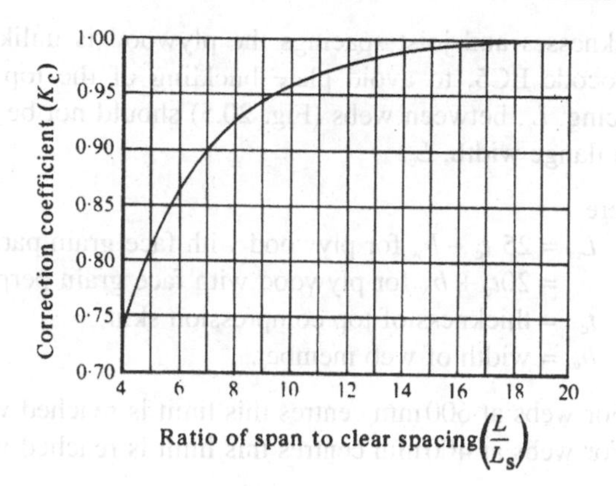

Fig. 20.4

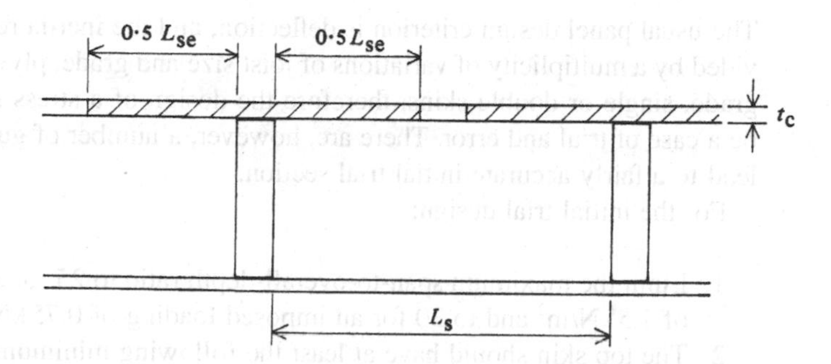

Fig. 20.5

In the earlier editions of this manual the design of stressed skin panels was based on Canadian practice. This took account of the full area of top and bottom plywood webs to determine the bending stresses and deflection according to normal engineering theory and then required the application of a modification factor to increase bending stresses and deflection to reflect the effect of shear lag. This correction factor is shown in Fig. 20.4 and is related to the ratio of span to clear web spacing.

In this edition the design method now relates to European practice which determines an effective top and bottom flange width such that bending stresses and deflection require no further modification.

In accordance with Eurocode EC5, no adjustment for shear lag in plywood flanges is required if the partial flange width, L_{se}, as shown in Fig. 20.5, is less than $0.1 \times$ span of panels irrespective of face grain orientation.

From this limitation, assuming 47 mm wide web members, it can be shown that no adjustment for shear lag is required if panel span is more than 5.53 m for webs at 600 mm centres and more than 3.53 m for webs at 400 mm centres.

20.3.2 Buckling of compression skin

The upper skin of a stress skin panel used as floor or roof decking is essentially in compression. In tests it has been shown that with normal proportions of plywood thicknesses and joist spacings the plywood is unlikely to buckle. According to Eurocode EC5, to avoid plate buckling of the top compression skin the clear spacing, L_s, between webs (Fig. 20.5) should not be greater than twice the effective flange width, L_{ef}

where

$L_{ef} = 25\ t_c + b_w$ for plywood with face grain parallel to webs
$= 20t_c + b_w$ for plywood with face grain perpendicular to webs
t_c = thickness of top compression skin
b_w = width of web member.

For webs at 600 mm centres this limit is reached when $t_c = 9.2$ mm
For webs at 400 mm centres this limit is reached when $t_c = 5.2$ mm.

20.4 SELECTING A TRIAL DESIGN CROSS SECTION

The usual panel design criterion is deflection, and the inertia required can be provided by a multiplicity of variations of joist size and grade, plywood thickness and grade, single or double skins, therefore the design of a stress skin panel tends to be a case of trial and error. There are, however, a number of guidelines which can lead to a fairly accurate initial trial section.

For the initial trial design:

1. Limit the maximum span-to-overall-depth ratio to 25 for an imposed loading of 1.5 kN/m^2 and to 30 for an imposed loading of 0.75 kN/m^2.
2. The top skin should have at least the following minimum thicknesses:
   9–12 mm for 0.75 kN/m^2 imposed loading
   12–18 mm for 1.5 kN/m^2 imposed loading.
3. The bottom tension skin is usually thinner than the compression skin, 6.5 or 9 mm sanded, good face quality if the soffit requires an architecturally improved finish, or unsanded if no special finish is required.
4. Spacing of web members should not normally exceed
   0.6 m for imposed loading of 0.75 kN m^2
   0.4 m for imposed loading of 1.5 kN m^2
   0.3 m for imposed loading more than 1.5 kN m^2.
   Spacings of around 0.4 m are most common and web members are normally 47 mm finished thickness.
5. The design criterion for long span panels is usually deflection. From section 5.3.2 it can be seen that the EI value of the trial section should be $EI = 4.34\,FL^2$ to limit bending deflection to 0.003 of span.

20.5 PERMISSIBLE STRESSES

20.5.1 Joist webs

If the web members are no more than 610 mm apart, with at least four members, they may be considered as part of a load-sharing system. The load-sharing modification factor ($K_8 = 1.1$) and E_{mean} may be used. Permissible stresses for the webs

Table 20.1 Permissible stresses for C24 strength class

Stress type	Grade stress (N/mm²)	Load–duration factor K_3		Load-sharing factor K_8	Permissible stress (N/mm²)
Bending	7.5	Long term	1.00	1.1	8.25*
		Medium	1.25	1.1	10.3*
Shear	0.71	Long term	1.00	1.1	0.78
		Medium	1.25	1.1	0.98
Modulus of elasticity	10 800				

*Not modified by depth factor K_7.

Table 20.2

Web section	47 × 97	47 × 122	47 × 147	47 × 195	47 × 195	47 × 220
EI (kNm²)	38.6	76.8	135	208	314	450
Depth factor K_7	1.132	1.104	1.082	1.064	1.049	1.035

are given in Table 20.1 for C24 strength class timber. The *EI* capacities and K_7 depth factors of commonly used web members are given in Table 20.2.

20.5.2 Canadian softwood plywood flanges

Permissible stresses for Canadian softwood plywood, face grain parallel to webs, for service classes 1 and 2 are taken from Table 44 of BS 5268-2 and repeated here as Table 20.3.

20.6 SELF-WEIGHT OF PANEL ELEMENTS

The designer can determine self-weight from the values given in Table 20.4 plus an allowance of around 10% for blocking, splice plates, glue and nails.

20.7 TYPICAL DESIGN FOR DOUBLE-SKIN PANEL

Consider the case of a double-skin roof panel spanning 4.2 m supporting a medium-term uniformly imposed loading of 0.40 kN/m² dead (excluding self-weight) and 0.6 kN/m² imposed. Panels are to be 1.2 m wide.

Estimate the trial section from the guiding principles discussed in section 20.4.

$$\text{Estimated overall depth} = 4200/30 = 140\text{ mm}$$

Table 20.3 Permissible stresses for Canadian softwood plywood

Stress type	Duration of load	Permissible stress in N/mm² for nominal thicknesses in mm Minimum thicknesses in round brackets				
		9.5 (9.0)	12.5 (12.0)	15.5 (15.0)	18.5 (18.0)	20.5 (20.0)
Tensile	Long term	4.78	4.53 [4]	4.38	4.92 [7]	4.43 [7]
	Medium	5.97	5.66 [4]	5.47	6.15 [7]	5.53 [7]
Compression	Long term	6.25	5.96 [4]	5.76	6.45 [7]	5.81 [7]
	Medium	7.81	7.45 [4]	7.20	8.08 [7]	7.26 [7]
Bearing on face	Long term	1.88	1.88	1.88	1.88	1.88
	Medium	2.35	2.35	2.35	2.35	2.35
Rolling shear†	Long term	0.39	0.39 [4]	0.51	0.39 [6]	0.39 [5]
	Medium	0.49	0.49 [4]	0.64	0.49 [6]	0.49 [5]
Modulus of elasticity in tension and compression	–	2600	2470 [4]	2390	2670 [7]	2410 [7]

Properties based on 'full area method'.
Where there are more than one layup for the same nominal ply thickness only the lower value is given with the appropriate veneer count indicated in brackets [ ].

Table 20.4

Canadian softwood plywood skins		Finnish Birch-faced plywood skins		Timber joists	
Nominal thickness (mm)	Weight (kN/m²)	Nominal thickness (mm)	Weight (kN/m²)	Size (mm)	Weight (kN/m²)
9.5	0.054	9	0.065	47 × 122	0.029
12.5	0.073	12	0.084	47 × 145	0.034
15.5	0.091	15	0.104	47 × 170	0.040
18.5	0.108	18	0.123	47 × 195	0.046
20.5	0.120	21	0.143	47 × 220	0.052

Try cross-section arrangement shown in Fig. 20.6.

Top skin, assume 12.5 mm Canadian softwood plywood unsanded. (Minimum thickness = 12.0 mm)

Lower skin, assume 9.5 mm Canadian softwood plywood unsanded. (Minimum thickness = 9.0 mm)

$$\text{Minimum web joist depth} = 140 - 12.0 - 9.0 = 119\,\text{mm}$$

For web joists try 47 × 122 C24 strength class timber.

Determine effective flange width according to EC5 with plywood outer ply grain direction parallel to webs.

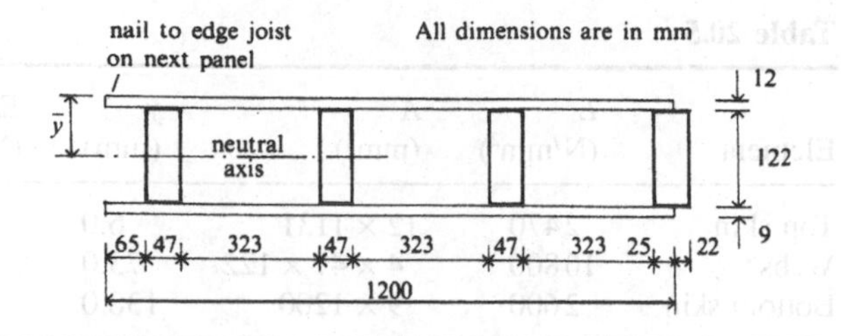

Fig. 20.6

Referring to Fig. 20.5:
For shear lag, partial flange width, L_{se}, must not be greater than

$$0.1 \times \text{span} = 0.1 \times 4200 = 420\,\text{mm}$$

To avoid plate buckling, partial flange width, L_{se}, must not be greater than

$$25t_c = 25 \times 12 = 300\,\text{mm}$$

For top compression skin,

$$\text{effective flange width} = L_{ef} = L_{se} + b_w = 300 + 47 = 347\,\text{mm}$$

To avoid plate buckling

$$L_s \leq 2 \times \text{effective flange width} = 694\,\text{mm}$$

$L_s = 323\,\text{mm}$, therefore plate buckling requirement is satisfied.

With $L_{se} = 300\,\text{mm}$,

$$\text{overall top skin effective width} = 1200 - 3 \times (323 - 300) = 1131\,\text{mm}$$

For the bottom tension skin the buckling limit is not applicable and the bottom skin effective width would be the lesser of shear lag limit of 420 mm or the actual width of 323 mm. Therefore adopt full skin width for lower tension flange.

$$\begin{aligned}\text{Self-weight of the panel per m}^2 &= \text{top skin} + \text{webs} + \text{bottom skin}\\ &= 0.073 + (4 \times 0.029/1.2) + 0.054\\ &= 0.22\,\text{kN/m}^2\end{aligned}$$

Applied load = dead + imposed + self-weight = $0.4 + 0.6 + 0.22 = 1.22\,\text{kN/m}^2$
Total load on panel = $1.22 \times 4.2 \times 1.2 = 6.15\,\text{kN}$

$$\text{Shear} = \frac{6.15}{2} = 3.07\,\text{kN}$$

$$\text{Bending moment} = \frac{6.15 \times 4.2}{8} = 3.23\,\text{kN m}$$

Deflection of panel

To limit bending deflection to $0.003 \times$ span,

$$EI \cong 4.34\,WL^2 = 4.34 \times 7.03 \times 4.2^2 = 538\,\text{kN m}^2$$

Table 20.5

Element	E (N/mm^2)	A (mm^2)	y (mm)	EA (N × 10^6)	EAy (N mm × 10^6)
Top skin	2470	12 × 1131	6.0	33.5	201
Webs	10 800	4 × 47 × 122	73.0	247.7	18 082
Bottom skin	2600	9 × 1200	138.0	28.1	3875
Σ				309.3	22 158

E values and actual thicknesses are from Tables 20.1 and 20.3.

Table 20.6

Element	EA (N × 10^6)	h_x (mm)	$(EA)h_x^2$ (kN m)
Top skin	33.50	65.60	144
Webs	247.70	1.4	–
Bottom skin	28.1	66.40	124
Σ			268

To calculate the actual *EI* value of the panel it is first necessary to locate the neutral axis which is distance $\bar{y}$ from the top surface (see Fig. 20.6).

$$\bar{y} = \frac{\sum EAy}{\sum EA}$$

where ΣEA = product of the E value and area A of each element in the panel
y = distance of the centroid of each element from the top surface reference plane. (Any reference plane could be used, but the top or bottom surface is most convenient.)

From Table 20.5,

$$\bar{y} = \frac{22158}{309.3} = 71.6\,\text{mm}$$

The self-*EI* capacity of the top and bottom skins about their neutral axis is small and can be disregarded with little loss of accuracy. Therefore the bending rigidity of total panel is:

$$EI = EI_{\text{webs}} + \sum (EA)h_x^2$$

where h_x = distance from the neutral axis to the centroid of the element.
From Tables 20.2 and 20.6,

$$EI = (4 \times 76.8) + 268 = 575\,\text{kN}\,\text{m}^2$$

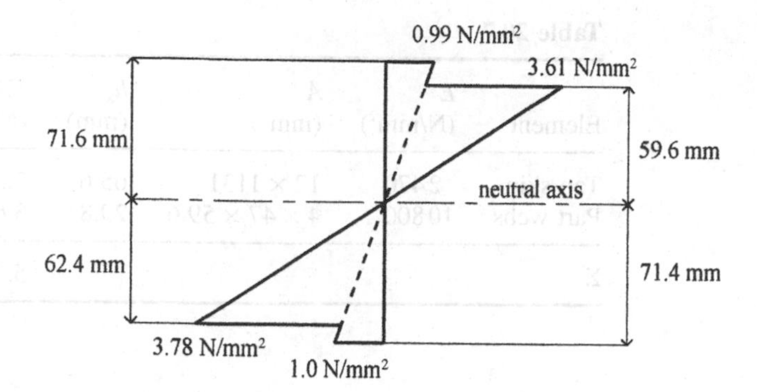

Fig. 20.7

$$\text{Deflection} = \frac{5 \times 6.15 \times 4.2^3}{384 \times 575} = 0.010\ \text{m}$$

Allowable deflection = 0.003 × span = 0.014 m

Bending stresses

The bending stress at any fibre of an element is:

$$\sigma_{m,a} = \frac{MyE}{EI}$$

where M = bending moment on the panel
y = distance from the neutral axis to the fibre under consideration
E = E value of the element under consideration
EI = bending rigidity of the full panel.

At the extreme fibre of the top skin ($y = 71.6$ mm):

$$\sigma_{m,a} = \frac{3.23 \times 10^6 \times 71.6 \times 2470}{575 \times 10^9} = 0.99\ \text{N/mm}^2 < 7.45\ \text{N/mm}^2$$

[Table 20.3, plywood mainly in compression]

At the extreme fibre of the bottom skin ($y = 143 - 71.6 = 71.4$ mm):

$$\sigma_{m,a} = \frac{3.23 \times 10^6 \times 71.4 \times 2600}{575 \times 10^9} = 1.0\ \text{N/mm}^2 < 5.97\ \text{N/mm}^2$$

[Table 20.3, plywood mainly in tension]

At the upper extreme fibre of the web joist ($y = 71.6 - 12 = 59.6$ mm):

$$\sigma_{m,a} = \frac{3.23 \times 10^6 \times 59.6 \times 10\,800}{575 \times 10^9} = 3.61\ \text{N/mm}^2$$

Allowable bending stress (from Tables 20.1 and 20.2) = 10.3 × 1.104
= 11.3 N/mm²

Table 20.7

Element	E (N/mm^2)	A (mm^2)	h_x (mm)	EAh_x (N mm × 10^9)
Top skin	2470	12 × 1131	65.6	2.20
Part webs	10 800	4 × 47 × 59.6	29.8	3.61
Σ				5.81

Table 20.8

Element	E (N/mm^2)	A (mm^2)	h_x (mm)	EAh_x (N mm × 10^9)
Part webs	10 800	4 × 47 × 62.4	31.2	3.95
Bottom skin	2600	9 × 1200	66.4	1.86
Σ				5.81

At the lower extreme fibre of the web joist ($y = 71.4 - 9 = 62.4$ mm):

$$\sigma_{m,a} = \frac{3.23\times10^6 \times 62.4\times10\,800}{575\times10^9} = 3.78\ \text{N/mm}^2 < 11.3\ \text{N/mm}^2$$

Therefore all these bending stresses are acceptable.

Horizontal shear stresses in web members

The horizontal shear stress at the neutral axis is:

$$\tau = \frac{F_v E_s}{(EI)t}$$

where F_v = applied shear force (N)
E_s = product of moment of elasticity and first moment of area about neutral axis (N mm)
EI = bending rigidity of panel (N mm)
t = total thickness of webs (mm).

Determine the value of $E_s = \Sigma EAh_x$ taking into consideration those elements (or parts of elements) which occur either above or below the neutral axis.

By tabulation, consider parts above the neutral axis (see Table 20.7):

Therefore: $E_s = 5.81$ N mm × 10^9

Alternatively the same result is determined by considering parts below the neutral axis (see Table 20.8):

Therefore: $E_s = 5.81$ N mm × 10^9

$$\text{Maximum horizontal shear stress} = \tau = \frac{F_v E_s}{(EI)t}$$

$$= \frac{3.07 \times 10^3 \times 5.81 \times 10^9}{575 \times 10^9 \times (4 \times 47)} = 0.17\ \text{N/mm}^2$$

Permissible horizontal shear stress (from Table 20.1) = 0.98 N/mm²
Satisfactory.

Rolling shear stress between ply skins and web members

The rolling shear stress at the junction of the web joists and the flanges is:

$$\tau_r = \frac{F_v E_s}{(EI)t}$$

where F_v = applied shear force (N)
E_s = product of moment of elasticity and first moment of area of relevant flange about neutral axis (N mm)
EI = bending rigidity of panel (N mm)
t = total thickness of webs (mm).

For a known panel arrangement and applied shear force F_v, EI and t are constant and the maximum rolling shear stress will occur at the junction where E_s (for skin only) is the larger. In this case it can be seen that the top skin is more critical (2.20 > 1.86 in Tables 20.7 and 20.8 respectively).

In Fig. 20.6, assuming that the shear in the panel is transferred uniformly between webs and skins, the rolling shear stress is over a contact width of

$$t = (3 \times 47) + 25 = 166\ \text{mm}$$

Therefore:

$$\text{Rolling shear stress, } \tau_r = \frac{3070 \times 2.2 \times 10^9}{575 \times 10^9 \times 166} = 0.07\ \text{N/mm}^2$$

When determining the permissible rolling shear stress, clause 4.7 of BS 5268-2 requires the application of stress concentration modification factor $K_{37} = 0.5$ and assuming that bonding pressure is achieved by nails, clause 6.10.1.3 of BS 5268-2 requires the application of the nail/glue modification factor $K_{70} = 0.9$. As the spacing of web members is less than 600 mm, the load-sharing factor $K_8 = 1.1$ is applicable.

$$\text{Permissible rolling shear stress} = 0.49\ \text{(Table 20.3)} \times 0.5 \times 0.9 = 0.22\ \text{N/mm}^2$$

Satisfactory

20.8 SPLICE PLATES

When full-length plywood sheets are not available, it will be necessary to introduce splices in the plywood. Although the stress skin panel is in bending, the stress in the skin or skins approximates closer to pure compression or tension and the splice plates should be designed accordingly. Although splices may not occur at the position of maximum moment, it is convenient for the purpose of design to assume that they do.

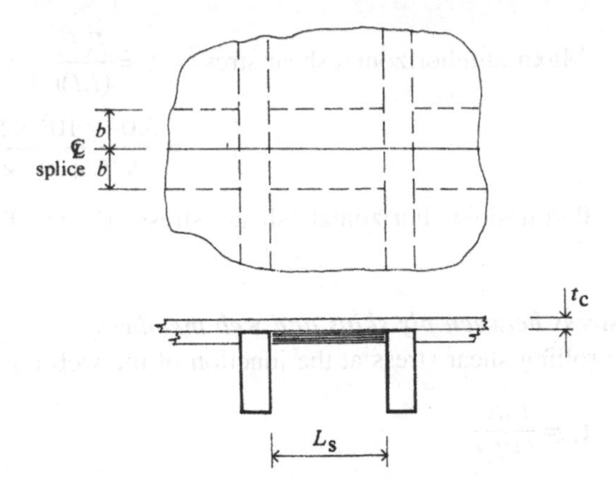

Fig. 20.8

The longitudinal members also act as splices and it is therefore only necessary to transfer through the plywood splices the force developed in the area of the ply skin between longitudinal members. The maximum force to be transferred will be expressed as:

$$\text{Force} = \sigma_{s,a} \times t_c \times L_s$$

where $\sigma_{s,a}$ = applied tensile or compression stress in plywood skin
t_c = skin thickness
L_s = spacing between longitudinals.

This force must be resisted by the cross-sectional area of the splice plate. It is usual to make the splice plate of width L_s (Fig. 20.8) and fix it in the same direction as the skin. If the skin is stressed to only a small proportion of the permissible stress, some reduction in plywood splice thickness to a value less than t_c may be acceptable. In practice, it is usual to specify the splice plate as the same thickness as the skin and no stress check is then required.

The force must also be transferred across the joint either by mechanical fasteners or glue. In the case of gluing, pressure may be achieved by clamps or nailing, the former being preferred, particularly for the lower skin if the soffit is of architectural quality.

The glue line stress should not exceed the permissible rolling shear stress. In this case the K_{37} factor of 0.5 for stress concentration is not applicable and the K_{70} of 0.9 will apply only if bonding pressure is achieved by nails or staples.

The half-length splice plate dimension b (Fig. 20.8) is given by the expression:

$$b = \frac{\sigma_{s,a} t_c}{\tau_{r,adm}}$$

where $\tau_{r,adm}$ = permissible rolling shear stress.

Consider a splice plate in the 12 mm thick top skin of panel in the example in section 20.7.

Compression stress in top skin = 0.99 N/mm^2

From Table 20.3,

$$\tau_{r,\,adm} = 0.49 \times 0.9 = 0.44\,N/mm^2$$

$$b = \frac{0.99 \times 12}{0.44} = 27\text{ mm}$$

and total minimum splice plate length required = $2b$ = 54 mm

Consider a splice plate in the 9 mm thick bottom skin of panel in the example in section 20.7

Tension stress in bottom skin = 1.0 N/mm²

From Table 20.3,

$$\tau_{r,\,adm} = 0.49 \times 0.9 = 0.44\,N/mm^2$$

$$b = \frac{1.0 \times 9}{0.44} = 20\text{ mm}$$

and total minimum splice plate length required = $2b$ = 40 mm

These examples have produced short length splice plates resulting from the relatively low stresses in the top and bottom skins. Many designs will give much higher stress levels, and longer splice plates should be anticipated.

20.9 TYPICAL DESIGN FOR SINGLE-SKIN PANEL

To compare the performance of a single-skin panel (stiffened panel) to a double-skin panel recalculate the panel assessed in section 20.7, but with the lower skin removed. The new section is shown in Fig. 20.9.

Self-weight of the panel per m² = top skin + webs

$$= 0.073 + \frac{4 \times 0.029}{1.2} = 0.17\text{ kN/m}^2$$

Applied load = dead + imposed + self-weight = 0.4 + 0.6 + 0.17 = 1.17 kN/m²

Total load on panel = 1.17 × 4.2 × 1.2 = 5.9 kN

$$\text{Shear} = \frac{5.9}{2} = 2.95\text{ kN}$$

$$\text{Moment} = \frac{5.9 \times 4.2}{8} = 3.10\text{ kN m}$$

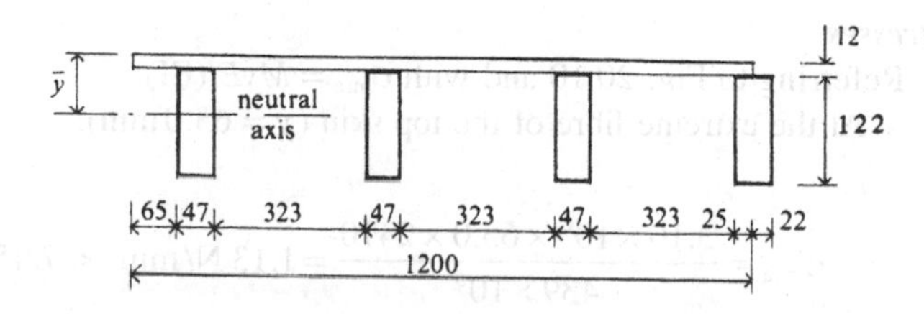

Fig. 20.9

Table 20.9

Element	E (N/mm^2)	A (mm^2)	y (mm)	EA (N × 10^6)	EAy (N mm × 10^6)
Top skin	2470	12 × 1131	6.0	33.5	201
Webs	10800	4 × 47 × 122	73.0	247.7	18082
Σ				281.2	18283

Table 20.10

Element	EA (N × 10^6)	h_x (mm)	$(EA)h_x^2$ (kN m)
Top skin	33.9	59.0	116.6
Webs	247.7	7.8	15.1
Σ			131.7

Deflection

By tabulation, determine location of neutral axis $\bar{y}$ distance of which is centroid of each element from top surface reference plane.

From Table 20.9,

$$\bar{y} = \frac{18\,283}{281.2} = 65.0 \text{ mm}$$

The self-EI capacity of the top about its neutral axis is small and can be disregarded with little loss of accuracy. Therefore the bending rigidity of total panel is:

$$EI = EI_{\text{webs}} + \sum (EA)h_x^2$$

where h_x = distance from neutral axis to centroid of the element.

From Tables 20.2 and 20.10,

$$EI = (4 \times 76.8) + 131.7 = 439 \text{ kN m}^2$$

$$\text{Bending deflection} = \frac{5 \times 5.9 \times 4.2^3}{384 \times 439} = 0.013 \text{ m}$$

which is approximately 1.24 times that for the double-skin panel.

Bending stresses

Referring to Fig. 20.10 and with $\sigma_{m,a} = MyE/(EI)$.

At the extreme fibre of the top skin ($y = 65.0$ mm):

$$\sigma_{m,a} = \frac{3.10 \times 10^6 \times 65.0 \times 2470}{439 \times 10^6} = 1.13 \text{ N/mm}^2 < 7.45 \text{ N/mm}^2$$

[Table 20.3, plywood mainly in compression]

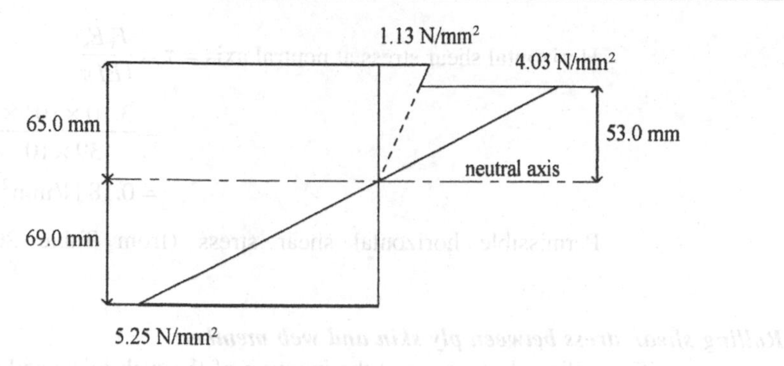

Fig. 20.10

Table 20.11

Element	E (N/mm²)	A (mm²)	h_x (mm)	EAh_x (N mm × 10⁹)
Top skin	2470	12 × 1131	59.0	1.98
Part webs	10800	4 × 47 × 53.0	26.5	2.85
Σ				4.83

At the upper extreme fibre of the web joist ($y = 65.0 - 12 = 53.0$ mm):

$$\sigma_{m,a} = \frac{3.10 \times 10^6 \times 53.0 \times 10\,800}{439 \times 10^9} = 4.03\ \text{N/mm}^2$$

Allowable bending stress (from Tables 20.1 and 20.2) = 10.3 × 1.104
= 11.3 N/mm²

At the lower extreme fibre of the web joist ($y = 122 - 53.0 = 69.0$ mm):

$$\sigma_{m,a} = \frac{3.10 \times 10^6 \times 69.0 \times 10\,800}{439 \times 10^9} = 5.25\ \text{N/mm}^2 < 11.3\ \text{N/mm}^2$$

Therefore all these bending stresses are acceptable.

Horizontal shear stresses

Determine the value of $E_s = \Sigma EAh_x$ taking into consideration those elements (or parts of elements) which occur either above or below the neutral axis.

By tabulation, consider parts above the neutral axis (see Table 20.11):

$E_s = 4.83$ N mm × 10⁹

Alternatively:

E_s (for case below neutral axis) = 10 800 × (4 × 47 × 69.0) × 34.5
= 4.83 × 10⁹ N mm

$$\text{Horizontal shear stress at neutral axis} = \tau = \frac{F_v E_s}{(EI)t}$$

$$= \frac{3.10 \times 10^3 \times 4.83 \times 10^9}{439 \times 10^9 \times (4 \times 47)}$$

$$= 0.18\ \text{N/mm}^2$$

Permissible horizontal shear stress (from Table 20.1) = 0.98 N/mm^2
Satisfactory

Rolling shear stress between ply skin and web members

The rolling shear stress at the junction of the web joists and the ply skin is:

$$\tau_r = \frac{F_v E_s}{(EI)t} = \frac{2950 \times 1.98 \times 10^9}{439 \times 10^9 \times 166} = 0.08\ \text{N/mm}^2$$

Permissible rolling shear stress = 0.49 (Table 20.3) × 0.5 × 0.9 = 0.22 N/mm^2
Satisfactory

Conclusion

In this instance removing the lower skin increases deflection by 24% and increases maximum bending stress by 36%.

Chapter 21
Trusses

21.1 INTRODUCTION

The range of truss profiles available using timber construction is probably far greater than with any other structural material. There may be a historical basis for the use of timber for trusses, or possibly the relative ease with which unusual truss shapes can be fabricated and assembled. Profiles regarded as 'traditional' are often specified for architectural or restoration purposes so the engineer needs to be familiar with both modern and traditional forms of truss design.

The structural function of a truss is to support and transfer loads from the points of application (usually purlins) to the points of support as efficiently and as economically as possible. The efficiency depends on the choice of a suitable profile consistent with the architectural requirements and compatible with the loading conditions. Typical 'idealized' truss profiles for three loading conditions are sketched in Fig. 21.1.

With a symmetrical system of loading (particularly important in the second case in Fig. 21.1, which is a four-pin frame and therefore unstable) the transfer of loading is achieved without internal web members, because the chord profile matches the bending moment of the simple span condition. Unfortunately, it is seldom possible to use a profile omitting internal members, because unbalanced loading conditions can nearly always occur from snow, wind or permanent loadings. Unbalanced conditions can also occur due to manufacturing and erection tolerances (see also section 3.12), nevertheless, the engineer should try to use a truss profile closely related to the idealized profile (the moment diagram), adding a web system capable of accommodating unbalanced loading. In this way the forces in the internal members and their connections are minimized, with consequent design simplicity and economy.

Undoubtedly the engineer will encounter cases in which the required architectural profile is at conflict with the preferred structural profile, therefore high stresses may be introduced into the web system and the connections. Economy must then be achieved by adopting the most suitable structural arrangement of internal members in which it is necessary to create an economical balance between materials and workmanship. The configuration of internal members should give lengths between node points on the rafters and ceiling ties such as to reduce the numbers of joints. Joints should be kept to a sensible minimum because the workmanship for each is expensive, and also the joint slip at each (except with glued joints) generally adds to the overall deflection of the truss (section 21.6.2). On the other hand, the slenderness ratio of the compression chords and the internal struts

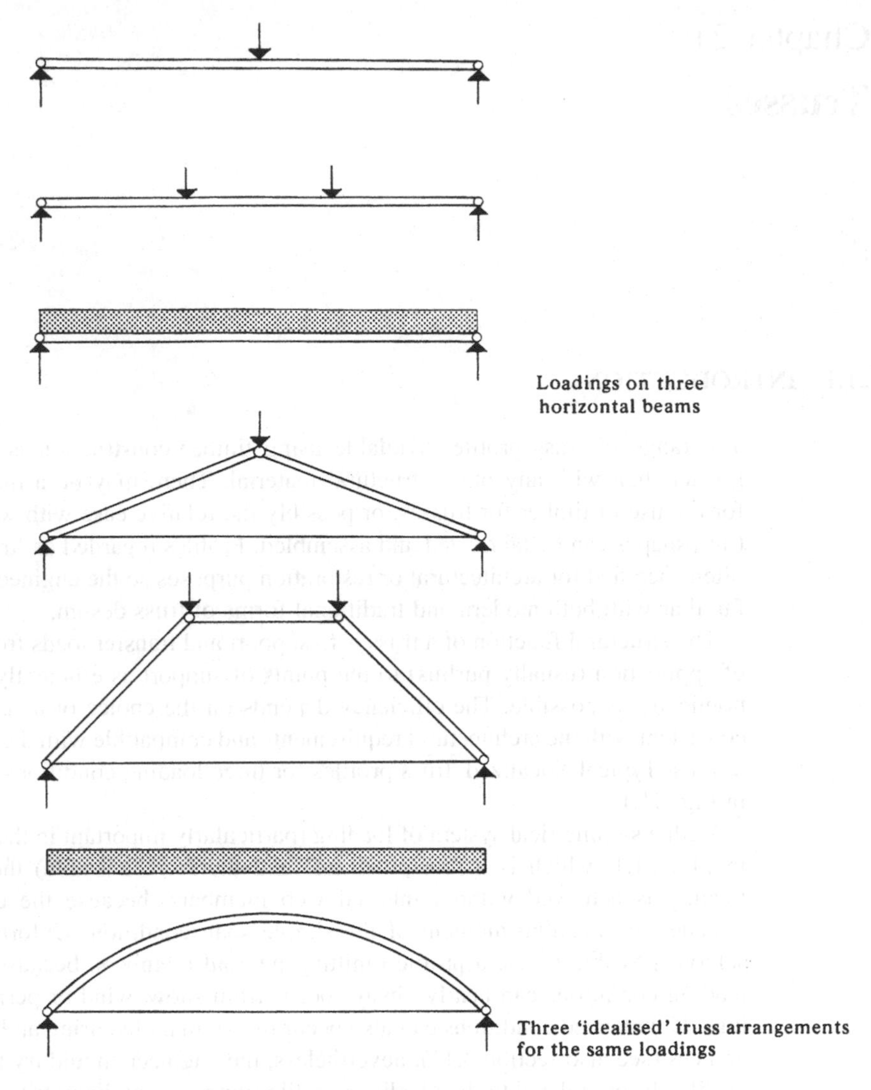

Fig. 21.1

must not be excessive, local bending on the chords must not be too large, and the angle between internal diagonals and the chords must not be too small.

The engineer is usually influenced by architectural considerations, the type and weight of roof material, support conditions, span and economy, and probably chooses from three basic truss types: pitched (mono- or duo-pitch), parallel-chord or bowstring trusses (Fig. 21.2).

The most common form for both domestic and industrial uses is the pitched truss. The shape fits the usual moment diagram reasonably well and is compatible with traditional roofing materials, such as tiles for domestic uses and corrugated sheeting for industrial applications. A portion of the applied loading is transferred directly through the top chord members to the points of support, while the web members transfer loads of relatively small to medium magnitude, and the joints

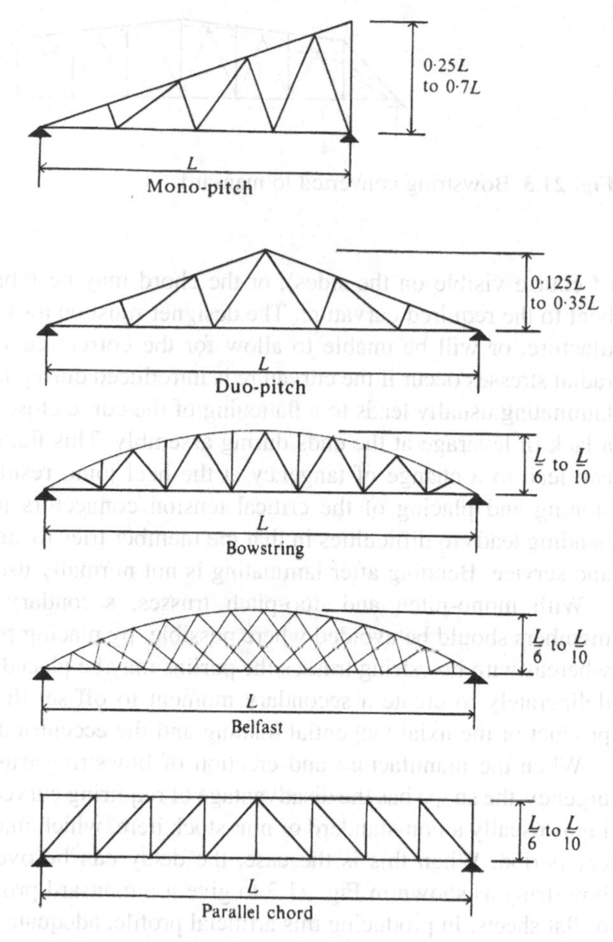

Fig. 21.2 General span-to-depth ratios for basic profiles.

can usually be designed to develop these loads with little difficulty. Mono-pitch trusses are suitable generally for spans only up to around 9 m. Above this span the vertical height is usually too large for architectural reasons, even if the truss slope is reduced below that suitable for tiling. Duo-pitch domestic trusses span up to around 12 m with duo-pitch industrial trusses spanning up to around 15 m, above which span they become difficult to transport unless fabricated in parts (Fig. 21.8).

For large-span industrial uses, bowstring trusses (Fig. 21.2) can be very economical. These may be regarded as the current alternative to the traditional all-nailed 'Belfast truss' (Fig. 21.2). With uniform loading and no large concentrated loads the arched top chord profile supports almost all of the applied loading, and spans in excess of 30 m are not uncommon. A parabolic profile is the most efficient theoretical choice to support uniform loading, but practical manufacturing considerations usually make it more convenient or necessary to adopt a circular profile for the top chord member. The top chord member is usually laminated (not necessarily with four or more members), using either clamp pressure or nail pressure for assembly. The curvature may be introduced while laminating (laminations

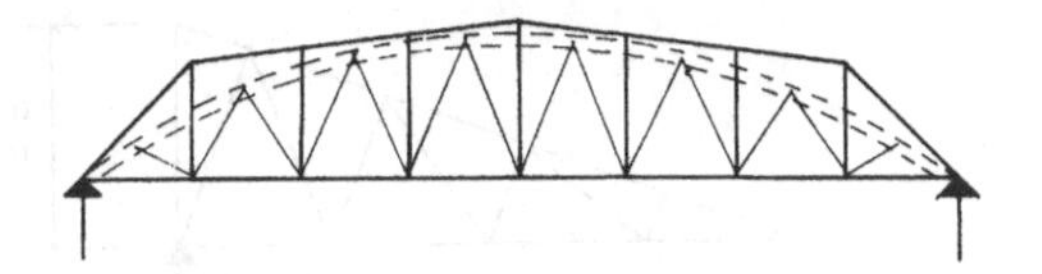

Fig. 21.3 Bowstring converted to mansard.

of course visible on the sides), or the chord may be fabricated straight and then bent to the required curvature. The designer must be aware of the method of manufacture, or will be unable to allow for the correct curvature stresses. The least radial stresses occur if the curvature is introduced during laminating. Bending after laminating usually leads to a flattening of the curve close to the heel joints, due to a lack of leverage at the ends during assembly. This flattening, if not recognized, can lead to a change of tangency at the heel joint, resulting in incorrect dimensioning and placing of the critical tension connectors in the tie member. Also, bending leads to difficulties in that the member tries to straighten during assembly and service. Bending after laminating is not normally used.

With mono-pitch and duo-pitch trusses, secondary bending on the chord members should be avoided where possible, by placing purlins at the node points, whereas with bowstring trusses the purlins may be placed between the node points deliberately to create a secondary moment to off-set the moment caused by the product of the axial tangential loading and the eccentricity of the chord.

When the manufacture and erection of bowstring trusses is a matter of some urgency, the shape has the disadvantage of requiring curved sheeting. Curved sheeting is usually a non-standard or non-stock item, which may have a very long delivery period. When this is the case, the delay can be overcome by adding to the bowstring as shown in Fig. 21.3 to give it a mansard profile, and revert to the use of flat sheets. In producing this artificial profile, adequate lateral restraint must still be provided to the curved compression chord, which is no longer restrained directly by the purlins.

Parallel-chord trusses are frequently specified as an alternative to ply web or glulam beam on long spans where the beams may be uneconomical. The loads in the web members are frequently very large, which causes some difficulty in providing adequate joints. The choice of web configuration is between the Howe (diagonals in compression, Table 21.17) the Pratt (diagonals in tension, Table 21.20) and the Warren type (diagonals in alternate compression and tension, Table 21.23).

When a parallel-chord truss is joined to a timber or steel column with connections at both the top and bottom chords (Fig. 21.4), this gives fixity in a building subject to sway (Figs 23.5–23.8) and a Pratt truss would be favoured.

When it is required to minimize the height of the perimeter wall, the Howe truss is favoured (Fig. 21.5).

As an indication of the difference in magnitude of forces in internal members dictated by the choice of truss profile, coefficients are presented in Fig. 21.6 for the three basic types at a typical span-to-depth ratio. Each of the three basic profiles may have raised bottom chords to give extra central clearance. This can be particularly useful in storage buildings with central access. The guide

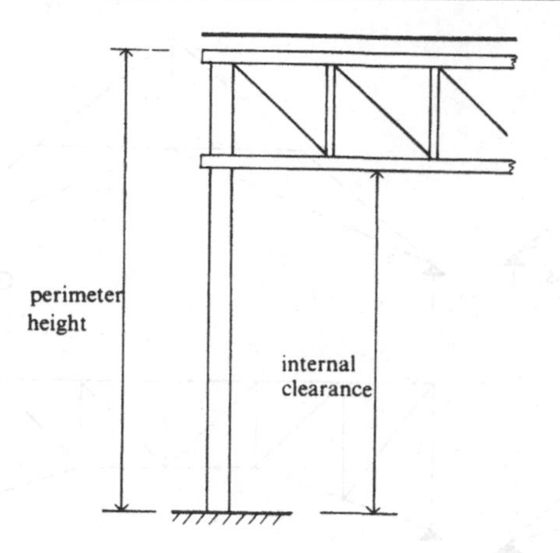

Fig. 21.4

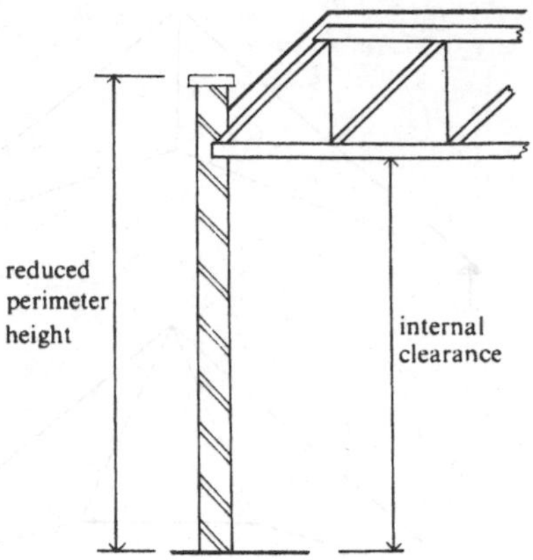

reduced
perimeter
height
internal
clearance

Fig. 21.5

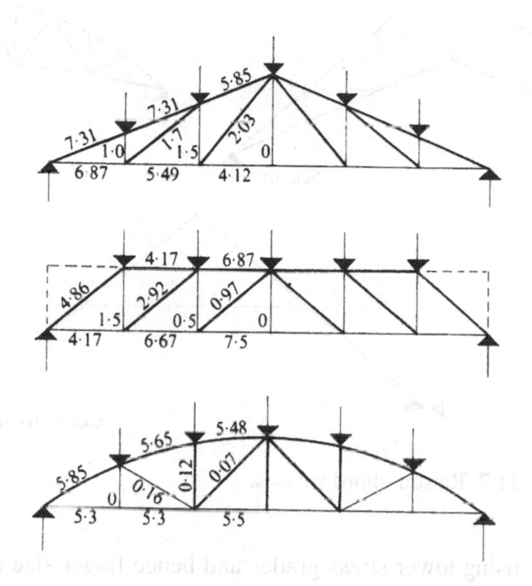

Fig. 21.6

span-to-central-depth ratios given in Fig. 21.2 should be maintained for maximum economy in design. Modified pitched, parallel-chord and bowstring trusses are sketched in Fig. 21.7, plus some traditional configurations. In practice, the collar beam is used only for small spans with steep slopes. If using a raised chord truss (and particularly if using the collar beam type), the designer should consider the possibility of thrusts occurring at the support points due to deflection of the framework.

The larger the span the more necessary it may be to use the smaller span-to-depth ratios. Although shallower depth trusses may be preferred aesthetically, they tend to have large deflections which may create secondary stresses. Deflection can be minimized by:

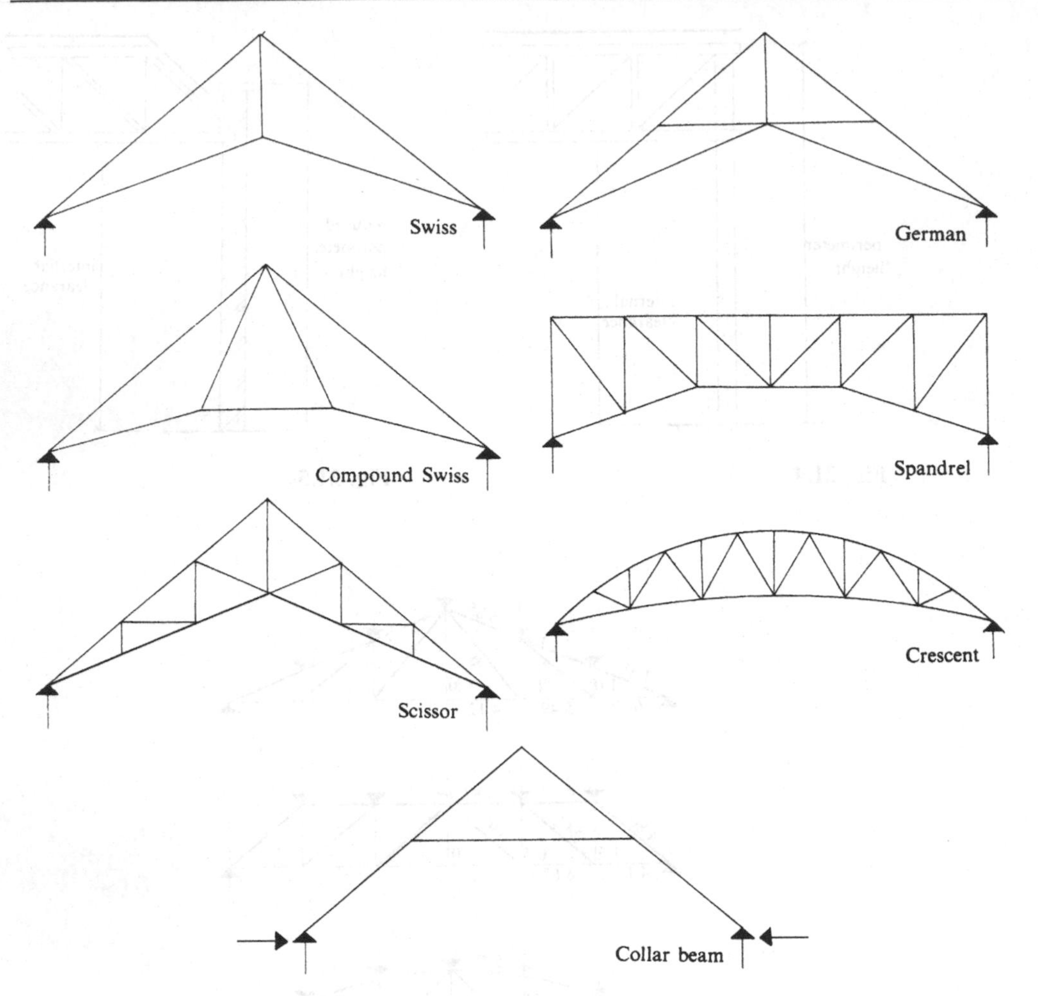

Fig. 21.7 Raised chord trusses.

- using lower stress grades and hence larger size members
- keeping the number of joints and mechanically jointed splices to a minimum, and
- using fastenings with low slip characteristics.

The use of lower-grade material may be regarded initially as an uneconomical proposition as this makes larger sizes necessary. However, in many cases the spacings and edge distances for connector groups call for the use of timber sections larger than those required to satisfy the axial loadings and moments. This may invalidate the apparent benefits of a higher stress grade. From the values in Chapter 18 it can be seen that connector capacities do not increase to the same extent as the efficiency increase of higher stress grades, and the required minimum spacings and edge distances do not reduce pro rata to the increase in grade stress. These points encourage the use of lower stress grades although, of course, the strength of the net timber section is less compared to the same size of a higher stress grade.

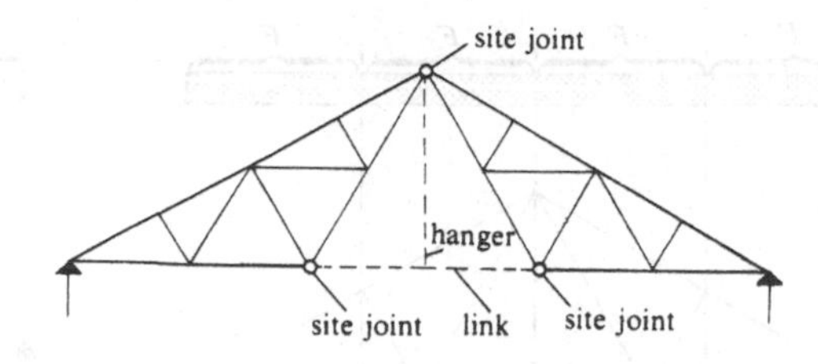

Fig. 21.8

To obtain an indication of the likely effect on the deflection of a truss by increasing the strength class from C16 to C24 compare the AE values of the tension members assuming members fully stressed. The ratios show that the strain in the C24 members leads to approximately 13% more deflection of the truss than if using C16 members. Providing the slenderness ratio of the compression members is quite small, the AE ratio in compression is similar to that in tension, and it can be assumed that a truss designed to use the minimum amount of C24 material deflects approximately 13% more than a truss designed for the minimum amount of C16 material, providing that there is no slip at joints. However, if the same type of mechanical connector is used in each truss, the slip effect is constant for each, which reduces the difference in deflection from 13% to approximately 7%.

Trusses (not trussed rafters) used on spans similar to domestic applications are usually spaced at 1.5 to 2 m centres, perhaps with solid purlins supporting intermediate rafters. For industrial uses where lighter-weight roof specifications can be expected, trusses are spaced at 5 to 6 m centres with solid or composite purlins at 0.8 to 1.8 m spacings supporting corrugated sheeting. Economy usually results if truss spacing increases with truss span.

Transportation is frequently a limiting factor with deep or long-span trusses. Trusses deeper than 3 m, or longer than 20 m, require special attention. The transport problem can usually be overcome by a partial or complete breakdown of the truss. For example, bowstring trusses may have the main members spliced at midspan, and in many cases the entire assembly can be carried out on site, although it is preferable to carry out an initial assembly in the works to ensure correct fit, then break down for transport. Pitched trusses, especially those of large span, can be fabricated in two halves and linked together on site, with a loose centre tie perhaps with an optional hanger (Fig. 21.8). Section 28.3 gives guidance on the maximum size of components which can be accommodated on normal lorries without special transport arrangements having to be made.

21.2 LOADING ON TRUSSES

Figure 21.9 shows a typical four-panel truss supporting a uniformly distributed load totalling $4F$. It is common and accepted practice to assume equal point loads at each node point from this UDL in calculating the axial loading in the truss members. If each rafter is isolated, as shown in Fig. 21.10, and treated as a member continuous over two bays without sinking supports, this has the effect of reducing the overall moment on the truss by reducing the central point load (Fig. 21.11).

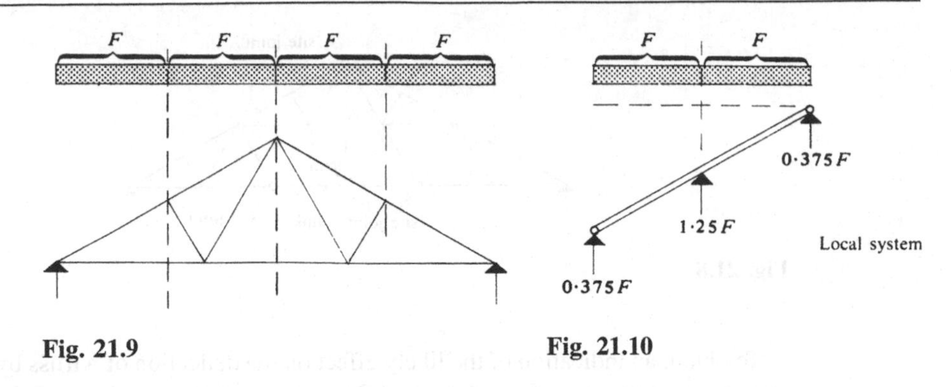

Fig. 21.9

Fig. 21.10

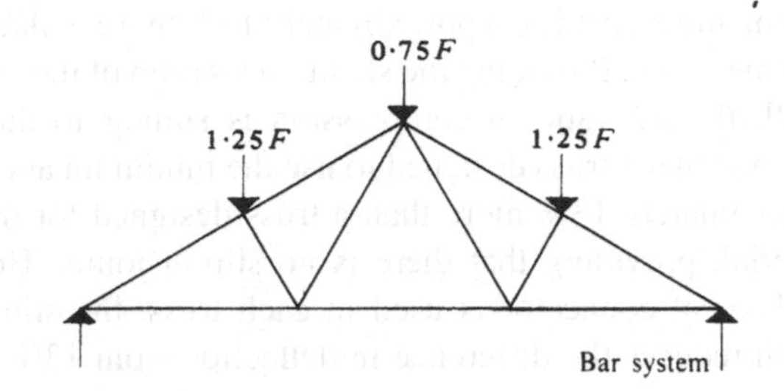

Fig. 21.11

Whether or not the designer takes account of this in normal truss design, it should be realized that, if secondary rafters are continuous over a purlin supported at mid-rafter and at the ridge, the load transferred to the truss will approach that shown in Fig 21.11.

With a truss having six panels, the loads at the intermediate nodes will be $1.1F$ and the apex load will be $0.8F$. With trusses of more than six bays it is usually acceptable to disregard any such increase in the intermediate node point loads.

21.3 TYPES OF MEMBERS AND JOINTS

21.3.1 Mono-chord and duo-chord trusses

It is usual for the mono-chord truss to use chords and internal members in one plane with mild steel gusset plates or perhaps even thick (say 14 gauge) punched metal plate fasteners on each side at all connections (Fig. 21.12). The design is similar to a steelwork truss, and has the advantage that all connectors are loaded parallel to grain. With lightly loaded members bolted joints are adequate, single-sided toothed plates may be used for medium loads and shear plate connectors used for heavily loaded members. This type of truss is extremely economical although the steel plates placed on the outer faces may not be visually acceptable for certain applications. Where appearance is a prime consideration the double-chord truss (Fig. 21.13) should be considered.

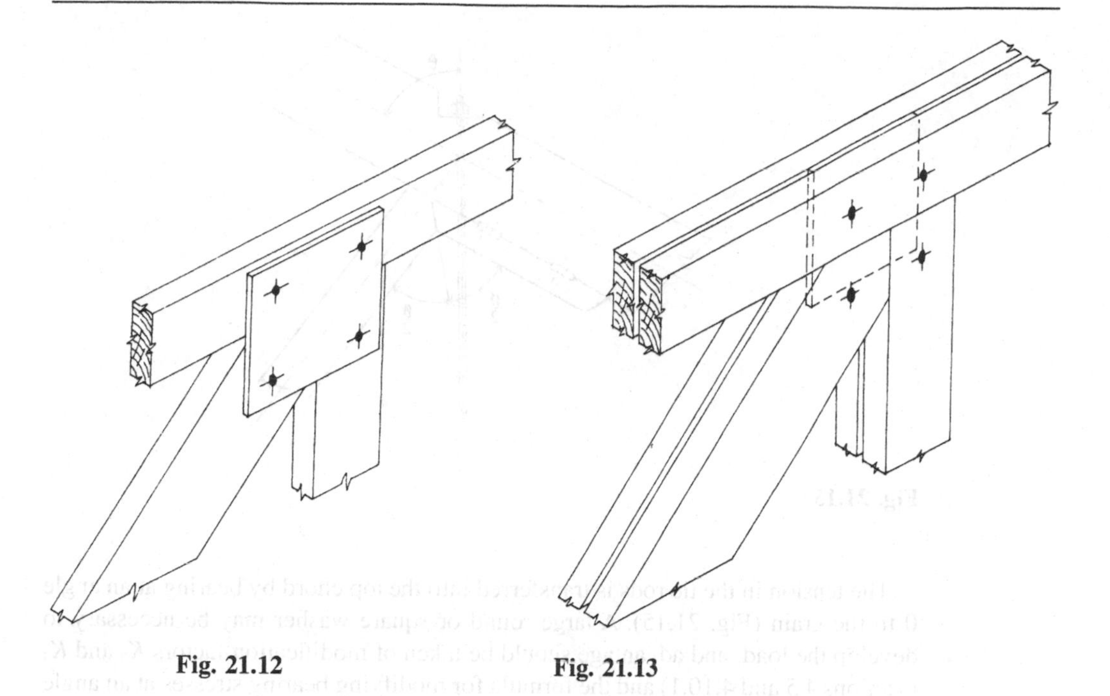

Fig. 21.12 Fig. 21.13

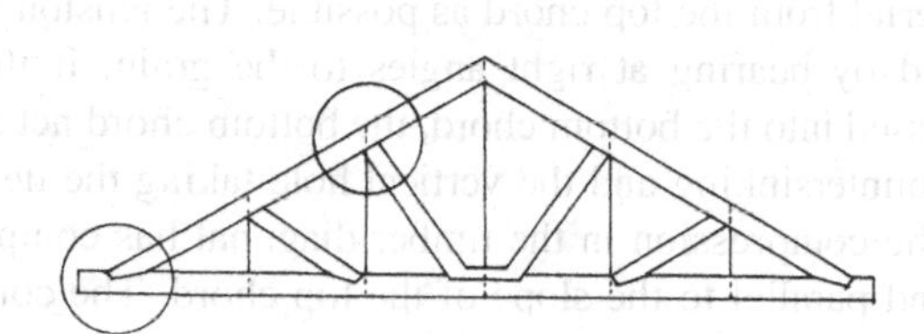

Fig. 21.14

The duo-chord truss, shown in Fig. 21.13, uses twin members for the chords and web members placed in the same two planes with mild steel plates between the members. Washers must be placed under the heads and nuts of bolts, circular washers being preferred.

Large-span trusses may use laminated members to give sections with a larger area and/or improved appearance.

21.3.2 Rod-and-block assemblies

Rod-and-block assemblies (Fig. 21.14) are not used to any great extent for truss design in the UK. This system uses steel rods for the internal tension members, and as such may be the choice for buildings where a traditional appearance is desirable. Because the steel rods will not take compression, this type of construction is not permitted when the wind loading creates reversal of stress in the internal members.

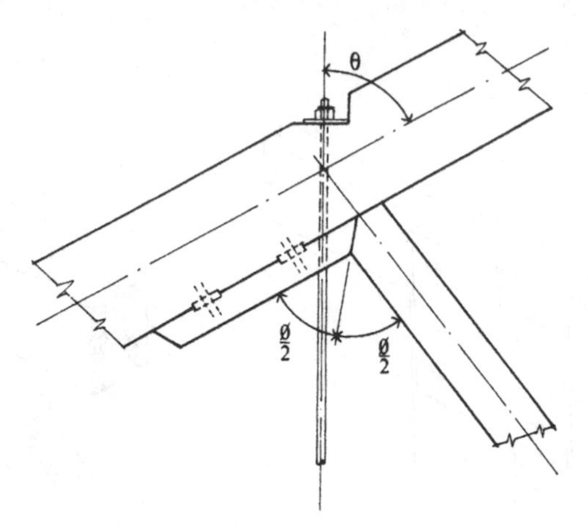

Fig. 21.15

The tension in the tie rods is transferred into the top chord by bearing at an angle θ to the grain (Fig. 21.15). A large round or square washer may be necessary to develop the load, and advantage should be taken of modification factors K_3 and K_4 (sections 4.5 and 4.10.1) and the formula for modifying bearing stresses at an angle to grain (section 14.10) to keep the bearing area to a minimum and remove as little material from the top chord as possible. The tension is transferred into the bottom chord by bearing at right angles to the grain. If the bolt head and washer are recessed into the bottom chord, the bottom chord net area should allow for the area of countersinking and the vertical hole taking the tie rod.

The compression in the timber diagonal has components of load perpendicular to and parallel to the slope of the top chord. The component perpendicular to the slope is taken in bearing, while the component parallel to the slope is taken first in bearing by the bearing block and is then transferred in shear (either by connectors or a glue line) into the top chord (Fig. 21.15). Maximum bearing pressure is developed by ensuring that the bearing line between the internal strut and the bearing block bisects the angle ϕ between them, giving the same angle of load to grain for both the strut and the bearing block.

The heel joint of a rod-and-block truss requires careful detailing. It is a notched joint and the bottom tie member is weakened by the notch and can be further weakened by eccentricities if the member forces and support reaction do not intercept at one point. A typical traditional heel detail is shown in Fig. 21.16.

Traditionally, the following conditions would have been intended when designing the heel joint:

1. The bearing between the compression chord and the ceiling tie would be perpendicular to the line of the compression chord, with a gap between the mitre angle (line AB) of approximately 10 mm at the extreme end of the mitre, to prevent bending or thrust being applied to the ceiling tie along this line. The compression load should align with the centre of the bearing area.
2. The lines of the thrust, tension and reaction would coincide at one point 'O' so that the tension force is concentric with the net section A–A below the step.

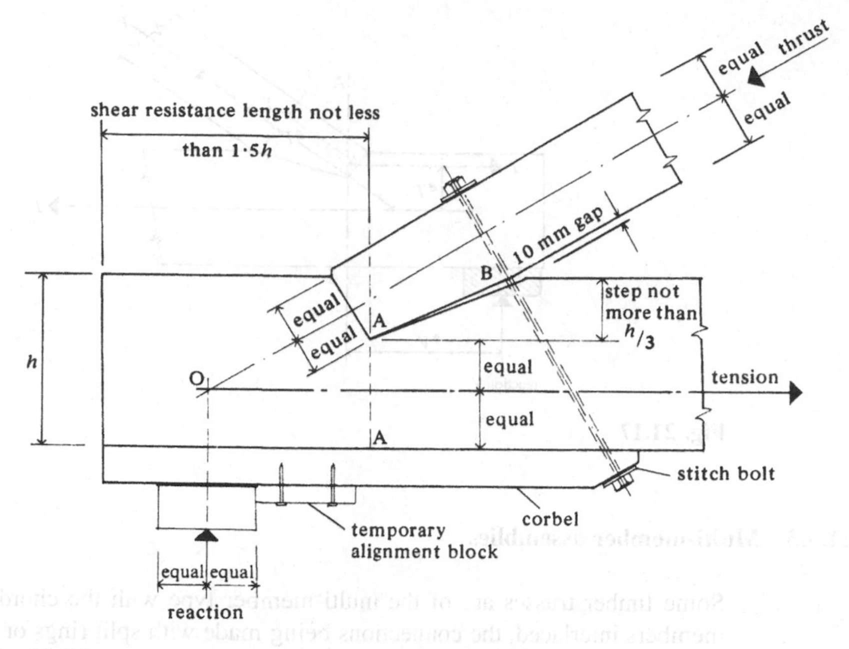

Fig. 21.16

3. Adequate length of tie would be provided beyond the step joint, to give adequate shear resistance to resist the component of thrust from the top chord parallel to the grain (numerically equal to the tension in the ceiling tie). An adequate depth of member would be provided at section A–A to resist the shear from the vertical component of thrust from the top chord (numerically equal to the end reaction).
4. A corbel would be added underneath the tie member to provide a bearing for the stitch bolt. The corbel would be attached to the tie to transfer a shear component from the stitch bolt should shearing commence beyond the step joint. The stitch bolt was introduced because the structure would otherwise be totally dependent on the shear resistance of the timber beyond the step. Its inclusion also helped to locate the chord member.
5. A bearing pad would be fixed underneath the ceiling tie, to ensure that the reaction occurred directly below point 'O'.

Modern designs are not as elaborate as that shown in Fig. 21.16 and are usually of the form shown in Fig. 21.17.

The relationship of the support point to the intersect of rafter and tie forces is often predetermined architecturally so that the design cannot avoid significant eccentricities. The most frequent problem encountered will be a large moment at A–A caused by the location of the end support. This moment will often dictate the size of the bottom tie member.

The net section at A–A of the horizontal tie member must be designed for a moment $M_L = Ve_V - Te_T$.

The sloping rafter member also takes into account the moment Pe_p resulting from the eccentric end bearing.

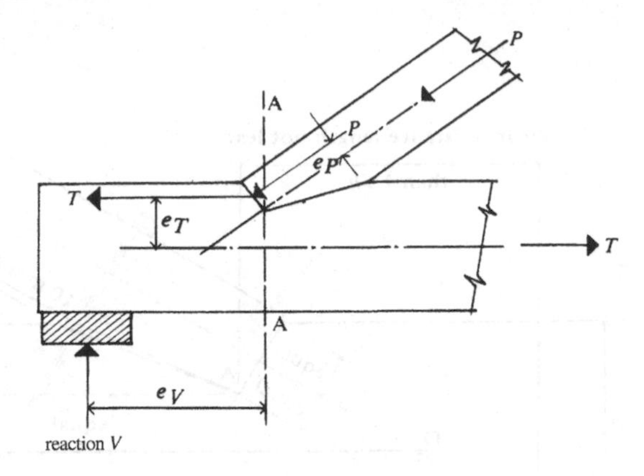

Fig. 21.17

21.3.3 Multi-member assemblies

Some timber trusses are of the multi-member type with the chords and internal members interlaced, the connections being made with split rings or toothed plates (although currently the duo-chord truss shown in Fig. 21.13 is probably most favoured architecturally). The Timber Research and Development Association have produced a comprehensive range of standard designs for domestic and industrial use, and certain tile manufacturers also issue design guides for domestic trusses using multi-member assemblies.

With a short-span or lightly loaded truss, or a truss with lightly loaded internals (bowstring or Warren), it is possible to use a single solid member for the internal compression members with twin-member chords, but only if an eccentricity is accepted at joints (see Fig. 21.18(a).) It is more usual to use twin members for internal struts and single members for internal ties leading to the type of joint assemblies sketched at (b) and (c) in Fig. 21.18.

When there is an eccentricity at a joint or joints, the effect of eccentricity and secondary moments should be considered (Fig. 21.19). Unless an external load is applied at the connection, the transverse components of the compression and tension members are the same, and if equal to V, give a local moment of Ve which produces a moment into the bottom chord member of $M_L = Ve/2$.

The seven-member arrangement sketched at (c) in Fig. 21.18 may occur in large-span parallel-chord trusses, particularly those involving cantilevers, and in eight-panel Fink trusses (Table 21.10) at the centre of the rafter. The direction and magnitude of the load as it affects each part of the connection is determined by considering each interface. This is best illustrated by the example given below.

Consider the members 1, 4 and 12 of the eight-panel Howe truss, for which coefficients are given in Table 21.19, assuming the angle between diagonal and vertical (and horizontal) to be 45°. Assume a point load of W at each upper-panel node point, and no loading on the bottom chord. The equilibrium diagram for the upper, outermost joint is shown in Fig. 21.20.

Members 1 and 4 are in compression and are twin members. Member 12 is a single section. Symmetry is obviously desirable in the assembly of the truss, there-

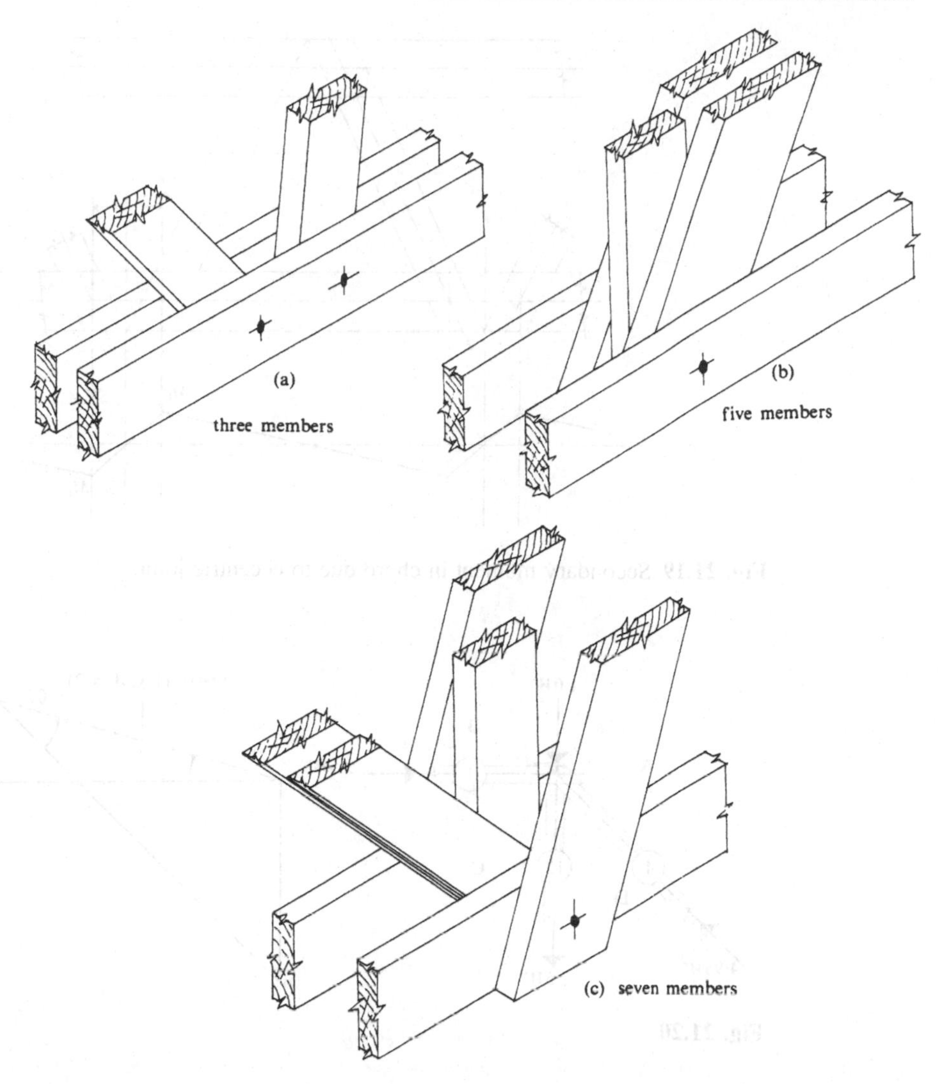

Fig. 21.18

fore member 12 is placed on the centreline. There is then a choice for the positioning of member 1 and member 4 which will influence the joint design and economy.

Firstly consider the arrangement illustrated in Fig. 21.21. The two parts of member 1 are placed at the outer positions (1) and (5) and the two parts of member 4 are placed in positions (2) and (4). The joint has to be considered on each interface, and at this stage the designer must decide how the vertical external loading is to be applied. Assume it is applied through brackets or by direct bearing on the top chord (i.e. members 1) and analyze the loading in two parts.

Loads to the left of interface B_L (Fig. 21.21)

The $0.5W$ external load is applied at the top of the half of member 1 to the left of interface B_L. Because of this external load, the load in the connector at interface

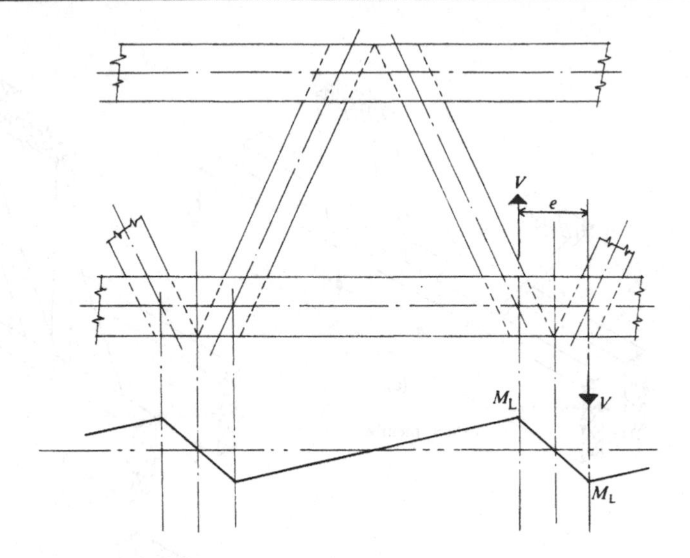

Fig. 21.19 Secondary moment in chord due to eccentric joint.

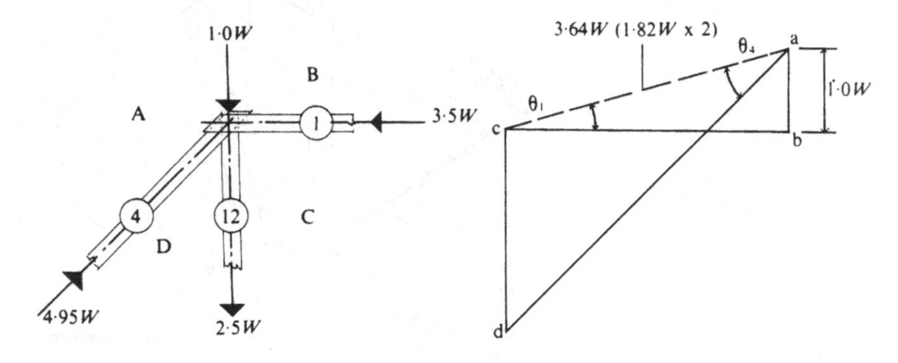

Fig. 21.20

B_L cannot be taken direct from the vector diagram in Fig. 21.20 except by joining points c and a. The loads in this connector to the left of interface B_L are $0.5W$ vertical and $1.75W$ horizontal (i.e. half $3.5W$) giving $1.82W$ at an angle of 16° to member 1 and 29° to member 4 (see Fig. 21.22).

If either single-sided or double-sided connector units are used, the part in member 1 acts at an angle of 16° and the part in member 4 at an angle of 29° to grain. As part of the loading is taken by the connecting bolt, whichever type of connector is used, it is important to use the correct diameter bolt.

Loads to the left of interface A_L (Fig. 21.21)

The connector unit at this interface is being acted upon by the $0.5W$ and $1.75W$ loads plus $2.475W$ from member 4 which, from Fig. 21.20, can be seen to lead to a resultant of $0.5 \times 2.5W = 1.25W$ acting vertically. The connector acts at an angle of 45° to the grain in member 4 and parallel to the grain in member 12 (see

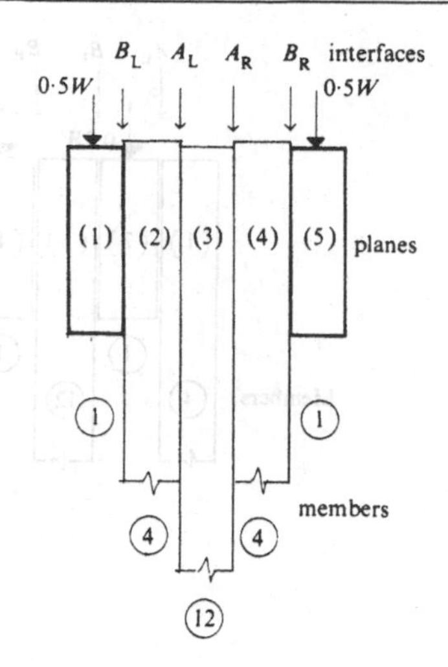

Fig. 21.21

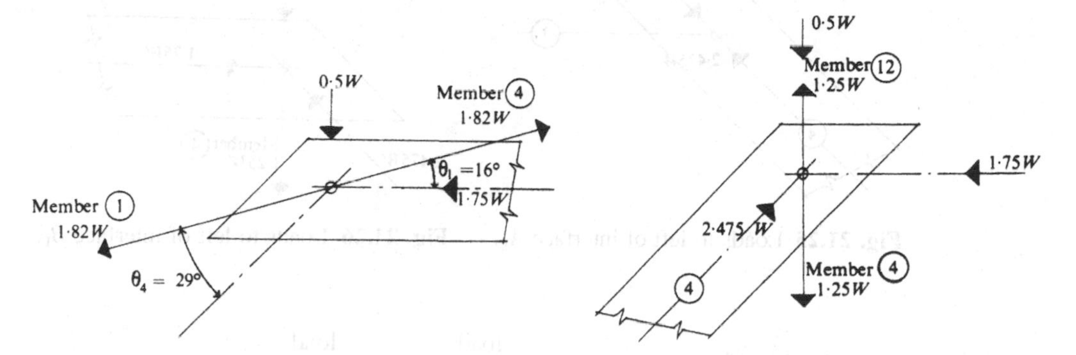

Fig. 21.22 Loads to left of interface B_L. **Fig. 21.23** Loads to left of interface A_L.

Fig. 21.23). An easier method of looking at this interface would be to cut the truss on its vertical axis and see how half the load in member 12 affects this interface.

Secondly, consider the arrangement illustrated in Fig. 21.24. The two parts of member 4 are placed at the outer positions (1) and (5) and the two parts of member 1 are placed in positions (2) and (4). The external point load is applied in two halves to the top of the chord 1. The load to the left of interface A_L (Fig. 21.25) is $2.475W$, applied parallel to member 4 and at 45° to member 1.

The loads to the left of interface B_L are shown in Fig. 21.26, and result in a load of $1.25W$ parallel to the grain in member 12 and perpendicular to grain in member 1. These loads could also be derived by considering the truss cut on its vertical axis and calculating how half the load in member 12 affects this interface.

At each interface, the number and size of connectors required to develop the load is dictated by the maximum angle of load to grain and can be expressed in general terms as:

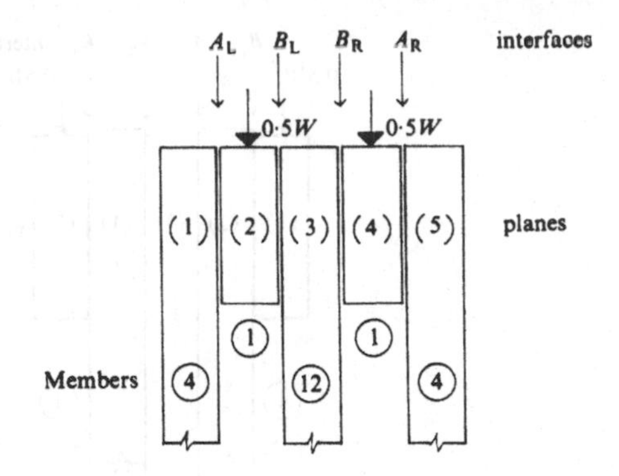

Fig. 21.24

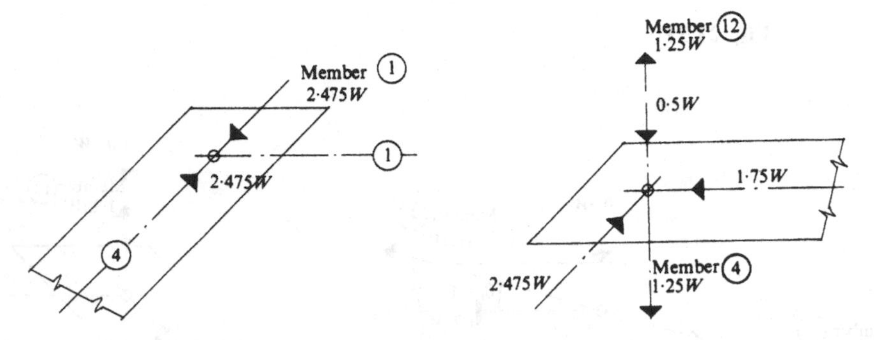

Fig. 21.25 Loads to left of interface A_L.

Fig. 21.26 Loads to left of interface B_L.

$$\text{Number of connectors} = \frac{\text{load}}{\text{connector capacity}} = \frac{\text{load}}{\overline{F}_0 \, K_\alpha}$$

where $\overline{F}_0$ = capacity of one connector parallel to grain
K_α = coefficient for angle of load to grain given in Table 18.36 for split rings and shear plate connectors.

For the sake of discussion the values for bolts and toothed plate connectors may be considered to be similar to $\overline{F}_0 K_\alpha$ The connector requirements for the two joint assemblies shown in Figs 21.21 and 21.24 can now be compared to determine which assembly is the most efficient.

The summation in Table 21.1 shows that with the chords in the inner positions, $(4.78 - 3.52) \times 100/3.52 = 36\%$, more connector capacity is required. The designer will not wish to work through the various combinations before deciding upon the optimum arrangement. A rule to follow is to assemble the member so that the angle between adjacent members is kept to a minimum.

In a multi-member joint it is inevitable that certain members will be loaded at an angle to the grain and that members such as member 4 in Fig. 21.21 and member 1 in Fig. 21.24 will receive a load from a different angle on each face. The entire joint assembly is in equilibrium when looked at in elevation, the components

Table 21.1

Interface	With the chords at the outer positions (Fig. 21.21) Load	Maximum angle to grain	K_α	Required number of connectors	With the chords at the inner positions (Fig. 21.24) Load	Maximum angle to grain	K_α	Required number of connectors
A	$1.25W$	45°	0.824	$1.52W/\bar{F}_0$	$2.475W$	45°	0.824	$3.00W/\bar{F}_0$
B	$1.82W$	29°	0.908	$2.00W/\bar{F}_0$	$1.25W$	90°	0.700	$1.78W/\bar{F}_0$
				$3.52W/\bar{F}_0$				$4.78W/\bar{F}_0$

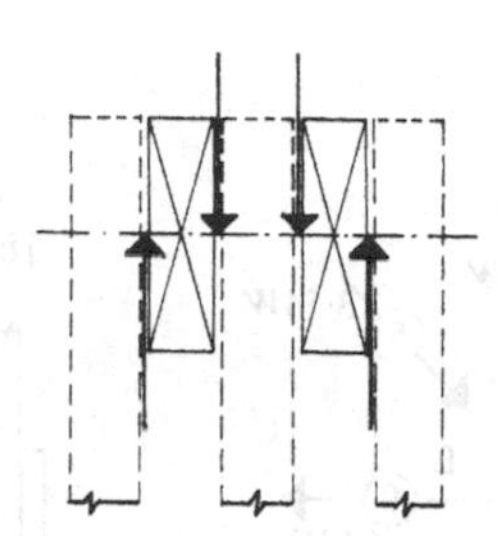

Fig. 21.27

perpendicular to the grain on each face being of equal magnitude but acting in opposite directions. This leads to a cleavage action within the thickness of the member which is illustrated in Fig. 21.27.

Although this action is recognized, no design method has been evolved to calculate the effect through the thickness of the timber, and the best guidance one can give to the designer is to limit the effect. The cleaving tendencies of the alternative joint assemblies (Figs 21.21 and 21.24) are seen in Fig. 21.28 by examining the magnitude of the components perpendicular to the grain.

With the chords in the inner planes, two adjacent faces are at 90°, which is bad from the point of view of efficiency of connectors (Table 21.1), and also the cleavage load of $1.75W$ perpendicular to the grain of member 1 is much larger than either of the cleavage loads with the chords in the outer planes. The arrangement with the chords in the outer planes is therefore preferable from these aspects and also tends to give a better appearance.

Figure 21.22 illustrates how the effect of an external load being applied at a node point throws the line of action of load on the connectors off the axes of members. When no external load occurs, the direction of the connector loadings at a joint is in line with the member axes. Consider, for example, a lower chord joint in which members 8, 9, 12 and 5 meet in the Howe truss illustrated in Table 21.19. The loads in members and the vector diagram are shown in Fig. 21.29.

Consider twin chord members 8 and 9 continuous through the joint and in the outer planes (1) and (5) with twin web member 12 in planes (2) and (4) and single web member 5 in the central plane (3). The net horizontal force to be

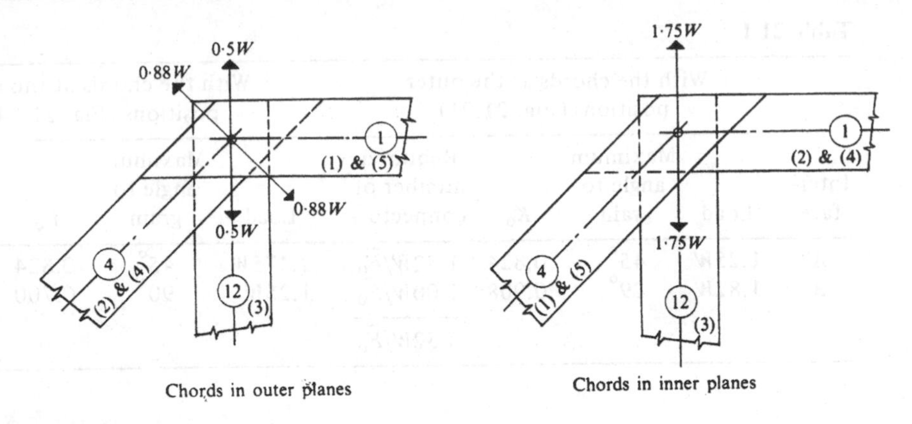

Fig. 21.28

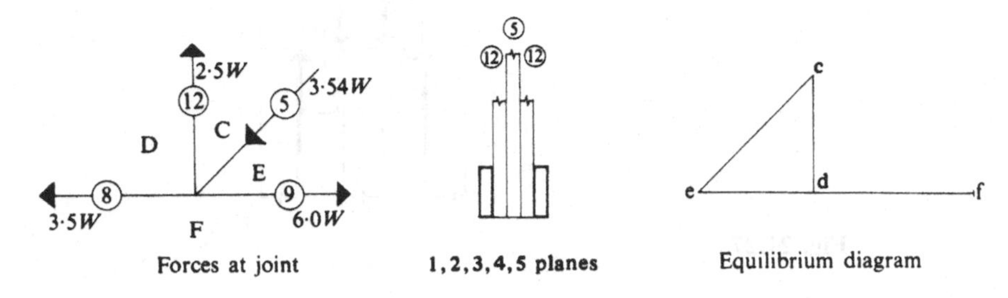

Fig. 21.29

developed at the joint from members 8 and 9 is $6.0W - 3.5W = 2.5W$. This force must firstly be transferred between members 8 and 9 and member 12 acting at 0° to grain in members 8 and 9 and at 90° to grain in member 12. The resulting component force of $3.54W$ is transferred between member 12 and member 5 acting at 45° to member 12 and at 0° to member 5.

Alternatively consider twin chord members 8 and 9 continuous through the joint and in planes (1) and (5) with twin web member 5 in planes (2) and (4) with single web member 12 in the central plane (3). The net horizontal force to be developed is $2.5W$ as before but on this occasion it is transferred between members 8 and 9 and member 5 acting at 0° to members 8 and 9 and 45° to grain in member 5. The resulting component force of $2.5W$ is transferred between members 5 and 12 acting at 45° to grain in each member. This alternative arrangement avoids forces being transferred at 90° to grain and simplifies the connector design.

21.4 DESIGN OF A PARALLEL-CHORD TRUSS

21.4.1 Loading

Determine the member sizes and principal joint details for the 12 m span parallel-chord truss shown in Fig. 21.30, which is carrying a medium-term UDL of

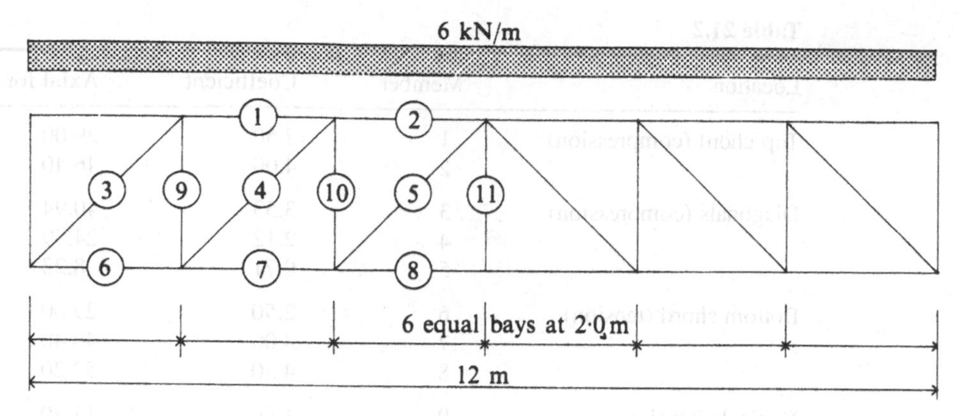

Fig. 21.30

5.8 kN/m applied to the top chord by joisting at 0.6 m centres. Use C24 timber in service classes 1 and 2. Neglect dead weight of truss.

The top and bottom chords are spaced 2.0 m vertically apart on centrelines and are twin members in the outer planes (1) and (5). The diagonals are twin members in planes (2) and (4) and the verticals are a single member in plane (3).

Load per panel $W = 2 \times 5.8 = 11.6$ kN = node point loading to top chord. Bending moments occur in the top chord at and between the node points. Figure 28.10 gives coefficients for these on two-, three- and four-span continuous beams. Although this is a six-span continuous chord it is sufficiently accurate (see section 21.2) to take the values from Fig. 28.10 as $0.107WL_N$ at the node points and $0.03WL_N$ between node points, where L_N is the distance between node points.

At node points

$$M = 0.107WL_N = 0.107 \times 11.6 \times 2 = 2.48 \text{ kN m}$$

Between node points

$$M = 0.036WL_N = 0.036 \times 11.6 \times 2 = 0.835 \text{ kN m}$$

The axial loads are tabulated in Table 21.2 calculated from the coefficients given in Table 21.18 with $\alpha = H/L = 2/12 = 0.167$.

21.4.2 Design of members

Whether to design the member first and then the joint, or vice versa, is a matter of individual preference. Most designers first carry out the design of members, realizing that some sizes may have to be increased or changed when the requirements of the joints are established. It is unwise for an engineer to finalize drawings, cost, etc., on the basis of the design of members without first having established that the connections can be accommodated. Connector requirements frequently dictate the required member sizes. To design tension members the net section applies and it is necessary to anticipate connector requirements. In this design it is assumed that 64 mm diameter split rings are the preferred choice.

Table 21.2

Location	Member	Coefficient	Axial force (kN)
Top chord (compression)	1	2.50	29.00
	2	4.00	46.40
Diagonals (compression)	3	3.53	40.94
	4	2.12	24.59
	5	0.71	8.23
Bottom chord (tension)	6	2.50	29.00
	7	4.00	46.40
	8	4.50	52.20
Verticals (tension)	9	1.50	17.40
	10	0.50	5.80
	11	0.00	0.00

Chapter 13 details the general design recommendations for tension members. The bottom chord has a maximum tension of 52.2 kN, and being located in the outer planes, connectors occur only in one face of the member.

From Table 13.1 twin 47 × 169 members (C16 strength class) with a single connector and bolt ($N_b = 1$ and $N_c = 1$) have a medium-term capacity, based on net area, of

$$2 \times 22.8 \times 1.25\,(K_3) = 57\,\text{kN}$$

From the footnote to Table 13.1, adjust to C24 capacity at 57 × 1.41 = 80.4 kN.

The verticals are single tension members with connectors on both faces and the maximum tension is 17.4 kN. From Table 13.1 it can be seen that one 47 × 145 member ($N_b = 1$ and $N_c = 2$) has a medium-term capacity, based on net area, of

$$17.3 \times 1.25(K_3) \times 1.41 = 30.4\,\text{kN}$$

Section 16.6 gives general design recommendations for spaced compression members in triangulated frameworks. The top chord member 2 has a bending moment of 2.48 kN m at the node position and 0.835 kN m at the centre of members with a maximum axial compression of 46.4 kN. With the principal axes as shown in Fig. 21.31, the effective lengths of the chords against compression are

$$L_{ex} = 0.85 \times 2.0 = 1.7\,\text{m} \quad \text{and} \quad L_{ey} = L_{ew} = 0.6\,\text{m}$$

the latter being set by the centres of joisting. The chord has to be designed on a trial-and-error basis.

Top chord members (see sections 16.4, 16.5 and 16.6)

In certain cases it may be necessary for the designer to consider two positions on member 2: (a) in the centre with $L_{ex} = 1.7$ m and a bending moment of 0.86 kN m and (b) at the node point with effective length equal to the distance between the points of contraflexure to each side of the node and a bending moment of 2.6 kN m. In this example the latter will be considered to be the critical condition

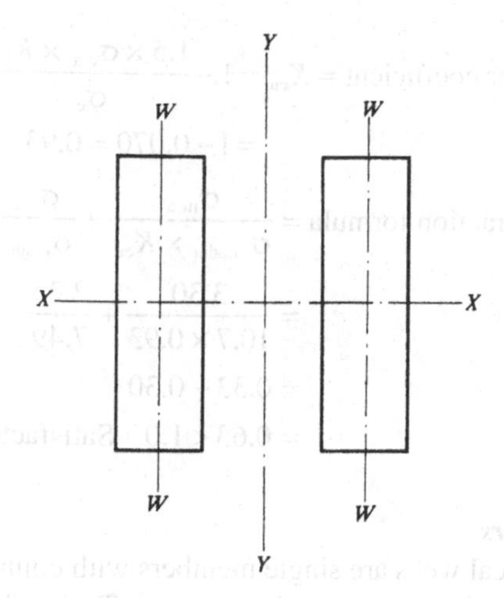

Fig. 21.31

with L_{ex} = 1.7 m. The member may be considered to have adequate lateral restraint for bending on the x–x axis to justify the full grade stress.

Try twin 47 × 219 C24 spaced 141 mm apart by three internal members each of 47 mm thickness

$$\lambda_x = \frac{1700\sqrt{12}}{219} = 27$$

$$\lambda_w = \frac{600\sqrt{12}}{47} = 44$$

By inspection, $\lambda_x < \lambda_w$.

Taking unmodified value of E_{min} to determinine K_{12} and medium-term value of σ_c:

$$\frac{E}{\sigma_c} = \frac{7200}{7.9 \times 1.25} = 7129 \quad \text{and} \quad \lambda_{max} = 44$$

From Table 15.4, $K_{12} = 0.759$

$$\sigma_{c,a} = \frac{46400}{2 \times 47 \times 219} = 2.25\ \text{N/mm}^2$$

$$\sigma_{c,adm} = \text{grade stress} \times K_3 \times K_{12} = 7.9 \times 1.25 \times 0.759 = 7.49\ \text{N/mm}^2$$

$$Z_x = \frac{2 \times 47 \times 219^2}{6} = 0.751 \times 10^6\ \text{mm}^3$$

$$\sigma_{m,a} = \frac{M}{Z} = \frac{2.48 \times 10^6}{0.751 \times 10^6} = 3.30\ \text{N/mm}^2$$

$$\sigma_{m,adm} = \text{grade stress} \times K_3 \times K_7 \times K_8 = 7.5 \times 1.25 \times 1.035 \times 1.1 = 10.7\ \text{N/mm}^2$$

$$\text{Euler critical stress} = \sigma_e = \frac{\pi^2 E}{\lambda^2} = \frac{\pi^2 \times 7200}{44^2} = 36.7\ \text{N/mm}^2$$

$$\text{Euler coefficient} = K_{eu} = 1 - \frac{1.5 \times \sigma_{c,a} \times K_{12}}{\sigma_e} = 1 - \frac{1.5 \times 2.25 \times 0.759}{36.7}$$
$$= 1 - 0.070 = 0.93$$

$$\text{Interaction formula} = \frac{\sigma_{m,a}}{\sigma_{m,adm} \times K_{eu}} + \frac{\sigma_{c,a}}{\sigma_{c,adm}}$$
$$= \frac{3.30}{10.7 \times 0.93} + \frac{2.25}{7.49}$$
$$= 0.33 + 0.30$$
$$= 0.63 < 1.0 \quad \text{Satisfactory.}$$

Internal web members

The vertical webs are single members with connectors on each face and maximum tension is 17.4 kN in member 9. From Table 13.1 (with $N_b = 1$ and $N_c = 2$) use a 47 × 169 C24 with capacity of 20.9 × 1.41 = 29.5 kN.

The maximum compression force in diagonals is 40.94 kN and the actual length is 2.8 m. $L_{ex} = L_{ey} = 2.8$ m but L_{ew} may be less depending on whether or not packing pieces are fitted. Check suitability of twin 47 × 145 C24 members using two bolted packs located at third points of overall length. The diagonals will have a 47 mm gap between them to accommodate the vertical internal members.

$$\lambda_x = \frac{2800\sqrt{12}}{145} = 67\ (<180)$$

Determine radius of gyration about y–y axis

$$I_y = \frac{(141^3 - 47^3) \times 145}{12} = 32.62 \times 10^6\ \text{mm}^4$$

$$A = 2 \times 47 \times 145 = 13\,630\ \text{mm}^2$$

$$i_y = \frac{\sqrt{I_y}}{A} = \sqrt{2393} = 48.9\ \text{mm}$$

Alternatively, referring to section 16.4

$$i_y \cong \frac{2b + (5c/3)}{\sqrt{12}}$$

where b = member width = 47 mm
c = spacing between members = 47 mm

$$i_y \cong \frac{2 \times 47 + (5 \times 47)/3}{3.464} \cong 49.7\ \text{mm}$$

Try bolted packs, with $c/b = 1$, $K_{13} = 2.4$ from Table 16.2

$$\lambda_y = \frac{L_e \times K_{13}}{i_y} = \frac{2800 \times 2.4}{48.9} = 137\ (<180)$$

$$\lambda_w = \frac{933\sqrt{12}}{47} = 69$$

Taking unmodified value of E_{min} to determinine K_{12} and medium-term value of σ_c:

$$\frac{E}{\sigma_c} = \frac{7200}{7.9 \times 1.25} = 729 \quad \text{and} \quad \lambda_{max} = 137$$

From Table 15.4, $K_{12} = 0.210$

$$\sigma_{c,a} = \frac{40\,940}{2 \times 47 \times 145} = 3.00\ \text{N/mm}^2$$

$$\sigma_{c,adm} = \text{grade stress} \times K_3 \times K_{12} = 7.9 \times 1.25 \times 0.210$$

$$= 2.07\ \text{N/mm}^2 < 3.00\ \text{N/mm}^2$$

Present choice of members/grade/pack connections is inadeqate by factor of 3.00/2.07 = 1.45. The following options should be considered in order to give $\sigma_{c,adm} > \sigma_{c,a}$:

1. Use a larger section, say 145 × 1.45 = 210 mm.
2. Improve pack performance, changing from bolts to connectors or gluing, in order to reduce λ_y.

Adopting the latter option try connectored packs such that $K_{13} = 1.8$

$$\lambda_y = \frac{L_e \times K_{13}}{i_y} = \frac{2800 \times 1.8}{48.9} = 103$$

From Table 15.4, $K_{12} = 0.335$

$$\sigma_{c,adm} = \text{grade stress} \times K_3 \times K_{12} = 7.9 \times 1.25 \times 0.335 = 3.31\ \text{N/mm}^2$$

$$> 3.11\ \text{N/mm}^2$$

Therefore connectored packs will permit use of 2/47 × 145 C24.

Alternatively, use glued packs such that $K_{13} = 1.1$

$$\lambda_y = \frac{L_e \times K_{13}}{i_y} = \frac{2800 \times 1.1}{48.9} = 63$$

$$\lambda_{max} = \lambda_w = 69$$

From Table 15.4, $K_{12} = 0.611$

$$\sigma_{c,adm} = \text{grade stress} \times K_3 \times K_{12} = 7.9 \times 1.25 \times 0.611$$

$$= 6.03\ \text{N/mm}^2 > 3.11\ \text{N/mm}^2$$

Note the significant improvement in the permissible stress provided by glued packs.

Check spacer blocks in diagonals for geometric limits and fixing requirements. The normal rules for spaced columns does not apply to the ends of compression members in triangulated frameworks (see BS 5268-2, clause 2.11.11c), the interface connections of the member joints being considered an acceptable alternative.

Intermediate packings (see section 16.4) should be at least 230 mm long and should be designed to transmit, between the abutting face of the packing and one adjacent shaft, a shear force of 0.65 P*b*/*na*, where

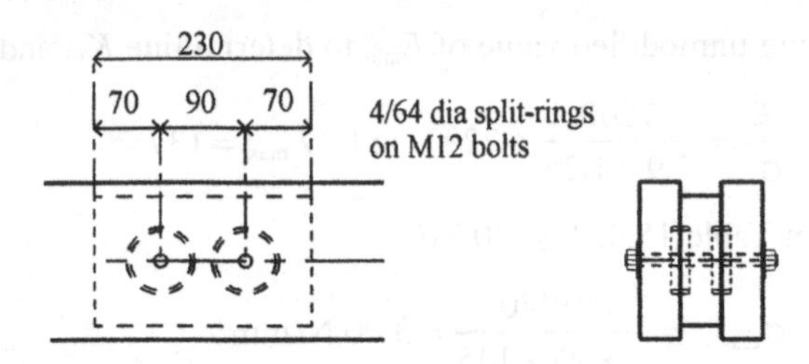

Fig. 21.32

P = total applied axial load = 42.4 kN
b = thickness of each shaft = 47 mm
n = number of shafts = 2
a = distance between centres of adjacent shafts = 94 mm.

Then

$$\text{Shear force} = \frac{0.65 \times 42.4 \times 47}{2 \times 94} = 6.89\,\text{kN}$$

Connectored packs

Provide 230 mm long pack with connectored pack as shown in Fig. 21.32.
From Table 18.39,

minimum loaded end distance = 70 mm ($K_C = 0.625$)
minimum spacing = 89 mm, say 90 mm ($K_S = 0.75$)

K_{64} is lesser of K_C and K_S

Basic connector load from Table 18.35 = 10.1 kN
connector capacity in pack = number of connectors × basic capacity × K_{64}
= 4 × 10.1 × 0.625 = 25.2 kN Satisfactory.

Glued packs

$$\text{Glue line stress} = \frac{6890}{230 \times 145} = 0.21\,\text{N/mm}^2$$

Allowable glue line stress (medium term)
= permissible medium-term shear stress parallel to grain
= 0.6 × 1.25 (K_3) = 0.75 N/mm² Satisfactory.

Screws or bolts may be used to give adequate glue line pressure. There should be at least 4 screws or bolts per packing spaced so as to provide a uniform pressure over the area of the packing.

The overall design may be simplified by using the same section and packing specificaton for lighter loaded diagonals and vertical members. This allows possible repetition of joint details and consequently simplifies material ordering and fabrication.

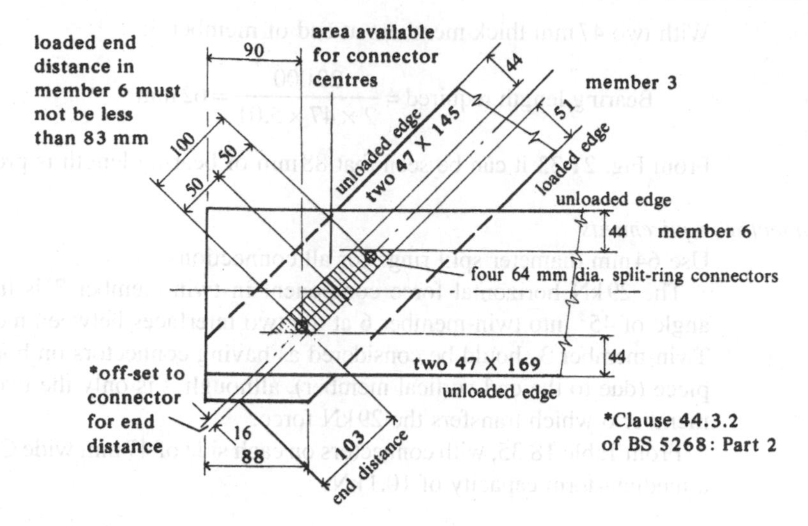

Fig. 21.33

21.4.3 Heel joint

The total load on the truss is 5.8 × 12 = 69.6 kN giving 34.8 kN reaction at each end.

The last vertical has 5.8 kN compression from the UDL on the last bay of the top chord. This 5.8 kN force can conveniently by-pass the joint connection by being extended below the bottom tie member to the truss support. The end diagonal therefore transfers a net vertical load of 29 kN to the truss support.

The vertical component of the load in diagonal member 3 can be isolated from the connector requirements by also extending member 3 below the horizontal member 6 (Fig. 21.33). This extension takes the 29 kN net vertical component of force direct to the support point and provides an improved end distance for the horizontal component of force resisted by member 6 acting at 45° to member 3. If the truss bottom tie member is suppported directly onto the support then the vertical component of load must be transferred through the connections, this will greatly increase the number of connectors required and may prove difficult if not impossible to accommodate within the interface area available.

Bearing at support

Consider bearing of twin members 3 at 45° to grain in C24 timber and refer to section 14.10. With no wane permitted the higher of the two listed values for compression perpendicular to grain may be adopted.

$$\sigma_{c,adm,90} = \text{grade stress} \times K_3 = 2.4 \times 1.25 = 3.00\ \text{N/mm}^2$$

$$\sigma_{c,adm,0} = \text{grade stress} \times K_3 = 7.9 \times 1.25 = 9.87\ \text{N/mm}^2$$

with $\alpha = 45°$, $\sin\alpha = 0.707$

$$\sigma_{c,adm,\alpha} = 9.87 - (9.87 - 3.00) \times 0.707 = 5.01\ \text{N/mm}^2$$

With two 47 mm thick members at end of member 3:

$$\text{Bearing length required} = \frac{29\,000}{2 \times 47 \times 5.01} = 62\,\text{mm}$$

From Fig. 21.33 it can be seen that 88 mm of bearing length is provided.

Connector requirements

Use 64 mm diameter split rings for all connections.

The 29 kN horizontal force component in twin-member 3 is transferred at an angle of 45° into twin-member 6 at the two interfaces between members 3 and 6. Twin-member 3 should be considered as having connectors on both faces of each piece (due to the end vertical member), although it is only the two faces abutting member 6 which transfers the 29 kN force.

From Table 18.35, with connectors on each side of 47 mm wide C24 timber adopt a medium-term capacity of 10.1 kN.

From Table 18.36, with $\alpha = 45°$, $K_\alpha = 0.824$
Modified connector capacity = 10.1 × 0.824 = 8.32 kN

$$\text{Number of connectors required} = \frac{29}{8.32} = 3.48, \text{ use 4 connectors.}$$

There are two interfaces through which to transfer the load, so two connectors are used at each face on two bolts. Four connectors give the percentage of allowable value (PAV) used as

$$\text{PAV} = \frac{3.48}{4} = 0.87$$

Connector spacing

In considering the actual end and edge distances and the spacing of connectors, BS 5268-2, clause 6.8.5, calls for the spacing factor K_{64} to be taken as the lowest of values K_S, K_C and K_D. However, in no case should the actual distances and spacing be less than the minimum tabulated values.

Having chosen the connector size and number required and having determined the PAV value in member 3 as 0.87, it can be seen that the connectors must be positioned such that end distance, edge distance and spacing provide, in each case, a value for K_{64} of at least 0.87. This requires K_S (spacing), K_C (end distance) and K_D (edge distance) to each have a value of 0.87 or more. Suitable spacings can be determined from Tables 93, 95 and 96 of BS 5268-2 or from the diagrams in Table 18.39 of this manual.

Unloaded edge distance: member 3

The unloaded edge distances in members 3 and 6 must be at least 44 mm from Table 18.39 which is independent of K_S, K_C and K_D.

Loaded edge distance: member 3

From Table 18.39 (or BS 5268-2, Table 90) the loaded edge distance in member 3 (loaded at 45°) requires a standard edge distance, $D_{st} = 70\,\text{mm}$ ($K_D = 1.0$) and with a minimum edge distance, D_{min} = 44 mm (K_D = 0.83). For intermediate values of edge distance, K_D is interpolated linearly. From K_D value graph in Table 18.39 with K_D = 0.87, the required edge distance must be at least 50 mm (51 mm is provided).

Unloaded end distance: member 3

Consider the unloaded end distance to member 3 loaded at 45° to grain.

From Table 18.39, C_{st} = 121 mm (K_C = 1.0) and C_{min} = 67 mm (K_C = 0.625). From the K_C value graph in Table 18.39, with K_C = 0.87, the required end distance must be at least 103 mm (103 mm is provided).

Loaded end distance: member 6

The loaded end distance to twin-member 6 requires a re-assessment of the percentage of allowable value used (PAV) as it applies to member 6, in which the connectors are loaded parallel to the grain on one face only.

Basic medium-term capacity from Table 18.35 = 10.7 kN.

$$\text{The number of connectors required} = \frac{29}{10.7} = 2.71, \text{ use 4 connectors.}$$

$$\text{PAV} = 2.71/4 = 0.68$$

From Table 18.39, C_{st} = 140 mm (K_C = 1.0) and C_{min} = 70 mm (K_C = 0.625). From the K_C value graph in Table 18.39, with K_C = 0.68, the required end distance must be at least 80 mm (90 mm is provided).

Spacing between connectors in member 3

Consider member 3 with loading at 45° to grain and with connector axis θ at 0° to grain.

From Table 18.39, C_{st} = 108 mm (K_C = 1.0) and C_{min} = 89 mm (K_C = 0.75). From the K_S value graph in Table 18.39, with K_S = 0.87, the required parallel to grain spacing must be at least 98 mm (100 mm is provided).

Spacing between connectors in member 6

Consider member 6 with loading parallel to grain, $\alpha = 0°$ and with connector axis θ at 45° to grain.

From Table 18.39, S_{st} = 112 mm (K_S = 1.0)
From Table 18.38, S_{min} = 89 mm (K_S = 0.75)
Actual spacing is 100 mm.

From the formula at the foot of Table 18.39 determine K_S:

$$K_S = 0.75 + 0.25[(100 - 89)/(112 - 89)] = 0.87 > 0.68 \quad \text{Satisfactory.}$$

K_S is calculated by way of example although it is obvious that the minimum value of 0.75 is satisfactory.

With the all above parameters determined, the joint is set out to give a convenient arrangement for the connectors, having regard to the required connector positioning, convenience of detailing and ease of manufacture. The first step is to superimpose the permitted end and edge distances on the member arrangement

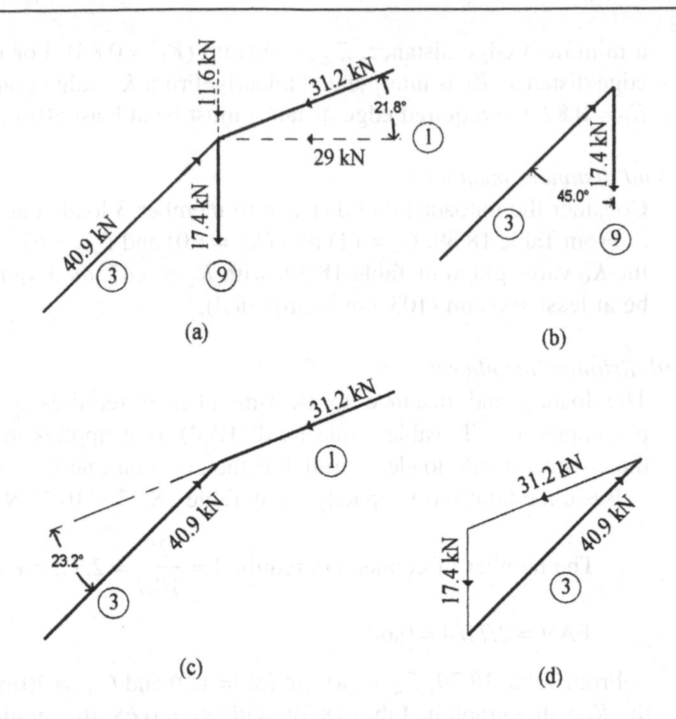

Fig. 21.34

to give an area within which the centres of the connectors can be placed. This is shown hatched in Fig. 21.33. In this case it is convenient to place the connectors on the centreline of the diagonal, 50 mm to each side of the intersection point of the axes.

21.4.4 Top chord joint

Consider the joint at the junction of members 1, 3 and 9. This joint is typical of the top chord joints, all of which are influenced by the applied vertical node point load, giving a direction of connector loading which in most cases does not coincide with the axes of the members.

The loads at the joint are summarized in Fig. 21.34(a). It should be noted that the 11.6 kN vertical load enters the joint as a shear on the two halves of member 1, and it is the vector sum of this shear and the axial load in member 1 which must be resisted by the connectors joining member 1 to member 3. The components of shear and axial load in member 1 are shown in Fig. 21.34 as dotted lines, and the vector and its angle as a full line.

Referring to Fig. 21.34(b), member 9 transfers 17.4 kN at 45° into member 3.

From Table 18.36, with $\alpha = 45°$, $K_\alpha = 0.824$

$$\text{Number of connectors required} = \frac{17.4}{10.1 \times 0.824} = 2.1$$

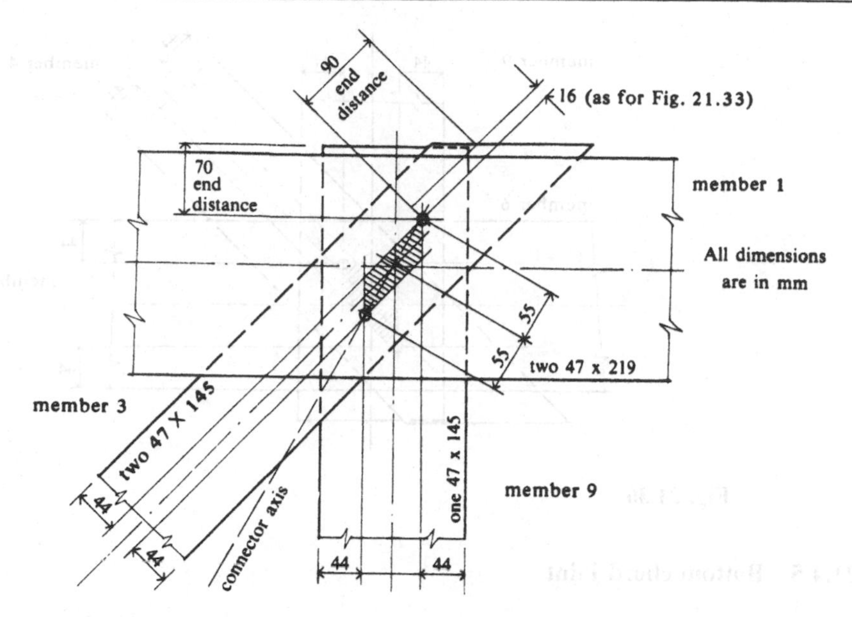

Fig. 21.35

Using four connectors, PAV = 2.10/4 = 0.525 which, from Table 18.39, permits minimum loaded endspacing and edge distances of C_{min} = 70 mm, S_{min} = 89 mm and D_{min} = 44 mm.

Referring to Fig. 21.34(c), member 1 transfers 31.2 kN at 23.2° into member 3.

From Table 18.36, with $\alpha = 23.2°$, $K_\alpha = 0.938$

$$\text{The number of connectors required} = \frac{31.2}{10.1 \times 0.938} = 3.3$$

Using four connectors, PAV = 3.3/4 = 0.82
From Table 18.39,

Unloaded edge distance = 44 mm
Loaded edge distance with $K_D = 0.82 < 0.825 = D_{min} = 44$ mm

Unloaded end distance to member 3 with $K_C = 0.82$ is interpolated between 0° (C_{st} = 102 mm) and 45° (C_{st} = 121 mm) to give 90 mm.

The actual spacing cannot be determined until the parallelogram of connector area is set out and θ determined, although it can be seen from the graph for K_S in Table 18.39 that spacing must be between 89 mm and 108 mm.

The required edge distances may be set out on the member sizes as determined in section 21.4.3. The available parallelogram of area is large enough (see Fig. 21.35) for the connectors to be spaced 110 mm apart, which exceeds 108 mm (S_{st}) therefore no further detailed check is required.

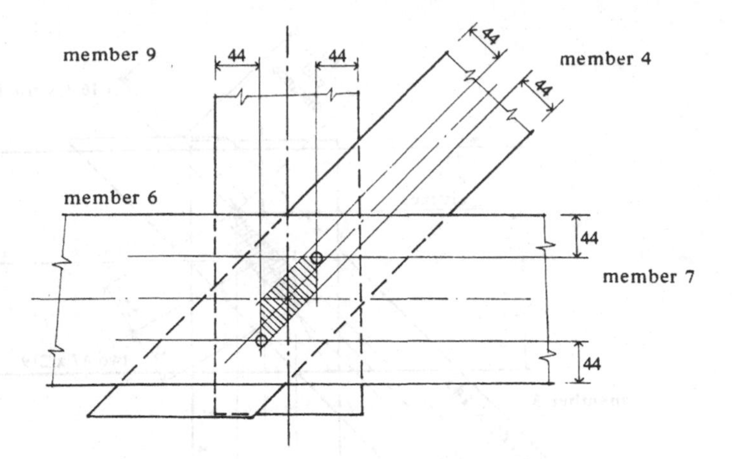

Fig. 21.36

21.4.5 Bottom chord joint

Consider the joint at the junction of members 9, 4, 6 and 7. This joint is typical of the bottom chord joints having no external node point load.

The joints between members 9 and 4 have the same requirements as those between 9 and 3, which have already been established.

The joint between members 4 and 6–7 transfers a horizontal load equal to the differential between 6 and 7 (i.e. 46.4 – 29.0 = 17.4 kN), which is at 45° to the grain of member 4.

From Table 18.36, with $\alpha = 45°$, $K = 0.824$

The number of connectors required $= \dfrac{17.4}{10.1 \times 0.824} = 2.1$

Using four connectors, PAV $= \dfrac{2.1}{4} = 0.52$

The joint is shown in Fig. 21.36. The area available for the centres of connectors is a six-sided figure in this case. The connectors are placed on the longest diagonal to give a maximum spacing of 100 mm.

At PAV = 0.52, $S_{min} = 89$ mm   Acceptable.

21.4.6 Tabular method of joint analysis

In sections 21.4.3–21.4.5, the three main joints were investigated and suitable connector arrangements derived. There may have been a number of alternative solutions which would have been equally acceptable. As a general rule, it will be found best to investigate the heavily loaded joints first. The remaining joints tend to be simpler to design and to show some repetition. It is not usual to mix connectors of different sizes in a given truss although there may be occasions where the connector size may change from joint to joint.

In practice, the design of joints is made easier if each joint is set out to a reasonably large scale. Various possible arrangements should be investigated to find the one most suited to detailing and the workshop.

It is convenient to establish the number of connectors, the percentage of allowable value used and the spacing requirement in tabular form. A typical format for the truss under consideration is given in Table 21.3. Because the angle of connector axis to grain, θ, is not known prior to the detailing of the joint, spacing S is tabulated assuming $\theta = 0°$. Then, if this spacing can be provided, no further calculation is required. If, however, the tabulated value of S cannot be accommodated, further calculations with values of θ derived from the joint detail may indicate a possible reduction in spacing, otherwise at least one member width must be increased.

21.5 BOWSTRING TRUSSES

21.5.1 Introduction

In bowstring trusses, the upper chord is usually circular (as distinct from the ideal parabolic shape), which gives a truss profile closely resembling the bending moment diagram for a UDL. With balanced loading, the internal members take little load and are usually very small considering the span. Because of this, single internal members can be used on the vertical centreline of the truss, plane (2), with the top and bottom chords each constructed in two halves and placed in the outer planes (1) and (3). The resulting secondary moments on the chords caused by eccentricity of connections can usually be disregarded under balanced loading but must be considered for unbalanced loading.

The truss is loaded either by purlins placed (a) at node points or (b) between node points to cancel out part of the bending moment induced by the curvature. The bending stress in the top chord can vary significantly as a result of the placing of the purlins or secondary system.

The axial loading in members can be derived by the construction of a vector diagram, although in doing so a large scale must be used to obtain any degree of accuracy (particularly for the web members). The axial loading in the top chord is fairly consistent across the whole span of the truss and can be established mathematically. The curvature of the top chord results in a bending moment being induced into the member equal to the product of the axial load and its eccentricity from a line joining the adjacent node points. This moment must be added algebraically to any moment induced in the member by the local loading from purlins to give the net moment on the member.

The moment in the top chord = $M = M_0 - F_N e$

where M_0 = the mid-panel moment due to localized purlin loading
F_N = the axial load tangentially at mid node points
e = the eccentricity to the line of action of F_N from the node points.

The eccentricity is determined from the radius of the top chord and the distance between node points:

$$e = \frac{(L_N)^2}{8r}$$

Table 21.3

Joint	Member to member	Conn. load (kN)	Member	α	K_α	Conns. one side or two	Medium term $\bar{F}_0$* (kN)	Number of conns. required	Use	PAV	Required spacing: End distance† C (mm)	Required spacing: Spacing S (mm)	Required spacing: Loaded edge distance D (mm)	Type of connector
3 6	3 6	29	6 3	0° 45°	1.00 0.824	1 2	10.7 10.1	2.73 3.50	4	0.68 0.875	80 103	89 98.5	– 51	64 mm dia. split rings
1 3 9	9 3	17.4	9 3	0° 45°	1.00 0.824	2 2	10.1 10.1	1.73 2.10	4	0.43 0.525	70 67	89 89	44 44	64 mm dia. split rings
	3 1	31.2	3 1	23° 22°	0.938 0.943	2 1	10.1 10.7	3.31 3.11	4	0.83 0.78	97 –	109 97	44 44	64 mm dia. split rings
9 4 6/7	9 4	17.4	9 4	0° 45°	1.00 0.824	2 2	10.1 10.1	1.73 2.10	4	0.43 0.525	70 67	89 89	44 44	64 mm dia. split rings
	4 6/7	17.4	4 6/7	45° 0°	0.824 1.00	2 1	10.1 10.7	2.10 1.64	4	0.525 0.41	67 –	89 89	44 44	64 mm dia. split rings

10 4	10	5.8	10	0°	1.0	2	10.1	0.58	2	0.29	70	–	44	64 mm dia. split rings
	4		4	45°	0.824	2	10.1	0.70		0.35	67	–	44	
1/2	4	21.0	4	11°	0.984	2	10.1	2.11	4	0.53	67	89	44	64 mm dia. split rings
	1/2		1/2	34°	0.882	1	10.7	2.23		0.56	–	89	44	
10 5	10	5.8	10	0°	1.0	2	10.1	0.58	2	0.29	70	–	44	64 mm dia. split rings
	5		5	45°	0.824	2	10.1	0.70		0.35	67	–	44	
7/8	5	5.8	5	45°	0.824	2	10.1	0.70	2	0.35	67	–	44	64 mm dia. split rings
	7/8		7/8	0°	1.0	1	10.7	0.55		0.27	–	–	44	
5	5	8.2	5	0°	1.0	1	10.7	0.78	2	0.39	64	–	44	64 mm dia. split rings
2	2		2	45°	0.824	1	10.7	0.95		0.47	–	–	44	

* F_o is the connector capacity from Table 18.35 at $\alpha = 0°$.
† Loaded or unloaded end distance.

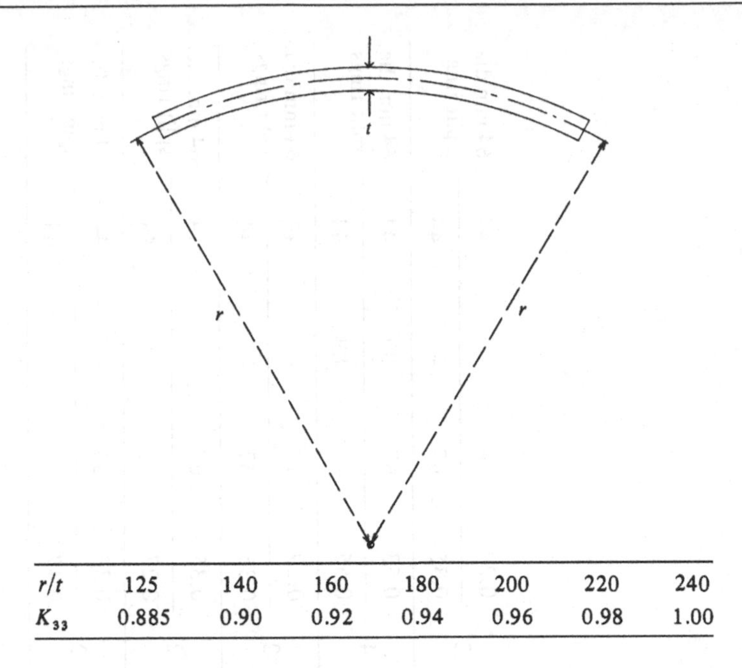

r/t	125	140	160	180	200	220	240
K_{33}	0.885	0.90	0.92	0.94	0.96	0.98	1.00

Fig. 21.37

where L_N = the distance between node points
r = the radius of the chord.

The chord should be fabricated as a curved laminated beam of rectangular cross section. The curvature and lamination thickness should be such that:

$$\frac{r}{t} > \frac{E_{mean}}{70} \quad \text{(for both softwood and hardwood)}$$

where r = radius of curvature
t = lamination thickness
E_{mean} = mean modulus of elasticity for the strength class.

Typically for C24 timber this would give a minimum value of $r/t = 10\,800/70 = 154$. With 45 mm thick laminations the minimum radius of curvature $= r = 154 \times 45 = 6930$ mm. When $r/t < 240$ the bending, tension and compression stresses must be modified by factor K_{33}, where

$$K_{33} = 0.76 + 0.001(r/t) \quad \text{(with } K_{33} \text{ not greater than 1.0)}$$

Values of K_{33} are given in Fig. 21.37.

In curved beams where the ratio of the minimum mean radius of curvature, r_{mean}, to the overall beam depth, h, is less than or equal to 15, the bending stress, σ_m, induced by a moment, M, should be taken as:

(a) in the extreme fibre on the concave side,

$$\sigma_m = \frac{K_{34}M}{Z}$$

where $K_{34} = 1 + \left(0.5\frac{h}{r_{mean}}\right)$ for $\frac{r_{mean}}{h} \le 10$

or $= 1.15 - \left(0.01\frac{r_{mean}}{h}\right)$ for $10 < \frac{r_{mean}}{h} \le 15$

Z = section modulus = $bh^2/6$.

(b) in the extreme fibre on the convex face,

$$\sigma_m = \frac{M}{Z}$$

The bending moment in a curved member will produce a radial stress which should be calculated as:

$$\sigma_r = \frac{3M}{2bhr_{mean}}$$

where σ_r = the radial stress perpendicular to grain
M = the bending moment
r_{mean} = the radius of curvature to the centre of the member
b = the breadth of the member
h = the depth of the member.

When the bending moment tends to increase the radius of curvature (i.e. to flatten the arc) the radial stress, σ_r, will be tension perpendicular to grain and should not be greater than one-third the permissible shear parallel to the grain stress with laminating factors K_{19} disregarded, i.e. for C24 timber, long-term duration

$$\sigma_{r,adm} = \frac{0.71}{3} = 0.23\ \text{N/mm}^2$$

When the bending moment tends to reduce the radius of curvature (i.e. create a tighter curve) the radial stress, σ_r, will be compression perpendicular to grain and should not be greater than 1.33 times the compression perpendicular to the grain stress for the strength class, i.e. for C24 timber, long-term duration

$$\sigma_{r,adm} = 1.33 \times 1.9 = 2.52\ \text{N/mm}^2$$

21.5.2 Preferred geometry for a bowstring truss

Some simplification of design is obtained with a bowstring truss if a preferred geometrical arrangement is adopted as shown in Fig. 21.38. The span-to-midspan height ratio should be approximately 8 to 1 to avoid large deflections on the one hand and excessive height and curvature on the other. It is usually convenient to adopt a radius for the top chord equal to the span of the truss ($L = r$), so that the tangent angle at the support is 30°.

The rise of the truss is $H = r - r\cos 30° = 0.134r = 0.134L$.

Provide a camber of $0.009L$ to the bottom chord. This gives a net truss height

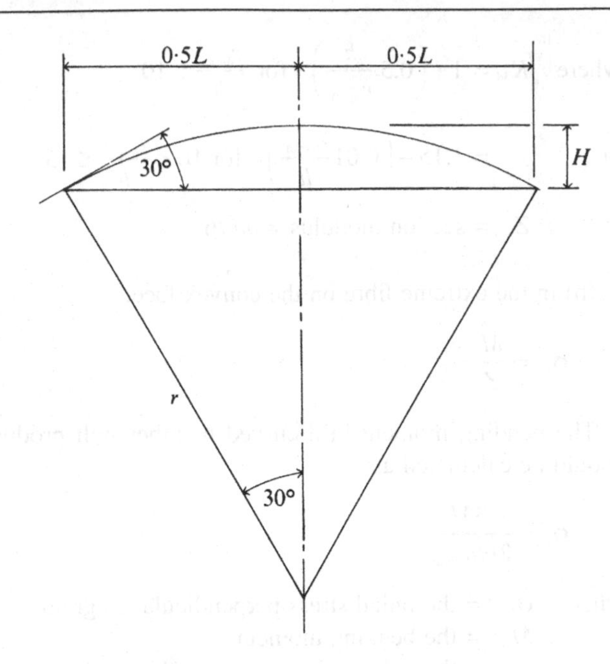

Fig. 21.38

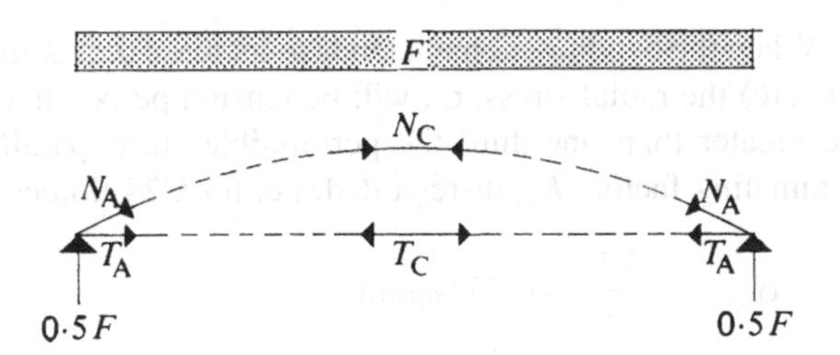

Fig. 21.39

of $0.134L - 0.009L = 0.125L$. The span-to-height ratio is then 8 to 1 (i.e. the distance between the centrelines of the upper and lower chords is $L/8$ at midspan).

$$\text{Top chord arc length} = \frac{60 \times 2\pi r}{360} = 1.05r = 1.05L$$

Referring to Fig. 21.39, with a UDL of F applied to the full span of the truss, the axial loads are resolved as:

Heel joint axial loads, $N_A = F$ and $T_A = 0.866F$
Midspan axial loads, $N_C = F$ and $T_C = F$.

21.5.3 Example of bowstring truss design

Consider 16 m span bowstring trusses at 6 m centres carrying 0.50 kN/m^2 dead load and 0.60 kN/m^2 imposed load. Purlins are placed at the node points (Fig. 21.40). The distance between centrelines of top and bottom chord at midspan is 2.0 m.

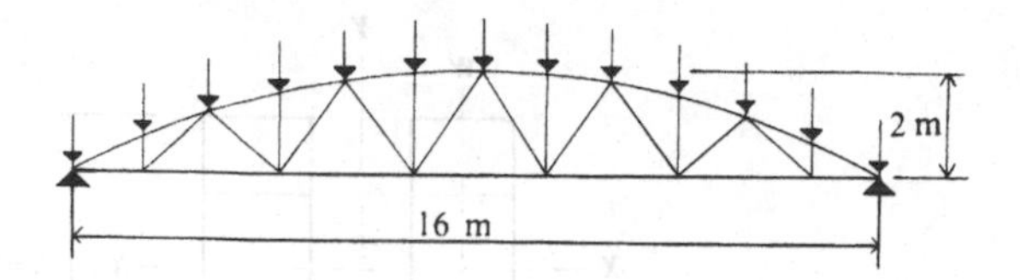

Fig. 21.40

The arc length is $16 \times 1.05 = 16.8\,m$, which is divided into twelve equal lengths of 1.4 m. Note that the division is along the arc and not on true plan.

The total loading on the truss is:

F (dead) $= 16.8 \times 6 \times 0.5 = 50.4\,kN$
F (imposed) $= 16 \times 6 \times 0.6 = 57.6\,kN$
Total medium-term load $= F = 50.4 + 57.6 = 108\,kN$

Top chord

Maximum axial load in top chord $= N_C = F = 108\,kN$

$$\text{Eccentricity on panel length} = e = \frac{(L_N)^2}{8r} = \frac{1.4^2}{8 \times 16} = 0.0153\,m$$

Moment due to eccentricity $= 108 \times 0.0153 = 1.65\,kN\,m$

$$\frac{r_{mean}}{h} = \frac{16000}{180} = 89$$

As r_{mean}/h is greater than 15 there is no need to increase the bending stress by K_{34}

$$\frac{r}{h} = \frac{16000}{180} = 89 > 15, \quad \text{therefore } K_{34} = 1.0$$

Design moment in top chord $= 1.65 \times K_{34} = 1.65 \times 1.0 = 1.65\,kN\,m$

For the top chord try two $65 \times 180\,mm$ C24 glulam (4 laminations at 45 mm) spaced 60 mm apart (Fig. 21.41) and laminated to curvature.

Radius of gyration

$$i_y \cong \frac{2b + (5c/3)}{\sqrt{12}} = \frac{2 \times 65 + (5 \times 60)/3}{3.464} = 66\,mm \quad \text{(see section 16.4)}$$

With restraint about both the x–x and y–y axes at 1.4 m centres and with packs at midpanel length to restrain the w–w axis at 0.7 m centres, the various slenderness ratios are:

$$\lambda_x = \frac{1400\sqrt{12}}{180} = 27$$

$$\lambda_y = \frac{1400}{66} = 21$$

$$\lambda_w = \frac{700\sqrt{12}}{65} = 37$$

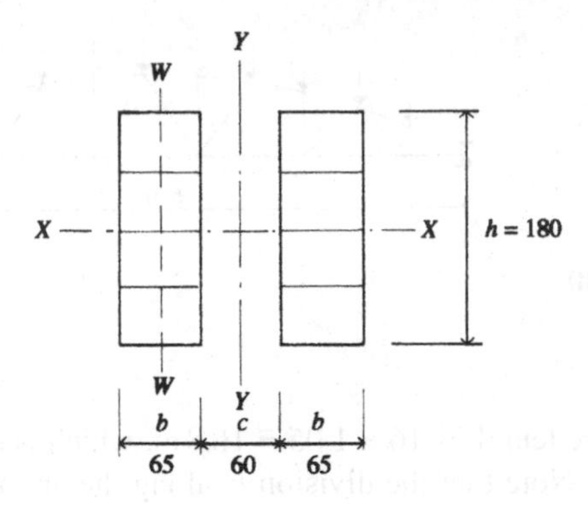

Fig. 21.41

From section 14.4 the E and σ_c value for entry into Table 17.1 to determine K_{12} are:

$$E = E_{mean}(\text{C24 grade}) \times K_{20} = 10\,800 \times 1.07 = 11\,556 \text{ N/mm}^2$$

and

$$\sigma_c(\text{medium term}) = \text{C24 grade stress} \times K_3 \times K_{17}$$
$$= 7.9 \times 1.25 \times 1.04 = 10.3 \text{ N/mm}^2$$

With

$$\frac{E}{\sigma_c} = \frac{11556}{10.3} = 1120 \quad \text{and} \quad \lambda_{max} = 37, \quad K_{12} = 0.82$$

$$\frac{r}{t} = \frac{16000}{45} = 356 > 240, \quad \text{therefore } K_{33} = 1.0$$

Then

$$\sigma_{c,adm} = \text{C24 grade stress} \times K_3 \times K_{12} \times K_{17} \times K_{33}$$
$$= 7.9 \times 1.25 \times 0.82 \times 1.04 \times 1.0 = 8.42 \text{ N/mm}^2$$

$$\sigma_{c,a} = \frac{10\,800}{2 \times 65 \times 180} = 4.61 \text{ N/mm}^2$$

$$\sigma_{m,adm} = \text{C24 grade stress} \times K_3 \times K_7 \times K_{15}$$
$$= 7.5 \times 1.25 \times 1.058 \times 1.26 = 12.5 \text{ N/mm}^2$$

Section modulus

$$Z_x = \frac{2 \times 65 \times 180^2}{6} = 702\,000 \text{ mm}^3$$

$$\frac{r_{mean}}{h} = \frac{16000}{180} = 89$$

As r_{mean}/h is greater than 15 there is no need to increase the bending stress by K_{34}.

Then

$$\sigma_{m,a} = \frac{M}{Z_x} = \frac{1.65 \times 10^6}{702\,000} = 2.35 \text{ N/mm}^2$$

Euler critical stress is

$$\sigma_{eu} = \frac{\pi^2 E}{\lambda^2} = \frac{9.872 \times 11556}{37^2} = 83.3 \text{ N/mm}^2$$

Euler coefficient is

$$K_{eu} = 1 - \frac{(1.5 \times \sigma_{c,a} \times K_{12})}{\sigma_{eu}}$$

$$= 1 - \frac{1.5 \times 4.61 \times 0.82}{83.3} = 1 - 0.068 = 0.932$$

$$\text{Interaction fromula} = \frac{\sigma_{m,a}}{\sigma_{m,adm} \times K_{eu}} + \frac{\sigma_{c,a}}{\sigma_{c,adm}}$$

$$= \frac{2.35}{12.5 \times 0.932} + \frac{4.61}{8.42}$$

$$= 0.20 + 0.55 = 0.75 < 1.0 \quad \text{Satisfactory.}$$

$$\text{Radial stress} = \frac{3M}{2bhr_{mean}}$$

$$= \frac{3 \times 1.65 \times 10^6}{2 \times 130 \times 180 \times 16000} = 0.007 \text{ N/mm}^2$$

The moment induced into the top chord, taking the force line between node points, tends to reduce the radius of curvature. Therefore the permissible radial stress = 1.33 × permissible compression perpendicular to grain

$$\sigma_{r,adm} = 1.33 \times 1.9 \times 1.25\,(K_3) = 3.16 \text{ N/mm}^2 \quad \text{Acceptable.}$$

Web members

It can be shown that the forces in the internal web members under balanced loading are extremely small for this form of truss, whereas unbalanced loading on one half of the span can produce larger axial compression loads in the members near to truss midspan (shown dotted in Fig. 21.42). The axial load is approximately three times the imposed panel point loading, the exact value depending on the number of bays.

The axial force in midspan web member may be estimated as follows:

Midspan vertical height = 2.0 m
Half bottom chord bay width = 16/12 = 1.33 m
Length of diagonal web member = $\sqrt{(2^2 + 1.33^2)} = \sqrt{(5.76)} = 2.4$ m
Take imposed load on left half truss = 57.6/2 = 28.8 kN
Reaction at right-hand edge = 28.8/4 = 7.2 kN
Shear on truss at midspan = 7.2 kN
Axial load in web member = 7.2 × 2.4/2.0 = 8.64 kN

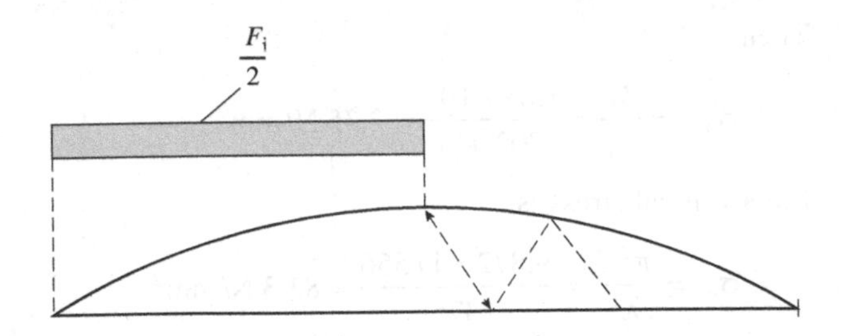

Fig. 21.42

A 60 × 145 mm C24 member would prove satisfactory.

The designer can see how important it is to consider unbalanced loading. Had only balanced loading been considered, this longest of the web members would have been calculated as carrying no load at all.

21.6 DEFLECTION OF TRUSSES

21.6.1 Elastic deflection

The deflection of a triangulated framework results from a combination of axial shortening of the compression members, lengthening of tension members and slip at the joints.

Deflection due to axial strain may be determined by computer analysis or calculated from the strain energy formula:

$$\delta_e = \sum \frac{FUL}{AE}$$

where δ_e = elastic deflection at a selected node point of the truss
F = load in each member of the truss caused by the applied loading (+ for compression, − for tension is the usual sign convention)
U = load in each member of the truss caused by a unit load placed at the node point for which the deflection is required and in the direction in which the deflection is to be calculated (+ for compression, − for tension is the usual sign convention)
L = the actual length to node points of each member
A = the area of each member (not the net area)
E = the modulus of elasticity of each member.

Values of F and U are determined by normal structural analysis. Coefficients for various truss configurations are tabulated in section 21.7 and also in section 20.8.

The elastic deflection of the truss designed in section 21.4 is calculated in Table 21.4, in the form usual with strain-energy methods. The unit load is placed at the centre of the top chord, therefore member 11 need not be considered.

If the load in a member due to the unit load is of the same sign as the load in the member due to the applied loading, the deformation of this member adds to

Table 21.4

Member	F (kN)	U	L (m)	A (mm^2)	E (N/mm^2)	FUL/AE (m)
1	+29.0	0.50	2.00	20600	8208	0.000171
2	+46.4	1.00	2.00	20600	8208	0.000549
3	+40.94	0.71	2.83	13630	8208	0.000735
4	+24.59	0.71	2.83	13630	8208	0.000442
5	+8.23	0.71	2.83	13630	8208	0.000148
6	−29.0	−0.71	2.00	15880	8208	0.000316
7	−46.4	−1.00	2.00	15880	8208	0.000712
8	−52.2	−1.50	2.00	15880	8208	0.001201
9	−17.4	−0.50	2.00	7980	7200	0.000304
10	−5.8	−0.50	2.00	7980	7200	0.000101

For half truss, $\Sigma = 0.004\,679$ m.

the external deflection being considered, whereas, if the signs are opposite, the deformation of the member reduces the external deflection.

The product FU is always positive in this case, and only half the truss is tabulated.

The total elastic deflection at midspan = $2 \times 0.0047 = 0.0094$ m.

The deflection has been determined assuming E_{min} at 7200 N/mm^2 for single members and $E_{min} \times K_9 = 7200 \times 1.14 = 8208$ N/mm^2 for twin members. This is a conservative interpretation of BS 5268-2, clause 2.10.7, which recommends that solid timber members acting alone should use the minimum modulus of elasticity. However, the construction of the truss will involve some 17 timbers for the web members and 12 timbers for the top and bottom chord, the latter assumed to be spliced at approximate third points of span to keep timbers within available lengths. This gives some 29 timber members influencing the stiffness of the truss, so it would appear reasonable to suggest that the adoption of the mean modulus of elasticty is more appropriate.

Using E_{mean} will reduce the elastic deflection to

$$\frac{0.0094 \times 8208}{10\,800} = 0.0071 \text{ m}$$

21.6.2 Slip deflection

An important cause of unanticipated deflection in trusses is caused by the slip which may occur at joints due to the settling of bolts/connectors in the holes/dappings provided. It is not possible to predict exactly the amount of slip at each joint, and hence not possible to predict exactly the overall deflection of a truss caused by slip deflection, but experience has shown that the following procedure gives anticipated deflections borne out fairly well in practice.

Table 21.5

Connector type	Slip (m)
Glue	nil
Bolt*	0.0026
Tooth plates*	0.0026
64 mm diameter split ring	0.0008
104 mm diameter split ring	0.0010
67 mm diameter shear plate*	
timber-to-timber	0.0026
timber-to-steel	0.0018
104 mm diameter shear plate*	
timber-to-timber	0.0031
timber-to-steel	0.0026

* An allowance of 0.0016 m is included for the usual bolt-hole tolerance.

Approximate slip values for various connector types are given in Table 21.5 which should be adequate for normal loading conditions. If the engineer can establish more accurate joint slip characteristics and control the assembly, including moisture content, the performance on site will match the calculated deflections more closely. Where the truly permanent load at a connector is a high percentage (say 70% or more) of the permissible long-term loading it may be advisable to build in an additional estimate for 'slip' to take account of indentation.

$$\text{Slip deflection} = \sum U\Delta_s$$

where U = unitless load in each member due to unit load applied at the point where deflection is being considered (i.e. usually midspan)

Δ_s = the total change in member length due to joint slip.

The midspan deflection of the truss in section 21.4 due to connector slip is calculated in Table 21.6. Split rings of 64 mm diameter are used with an allowance for slip at one end of ±0.0008 m or ±0.0016 m if slip occurs at both ends. Only half the truss is tabulated.

Values in parentheses assume the introduction of top and bottom chord splices to permit the use of available solid timbers which are usually limited to approximately 4.8 m lengths.

$$\text{Total slip deflection} = 2 \times 0.009\,176$$
$$\cong 0.018\,\text{m, of which 35\% is due to chord splices}$$
$$\text{Total deflection of the truss} = \text{elastic deflection} + \text{slip deflection}$$
$$= 0.007 \text{ (assuming } E_{mean}) + 0.018 = 0.025\,\text{m}$$

Note that the slip deflection is significantly more than the elastic deflection in this example. To reduce slip deflection the designer should consider minimizing the number of internal web members and top and bottom chord splices.

Table 21.6

Member	U	No. of slip ends	Δ_s (m)	$U\Delta_s$ (m)
1	+0.5	1	+0.000 800	+0.000 400
2	+1.0	(2)	(+0.001 600)	(+0.001 600)
3	+0.71	2	+0.001 600	+0.001 136
4	+0.71	2	+0.001 600	+0.001 136
5	+0.71	2	+0.001 600	+0.001 136
6	−0.71	1	−0.000 800	+0.000 568
7	−1.0	(2)	(−0.001 600)	(+0.001 600)
8	−1.5	0	0	0
9	−0.5	2	−0.001 600	+0.000 800
10	−0.5	2	−0.001 600	+0.000 800

For half truss, Σ = 009 176 m.

Table 21.7

Symmetrical flat and pitched trusses	Bowstring trusses
$\delta_t = \dfrac{L^2}{4290H}\left[\dfrac{L}{26.6}+1\right]$	$\delta_t = \dfrac{L^2}{19\,000H}\left[\dfrac{L}{6}+1\right]$

δ_t = the total deflection under total loading (m).
L = the span of the truss (m).
H = the height (m) of the truss at mid span.

21.6.3 Standard deflection formulae

The American Institute of Timber Construction Standards quotes values for the total deflection of flat, pitched and bowstring trusses which are purely empirical, being independent of loading and member size. These are reproduced in Table 21.7 in metric form.

Use the formula in Table 21.7 for flat trusses to check the calculated deflection of the example in section 21.6.2

$$\text{Deflection} = \frac{12^2}{4290 \times 2}\left[\frac{12}{26.6}+1\right] = 0.0244\,\text{m compared to } 0.025\,\text{m}$$

The agreement is good, particularly as the formula makes no allowance for the E value used or the degree to which the truss is stressed or jointed.

21.7 COEFFICIENTS OF AXIAL LOADING

To save design-office time, Tables 21.8–21.25 give coefficients by which the rafter node point loading should be multiplied to arrive at the axial loading in the members. Coefficients are also tabulated for a unit load placed at the top chord of parallel-chord trusses, except in the case of Tables 21.23 and 21.25, where the unit load is placed at the centre of the bottom chord. In the case of Tables 21.11–21.25, coefficients are also given relating to ceiling node point loading.

Table 21.8

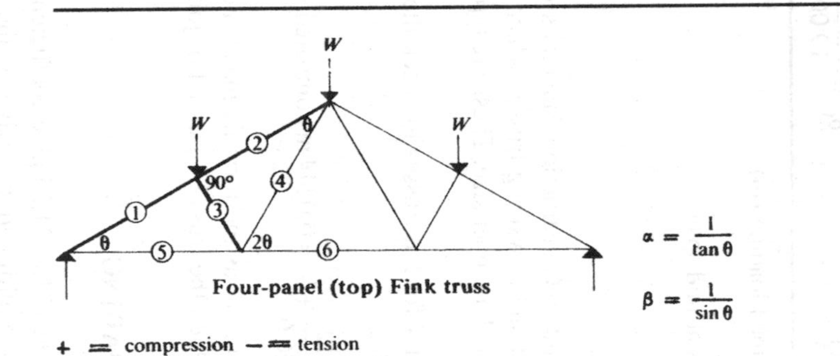

Four-panel (top) Fink truss

$$\alpha = \frac{1}{\tan\theta}$$

$$\beta = \frac{1}{\sin\theta}$$

+ = compression − = tension

Member	Compression or tension	θ = 15° W	15° Unit	17½° W	17½° Unit	20° W	20° Unit	22½° W	22½° Unit	25° W	25° Unit	27½° W	27½° Unit	30° W	30° Unit	32½° W	32½° Unit	35° W	35° Unit	General formulae W	General formulae Unit
1	+	5.80	1.93	5.00	1.66	4.39	1.46	3.92	1.30	3.55	1.18	3.25	1.08	3.00	1.00	2.79	0.93	2.61	0.87	1.5β	0.5β
2	+	5.54	1.93	4.69	1.66	4.04	1.46	3.54	1.30	3.13	1.18	2.79	1.08	2.50	1.00	2.25	0.93	2.04	0.87	$\frac{3\alpha^2 + 1}{2\beta}$	0.5β
3	+	0.97	0	0.95	0	0.94	0	0.92	0	0.91	0	0.89	0	0.87	0	0.84	0	0.82	0	α/β	0
4	−	1.87	0	1.59	0	1.37	0	1.21	0	1.07	0	0.96	0	0.87	0	0.78	0	0.71	0	0.5α	0
5	−	5.60	1.87	4.76	1.59	4.12	1.37	3.62	1.21	3.22	1.07	2.88	0.96	2.60	0.87	2.35	0.78	2.14	0.71	1.5α	0.5α
6	−	3.73	1.87	3.17	1.59	2.75	1.37	2.41	1.21	2.14	1.07	1.92	0.96	1.73	0.87	1.57	0.78	1.43	0.71	α	0.5α
	α	3·73		3·17		2·74		2·41		2·14		1·92		1·73		1·57		1·43			
	β	3·86		3·33		2·92		2·61		2·36		2·17		2·00		1·86		1·74			

Table 21.9

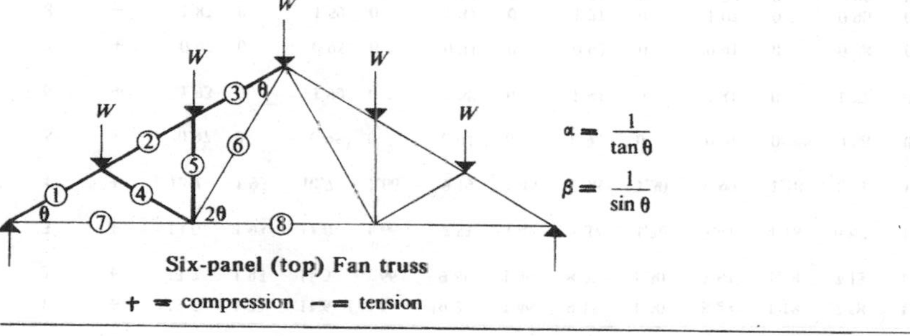

Six-panel (top) Fan truss

+ = compression − = tension

Member	Compression or tension	θ = 15° W	15° Unit	17½° W	17½° Unit	20° W	20° Unit	22½° W	22½° Unit	25° W	25° Unit	27½° W	27½° Unit	30° W	30° Unit	32½° W	32½° Unit	35° W	35° Unit	General formulae W	General formulae Unit
1	+	9.66	1.93	8.31	1.66	7.31	1.46	6.53	1.30	5.92	1.18	5.41	1.08	5.00	1.00	4.65	0.93	4.36	0.87	2.5β	0.5β
2	+	8.19	1.93	7.01	1.66	6.10	1.46	5.40	1.30	4.83	1.18	4.39	1.08	4.00	1.00	3.67	0.93	3.39	0.87	$\frac{13\beta^2 - 4}{6\beta}$	0.5β
3	+	9.14	1.93	7.69	1.66	6.60	1.46	5.75	1.30	5.06	1.18	4.48	1.08	4.00	1.00	3.58	0.93	3.22	0.87	$\frac{5\alpha^2 + 1}{2\beta}$	0.5β
4	+	1.54	0	1.38	0	1.27	0	1.18	0	1.11	0	1.05	0	1.00	0	0.95	0	0.91	0	$\frac{\alpha\sqrt{(\alpha^2 + 9)}}{3\beta}$	0
5	+	1.54	0	1.38	0	1.27	0	1.18	0	1.11	0	1.05	0	1.00	0	0.95	0	0.91	0	$\frac{\alpha\sqrt{(\alpha^2 + 9)}}{3\beta}$	0
6	−	3.73	0	3.17	0	2.75	0	2.41	0	2.14	0	1.92	0	1.73	0	1.57	0	1.43	0	α	0
7	−	9.33	1.87	7.93	1.59	6.87	1.37	6.04	1.21	5.36	1.07	4.80	0.96	4.33	0.87	3.92	0.78	3.57	0.71	2.5α	0.5α
8	−	5.60	1.87	4.76	1.59	4.12	1.37	3.62	1.21	3.22	1.07	2.88	0.96	2.60	0.87	2.35	0.78	2.14	0.71	1.5α	0.5α

	15°	17½°	20°	22½°	25°	27½°	30°	32½°	35°
α	3·73	3·17	2·74	2·41	2·14	1·92	1·73	1·57	1·43
β	3·86	3·33	2·92	2·61	2·34	2·17	2·00	1·86	1·74

Table 21.10

Eight-panel (top) Fink truss

+ = compression − = tension

$\alpha = \dfrac{1}{\tan\theta}$

$\beta = \dfrac{1}{\sin\theta}$

Member	Compression or tension	θ = 15° W	15° Unit	17½° W	17½° Unit	20° W	20° Unit	22½° W	22½° Unit	25° W	25° Unit	27½° W	27½° Unit	30° W	30° Unit	32½° W	32½° Unit	35° W	35° Unit	General formulae W	General formulae Unit
1	+	13.5	1.93	11.6	1.66	10.2	1.46	9.15	1.30	8.28	1.18	7.58	1.08	7.00	1.00	6.51	0.93	6.10	0.87	3.5β	0.5β
2	+	13.2	1.93	11.3	1.66	9.88	1.46	8.75	1.30	7.84	1.18	7.13	1.08	6.50	1.00	5.97	0.93	5.51	0.87	$\dfrac{3.5\beta^2-1}{\beta}$	0.5β
3	+	13.0	1.93	11.0	1.66	9.53	1.46	8.37	1.30	7.41	1.18	6.67	1.08	6.00	1.00	5.43	0.93	4.94	0.87	$\dfrac{3.5\beta^2-2}{\beta}$	0.5β
4	+	12.7	1.93	10.7	1.66	9.19	1.46	7.98	1.30	6.99	1.18	6.21	1.08	5.50	1.00	4.90	0.93	4.36	0.87	$\dfrac{3.5\beta^2-3}{\beta}$	0.5β
5	+	0.97	0	0.95	0	0.94	0	0.92	0	0.91	0	0.88	0	0.86	0	0.84	0	0.82	0	α/β	0
6	+	1.93	0	1.90	0	1.88	0	1.84	0	1.81	0	1.77	0	1.73	0	1.69	0	1.64	0	$2\alpha/\beta$	0
7	+	0.97	0	0.95	0	0.94	0	0.92	0	0.91	0	0.88	0	0.86	0	0.84	0	0.82	0	α/β	0
8	−	1.87	0	1.59	0	1.37	0	1.21	0	1.07	0	0.96	0	0.87	0	0.78	0	0.71	0	0.5α	0
9	−	1.87	0	1.59	0	1.37	0	1.21	0	1.07	0	0.96	0	0.87	0	0.78	0	0.71	0	0.5α	0
10	−	3.73	0	3.17	0	2.75	0	2.41	0	2.14	0	1.92	0	1.73	0	1.57	0	1.43	0	1.0α	0
11	−	5.6	0	4.76	0	4.12	0	3.62	0	3.22	0	2.88	0	2.60	0	2.35	0	2.14	0	1.5α	0
12	−	13.1	1.87	11.1	1.59	9.62	1.37	8.45	1.21	7.51	1.07	6.72	0.96	6.06	0.87	5.49	0.78	5.00	0.71	3.5α	0.5α
13	−	11.2	1.87	9.51	1.59	8.24	1.37	7.24	1.21	6.43	1.07	5.76	0.96	5.20	0.87	4.71	0.78	4.28	0.71	3α	0.5α
14	−	7.46	1.87	6.34	1.59	5.49	1.37	4.83	1.21	4.29	1.07	3.84	0.96	3.46	0.87	3.14	0.78	2.86	0.71	2α	0.5α

	15°	17½°	20°	22½°	25°	27½°	30°	32½°	35°
α	3·73	3·17	2·74	2·41	2·14	1·92	1·73	1·57	1·43
β	3·86	3·33	2·92	2·61	2·36	2·17	2·00	1·86	1·74

Table 21.11

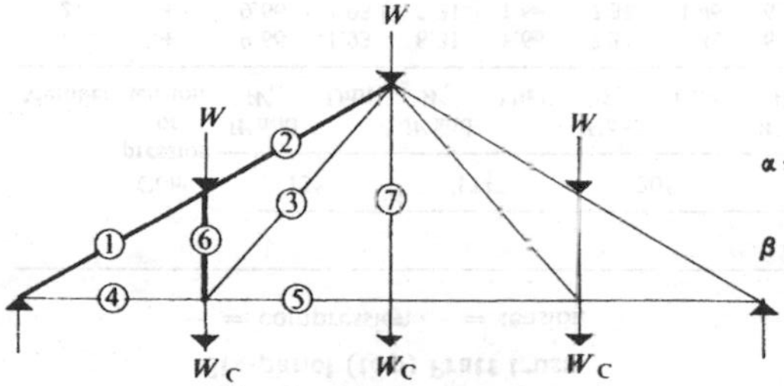

W W W
1 2 3 4 5 6 7
W_C W_C W_C

Four-panel (top) Pratt truss

+ = compression − = tension

$$\alpha = \frac{1}{\tan\theta}$$

$$\beta = \frac{1}{\sin\theta}$$

Member	Compression or tension	Coefficients for all slopes: W	W_c	Unit
6	+	1	0	0
7	−	0	1	0

Member	Compression or tension	θ = 15°: W and W_c	15°: Unit	17½°: W and W_c	17½°: Unit	20°: W and W_c	20°: Unit	22½°: W and W_c	22½°: Unit	25°: W and W_c	25°: Unit	27½°: W and W_c	27½°: Unit	30°: W and W_c	30°: Unit	32½°: W and W_c	32½°: Unit	35°: W and W_c	35°: Unit	General formulae: W and W_c	General formulae: Unit
1	+	5.80	1.93	5.00	1.56	4.39	1.46	3.92	1.30	3.55	1.18	3.25	1.08	3.00	1.00	2.79	0.93	2.61	0.87	1.5β	0.5β
2	+	5.80	1.93	5.00	1.56	4.39	1.46	3.92	1.30	3.55	1.18	3.25	1.08	3.00	1.00	2.79	0.93	2.61	0.87	1.5β	0.5β
3	−	2.12	0	1.87	0	1.70	0	1.57	0	1.47	0	1.39	0	1.32	0	1.27	0	1.23	0	$0.5\sqrt{(\alpha^2 + 4)}$	0
4	−	5.60	1.87	4.76	1.59	4.12	1.37	3.62	1.21	3.22	1.07	2.88	0.96	2.60	0.87	2.35	0.78	2.14	0.71	1.5α	0.5α
5	−	3.73	1.87	3.17	1.59	2.75	1.37	2.41	1.21	2.14	1.07	1.92	0.96	1.73	0.87	1.57	0.78	1.43	0.71	α	0.5α
	α	3·73		3·17		2·74		2·41		2·14		1·92		1·73		1·57		1·43			
	β	3·86		3·33		2·92		2·61		2·36		2·17		2·00		1·86		1·74			

Table 21.12

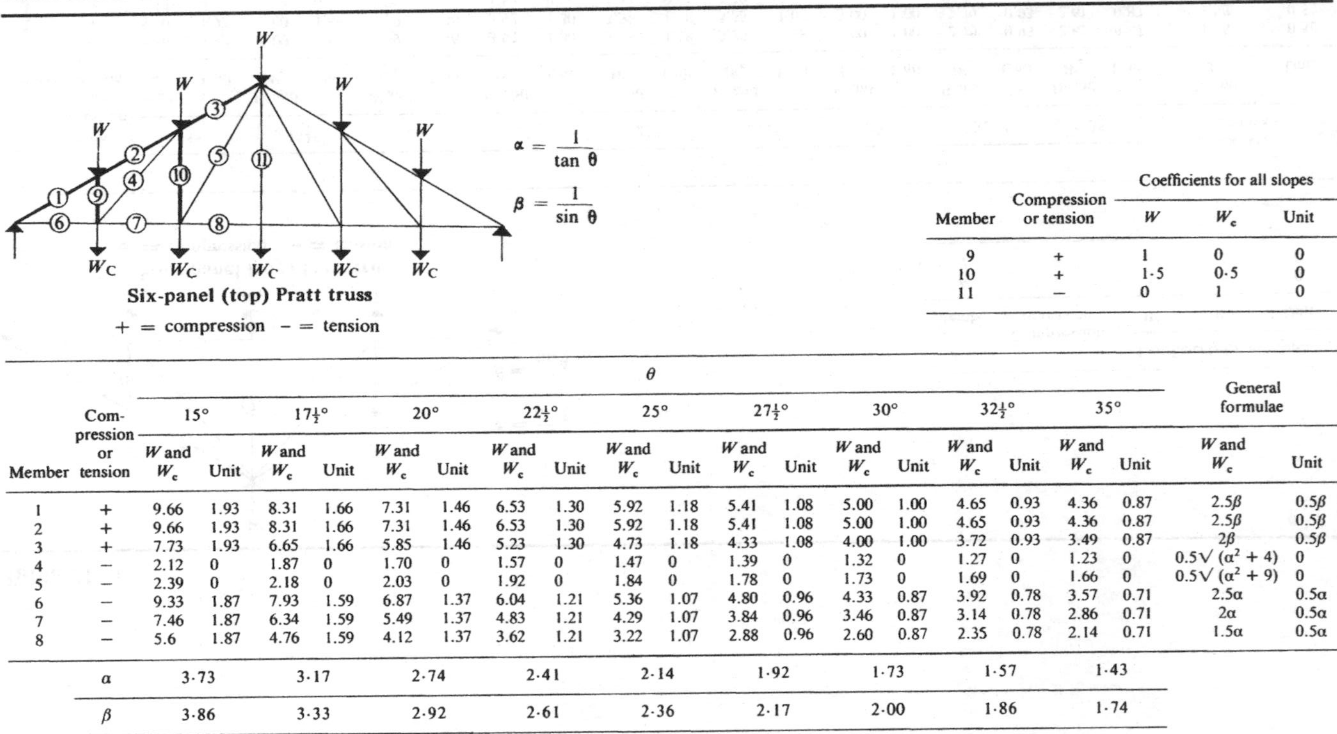

Six-panel (top) Pratt truss

+ = compression – = tension

$$\alpha = \frac{1}{\tan \theta}$$

$$\beta = \frac{1}{\sin \theta}$$

Member	Compression or tension	Coefficients for all slopes W	W_c	Unit
9	+	1	0	0
10	+	1·5	0·5	0
11	–	0	1	0

Member	Compression or tension	θ: 15° W and W_c	Unit	17½° W and W_c	Unit	20° W and W_c	Unit	22½° W and W_c	Unit	25° W and W_c	Unit	27½° W and W_c	Unit	30° W and W_c	Unit	32½° W and W_c	Unit	35° W and W_c	Unit	General formulae W and W_c	Unit
1	+	9.66	1.93	8.31	1.66	7.31	1.46	6.53	1.30	5.92	1.18	5.41	1.08	5.00	1.00	4.65	0.93	4.36	0.87	2.5β	0.5β
2	+	9.66	1.93	8.31	1.66	7.31	1.46	6.53	1.30	5.92	1.18	5.41	1.08	5.00	1.00	4.65	0.93	4.36	0.87	2.5β	0.5β
3	+	7.73	1.93	6.65	1.66	5.85	1.46	5.23	1.30	4.73	1.18	4.33	1.08	4.00	1.00	3.72	0.93	3.49	0.87	2β	0.5β
4	–	2.12	0	1.87	0	1.70	0	1.57	0	1.47	0	1.39	0	1.32	0	1.27	0	1.23	0	$0.5\sqrt{(\alpha^2 + 4)}$	0
5	–	2.39	0	2.18	0	2.03	0	1.92	0	1.84	0	1.78	0	1.73	0	1.69	0	1.66	0	$0.5\sqrt{(\alpha^2 + 9)}$	0
6	–	9.33	1.87	7.93	1.59	6.87	1.37	6.04	1.21	5.36	1.07	4.80	0.96	4.33	0.87	3.92	0.78	3.57	0.71	2.5α	0.5α
7	–	7.46	1.87	6.34	1.59	5.49	1.37	4.83	1.21	4.29	1.07	3.84	0.96	3.46	0.87	3.14	0.78	2.86	0.71	2α	0.5α
8	–	5.6	1.87	4.76	1.59	4.12	1.37	3.62	1.21	3.22	1.07	2.88	0.96	2.60	0.87	2.35	0.78	2.14	0.71	1.5α	0.5α
	α	3·73		3·17		2·74		2·41		2·14		1·92		1·73		1·57		1·43			
	β	3·86		3·33		2·92		2·61		2·36		2·17		2·00		1·86		1·74			

Table 21.13

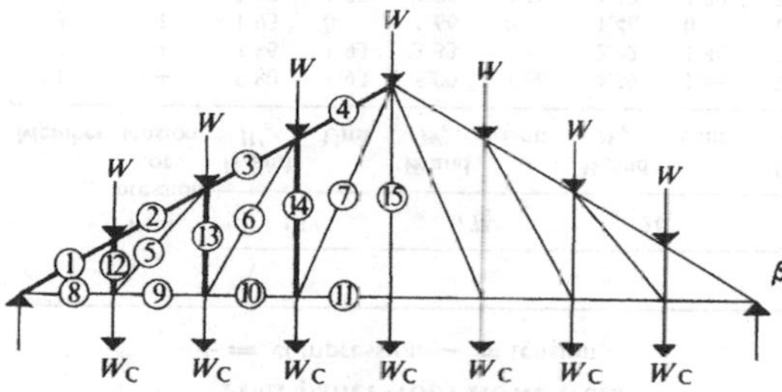

Eight-panel (top) Pratt truss

+ = compression — = tension

$\alpha = \frac{1}{\tan \theta}$

$\beta = \frac{1}{\sin \theta}$

Member	Compression or tension	Coefficients for all slopes: W	W_c	Unit
12	+	1	0	0
13	+	1·5	0·5	0
14	+	2	1	0
15	—	0	1	0

Member	Compression or tension	θ = 15°: W and W_c	Unit	17½°: W and W_c	Unit	20°: W and W_c	Unit	22½°: W and W_c	Unit	25°: W and W_c	Unit	27½°: W and W_c	Unit	30°: W and W_c	Unit	32½°: W and W_c	Unit	35°: W and W_c	Unit	General formulae: W and W_c	Unit
1	+	13.5	1.93	11.6	1.66	10.2	1.46	9.15	1.30	8.28	1.18	7.58	1.08	7.00	1.00	6.51	0.93	6.10	0.87	3.5β	0.5β
2	+	13.5	1.93	11.6	1.66	10.2	1.46	9.15	1.30	8.28	1.18	7.58	1.08	7.00	1.00	6.51	0.93	6.10	0.87	3.5β	0.5β
3	+	11.6	1.93	9.98	1.66	8.77	1.46	7.84	1.30	7.10	1.18	6.50	1.08	6.00	1.00	5.58	0.93	5.23	0.87	3β	0.5β
4	+	9.66	1.93	8.31	1.66	7.31	1.46	6.53	1.30	5.92	1.18	5.41	1.08	5.00	1.00	4.65	0.93	4.36	0.87	2.5β	0.5β
5	—	2.12	0	1.87	0	1.70	0	1.57	0	1.47	0	1.39	0	1.32	0	1.27	0	1.23	0	$0.5\sqrt{(\alpha^2 + 4)}$	0
6	—	2.39	0	2.18	0	2.03	0	1.92	0	1.84	0	1.78	0	1.73	0	1.69	0	1.66	0	$0.5\sqrt{(\alpha^2 + 9)}$	0
7	—	2.74	0	2.55	0	2.43	0	2.34	0	2.27	0	2.22	0	2.18	0	2.15	0	2.12	0	$0.5\sqrt{(\alpha^2 + 16)}$	0
8	—	13.1	1.87	11.1	1.59	9.62	1.37	8.45	1.21	7.51	1.07	6.72	0.96	6.06	0.87	5.49	0.78	5.00	0.71	3.5α	0.5α
9	—	11.2	1.87	9.51	1.59	8.24	1.37	7.24	1.21	6.43	1.07	5.76	0.96	5.20	0.87	4.71	0.78	4.28	0.71	3α	0.5α
10	—	9.33	1.87	7.93	1.59	6.87	1.37	6.04	1.21	5.36	1.07	4.80	0.96	4.33	0.87	3.92	0.78	3.57	0.71	2.5α	0.5α
11	—	7.46	1.87	6.34	1.59	5.49	1.37	4.83	1.21	4.29	1.07	3.84	0.96	3.46	0.87	3.14	0.78	2.86	0.71	2α	0.5α
	α	3·73		3·17		2·74		2·41		2·14		1·92		1·73		1·57		1·43			
	β	3·86		3·33		2·92		2·61		2·36		2·17		2·00		1·86		1·74			

Table 21.14

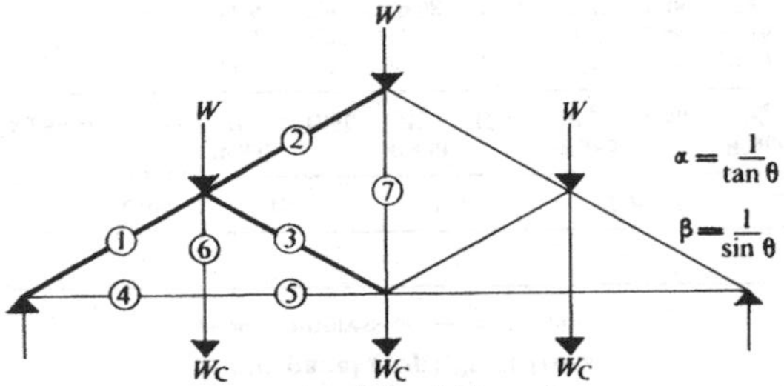

$\alpha = \frac{1}{\tan\theta}$

$\beta = \frac{1}{\sin\theta}$

Four-panel (top) Howe truss

+ = compression − = tension

Member	Compression or tension	Coefficients for all slopes: W	W_c	Unit
6	–	0	1·0	0
7	–	1·0	2·0	0

Member	Compression or tension	θ: 15°, W and W_c	15°, Unit	17½°, W and W_c	17½°, Unit	20°, W and W_c	20°, Unit	22½°, W and W_c	22½°, Unit	25°, W and W_c	25°, Unit	27½°, W and W_c	27½°, Unit	30°, W and W_c	30°, Unit	32½°, W and W_c	32½°, Unit	35°, W and W_c	35°, Unit	General formulae, W and W_c	General formulae, Unit
1	+	5.80	1.93	5.00	1.66	4.39	1.46	3.92	1.30	3.55	1.18	3.25	1.08	3.00	1.00	2.79	0.93	2.61	0.87	1.5β	0.5β
2	+	3.86	1.93	3.33	1.66	2.92	1.46	2.61	1.30	2.37	1.18	2.16	1.08	2.00	1.00	1.86	0.93	1.74	0.87	1.0β	0.5β
3	+	1.93	0	1.66	0	1.46	0	1.30	0	1.18	0	1.08	0	1.00	0	0.93	0	0.87	0	0.5β	0
4	–	5.60	1.87	4.76	1.59	4.12	1.37	3.62	1.21	3.22	1.07	2.88	0.96	2.60	0.87	2.35	0.78	2.14	0.71	1.5α	0.5α
5	–	5.60	1.87	4.76	1.59	4.12	1.37	3.62	1.21	3.22	1.07	2.88	0.96	2.60	0.87	2.35	0.78	2.14	0.71	1.5α	0.5α
	α	3·73		3·17		2·74		2·41		2·14		1·92		1·73		1·57		1·43			
	β	3·86		3·33		2·92		2·61		2·36		2·17		2·00		1·86		1·74			

Table 21.15

Six-panel (top) Howe truss

+ = compression − = tension

$\alpha = \dfrac{1}{\tan\theta}$

$\beta = \dfrac{1}{\sin\theta}$

Member	Compression or tension	Coefficients for all slopes: W	W_c	Unit
9	–	0	1·0	0
10	–	0·5	1·5	0
11	–	2	3	0

Member	Compression or tension	θ = 15°: W and W_c	Unit	17½°: W and W_c	Unit	20°: W and W_c	Unit	22½°: W and W_c	Unit	25°: W and W_c	Unit	27½°: W and W_c	Unit	30°: W and W_c	Unit	32½°: W and W_c	Unit	35°: W and W_c	Unit	General formulae: W and W_c	Unit
1	+	9.66	1.93	8.31	1.66	7.31	1.46	6.53	1.30	5.92	1.18	5.41	1.08	5.00	1.00	4.65	0.93	4.36	0.87	2.5β	0.5β
2	+	7.73	1.93	6.65	1.66	5.85	1.46	5.23	1.30	4.73	1.18	4.33	1.08	4.00	1.00	3.72	0.93	3.49	0.87	2β	0.5β
3	+	5.80	1.93	5.00	1.66	4.39	1.46	3.92	1.30	3.55	1.18	3.25	1.08	3.00	1.00	2.79	0.93	2.61	0.87	1.5β	0.5β
4	+	1.93	0	1.66	0	1.46	0	1.30	0	1.18	0	1.08	0	1.00	0	0.93	0	0.87	0	0.5β	0
5	+	2.12	0	1.87	0	1.70	0	1.57	0	1.47	0	1.39	0	1.32	0	1.27	0	1.23	0	$0.5\sqrt{(\alpha^2 + 4)}$	0
6	−	9.33	1.87	7.93	1.59	6.87	1.37	6.04	1.21	5.36	1.07	4.80	0.96	4.33	0.87	3.92	0.78	3.57	0.71	2.5α	0.5α
7	−	9.33	1.87	7.93	1.59	6.87	1.37	6.04	1.21	5.36	1.07	4.80	0.96	4.33	0.87	3.92	0.78	3.57	0.71	2.5α	0.5α
8	−	7.46	1.87	6.34	1.59	5.49	1.37	4.83	1.21	4.29	1.07	3.84	0.96	3.46	0.87	3.14	0.78	2.86	0.71	2α	0.5α
	α	3·73		3·17		2·74		2·41		2·14		1·92		1·73		1·57		1·43			
	β	3·86		3·33		2·92		2·61		2·36		2·17		2·00		1·86		1·74			

Table 21.16

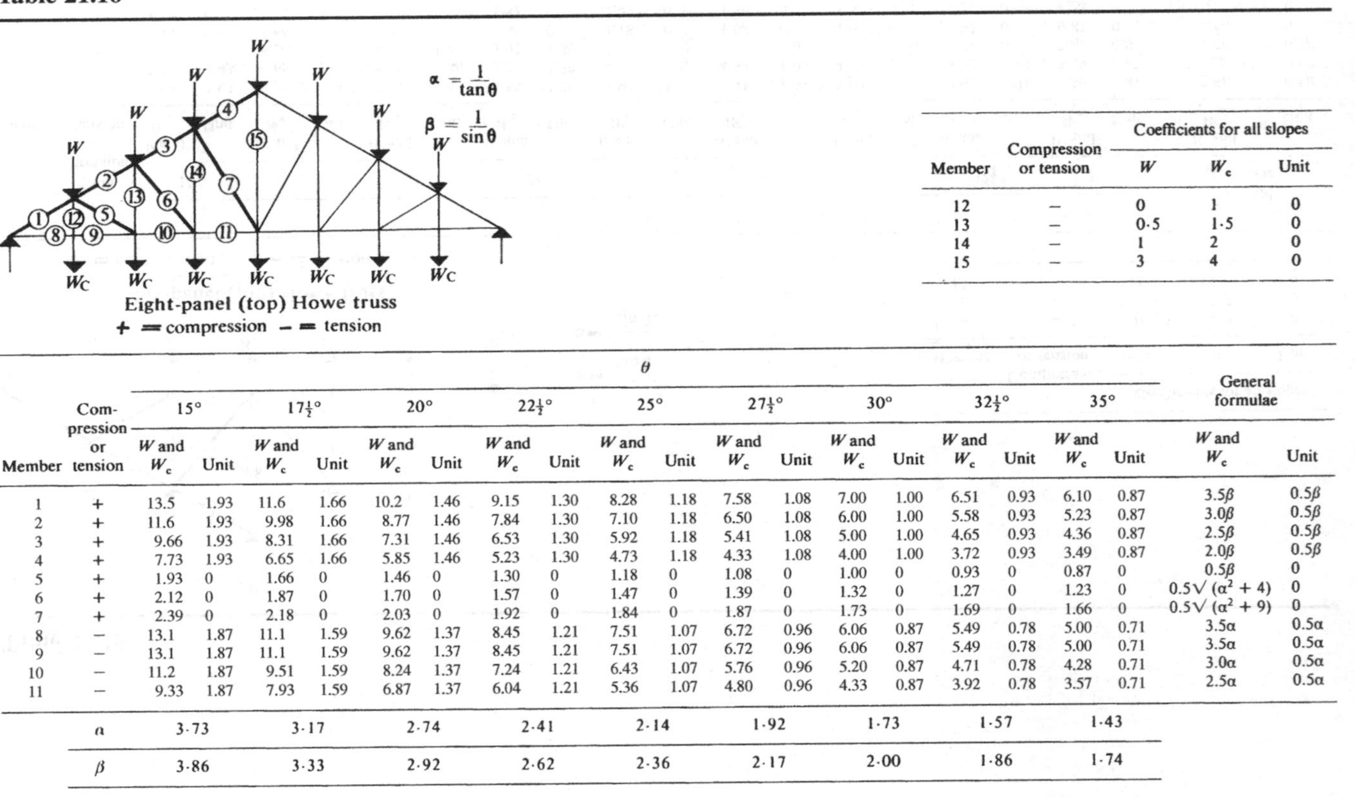

Eight-panel (top) Howe truss
+ = compression – = tension

Member	Compression or tension	Coefficients for all slopes: W	W_c	Unit
12	–	0	1	0
13	–	0·5	1·5	0
14	–	1	2	0
15	–	3	4	0

Member	Compression or tension	θ = 15°: W and W_c	Unit	17½°: W and W_c	Unit	20°: W and W_c	Unit	22½°: W and W_c	Unit	25°: W and W_c	Unit	27½°: W and W_c	Unit	30°: W and W_c	Unit	32½°: W and W_c	Unit	35°: W and W_c	Unit	General formulae: W and W_c	Unit
1	+	13.5	1.93	11.6	1.66	10.2	1.46	9.15	1.30	8.28	1.18	7.58	1.08	7.00	1.00	6.51	0.93	6.10	0.87	3.5β	0.5β
2	+	11.6	1.93	9.98	1.66	8.77	1.46	7.84	1.30	7.10	1.18	6.50	1.08	6.00	1.00	5.58	0.93	5.23	0.87	3.0β	0.5β
3	+	9.66	1.93	8.31	1.66	7.31	1.46	6.53	1.30	5.92	1.18	5.41	1.08	5.00	1.00	4.65	0.93	4.36	0.87	2.5β	0.5β
4	+	7.73	1.93	6.65	1.66	5.85	1.46	5.23	1.30	4.73	1.18	4.33	1.08	4.00	1.00	3.72	0.93	3.49	0.87	2.0β	0.5β
5	+	1.93	0	1.66	0	1.46	0	1.30	0	1.18	0	1.08	0	1.00	0	0.93	0	0.87	0	0.5β	0
6	+	2.12	0	1.87	0	1.70	0	1.57	0	1.47	0	1.39	0	1.32	0	1.27	0	1.23	0	$0.5\sqrt{(\alpha^2 + 4)}$	0
7	+	2.39	0	2.18	0	2.03	0	1.92	0	1.84	0	1.87	0	1.73	0	1.69	0	1.66	0	$0.5\sqrt{(\alpha^2 + 9)}$	0
8	–	13.1	1.87	11.1	1.59	9.62	1.37	8.45	1.21	7.51	1.07	6.72	0.96	6.06	0.87	5.49	0.78	5.00	0.71	3.5α	0.5α
9	–	13.1	1.87	11.1	1.59	9.62	1.37	8.45	1.21	7.51	1.07	6.72	0.96	6.06	0.87	5.49	0.78	5.00	0.71	3.5α	0.5α
10	–	11.2	1.87	9.51	1.59	8.24	1.37	7.24	1.21	6.43	1.07	5.76	0.96	5.20	0.87	4.71	0.78	4.28	0.71	3.0α	0.5α
11	–	9.33	1.87	7.93	1.59	6.87	1.37	6.04	1.21	5.36	1.07	4.80	0.96	4.33	0.87	3.92	0.78	3.57	0.71	2.5α	0.5α
	α	3·73		3·17		2·74		2·41		2·14		1·92		1·73		1·57		1·43			
	β	3·86		3·33		2·92		2·62		2·36		2·17		2·00		1·86		1·74			

Table 21.17

+ = compression − = tension $a = \frac{H}{L}$ $b = \sqrt{[\frac{1}{a}]^2 + 16}$

Member	Compression or tension	Constant coefficients W	W_c	Unit
6	–	0·5	1·5	0.5
7	–	0	1	0

Member	Compression or tension	Values of a: 0·15		0·20		0·25		0·30		0·35		0·40		General formulae	
		W and W_c	Unit	W and W_c	Unit	W and W_c	Unit	W and W_c	Unit	W and W_c	Unit	W and W_c	Unit	W and W_c	Unit
1	+	2·50	0·83	1·87	0·62	1·50	0·50	1·25	0·42	1·07	0·36	0·94	0·31	$\frac{3}{8a}$	$\frac{1}{8a}$
2	+	2·91	0·97	2·40	0·80	2·12	0·71	1·95	0·65	1·84	0·61	1·77	0·59	$\frac{3b}{8}$	$\frac{b}{8}$
3	+	0·97	0·97	0·80	0·80	0·71	0·71	0·65	0·65	0·61	0·61	0·59	0·59	$\frac{b}{8}$	$\frac{b}{8}$
4	–	2·50	0·83	1·87	0·62	1·50	0·50	1·25	0·42	1·07	0·36	0·94	0·31	$\frac{3}{8a}$	$\frac{1}{8a}$
5	–	3·33	1·67	2·5	1·25	2·00	1·00	1·67	0·83	1·43	0·71	1·25	0·62	$\frac{1}{2a}$	$\frac{1}{4a}$
Values of b		7·77		6·40		5·66		5·21		4·92		4·72			
Values of θ		31°		39°		45°		50°		55°		58°			

Table 21.18

Six-panel Howe truss

+ = compression – = tension $a = \frac{H}{L}$ $b = \sqrt{\left[\frac{1}{a}\right]^2 + 36}$

Member	Compression or tension	Constant coefficients W	W_c	Unit
9	–	1·5	2·5	0·5
10	–	0·5	1·5	0·5
11	–	0	1	0

		Values of a												General formulae	
	Com-pression or tension	0·10		0·15		0·167		0·20		0·25		0·30			
Member		W and W_c	Unit	W and W_c	Unit	W and W_c	Unit	W and W_c	Unit	W and W_c	Unit	W and W_c	Unit	W and W_c	Unit
1	+	4·17	0·83	2·78	0·56	2·50	0·50	2·08	0·42	1·67	0·33	1·39	0·28	$\frac{5}{12a}$	$\frac{1}{12a}$
2	+	6·67	1·67	4·44	1·11	4·00	1·00	3·33	0·83	2·67	0·67	2·22	0·56	$\frac{2}{3a}$	$\frac{1}{6a}$
3	+	4·86	0·97	3·74	0·75	3·53	0·71	3·25	0·65	3·00	0·60	2·86	0·57	$\frac{5b}{12}$	$\frac{b}{12}$
4	+	2·92	0·97	2·24	0·75	2·12	0·71	1·95	0·65	1·80	0·60	1·72	0·57	$\frac{b}{4}$	$\frac{b}{12}$
5	+	0·97	0·97	0·75	0·75	0·71	0·71	0·65	0·65	0·60	0·60	0·57	0·57	$\frac{b}{12}$	$\frac{b}{12}$
6	–	4·17	0·83	2·78	0·56	2·50	0·71	2·08	0·42	1·67	0·33	1·39	0·28	$\frac{5}{12a}$	$\frac{1}{12a}$
7	–	6·67	1·67	4·44	1·11	4·00	1·00	3·33	0·83	2·67	0·67	2·22	0·56	$\frac{2}{3a}$	$\frac{1}{6a}$
8	–	7·50	2·50	5·00	1·67	4·50	1·50	3·75	1·25	3·00	1·00	2·50	0·83	$\frac{3}{4a}$	$\frac{1}{4a}$
Values of b		11·7		8·97		8·48		7·81		7·21		6·86			
Values of θ		31°		42°		45°		50°		56°		61°			

Table 21.19

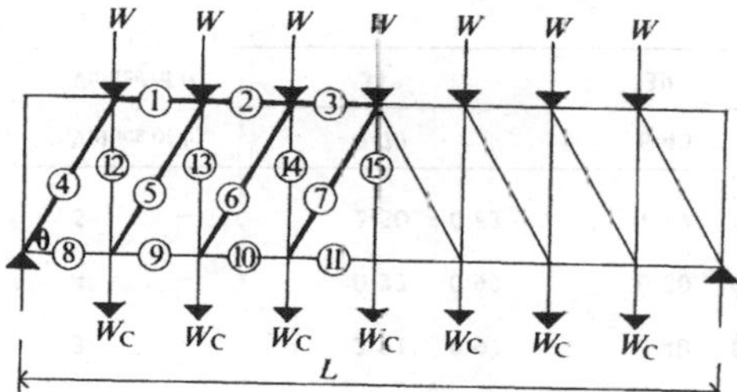

Eight-panel Howe truss

+ = compression − = tension $a = \frac{H}{L}$ $b = \sqrt{[\frac{1}{a}]^2 + 64}$

Member	Compression or tension	Constant coefficients W	Constant coefficients W_c	Constant coefficients Unit
12	−	2·5	3·5	0·5
13	−	1·5	2·5	0·5
14	−	0·5	1·5	0·5
15	−	0	1·0	0

Member	Compression or tension	a = 0·075: W and W_c	a = 0·075: Unit	a = 0·10: W and W_c	a = 0·10: Unit	a = 0·125: W and W_c	a = 0·125: Unit	a = 0·15: W and W_c	a = 0·15: Unit	a = 0·175: W and W_c	a = 0·175: Unit	a = 0·20: W and W_c	a = 0·20: Unit	General formulae: W and W_c	General formulae: Unit
1	+	5·83	0·83	4·37	0·62	3·50	0·50	2·92	0·42	2·50	0·36	2·19	0·31	$\frac{7}{16a}$	$\frac{1}{16a}$
2	+	10·0	1·67	7·50	1·25	6·00	1·00	5·00	0·83	4·28	0·71	3·75	0·62	$\frac{3}{4a}$	$\frac{1}{8a}$
3	+	12·5	2·22	9·37	1·67	7·50	1·33	6·25	1·11	5·36	0·95	4·69	0·83	$\frac{15}{16a}$	$\frac{1}{6a}$
4	+	6·80	0·97	5·60	0·80	4·95	0·71	4·55	0·65	4·30	0·61	4·13	0·59	$\frac{7b}{16}$	$\frac{b}{16}$
5	+	4·86	0·97	4·00	0·80	3·54	0·71	3·25	0·65	3·07	0·61	2·95	0·59	$\frac{5b}{16}$	$\frac{b}{16}$
6	+	2·92	0·97	2·40	0·80	2·12	0·71	1·95	0·65	1·84	0·61	1·77	0·59	$\frac{3b}{16}$	$\frac{b}{16}$
7	+	0·97	0·97	0·80	0·80	0·71	0·71	0·65	0·65	0·61	0·61	0·59	0·59	$\frac{b}{16}$	$\frac{b}{16}$
8	−	5·83	0·83	4·37	0·62	3·50	0·50	2·92	0·42	2·50	0·36	2·19	0·31	$\frac{7}{16a}$	$\frac{1}{16a}$
9	−	10·0	1·67	7·50	1·25	6·00	1·00	5·00	0·83	4·28	0·71	3·75	0·62	$\frac{3}{4a}$	$\frac{1}{8a}$
10	−	12·5	2·22	9·37	1·67	7·50	1·33	6·25	1·11	5·36	0·95	4·69	0·83	$\frac{15}{16a}$	$\frac{1}{6a}$
11	−	13·3	3·33	10·0	2·50	8·00	2·00	6·67	1·67	5·71	1·43	5·00	1·25	$\frac{1}{a}$	$\frac{1}{4a}$
Values of b		15·5		12·8		11·3		10·4		9·83		9·43			
Values of θ		31°		39°		45°		50°		55°		58°			

Table 21.20

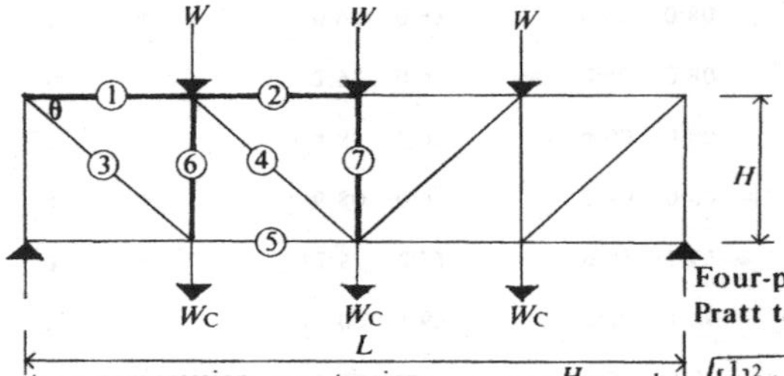

Four-panel Pratt truss

+ = compression − = tension $a = \frac{H}{L}$ $b = \sqrt{\left[\frac{1}{a}\right]^2 + 16}$

Member	Compression or tension	Constant coefficients W	W_c	Unit
6	+	1·5	0·5	0·5
7	+	1·0	0	1·0

		Values of a												General formulae	
		0·15		0·20		0·25		0·30		0·35		0·40			
Member	Compression or tension	W and W_c	Unit	W and W_c	Unit	W and W_c	Unit	W and W_c	Unit	W and W_c	Unit	W and W_c	Unit	W and W_c	Unit
1	+	2·50	0·83	1·87	0·62	1·50	0·50	1·25	0·42	1·07	0·36	0·94	0·31	$\frac{3}{8a}$	$\frac{1}{8a}$
2	+	3·33	1·67	2·50	1·25	2·00	1·00	1·67	0·83	1·43	0·71	1·25	0·62	$\frac{1}{2a}$	$\frac{1}{4a}$
3	−	2·91	0·97	2·40	0·80	2·12	0·71	1·95	0·65	1·84	0·61	1·77	0·59	$\frac{3b}{8}$	$\frac{b}{8}$
4	−	0·97	0·97	0·80	0·80	0·71	0·71	0·65	0·65	0·61	0·61	0·59	0·59	$\frac{b}{8}$	$\frac{b}{8}$
5	−	2·50	0·83	1·87	0·62	1·50	0·50	1·25	0·42	1·07	0·36	0·94	0·31	$\frac{3}{8a}$	$\frac{1}{8a}$
Values of b		7·77		6·40		5·66		5·21		4·92		4·72			
Values of θ		31°		39°		45°		50°		55°		58°			

Table 21.21

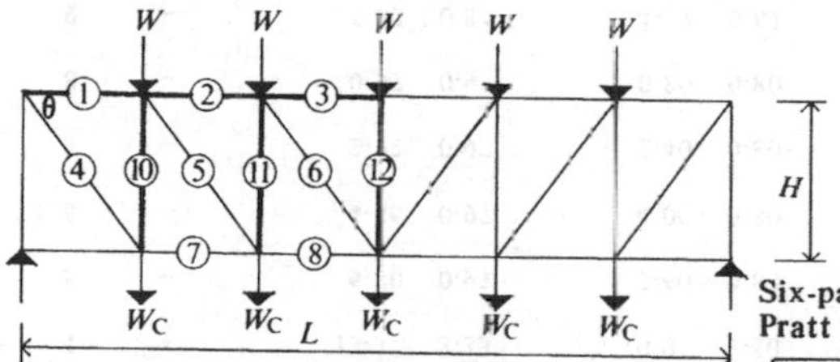

Six-panel Pratt truss

+ = compression − = tension $a = \frac{H}{L}$ $b = \sqrt{\left[\frac{1}{a}\right]^2 + 36}$

Member	Compression or tension	Constant coefficients: W	W_c	Unit
10	+	2·5	1·5	0·5
11	+	1·5	0·5	0·5
12	+	1·0	0	1·0

Member	Compression or tension	a = 0·10: W and W_c	Unit	0·15: W and W_c	Unit	0·167: W and W_c	Unit	0·20: W and W_c	Unit	0·25: W and W_c	Unit	0·30: W and W_c	Unit	General formulae: W and W_c	Unit
1	+	4·17	0·83	2·78	0·56	2·50	0·50	2·08	0·42	1·67	0·33	1·39	0·28	$\frac{5}{12a}$	$\frac{1}{12a}$
2	+	6·67	1·67	4·44	1·11	4·00	1·00	3·33	0·83	2·67	0·67	2·22	0·56	$\frac{2}{3a}$	$\frac{1}{6a}$
3	+	7·5	2·5	5·00	1·67	4·50	1·50	3·75	1·25	3·00	1·00	2·50	0·83	$\frac{3}{4a}$	$\frac{1}{4a}$
4	−	4·86	0·97	3·74	0·75	3·53	0·71	3·25	0·65	3·00	0·60	2·86	0·57	$\frac{5b}{12}$	$\frac{b}{12}$
5	−	2·92	0·97	2·24	0·75	2·12	0·71	1·95	0·65	1·80	0·60	1·72	0·57	$\frac{b}{4}$	$\frac{b}{12}$
6	−	0·97	0·97	0·75	0·75	0·71	0·71	0·65	0·65	0·60	0·60	0·57	0·57	$\frac{b}{12}$	$\frac{b}{12}$
7	−	4·17	0·83	2·78	0·56	2·50	0·71	2·08	0·42	1·67	0·33	1·39	0·28	$\frac{5}{12a}$	$\frac{1}{12a}$
8	−	6·67	1·67	4·44	1·11	4·00	1·00	3·33	0·83	2·67	0·67	2·22	0·56	$\frac{2}{3a}$	$\frac{1}{6a}$
Values of b		11·7		8·97		8·48		7·81		7·21		6·86			
Values of θ		31°		42°		45°		50°		56°		61°			

Table 21.22

Eight-panel Pratt truss

+ = compression − = tension $a = \frac{H}{L}$ $b = \sqrt{\left[\frac{1}{a}\right]^2 + 64}$

Member	Compression or tension	Constant coefficients W	W_c	Unit
12	+	3·5	2·5	0·5
13	+	2·5	1·5	0·5
14	+	1·5	0·5	0·5
15	+	1·0	0	1·0

Member	Compression or tension	Values of *a*: 0·075 W and W_c	0·075 Unit	0·10 W and W_c	0·10 Unit	0·125 W and W_c	0·125 Unit	0·15 W and W_c	0·15 Unit	0·175 W and W_c	0·175 Unit	0·20 W and W_c	0·20 Unit	General formulae W and W_c	General formulae Unit
1	+	5·83	0·83	4·37	0·62	3·50	0·50	2·92	0·42	2·50	0·36	2·19	0·31	$\frac{7}{16a}$	$\frac{1}{16a}$
2	+	10·0	1·67	7·50	1·25	6·00	1·00	5·00	0·83	4·28	0·71	3·75	0·62	$\frac{3}{4a}$	$\frac{1}{8a}$
3	+	12·5	2·22	9·37	1·67	7·50	1·33	6·25	1·11	5·36	0·95	4·69	0·83	$\frac{15}{16a}$	$\frac{1}{6a}$
4	+	13·3	3·33	10·0	2·50	8·00	2·00	6·67	1·67	5·71	1·43	5·00	1·25	$\frac{1}{a}$	$\frac{1}{4a}$
5	−	6·80	0·97	5·60	0·80	4·95	0·71	4·55	0·65	4·30	0·61	4·13	0·59	$\frac{7b}{16}$	$\frac{b}{16}$
6	−	4·86	0·97	4·00	0·80	3·54	0·71	3·25	0·65	3·07	0·61	2·95	0·59	$\frac{5b}{16}$	$\frac{b}{16}$
7	−	2·92	0·97	2·40	0·80	2·12	0·71	1·95	0·65	1·84	0·61	1·77	0·59	$\frac{3b}{16}$	$\frac{b}{16}$
8	−	0·97	0·97	0·80	0·80	0·71	0·71	0·65	0·65	0·61	0·61	0·89	0·59	$\frac{b}{16}$	$\frac{b}{16}$
9	−	5·83	0·83	4·37	0·62	3·50	0·50	2·92	0·42	2·50	0·36	2·19	0·31	$\frac{7}{16a}$	$\frac{1}{16a}$
10	−	10·0	1·67	7·50	1·25	6·00	1·00	5·00	0·83	4·28	0·71	3·75	0·62	$\frac{3}{4a}$	$\frac{1}{8a}$
11	−	12·5	2·22	9·37	1·67	7·50	1·33	6·25	1·11	5·36	0·95	4·69	0·83	$\frac{15}{16a}$	$\frac{1}{6a}$
Values of *b*		15·5		12·8		11·3		10·4		9·83		9·43			
Values of θ		31°		39°		45°		50°		55°		58°			

Table 21.23

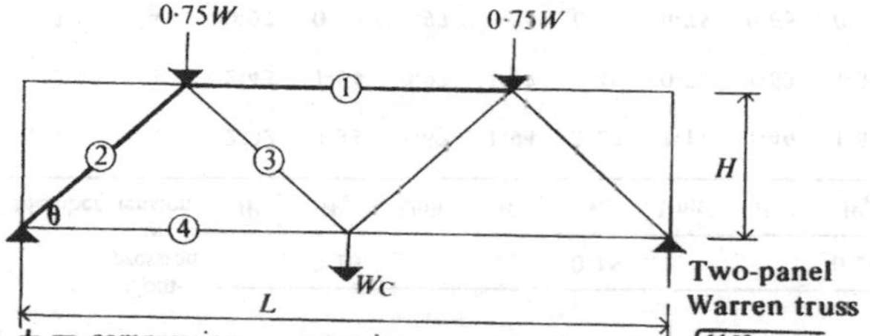

Two-panel Warren truss

+ = compression − = tension $a = \frac{H}{L}$ $b = \sqrt{\left[\frac{1}{a}\right]^2 + 16}$

Member	Com-pression or tension	Values of a: 0·15			0·20			0·25			0·30			0·35			0·40			General formulae		
		W	W_c	Unit	W	W_c	Unit	W	W_c	Unit	W	W_c	Unit	W	W_c	Unit	W	W_c	Unit	W	W_c	Unit
1	+	1·25	1·66	1·66	0·94	1·25	1·25	0·75	1·0	1·0	0·62	0·83	0·83	0·54	0·71	0·71	0·47	0·62	0·62	$\frac{3}{16a}$	$\frac{1}{4a}$	$\frac{1}{4a}$
2	+	1·46	0·97	0·97	1·20	0·80	0·80	1·06	0·71	0·71	0·98	0·65	0·65	0·92	0·61	0·61	0·88	0·59	0·59	$\frac{3b}{16}$	$\frac{b}{8}$	$\frac{b}{8}$
3	−	0	0·97	0·97	0	0·80	0·80	0	0·71	0·71	0	0·65	0·65	0	0·61	0·61	0	0·59	0·59	0	$\frac{b}{8}$	$\frac{b}{8}$
4	−	1·25	0·83	0·83	0·94	0·62	0·62	0·75	0·50	0·50	0·62	0·42	0·42	0·54	0·36	0·36	0·47	0·31	0·31	$\frac{3}{16a}$	$\frac{1}{8a}$	$\frac{1}{8a}$
Values of b		7·77			6·40			5·66			5·21			4·92			4·72					
Values of θ		31°			39°			45°			50°			55°			58°					

Table 21.24

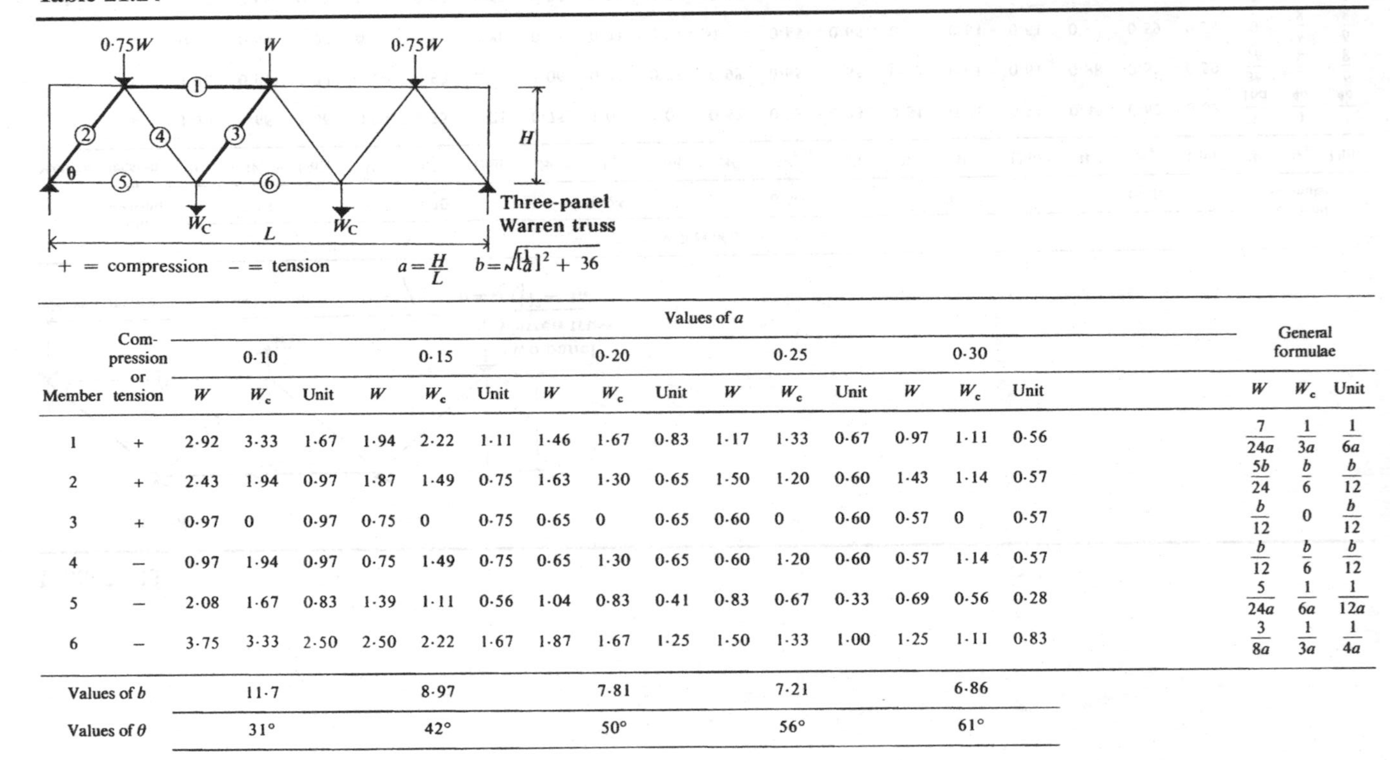

+ = compression − = tension $a = \frac{H}{L}$ $b = \sqrt{[\frac{1}{a}]^2 + 36}$

Member	Compression or tension	Values of a: 0·10 W	0·10 W_c	0·10 Unit	0·15 W	0·15 W_c	0·15 Unit	0·20 W	0·20 W_c	0·20 Unit	0·25 W	0·25 W_c	0·25 Unit	0·30 W	0·30 W_c	0·30 Unit	General formulae W	General formulae W_c	General formulae Unit
1	+	2·92	3·33	1·67	1·94	2·22	1·11	1·46	1·67	0·83	1·17	1·33	0·67	0·97	1·11	0·56	$\frac{7}{24a}$	$\frac{1}{3a}$	$\frac{1}{6a}$
2	+	2·43	1·94	0·97	1·87	1·49	0·75	1·63	1·30	0·65	1·50	1·20	0·60	1·43	1·14	0·57	$\frac{5b}{24}$	$\frac{b}{6}$	$\frac{b}{12}$
3	+	0·97	0	0·97	0·75	0	0·75	0·65	0	0·65	0·60	0	0·60	0·57	0	0·57	$\frac{b}{12}$	0	$\frac{b}{12}$
4	−	0·97	1·94	0·97	0·75	1·49	0·75	0·65	1·30	0·65	0·60	1·20	0·60	0·57	1·14	0·57	$\frac{b}{12}$	$\frac{b}{6}$	$\frac{b}{12}$
5	−	2·08	1·67	0·83	1·39	1·11	0·56	1·04	0·83	0·41	0·83	0·67	0·33	0·69	0·56	0·28	$\frac{5}{24a}$	$\frac{1}{6a}$	$\frac{1}{12a}$
6	−	3·75	3·33	2·50	2·50	2·22	1·67	1·87	1·67	1·25	1·50	1·33	1·00	1·25	1·11	0·83	$\frac{3}{8a}$	$\frac{1}{3a}$	$\frac{1}{4a}$
Values of b			11·7			8·97			7·81			7·21			6·86				
Values of θ			31°			42°			50°			56°			61°				

Table 21.25

Four-panel Warren truss

+ = compression − = tension $a = \frac{H}{L}$ $b = \sqrt{\left[\frac{1}{a}\right]^2 + 64}$

Member	Compression or tension	0·075 W	0·075 W_c	0·075 Unit	0·10 W	0·10 W_c	0·10 Unit	0·125 W	0·125 W_c	0·125 Unit	0·15 W	0·15 W_c	0·15 Unit	0·175 W	0·175 W_c	0·175 Unit	0·20 W	0·20 W_c	0·20 Unit	General formulae W	General formulae W_c	General formulae Unit
1	+	4·58	5·00	1·67	3·44	3·75	1·25	2·75	3·00	1·00	2·29	2·50	0·83	1·96	2·14	0·71	1·72	1·87	0·62	$\frac{11}{32a}$	$\frac{3}{8a}$	$\frac{1}{8a}$
2	+	6·25	6·67	3·33	4·69	5·00	2·50	3·75	4·00	2·00	3·12	3·33	1·67	2·70	2·86	1·43	2·34	2·50	1·25	$\frac{15}{32a}$	$\frac{1}{2a}$	$\frac{1}{4a}$
3	+	3·39	2·91	0·97	2·80	2·40	0·80	2·47	2·12	0·71	2·27	1·95	0·65	2·15	1·84	0·61	2·06	1·77	0·59	$\frac{7b}{32}$	$\frac{3b}{16}$	$\frac{b}{16}$
4	+	1·94	0·97	0·97	1·60	0·80	0·80	1·41	0·71	0·71	1·30	0·65	0·65	1·23	0·61	0·61	1·18	0·59	0·59	$\frac{b}{8}$	$\frac{b}{16}$	$\frac{b}{16}$
5	−	1·94	2·91	0·97	1·60	2·40	0·80	1·41	2·12	0·71	1·30	1·95	0·65	1·23	1·84	0·61	1·18	1·77	0·59	$\frac{b}{8}$	$\frac{3b}{16}$	$\frac{b}{16}$
6	−	0	0·97	0·97	0	0·80	0·80	0	0·71	0·71	0	0·65	0·65	0	0·61	0·61	0	0·59	0·59	0	$\frac{b}{16}$	$\frac{b}{16}$
7	−	2·92	2·50	0·83	2·19	1·87	0·62	1·75	1·50	0·50	1·46	1·25	0·41	1·25	1·07	0·36	1·09	0·94	0·31	$\frac{7}{32a}$	$\frac{3}{16a}$	$\frac{1}{16a}$
8	−	6·25	5·83	2·22	4·69	4·37	1·67	3·75	3·50	1·33	3·12	2·92	1·11	2·70	2·50	0·95	2·34	2·19	0·83	$\frac{15}{32a}$	$\frac{7}{16a}$	$\frac{1}{6a}$
Values of b		15·5			12·8			11·3			10·4			9·83			9·43					
Values of θ		31°			39°			45°			50°			55°			58°					

(Column groups 0·075 to 0·20 are values of a.)

Chapter 22
Structural Design for Fire Resistance

22.1 INTRODUCTION

Research into the effects of fire on timber has reached a stage at which the performance of timber can be predicted with sufficient accuracy for a design method to be evolved to calculate the period of fire resistance of even completely exposed load-bearing solid timber or glulam sections. This is presented in Section 4.1 of BS 5268: Part 4: 1978 'Method of calculating fire resistance of timber members'.

Also, considerable testing has been carried out on many combinations of standard wall components (load bearing and non-load bearing), and floor components of timber joists with various top and bottom skins. Fire resistance periods for the *stability*, *integrity* and *insulation* performance levels of these can be found in Building Regulations and various trade association and company publications. A design method to predict the stability, integrity and insulation performance levels of combinations without the necessity of a fire test is presented in BS 5268: Part 4: Section 4.2.

Stability refers to the ability of an element to resist collapse. *Integrity* relates to resistance to the development of gaps, fissures or holes which allow flames or hot gases to breach the integrity of an element of construction. *Insulation* relates to the resistance to the passage of heat which would cause the unheated face of an element of construction to rise to an unacceptable level. Not all elements are required to meet all three criteria. For example, an exposed beam or column would be required to meet only the stability criterion, whereas a wall or floor designed to contain an actual fire would usually be required to meet all three criteria. A component rated as 30/30/15 indicates stability for 30 minutes, integrity for 30 minutes and insulation for 15 minutes.

The intention of fire regulations in the UK is to save life, not property. The required fire resistance period quoted in regulations for a component or element represents a length of time during which it is considered that people can escape, and is related to the position in a structure, availability of fire applicances, access, etc. Although the regulations are not designed to save property, it will be obvious from the figures given below that, if a fire on a clad or exposed timber component or element is extinguished quickly, there is a chance that the structure will continue to be serviceable and may be able to be repaired insitu.

The fire resistance stability period for an element is basically the time during which it is required to support the design load while subjected to fire without failure taking place. In the case of beams, it is sensible for a deflection limit to be set and BS 5268: Part 4: Section 4.1 gives this as span/30.

22.2 PROPERTIES OF TIMBER IN FIRE

As defined by BS 476: Part 4, timber is 'combustible' but, as defined by BS 476: Part 5, timber is 'not easily ignitable' in the sizes normally used for building purposes.

Timber having a density in excess of 400 kg/m^3 (i.e. most structural softwoods and hardwoods) falls naturally into the class 3 'surface spread of flame' category of BS 476: Part 7 and can be upgraded by specialist surface or pressure treatments to class 1 or class 0.

Without a source of flame or prolonged pre-heating, timber will not ignite spontaneously until a temperature of around 450 °C to 500 °C is reached. It is worth realizing that this is around the temperature at which steel and aluminium lose most of their useful strength, hence the photographs one sees of timber members still standing after a fire and supporting tangled steel secondary members.

Once timber has ignited, as it burns it builds up a layer of charcoal on the surface and this, being a good thermal insulant, protects the timber immediately beneath the charred layer. Figure 22.1 shows a typical temperature plot through the section of a timber member which has been exposed in a furnace to a temperature of some 930 °C. The temperature immediately under the charcoal is less than 200 °C and the temperature in the centre of the member is less than 90 °C. Timber has a very low coefficient of thermal expansion and the strength properties of the uncharred part of the member are virtually unaffected, even during the fire period. After the fire is extinguished the properties of the residual section can be considered not to have been affected by the fire. Obviously this gives the possibility of developing a design method based on the strength of the residual section.

Another important positive property of the performance of timber in fire is that the charring rate when exposed to a standard fire test is sufficiently constant and predictable for values to be quoted. For the softwoods given in BS 5268-2 (except western red cedar) the notional charring rates are taken as 20 mm in 30 minutes and 40 mm in 60 minutes, with linear interpolation and extrapolation permitted for time periods between 15 and 90 minutes.

The 30 minute value for western red cedar is 25 mm.

The 30 minute value for oak, utile, keruing, teak, greenheart and jarrah is 15 mm.

The charring rates for glulam may be taken as the same as solid timber providing the adhesive used in assembly is resorcinol–formaldehyde, phenol–formaldehyde, phenol/resorcinol–formaldehyde, urea–formaldehyde, or urea/melamine–formaldehyde.

A column exposed to the fire on all four faces (including one which abuts or forms part of a wall which does not have the required fire resistance) should be assumed to char on all four faces at 1.25 times the rates given above. However, when a column abuts a construction which has a fire resistance at least as high as the required fire resistance of the column, it may be assumed that this prevents charring of the member at that point or surface (see Fig. 22.2).

The corners of a rectangular section become rounded when subjected to fire. The radius has been found to approximate to the depth of charring (see Fig. 22.3) but, for exposure periods of 30 minutes or less, if the least dimension of a rectangular *residual* section is not less than 50 mm, the rounding is considered insignificant and can be disregarded.

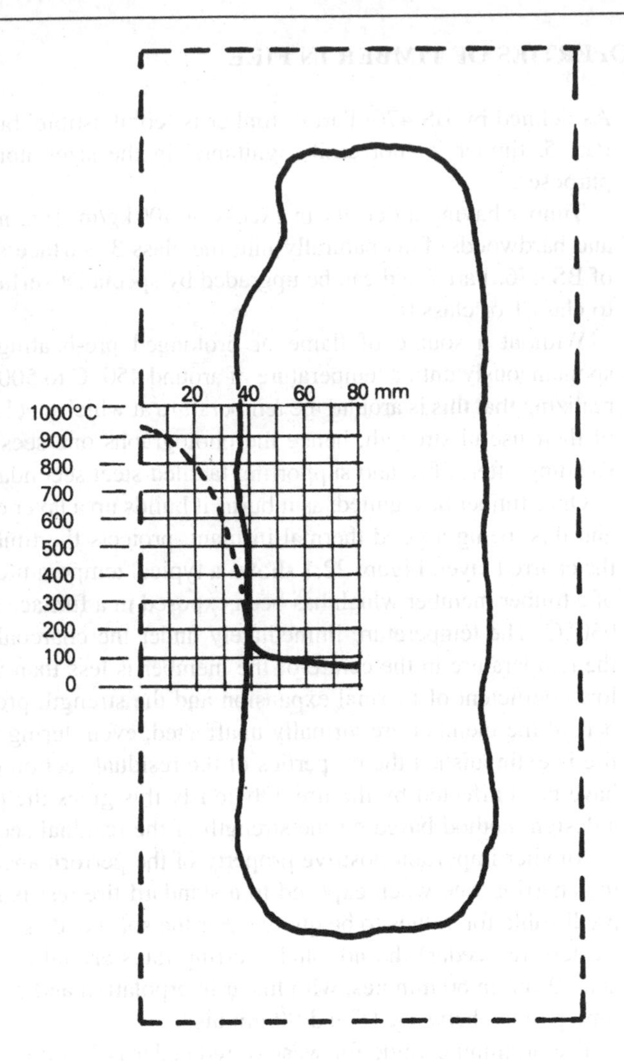

Fig. 22.1

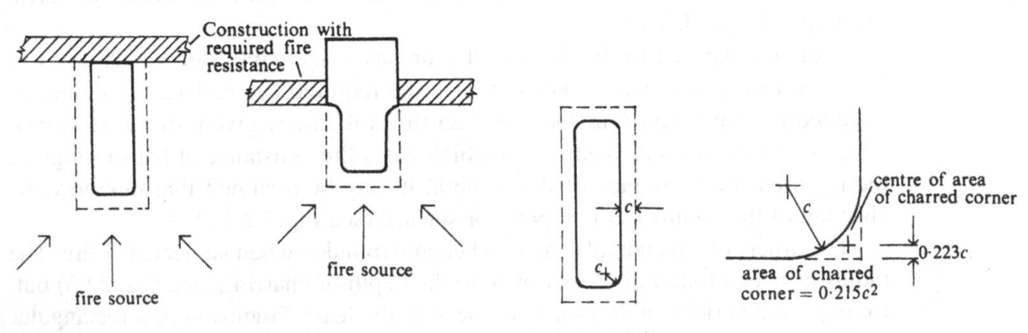

Fig. 22.2

Fig. 22.3

Metal fasteners within the charring line will conduct heat into the timber. They must either be fully protected, or charring of each individual piece of timber must be considered to act on all faces, even if two timber surfaces are butting.

22.3 DESIGN METHOD FOR THE RESIDUAL SECTION

Knowledge of the charring rate and the fire stresses enables a design method to be evolved to check whether the chosen section will withstand the design load for the period of required fire resistance. In the case of exposed sections, this is often referred to as the 'sacrificial timber' design method or the 'residual section' design method. The designer is referred to BS 5268: Part 4: Section 4.1 for detailed rules for this method.

In this method the load-bearing capacity of a flexural member should be calculated using the residual section and permissible stresses equal to the permissible long-term dry stresses times 2.25 when the minimum initial breadth (i.e. width not thickness) is 70 mm or more, and times 2.00 (instead of 2.25) when less than 70 mm.

The designer must judge the degree of load sharing which will remain at the end of the charring period in choosing values of K_8, E, etc., to use in calculations. The deflection should be limited to span/30. The notional charring rate is as detailed in section 22.2.

In checking a column, no restraint in direction (as against positional restraint) should be assumed in determining the effective length, unless this can be assured at the end of the charring period. Slenderness ratio L_e/i can be up to 250. The permissible compression stress parallel to the grain can be taken as 2.00 times the grade long-term dry stress modified for slenderness. The charring rate is as detailed in section 22.2.

The residual section of a tension member should be determined using a charring rate 1.25 times the notional rates given in section 22.2, and the permissible stress may be taken as 2.00 times the permissible long-term dry stress (including K_{14} based on the reduced greater dimension).

Where there is combined bending and compression (or combined bending and tension) one presumes that the size of the residual section should be calculated using a charring rate of 1.25 times the notional rate. The permissible stresses are calculated as above and combined by the usual interaction formula.

22.4 STRESS GRADE

As the outside of the section chars away, the knot area ratio changes, and this affects the stress grade to a certain extent. However, in the context of the accuracy of the residual section method, it is suggested that the designer assumes that the stress grade remains the same. Where there has been a shake, charring will tend to increase the stress grade. Charring will also usually reduce knot area ratios.

22.5 PLY WEB BEAMS

The residual section method is generally not appropriate in checking ply web beams when a fire resistance is required. When using ply web components for which a fire resistance is required, it is normal to protect them by a ceiling or cladding. In a special case a fire test might be used to validate a beam design which is to be exposed.

22.6 CONNECTIONS

The design method outlined in section 22.3 is for solid timber or glulam sections, and special consideration must be given to the resistance of connections, particularly if metal is included in the connection, and especially if the connection is required to develop a fixing moment. The effect of fire on a frame may be to induce failure by changing moment connections into hinges.

It may be possible to protect connections locally, either by adding material or ensuring that all the connection unit is behind the charring line. If even a part of a steel connection is within the charring area, it will tend to conduct heat into the section. The engineer may justify a particular connection by design on the net section or by test under conditions of fire.

BS 5268: Part 2: Section 4.1 gives certain 'deemed to satisfy' clauses for metal hangers protected by plasterboard.

22.7 TESTING FOR FIRE RESISTANCE

When the residual section method is not appropriate, or when desired, one can test for fire resistance. The test procedure for the UK is detailed in BS 476: Part 8. It is worth realizing that the fire test described in BS 476: Part 8 is arbitrary and does not necessarily reflect exactly what will occur in a real fire, although it has proved to be a useful yardstick for many years.

22.8 PROPRIETARY TREATMENTS FOR SURFACE SPREAD OF FLAME

There are several proprietary chemical fluid methods of treating timber to up-grade the surface spread of flame classification (see BS 476: Part 7) from class 3 to class 1 (even to class 0). Current formulations do not increase the fire resistance to any measurable extent.

22.9 CHECK ON THE FIRE RESISTANCE OF A GLULAM BEAM

Check whether an exposed 90 × 180 mm glulam beam (Fig. 22.4) of strength class C24 subject to a design bending moment of 3.0 kN m (4 kN UDL over 6 m), has a fire resistance of 30 minutes. The beam is not part of a load-sharing system. Full

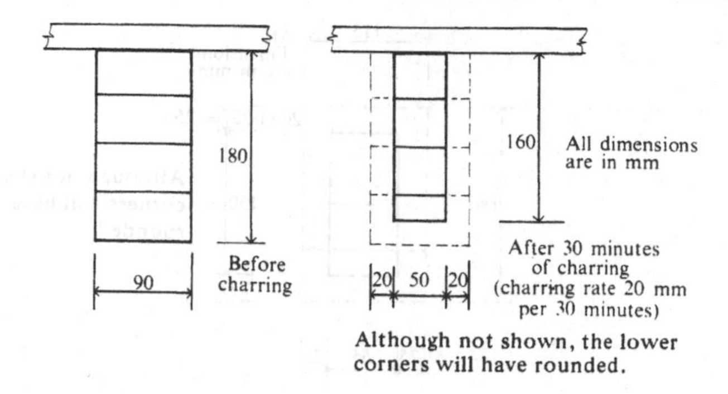

Fig. 22.4

lateral restraint may be assumed to be given by the decking, which also has a fire resistance of at least 30 minutes. Location is service class 2.

After 30 minutes, the residual width is 50 mm, therefore rounding can be disregarded (section 22.2).

$$Z \text{ of depleted section} = \frac{50 \times 160^2}{6} = 0.213 \times 10^6 \text{ mm}^3$$

Assume four laminations, even though one is badly charred:

$$\text{Permissible 'fire' bending stress} = 2.25 \times \sigma_{m,\text{grade}} \times K_{15} = 2.25 \times 7.5 \times 1.26 = 21.3 \text{ N/mm}^2$$

$$\text{Moment capacity} = 21.3 \times 0.213 \times 10^6 = 4.53 \times 10^6 \text{ N mm} = 4.53 \text{ kN m} > \text{applied moment of } 3.0 \text{ kN m}$$

Check that deflection is less than span/30 = 6000/30 = 200 mm

$$I = \frac{50 \times 160^3}{12} = 17.1 \times 10^6 \text{ mm}^4$$

$$E = E_{\text{mean}} \times K_{20} = 10800 \times 1.07 = 11556 \text{ N/mm}^2$$

$$EI = 198 \text{ kN m}^2$$

$$\text{Bending deflection} = \frac{5 \times F \times L^3}{384 \times EI} = \frac{5 \times 4 \times 6^3}{384 \times 198} = 0.0568 \text{ m} = 57 \text{ mm} < 200 \text{ mm Acceptable.}$$

22.10 CHECK ON THE FIRE RESISTANCE OF A GLULAM COLUMN

Check whether a C24 glulam column 115 × 180 mm (Fig. 22.5) exposed on all four sides has a fire resistance of 30 minutes.

Effective length = 2.6 m (before and after charring)
Axial load = 16 kN m
Bending moment = 2.0 kN m about *x-x* axis

The charring rate is 1.25 times 20 mm in 30 minutes. The size of the residual section is shown in Fig. 22.5. Location is service class 2.

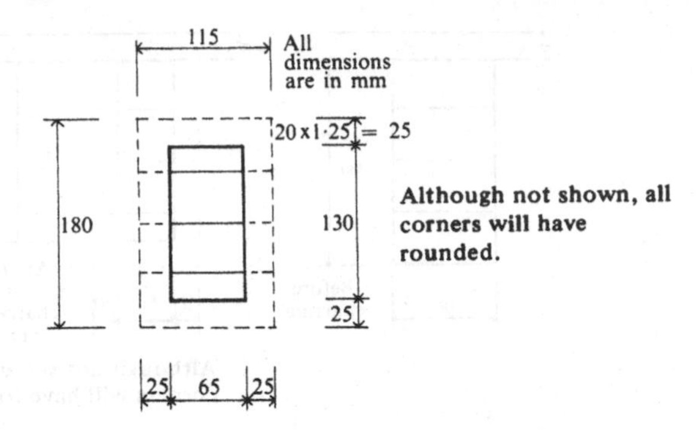

Fig. 22.5 Residual section after 30 minutes.

Because the lesser dimension of the residual section is more than 50 mm, rounding of corners can be disregarded.

The residual section is checked as a normal column but with the permissible compression stress increased by 2.00, and with the effective length determined assuming no directional restraint at ends.

Area of depleted section = 65 × 130 = 8450 mm^2

$$\text{Section modulus of depleted section} = \frac{65 \times 130^3}{6} = 0.183 \times 10^6 \text{ mm}^3$$

$$\sigma_{c,a} = \frac{16000}{8450} = 1.89 \text{ N/mm}^2$$

$$\sigma_{m,a} = \frac{2.0 \times 10^6}{0.183 \times 10^6} = 10.9 \text{ N/mm}$$

Refer to section 7.9.2 for depth factor K_7.

$$\sigma_{m,adm} = 2.0 \times \sigma_{m,grade} \times \underset{\text{(duration of load)}}{K_3} \times \underset{\text{(depth)}}{K_7} \times \underset{\text{(number of laminations)}}{K_{15}}$$

$$= 2.0 \times 7.5 \times \underset{\text{(long term)}}{1.0} \times 1.10 \times 1.26$$

$$= 20.8 \text{ N/mm}^2 > 10.9 \text{ N/mm}^2$$

$$\lambda_{max} = \frac{2600\sqrt{12}}{65} = 138$$

From Table 17.1, $K_{12} = 0.321$

$$\sigma_{c,adm} = 2.0 \times \sigma_{c,grade} \times \underset{\text{(duration of load)}}{K_3} \times K_{12} \times \underset{\text{(number of laminations)}}{K_{17}}$$

$$= 2.0 \times 7.9 \times 1.0 \times 0.321 \times 1.04$$
$$= 5.27 \text{ N/mm}^2 > 1.89 \text{ N/mm}^2$$

Check combined bending and compression:

$$\text{Euler stress} = \sigma_{eu} = \frac{\pi^2 E}{\lambda^2} = \frac{9.87 \times 11\,556}{138^2} = 6.0\ \text{N/mm}^2$$

$$\text{Euler coefficient} = K_{eu} = 1 - \frac{(1.5 \times \sigma_{c,a} \times K_{12})}{\sigma_{eu}}$$

$$= 1 - \frac{(1.5 \times 1.89 \times 0.321)}{6.0} = 1 - 0.15 = 0.85$$

$$\text{Combined ratio} = \frac{\sigma_{m,a}}{\sigma_{m,adm} \times K_{eu}} + \frac{\sigma_{c,a}}{\sigma_{c,adm}}$$

$$= \frac{10.9}{20.8 \times 0.85} + \frac{1.89}{5.27}$$

$$= 0.62 + 0.36 = 0.98 < 1.0 \quad \text{Satisfactory.}$$

The section provides 30 minutes fire resistance.

Chapter 23
Considerations of Overall Stability

23.1 GENERAL DISCUSSION

It is not sufficient to design individual components for what one may regard as their isolated loading conditions without considering how they are affected by, and how they affect the overall stability of, the building or frame in which they occur. For example, a horizontal beam which is also part of the horizontal bracing of a building, may have to carry axial compression or tension due to side wind on the building; a column which is part of a vertical bracing frame may have to carry tension or additional compression due to wind on the end of the building; the bottom tie of a truss, or the beam of a portal frame may have to take compression or additional tension from the action of side wind.

The usual practice is to design individual components for the gravity loading and subsequently to check those members that are called upon to carry additional loads from stability considerations. The stability loads are usually of short- or very short-term duration from wind, and consequently the component as designed for long- or medium-term loading is usually adequate to carry the additional stability loading, if restrained laterally.

It is dangerous to forget considerations of stability. The classic error in disregarding stability calculations is the 'four-pin frame'. This can occur if a simply supported beam is supported at each end by vertical columns which have no fixity at either end, in a building where the roof is neither provided with horizontal bracing nor acts as a horizontal diaphragm (Fig. 23.1). On an architect's drawing this looks sound, but failure occurs as sketched on the right of Fig. 23.1 unless the roof is braced or is a horizontal membrane, or unless fixity is provided at least at one joint.

Before proceeding with the calculations of any building, the designer must decide whether the design is to be based on the building being permitted to sway or being restrained from swaying. This question is absolutely fundamental to any building design, yet so often is not stated (although often it is understood).

23.2 NO SWAY CONDITION

When a designer has a choice it is normally more economical to design a building not to sway. This requires that the roof plane is either braced or acts as a horizontal diaphragm or membrane, and that the ends of the bracing or diaphragm are connected through stiffened end walls and/or cross-walls to the ground (Fig. 23.2).

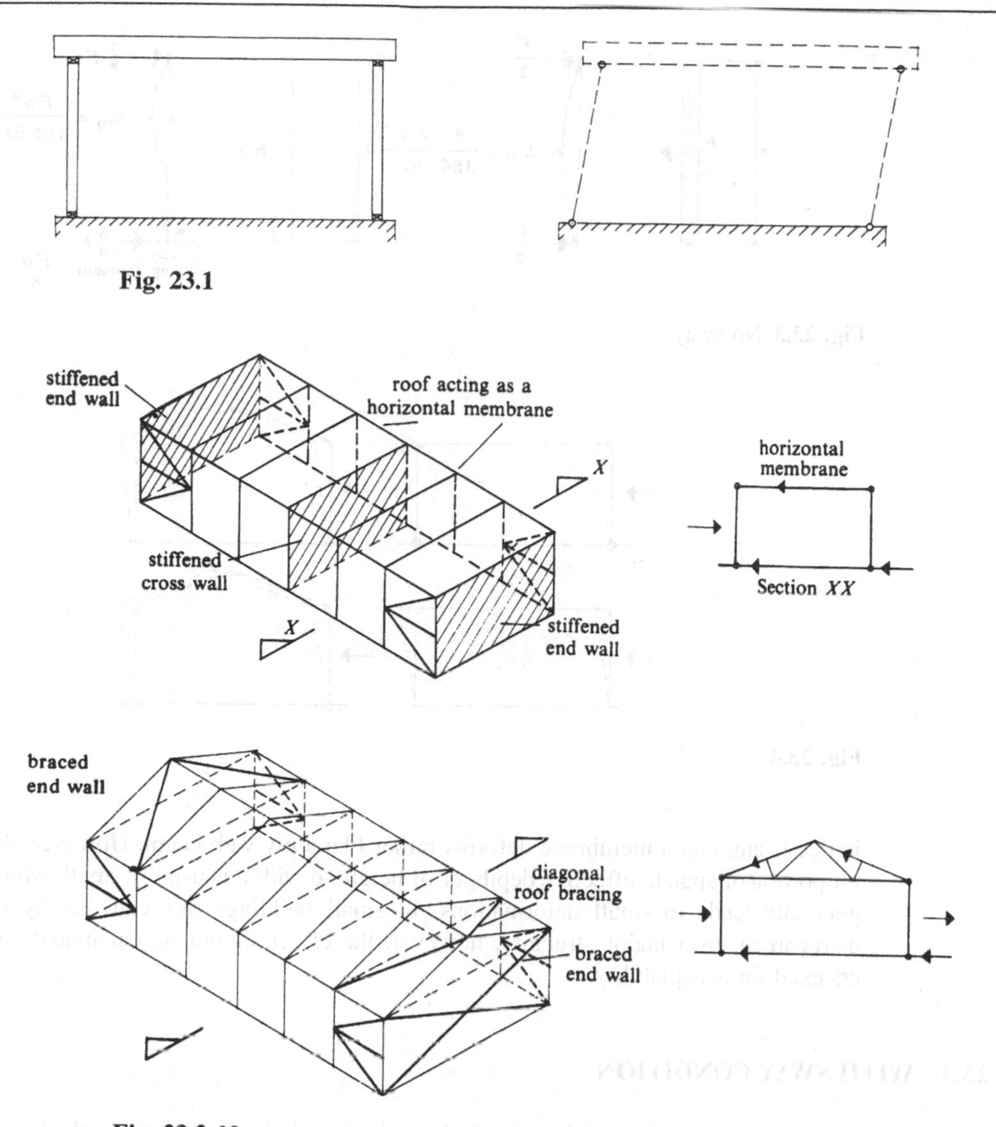

Fig. 23.1

Fig. 23.2 No cross sway.

With the type of construction sketched in Fig. 23.2, the columns can be hinged top and bottom because the overall stability comes from the bracing. If the columns are rather tall, the decision may be taken to fix the base to reduce deflection (Fig. 23.3).

Whether or not a building is designed to permit local sway, it is usually restrained from sway in the longitudinal direction by roof and side-wall bracing, diaphragms, or stiffened panels.

In deciding the basic design of a building, it is necessary to distinguish between 'sway' and 'no sway', but the designer should realize that even when a building can be designed for 'no sway' due to there being adequate bracing or membrane effect, the bracing or membrane deforms in itself. A bracing frame deforms like a

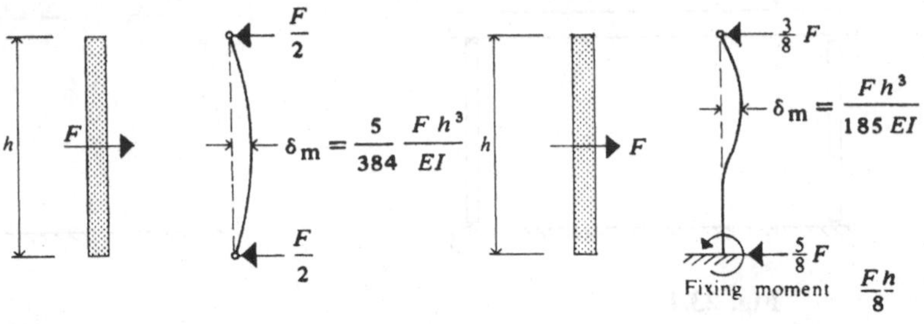

$$\delta_m = \frac{5}{384}\frac{Fh^3}{EI}$$

$$\delta_m = \frac{Fh^3}{185\,EI}$$

Fixing moment $\frac{Fh}{8}$

Fig. 23.3 No sway.

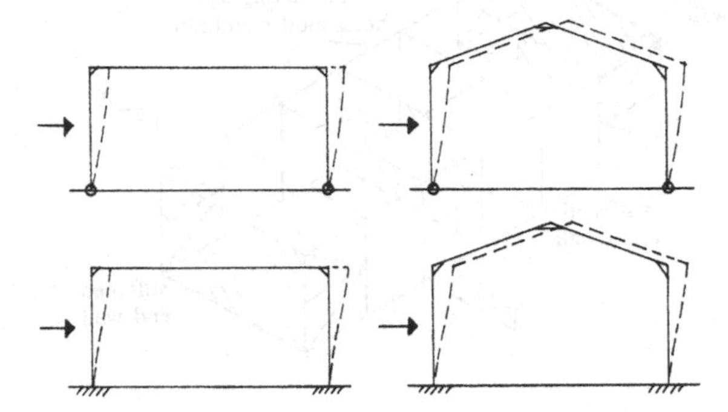

Fig. 23.4

lattice beam, and a membrane deforms rather like a ply web beam. However, the proportion of span to effective depth, or of height to width, is usually small, which generally leads to small deformations. In small buildings this can usually be disregarded. In a major structure, however, the effect should be calculated and checked for acceptability.

23.3 WITH SWAY CONDITION

With several types of building, the designer has no choice but to permit the building to sway and to design accordingly, by providing the necessary fixing moments and sufficient stiffness to limit sway. For example, a tower or mast can not be prevented from swaying (other than by guy ropes). A long narrow building without cross-walls, or split by expansion bays, must be designed for sway.

When designing a building for sway, it is quite usual to provide two-pin or fully fixed portal frames. The types of portal shown in Fig. 23.4 can be analysed by the Kleinlogel formulae given in the *Steel Designers' Manual* and with various computer software packages.

The frame sketched in Fig. 23.5 will sway as indicated under the action of side loading. Whether or not the columns have equal stiffness and whether or not the loading on each side is equal, the columns will sway to the same degree at the eaves. This sway is the algebraic sum of the deflection of a cantilever affected by the side loading and by the horizontal thrust at the base. A point load H at the end

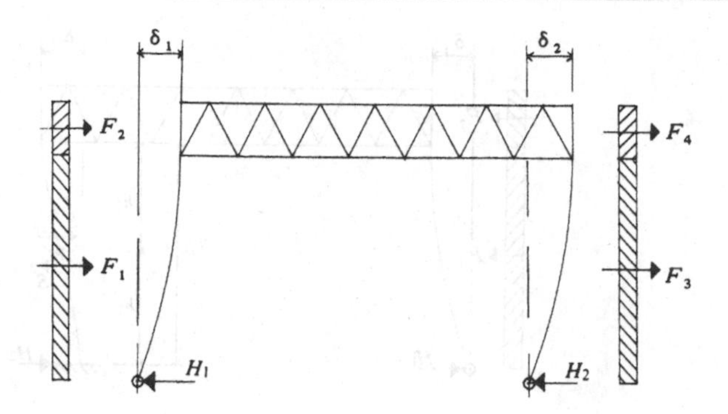

Fig. 23.5

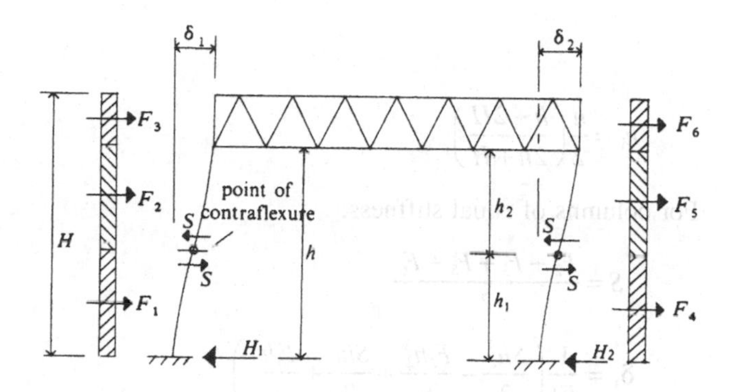

Fig. 23.6

of a cantilever causes a bending deflection at the end of $HL^3/3EI$. Similarly, a UDL of total F on a cantilever causes a deflection of $FL^3/8EI$.

$$H_2 = F_1 + F_2 + F_3 + F_4 - H_1$$

$$\delta_1 = \frac{H_1h^3}{3EI} - \frac{F_1h^3}{8EI}$$

$$\delta_2 = \frac{H_2h^3}{3EI} - \frac{F_3h^3}{8EI}$$

$$\delta_1 = \delta_2$$

where h = height to the underside of the cross-beam.

The force in the lower connection to the beam is found by taking moments about the top where the moment is zero. The force in the top connection is found by equating horizontal forces.

The frame sketched in Fig. 23.6 will sway as indicated. Even though the horizontal loadings on each side may differ, the points of contraflexure may be assumed to occur at the same level. The horizontal shear at the points of contraflexure will be the sum of the horizontal loadings above the level of the points of contraflexure shared to the columns in proportion to their stiffness (i.e. shared equally if columns are of equal stiffness).

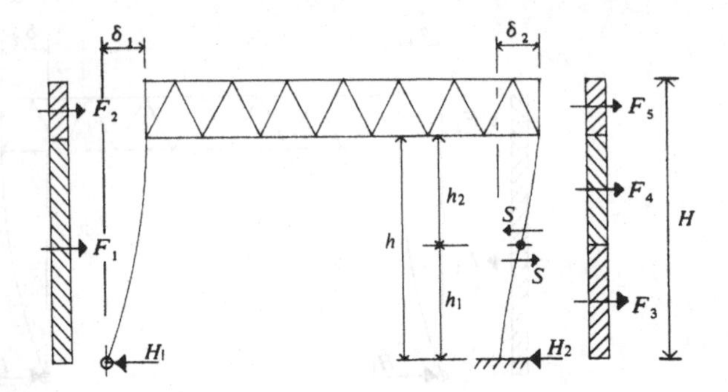

Fig. 23.7

$$h_1 = \frac{h}{2}\left(\frac{h+2H}{2h+H}\right)$$

For columns of equal stiffness:

$$S = \frac{F_2 + F_3 + F_5 + F_6}{2}$$

$$\delta_1 = \frac{1}{EI}\left(\frac{Sh_2^3}{3} - \frac{F_2h_2^3}{8} + \frac{Sh_1^3}{3} + \frac{F_1h_1^3}{8}\right)$$

$$\delta_2 = \frac{1}{EI}\left(\frac{Sh_2^3}{3} - \frac{F_5h_2^3}{8} + \frac{Sh_1^3}{3} + \frac{F_4h_1^3}{8}\right)$$

$$\delta_1 = \delta_2$$

$$H_1 = S - F_1$$

$$H_2 = S + F_4$$

The frame sketched in Fig. 23.7 will sway as indicated. h_1 is calculated as for Fig. 23.6.

For columns of equal stiffness:

$$S = \frac{F_2 + F_3 + F_5 + F_6}{2}$$

$$\delta_1 = \frac{H_1h^3}{3EI} - \frac{F_1h^3}{8EI}$$

$$\delta_2 = \frac{Sh_2^3}{3EI} - \frac{F_4h_2^3}{8EI} + \frac{Sh_1^3}{3EI} + \frac{F_3h_1^3}{8EI}$$

$$\delta_1 = \delta_2$$

$$H_2 = S + F_3$$

$$H_1 + H_2 = F_1 + F_2 + F_3 + F_4 + F_5$$

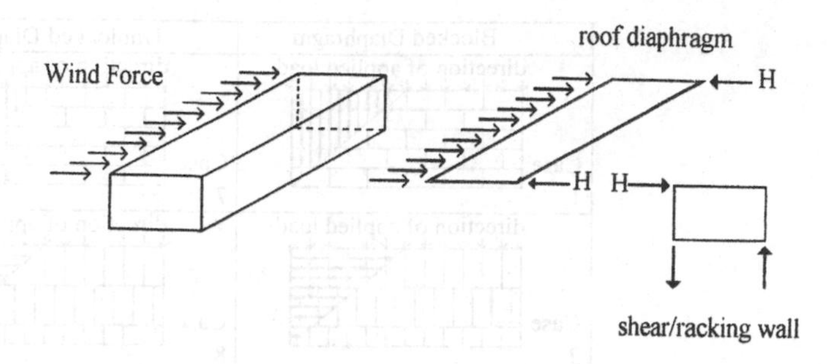

Fig. 23.8

23.4 DIAPHRAGM ACTION

As an alternative to the traditional braced frame approach to achieve stability, it is usual in timber design to take advantage of the inherent stiffness of the various claddings used to deck the floors and sheath the walls.

Figure 23.8 shows the essential elements for stabilizing a building using diaphragm action. The roof acts as a deep beam to transfer wind forces through the roof to appropriate shear walls and then to the foundations. Although the following considerations refer specifically to plywood diaphragms other wood-based materials such as chipboard may provide an adequate solution.

23.5 HORIZONTAL DIAPHRAGMS

23.5.1 General discussion

For large structures the adequacy of the diaphragm action should be checked. For smaller buildings, such as domestic dwellings, some diaphragm action will be inherent in the construction.

According to clause 6.5 of BS 5268-6.2 'Code of practice for timber framed walls – Buildings other than dwellings not exceeding four storeys':

Some horizontal diaphragms have sufficient strength to transmit lateral forces to supporting walls without the need for a further design consideration or calculations, provided that in the case of intermediate floors or flat roofs, a wood-based deck, sub-deck or lining is fixed directly to the upper or lower face of the joists, and in the case of pitched roofs, a plasterboard ceiling in combination with roof bracing as recommended in BS 5256-3.

Such diaphragms should have no horizontal dimension exceeding 12 m between support walls and a length to width ratio, l/w, not greater than 2, where l is the greatest horizontal dimension of the diaphragm and w is the smallest horizontal dimension.

BS 5268-2 gives no futher recommendations for diaphragms and it will be left to the designer to investigate the adequacy of diaphragm action outside of the scope of BS 5268-6.

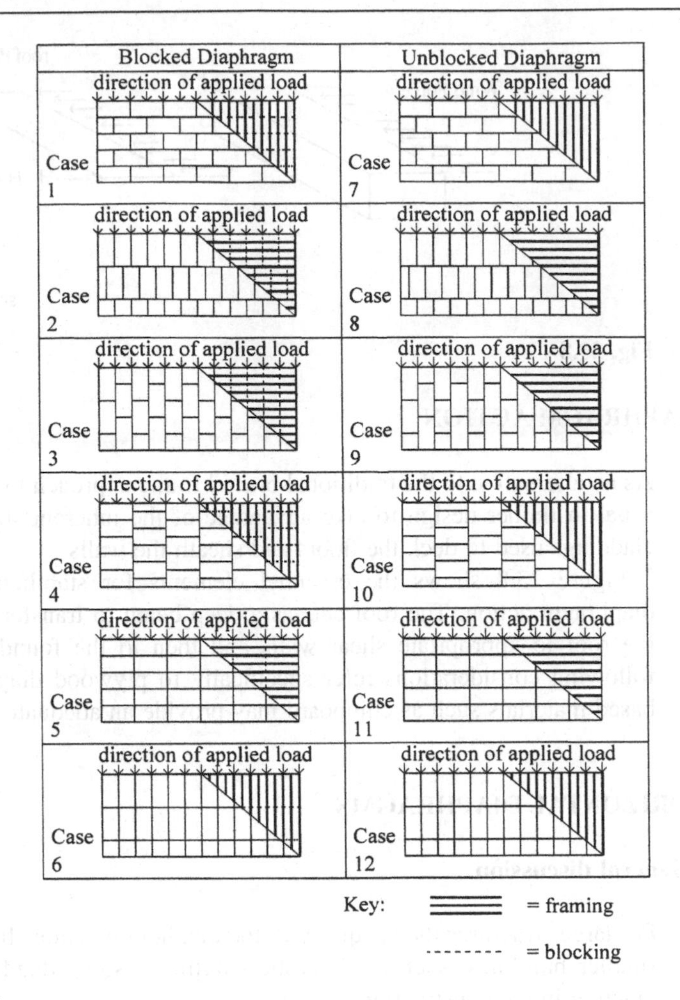

Fig. 23.9

In the absence of design guidance from BS 5268-2, the following recommendations are based on American practice with some adjustments appropriate to UK practice.

The span/width ratio for a horizontal diaphragm should not exceed 4:1. This limit is set in order to keep the deflection of the diaphragm within acceptable limits. Deflection is not normally critical and calculations are not usually required.

Horizontal diaphragms fall into two categories: (a) *blocked* and (b) *unblocked*. Six potential panel layouts are shown in Fig. 23.9.

- A 'blocked' diaphragm (cases 1–6 in Fig. 23.9) is one in which all four edges are adequately supported either by roof beams or lateral blocking pieces. Blocked diaphragms will be stronger than unblocked diaphragms because all four edges are nailed and the provision of blocking splices the sheets together and improves the continuity of the system.

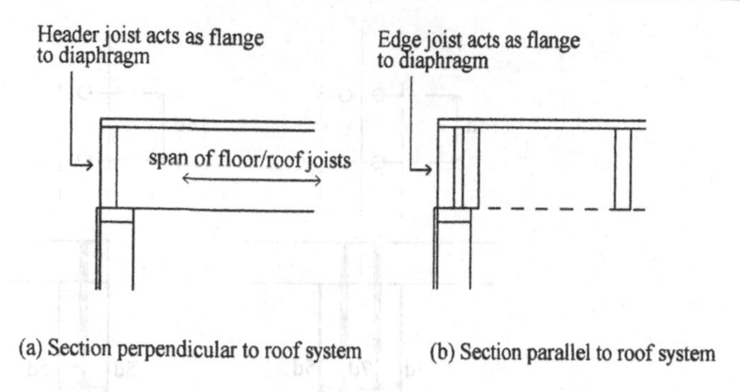

Fig. 23.10

- An 'unblocked' diaphragm (cases 7–12 in Fig. 23.9) is one in which only those edges which are supported directly on roof beams provide a shear transfer between panels. Panel edges which have tongue and groove joints are classed as unblocked because they have no support to receive nailing to transfer shear forces.

The shear resistance provided by the roof/floor sheathing depends on:

- plywoods sheathing species/grade
- sheathing layout
- width of supporting beam/joist system
- provision of lateral blocking
- nailing size and spacing.

The diaphragm may be considered to act as a deep beam to resist lateral forces transferred into suitable vertical shear walls (racking walls) to establish the overall stability of the structure. It follows that the diaphragm should have a top and bottom chord member, acting as a beam flange, to resist induced bending moments with adequate nailing to the plywood which acts as the web of the beam to resist shear forces. This is normally achieved by a header joist, sometimes referred to as a rim beam or the edge members of the support beam system, as illustrated in Fig. 23.10.

The bending moment and shear forces in the system are calculated as for a simple beam. The bending moment, M, is resisted by a couple, consisting of a tension force, T, and a compression force, C, spaced apart by the depth of the diaphragm, b.

$$T = C = M/b$$

It was shown in section 4.11 that for an ordinary beam the shear distribution over the depth of the beam is parabolic. For diaphragm design the shear is assumed to be uniformly distributed over the full depth.

$$\text{Applied average panel shear stress (N/mm}^2) = \frac{V}{bt}$$

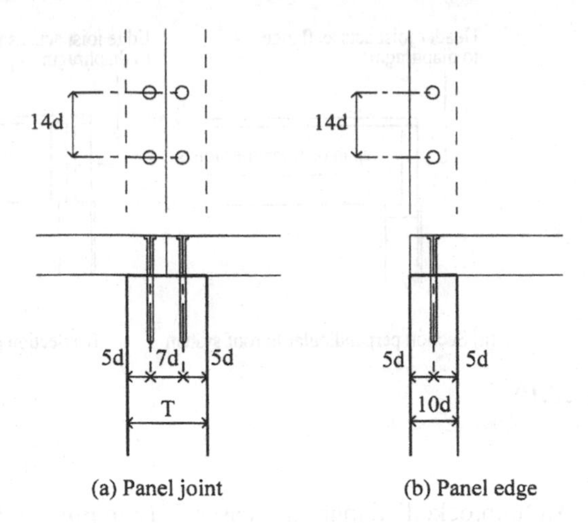

Fig. 23.11

Table 23.1

Nail diameter (mm)	Minimum thickness T (mm)	Minimum spacing S (mm)
2.65	45	38
3	51	42
3.35	57	47

where V = applied shear (N)
b = depth of diaphragm (mm)
t = thickness of plywood (mm).

Allowable panel shear stress for the plywood = $\tau_{grade} \times K_3$

where τ_{grade} = grade panel shear stress (N/mm^2) (see Tables 40–56 of BS 5268-2)
K_3 = modification factor for duration of load.

When the diaphragm is resisting a wind load, which is the usual condition, and assuming the diaphragm does not exceed 50 m in length or depth, the loading may be regarded as of very short-term duration with $K_3 = 1.5$.

The allowable shear in the diaphragm will frequently be limited by the lateral nail fastener load at the plywood sheet boundaries. Figure 23.11 indicates the required minimum spacing and edge distance for nails into support joists, blocking and headers. Table 23.1 shows the minimum widths of timbers to accommodate the most acceptable nails.

Table 23.2 lists the nail capacity for various ply thicknesses assuming plywood to be of grade I or better. Grade II plywood is unlikely to be used because of higher cost. Timber is assumed to be strength class C16 or better. Minimum boundary nailing should be 150 mm centres with intermediate members nailed at 300 mm centres.

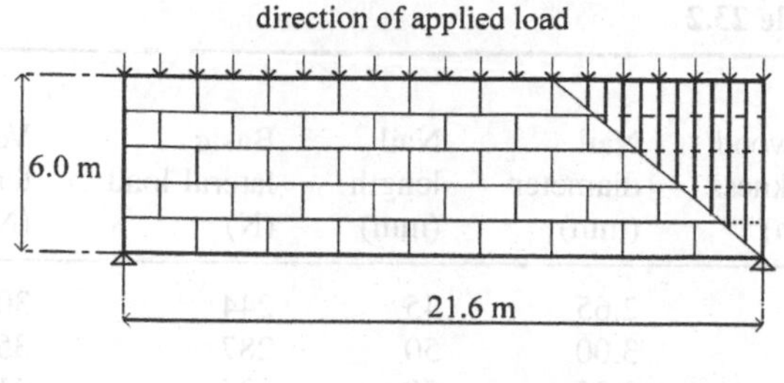

Fig. 23.12

23.5.2 Example of checking a horizontal diaphragm

Check the adequacy of the roof diaphragm sketched in Fig. 23.9 as (a) a blocked diaphragm and (b) an unblocked diaphragm. The roof is 21.6 m long and is supported by 406 mm deep I section joists, spanning 6.0 m and spaced at 600 mm centres. The flanges to the I section are 45 mm wide. The roof sheathing is 12.5 mm Canadian softwood plywood (12.0 mm minimum thickness) and is nailed with 2.65 mm diameter × 45 mm long nails at 150 mm centres at boundary and supported edges and at 300 mm centres elsewhere.

Horizontal loading at roof level from wind forces is 0.75 kN per metre run of building.

$$\text{Total wind load} = 0.75 \times 21.6 = 16.2\,\text{kN}$$

$$\text{End reaction} = \text{maximum shear} = V = 16.2/2 = 8.1\,\text{kN}$$

(a) ***Blocked diaphragm*** (case 1 in Fig. 23.9)

Panel shear stress

$$\text{Average shear stress in plywood} = \frac{V}{bt} = \frac{8100}{6000 \times 12} = 0.112\ \text{N/mm}^2$$

From Table 44 of BS 5268-2,

$$\text{grade panel shear} = \tau_{\text{grade}} = 1.72\ \text{N/mm}^2$$

$$\text{Allowable shear stress in plywood} = \tau_{\text{grade}} \times K_3 = 1.72 \times 1.5 = 2.58\ \text{N/mm}^2$$

It is seen that panel shear stress is not critical.

Boundary and edge nailing

$$\text{Applied lateral nail shear force} = \frac{V}{b} = \frac{8100}{6000} = 1.35\ \text{N/mm}$$

From Table 23.2, 2.65 mm diameter × 45 long nails at 150 mm centres provide an allowable load of 2.03 N/mm, therefore proposed nailing is adequate.

Table 23.2

Plywood thickness (mm)	Nail diameter (mm)	Nail length (mm)	Basic lateral load (N)	Very short-term load* (N)	Allowable boundary/panel edge load† (N/mm)
12	2.65	45	244	305	2.03
	3.00	50	287	358	2.38
	3.35	50	334	417	2.78
15	2.65	50	271	338	2.25
	3.00	65	314	392	2.61
	3.35	65	360	450	3.00
18	2.65	50	283	353	2.35
	3.00	65	245	431	2.87
	3.35	65	391	488	3.25
21	2.65	50	288	360	2.40
	3.00	65	360	450	3.00
	3.35	65	426	532	3.54

* The basic lateral load may be increased by modification factor K_{48} which has a value of 1.25 for very short-term load duration.

† The allowable boundary/panel edge load is for nailing spaced at 150 mm centres and is applicable to all 'blocked' diaphragm (cases 1–6 in Fig. 23.9). This load may be increased by reducing the spacing to not less than the minimum value given in Table 23.1.
For 'unblocked' diaphragms adopt 67% load for case 7, Fig. 23.9 and 50% load for cases 8–12, Fig. 23.9. (These load reductions are based on test and are not subject to a mathematical proof.)
Applied boundary load = V/b
where V = applied shear (N) and b = depth of diaphragm (mm).

Moment

$$\text{Bending moment} = \frac{16.2 \times 21.6}{8} = 43.7\,\text{kNm}$$

$$\text{Force in header/rim beam} = C = T = \frac{M}{b} = \frac{43.7}{6} = 7.28\,\text{kN}$$

Provide 45 × 406 Versa-Lam LVL rim beam (see section 10.3)

$$\text{Axial stress in rim beam} = \frac{\text{axial load}}{\text{area}} = \frac{7280}{45 \times 406} = 0.40\,\text{N/mm}^2,$$

which is well within acceptable limits.

A joint is required at any break in the rim beam. This normally takes the form of a backing plate, timber or steel, with sufficient fixings either nails or bolts to develop the 7.28 kN axial force.

(b) ***Unblocked diaphragm*** (case 7 in Fig. 23.9)

Panel shear stress and moment

Panel shear stress and moment are satisfacory by inspection.

Boundary and edge nailing

Applied lateral nail shear force = 1.35 N/mm, as before.

From Table 23.2, 2.65 mm diameter × 45 long nails at 150 mm centres provide an allowable load of 2.03 N/mm which must be reduced by 67% (see note† to Table 23.2).

Allowable load at 150 mm centres = 2.03 × 0.67 = 1.36 N/mm.

Therefore, proposed nailing is adequate.

23.6 VERTICAL SHEAR WALLS

23.6.1 General discussion

With timber construction it is very common to brace an external wall by cladding one face of vertical studs (spaced at not more than 610 mm centres) with a primary board material of category 1 material such as plywood, medium board, chipboard, tempered hardboard or OSB/3 or a category 2 material, i.e. bitumen-impregnated insulating board. The racking capacity may be complemented by a secondary board such as plasterboard on the inner stud face. The primary and secondary board materials are combined in racking resistance to act as a vertical diaphragm.

In addition to the racking and overturning effects applied to the racking wall, there is horizontal shear due to wind on the face of the panel which must be transmitted to the base (and top) usually by nails or ballistic nail fixings.

In considering horizontal loads acting at the base of a stud wall the designer may consider the restraining effect of friction between the stud wall/damp-proof course/base (providing of course that there is no residual uplift for the design case being considered). Tests have shown the coefficient of friction to be as high as 0.4 and a figure of 0.3 is frequently adopted. However, even if the designer decides to take advantage of friction, sufficient locating fixings will be used.

Narrow panels or panels with large openings may have to be disregarded in considering racking strength. A certain amount of engineering judgement is required.

23.6.2 Design considerations

The design of timber studded shear walls should comply with either:
BS 5268-6.1 'Dwelling not exceeding four storeys' or
BS 5268-6.2 'Buildings other than dwellings not exceeding four storeys'.

BS 5268-6.2 limits it scope to:

(1) buildings not more than four storeys high
(2) building overall height not more than 15.0 m

(3) for single-storey buildings a maximum panel height of 6.2 m
(4) for buildings more than one storey, a maximum panel height of 4.8 m

The prime considerations applicable to both standards are:

- Shielding effect of any masonry cladding which may relieve the timber frame of applied load.
- Basic racking capacity according to choice of primary and secondary board materials.
- Enhancing the basic racking capacity by increasing nail size and/or density.
- Enhancing the basic racking capacity by increasing the board thickness.
- Enhancing the basic racking capacity due to any net vertical load on the panel. The tabulated basic racking capacities assume zero net vertical load: If there is a net uplift (due to wind) the tabulated values still apply but provision must be made for vertical anchorage of the panels to the sub-structure.
- Enhancing the basic racking capacity by considering the racking contribution from any adjacent masonry veneer. This should be limited to 25% of the panel capacity.
- Dividing wall into suitable individual lengths as neccesary to suit production and to optimize racking resistance. Door openings should be regarded as structural breaks and the effective panel lengths should stop each side of a door opening.
- Enhancing the racking capacity by 10% to take account of the stiffening effect of corners, and the interaction of walls and floors through multiple fixings.
- Reducing the racking capacity attributable to the area of framed openings.

The designer is referred to the above standards for full design recommendations and the range of modification factors to be applied to both loading and racking capacity.

The following example provides a typical assessment of racking capacity for the gable wall resisting the lateral forces on the horizontal diaphragm in the previous example. It is left to the designer to verify the component parts of the calculation.

23.6.3 Determination of racking capacity

Consider the gable end wall, with door and window openings as shown in Fig. 23.13, resisting the racking force from the horizontal diaphragm in section 23.5.2. Panels have studs at 600 mm centres which are clad on the outer face with 9.5 mm plywood (category 1 material). The plywood is fixed to studs using 3.00 mm diameter nails at least 50 mm long at maximum spacing of 150 mm to perimeter of boards and 300 mm maximum internal spacing. The inner face of panels is lined with 12.5 mm plasterboard. The plasterboard is fixed to studs using 2.65 mm diameter nails at least 40 mm long at maximum spacing of 150 mm centres.

It is assumed that a suitable outer masonry veneer will contribute to the racking resistance of the wall. It is also assumed that the building, being 21.6 m long, is not a dwelling and is to be assessed in accordance with BS 5268-6.2.

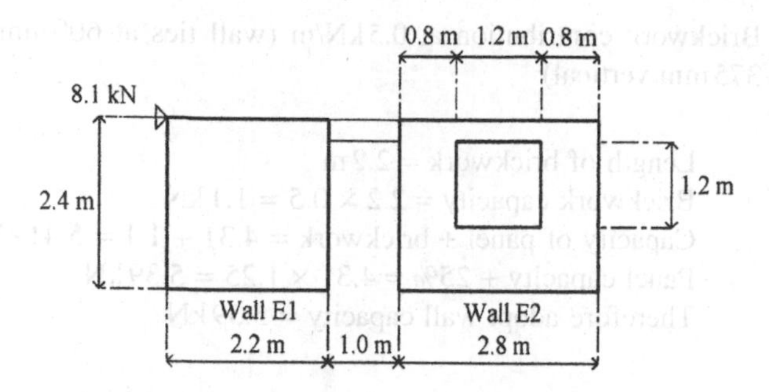

Fig. 23.13

From Table 2 of BS 5268-6.2:

$$\text{Basic racking resistance} = 1.68 + 0.28 = 1.96\,\text{kN/m}$$

Assume that the door effectively creates two separate panels E1 and E2.

$$\text{Racking resistance of wall} = R_b L K_m K_w$$

where R_b = basic racking resistance (kN/m)
L = wall length (m)
K_m = product of material modification factors, K_{201}, K_{202} and K_{203}
K_w = product of wall modification factors, K_{204}, K_{205}, K_{206} and K_{207}.

Wall E1

Wall length = $L = 2.2$ m; wall height = $h = 2.4$ m; then, $L/h = 0.917$

No framed openings and no net vertical load on panel.

Material modification factors are:

K_{201} (nail diameter) = 1.0

K_{202} (nail spacing) = 1.0

K_{203} (board thickness) = 1.0

therefore,

$$K_m = 1.0 \times 1.0 \times 1.0 = 1.0$$

Wall modification factors are:

K_{204} (shape factor) = 0.917

K_{205} (framed openings) = 1.0

K_{206} (vertical load) = 1.0

K_{207} (interaction) = 1.1

therefore,

$$K_w = 0.917 \times 1.0 \times 1.0 \times 1.1 = 1.0$$

$$\begin{aligned}\text{Panel capacity} &= \text{basic racking capacity} \times \text{length} \times K_m \times K_w \\ &= 1.96 \times 2.2 \times 1.0 \times 1.0 = 4.31\,\text{kN}\end{aligned}$$

Brickwork contribution at 0.5 kN/m (wall ties at 600 mm centres horizontal and 375 mm vertical):

Length of brickwork = 2.2 m
Brickwork capacity = $2.2 \times 0.5 = 1.1$ kN
Capacity of panel + brickwork = $4.31 + 1.1 = 5.41$ kN
Panel capacity + 25% = $4.31 \times 1.25 = 5.39$ kN
Therefore adopt wall capacity = 5.39 kN

Wall E2

Wall length = $L = 2.8$ m; wall height = $h = 2.4$ m; then, $L/h = 1.17$
Area of framed openings = $1.2 \times 1.2 = 1.44\,m^2$
Framed opening ratio = $1.44/(2.8 \times 2.4) = 0.214$

No net vertical load on panel.

Material modification factors are:

K_{201} (nail diameter) = 1.0

K_{202} (nail spacing) = 1.0

K_{203} (board thickness) = 1.0

therefore,

$K_m = 1.0 \times 1.0 \times 1.0 = 1.0$

Wall modification factors are:

K_{204} (shape factor) = 1.065

K_{205}(framed openings) = 0.521

K_{206} (vertical load) = 1.0

K_{207} (interaction) = 1.1

therefore,

$K_w = 1.065 \times 0.521 \times 1.0 \times 1.1 = 0.61$

Panel capacity = basic racking capacity × length × $K_m \times K_w$
= $1.96 \times 2.8 \times 1.0 \times 0.61 = 3.35$ kN

Brickwork contribution at 0.5 kN/m (wall ties at 600 mm centres horizontal and 375 mm vertical)

Length of brickwork = 1.6 m
Brickwork capacity = $1.6 \times 0.5 = 0.8$ kN
Capacity of panel + brickwork = $3.35 + 0.8 = 4.15$ kN
Panel capacity + 25% = $3.35 \times 1.25 = 4.18$ kN

Therefore, adopt wall capacity = 4.15 kN

Sum of panels E1 and E2 = $5.39 + 4.15 = 9.54$ kN > 8.1 kN Satisfactory.

If the racking summation is inadequate then the design must be re-assessed. Consider the following options (assuming framed opening sizes cannot be reduced).

1. Provide additional perimeter board nails in order to increase K_{202}.
2. Increase board thickness so as to increase K_{203}.
3. Look for any enhancement due to applied vertical load which is frequently ignored in a first assessment.
4. Convert any suitable internal partitions into racking walls.
5. Provide a higher category material for the primary board if category 1 material is not already specified.
6. Provide a higher category material for the secondary board. If neccesary provide a category 1 material. If the secondary board is then covered with plasterboard this layer of plasterboard should not be included in any determination of racking resistance.

Chapter 24
Preservation, Durability, Moisture Content

24.1 INTRODUCTION: PRESERVATION

The subject of preservation is covered extensively in many publications, and it is not intended that this chapter or manual should attempt to cover the subject in depth. The intention is to guide a designer in deciding when preservation is necessary, when there is a choice, or when preservation is unnecessary, and also to put forward several points of which a designer should be aware.

The notes are related to UK conditions, with the more commonly available softwoods in mind, and related to structural components rather than joinery.

As with structural design, the influence of the European standards has been significant. In particular the change from prescriptive treatment cycles, such as a specified preservative applied under a certain pressure regime for a minimum period of time, has been replaced by quantitative analysis of the preservative in the timber. The European standards do not cover all the processes and procedures employed in the UK so there are 'remnants' of British Standards and Codes of Practice that are still required.

24.2 DURABILITY

Each species of timber has a certain natural durability which is usually greater for the heartwood than the sapwood. Durability in this instance is a measure of the natural resistance to fungal decay (not insect attack). The Princes Risborough Laboratory of the Building Research Establishment evolved a classification for durability which is shown in Table 24.1. It is important to realize, however, that it is possible to refer to durability only in relative and indicative terms, and that the classifications refer to 51 mm × 51 mm pieces of heartwood in the ground. In general, larger pieces would have longer life.

It is also important to realize that even species with low natural durability will not decay if the moisture content in service is kept below 20–22%. Even occasional short periods when the moisture content is raised above 20–22% will not lead to decay, although discoloration (e.g. blue stain) may occur in the sapwood of some species. Even a piece of timber given the emotive description of 'perishable' (Table 24.1) will not decay if the service conditions lead to a moisture content of 20–22% or less. On the other hand, it is as well to realize that if a species with high natural durability (or properly preserved timber) is used and moisture is trapped (e.g. by a poor building detail or use of an unsuitable paint), the timber

Table 24.1 Durability classification

Grade of durability	Approximate life in contact with the ground (years)
Very durable	More than 25
Durable	15–25
Moderately durable	10–15
Non-durable	5–10
Perishable	Less than 5

Table 24.2

Timber in ground contact	Grade of durability of: unpreserved heartwood	Grade of durability of: unpreserved sapwood
European Redwood	Non-durable	Not known to the same extent as heartwood but usually close to or better than the boundary between non-durable and perishable
European Whitewood	Non-durable	
Hem-Fir	Non-durable	
Spruce-Pine-Fir	Non-durable	
Douglas-Fir-Larch	Moderately durable	
UK-grown Sitka Spruce	Non-durable	

will not decay but the trapped moisture may cause trouble in another way (e.g. paint may peel off). Therefore, the use of a durable timber or preserved timber should never be considered as an alternative to good detailing.

The sapwood of almost all timbers (not just softwoods) is either perishable or non-durable. Engineers should not fall into the trap of assuming that all available hardwoods are durable. Some of the cheaper hardwoods currently available (usually used for joinery rather than structural use) are less durable than several commonly available softwoods. Table 24.2 lists the durability rating of the heartwood and sapwood of commonly available softwoods used for structural purposes, mainly extracted from a *Handbook of Softwoods* by the Building Research Establishment.

As different biological agencies of timber deterioration occur in different end uses, it is essential that timber in a particular end use has a durability appropriate to that end use. BS EN 335-1 defines a hazard class system for different service environments in which wood and wood-based products are used. Consideration should be given to the most suitable preservative for the end use situation.

The assignment of a particular end use class to a component assumes good design and maintenance of the construction. If conditions arise during the service life of the component which result in unexpected wetting of the timber – for example as a result of design faults, condensation, failure of other materials, poor workmanship or lack of maintenance — the hazard class initially assigned to the component can change and therefore the recommendations for preservation can change.

Table 24.3 The four-class system of treatability

Class	Description	Explanation
1	Easy to treat	Easy to treat; sawn timber can be penetrated completely by pressure treatment without difficulty
2	Moderately easy to treat	Fairly easy to treat; usually complete penetration is not possible, but after 2 or 3 hours by pressure treatment more than 6 mm lateral penetration can be reached in softwoods and in hardwoods a large proportion of the vessels will be penetrated
3	Difficult to treat	Difficult to treat; 3 to 4 hours by pressure treatment may not result in more than 3 mm to 6 mm lateral penetration
4	Extremely difficult to treat	Virtually impervious to treatment; little preservative absorbed even after 3 to 4 hours by pressure treatment; both lateral and longitudinal penetration minimal

In addition to the end use classification, other service factors exist, based on safety and economic considerations, which the specifier or user of the treated timber should take into account when making a judgement on the need for preservative treatment. The extra costs of pre-treatment of the timber component, or the use of naturally durable timbers, should be balanced against the future cost of remedial treatments or replacement of failed components. For example, some timber components can have a short service life requirement or could be inexpensive to replace. In these circumstances preservation could be considered unnecessary. Alternatively, where damage by insects or fungi would be difficult or expensive to repair or where the associated weakening would endanger life or the integrity of the structure, then timber of adequate natural durability, or timber with preservative treatment, is recommended.

24.3 AMENABILITY TO PRESERVATIVE TREATMENT

The ease with which timber can be impregnated or treated with preservative can be an important consideration where it is decided to increase the natural durability. When correctly treated, non-durable timber can be made at least as durable as naturally durable timbers.

BS EN 350-2: 1994 provides a set of broad descriptions for classifying treatability based on general observations associated with vacuum/pressure treatment processes. A four-class system is used, as shown in Tabe 24.3.

The classification of the heartwood and sapwood of a few softwoods is given in Table 24.4. These classifications are related to pressure impregnation.

Table 24.4

Softwood	Heartwood	Sapwood
European whitewood	Resistant	Resistant
European redwood	Moderately resistant	Permeable
Douglas fir	Resistant	Resistant
Spruce–pine–fir*	Resistant	Resistant*
Hem–fir†	Resistant	Resistant†
UK Sitka spruce	Resistant	Resistant

* Amenability based on the spruce.
† Amenability based on the hemlock.

24.4 RISK AND AVOIDANCE

24.4.1 Risk and avoidance of decay

As stated in section 24.2, codes and standards are in agreement that there is little risk of decay, even in sapwood, if the timber is maintained at a moisture content in service of 20–22% or less. Therefore, in most cases of timber used inside a building there is rarely a 100% case for the use of preservation unless there is a local hazard. Also, by good practice and detailing, any risk of decay can be minimized with reasonable certainty for many other cases. However, over recent years it has become quite common to preserve many components where, strictly speaking, the risk of decay hardly warrants it. Where this happens the preservation can be regarded as an extra insurance against mischance.

In deciding on whether or not to preserve, and how to preserve, the specifier is faced with several codes, standards and regulations to consult. Some of these are written in general terms while others are very specific. Those likely to be most pertinent to a UK engineer are discussed in section 24.7.

24.4.2 Risk and avoidance of insect attack

Reference is made below to the four most common forms of insect attack encountered in the UK in softwoods. In dry timber, it is a fair assumption that only the house longhorn beetle, if present, could be significant regarding the risk of structural failure and, in the areas of England where it occurs, the Building Regulations call for preservation in roofs.

- *Furniture beetles* (*Anobium punctatum* De Geer). The furniture beetle is the most common form of insect attack in the UK. These insects can attack dried sapwood, and some heartwood.
- *Ambrosia beetles* (Pinhole borer). Although attack is less common in softwoods than hardwoods, the standing tree or recently felled logs of several softwoods can be attacked by Pinhole borer. Attack ceases when the timber is dry. Attack is more common in the sapwood than in the heartwood and takes the

Table 24.5

Species	Furniture beetles	House longhorn beetles	Pinhole borers	Wood wasps
European whitewood	✓	✓	✓	✓
European redwood	✓	✓	✓	✓
Douglas fir	*	✓	✓	*
Hem–fir	✓	*	✓	✓
Spruce–pine–fir	✓	✓	✓	*
UK Sitka spruce	✓	*	*	✓

✓ Species recorded as having been attacked by the insect indicated in the column.
* Not immune although not recorded as having been attacked.

form of circular holes or short tunnels up to 3 mm diameter. The holes are dark stained and contain no dust.

- *Wood wasps* (Siricidae). These insects attack the standing tree and logs. The attack dies when the timber is dried. The tunnels are circular and filled with tightly packed bore dust.
- *House longhorn beetles* (*Hylotrupes bajulus* L). There is one area in the Home Counties of England (delineated in B3 of the Building Regulations for England and Wales) where house longhorn beetles are a risk to softwood in roofs. They can attack dried softwood. Although the Building Regulations call for all softwoods in roofs in this area to be preserved, only four of the species listed in Table 24.5 are recorded as having been attacked (perhaps because the other two were not used before preservation became mandatory in the risk area).

Termites are not a hazard in the UK but must be considered if an engineer is producing a design for many other areas (e.g. the Middle East). There are many types. Although not all will climb, the designer must realize that timber can be attacked when in store on the ground.

In the area of the Home Counties where the house longhorn beetles are a risk, the Building Regulations require softwood to be treated with a suitable preservative. 'Deemed-to-satisfy' provisions are water-borne CCA in accordance with BS 4072, or organic solvent-based formulations in accordance with BS 5707 and other preservatives tested in accordance with BS EN 599-1 for which the supplier claims performance.

If it is desired to treat against insect attack (as well as decay) the specifier should ensure that the preservative contains an insecticide. Some of the preservatives used for preserving external joinery contain no insecticides because insect attack is not usually a problem. CCA and other copper-based preservatives are effective against the insects encountered in the UK.

24.5 TYPES OF PRESERVATIVE

A wide variety of preservative timber treatments are now available to the designer. Each is tailored to meet the requirements of particular end use situations. The application process controls the amount of preservative applied to the timber such that it will be *'fit for purpose'*, i.e. provide adequate performance in its eventual end use. Therefore, it is important that the designer ensures that treatment has been carried out according to the correct specification. One of the six 'Essential Requirements' of the Construction Products Directive 89/106/LEC (CPD) states that building products covered in the scope of the CPD should exhibit adequate 'Mechanical Stability and Strength' throughout their design life. There is therefore a clear legal onus on the designer of any structure encompassed by the CPD to take the necessary measures to ensure that the materials specified are to sufficient strength and durability to fulfil these requirements, and the need to apply the appropriate timber preservative to wood arises. While the designer/specifier continues to have considerable freedom in the choice and use of preservatives, the duty to safeguard the mechanical stability and strength of the structure is now more explicit.

The principal preservative types used in the UK for building components are:

1. Water-based products applied by vacuum pressure processes: traditionally dominated by chromated copper arsenate (CCA) based preservatives, alternatives that do not contain chrome or arsenic are now also available. These are based on copper in combination with organic biocides and are increasingly being used and specified.

   These products give a green colour to the timber. They are appropriate for all building applications, with the exception of joinery components, and can be used in external applications, e.g. cladding, without the need for a coating. They would not normally be used for joinery applications because the application process increases the moisture content in the timber considerably, and even after re-drying, there can be an increase in the cross section and grain raising.
2. Water-based products applied by double vacuum processes: introduced in the mid-1990s, these new systems are based on organic biocides whereas traditional solvent-based products contained metals, such as tin and zinc. While solvent-based products are still in use, particularly for joinery components, there has been a trend away from solvents to more environmentally acceptable water-based systems.

   These products are colourless and the appearance of the resulting treated timber is not significantly altered, although many treatments use a colour marker to show that the timber has been treated. If used in external building applications, it is necessary to apply and maintain a surface coating to achieve the desired service life given.
3. Solvent-based products applied by double vacuum processes: now more usually based on organic biocides, i.e. metal-free, solvent-based products are, in the main, restricted to joinery applications.

   Again the appearance of the resulting treated timber is not significantly altered, and when used in external building applications, a coating is required.

In internal building applications, desired service lives for all the products are in the region of 60 years. In external use, this figure is more usually in the region of 30 years.

24.6 ADDITIONAL NOTES ON PRESERVATION

24.6.1 Pre-treatment considerations

As far as possible, all cutting, machining, planing, notching and boring must be carried out before the timber is preserved. If some cutting after treatment is unavoidable, the cut surfaces should be given a thorough application ('swabbing') with a suitable end grain preservative.

The pre-treatment moisture content should be 28% or less to allow for effective treatment.

If a water-borne vacuum pressure process is to be used, it is important for a specifier to realize that the use of a water-borne process increases the moisture content of the timber very considerably, causes an increase in the cross section and will raise grain. Therefore, timber for building usually has to be re-dried before being used. With some components (e.g. small battens) it is usually acceptable to air dry the timber after preservation and before use, but for certain uses it will be necessary to kiln the timber before use.

Organic-solvent processes require timber to be at a moisture content of about 22% or less (preferably less) at time of treatment. They do not increase the moisture content of the timber, nor do they affect the dimensions/profile of the sections or raise the grain. Where an organic solvent is used to treat exterior timber it may be advantageous to use one containing a water repellent.

24.6.2 Post-treatment considerations

If timber treated with an organic solvent is to be painted it is essential to limit the amount of free solvent left in the timber to be compatible with the paint, and to leave sufficient time for the solvent to evaporate. This point is particularly important if the timber is to be factory finished.

24.6.3 Effect of preservatives on strength

Normally, preservatives do not affect the strength of timber to any measurable extent, but there are certain fire-retardant treatments that are said to reduce the strength. When using a specific fire-retardant the designer is advised to cheek on this point with the manufacturer (and on the compatibility with adhesives).

24.6.4 Compatibility of metal fixings and preservatives

Due to the diversity of products available in the market, it is important to check with the supplier as to the suitability of the various metal-fixing types.

24.6.5 Compatibility of glues and preservatives

It is usually relatively easy to glue preserved timber or preserve glued timber, providing certain precautions are taken. Many points are covered in Chapter 19, but if in doubt the designer should check with the manufacturers of the glue and the preservative. With certain fire-retardant solutions, particularly those including ammonia or inorganic salts, certain precautions must be taken. It may not be possible to glue at all on timber already treated with certain of these solutions and, if preservation is carried out after gluing, usually at least seven days must elapse between gluing and preservation. If the preservative contains any additive it is as well to check the compatibility with the adhesive.

24.6.6 Compatibility of paint and preservatives

It is usually possible to paint successfully on the surface of preserved timber, although with certain preservatives containing water-repellent waxes, there may be some difficulties, or some precautions that have to be taken. Once more the advice of the manufacturers should be sought.

24.6.7 Fire retardants

Special formulations are available which will raise the surface spread of flame classification of timber to class 1 when tested according to BS 476: Part 7, or achieve a Building Regulation rating of class 0, which requires a fire propagation index (I) not exceeding 12 and a sub-index (ii) not exceeding 6, when tested in accordance with BS 476: Part 6. They do not improve the fire resistance (time) of timber to any noticeable extent.

Fire-retardant impregnation formulations are water based and fall into one of the broad categories listed below.

The three types of system may be distinguished by properties which limit, or recommend, their use in specific circumstances. The variation in these properties is largely due to the nature of the chemical used in the formulation and, the complexity and degree of chemical reaction required in formulating them.

- *Type A* Interior type – based on simple inorganic salts. The common feature is that the treated timber or board product is sensitive to high humidity prolonged exposure, resulting in salt efflorescence and/or migration. Use is therefore restricted to interior environments where humidities do not exceed 70%. The salts are corrosive to all metals and structural performance can be reduced.
- *Type B* Humidity-resistant type – based on chemically reacted or blended organic and inorganic chemicals. The treated timber or board product is far less sensitive to high or fluctuating humidity, due to the lower solubility in water of the chemicals and, in some instances, a degree of chemical interaction. They can therefore be used in almost all interior situations and, in some cases, in certain weather-protected exterior situations.
- *Type C* Leach-resistant type – based on a polymeric resin system. The treated timber or board product can be used in all interior and exterior situations. The leach resistance is achieved by high temperature curing of the complex chemical system in the treated wood following impregnation and re-drying.

The above reference types are listed in the *British Wood Preserving and Damp Proofing Association Manual*, Section 2, 'Fire retardant treatments'.

24.6.8 Blue stain/anti-stain

The sapwood of certain species (e.g. European redwood) is susceptible to discoloration of the sapwood. There are two different types: sap stain, which occurs on freshly felled timber, and blue stain, which occurs in service. Some sawmills treat their sawn timber with anti-stain chemicals to prevent sap stain.

Staining organisms do not cause structural damage, they are merely disfiguring and unsightly; nor are they a sign of incipient decay. Blue stain in service will not occur or spread while the timber is at less than about 20–22% moisture content. It is mentioned here because it is quite common for a designer to be questioned about blue stain. It is not a structural defect, and is permitted by BS 4978 in stress-graded timber. However, some preservative treatments will protect against blue stain and the designer may wish to check with the manufacturer.

24.6.9 Size of preserving plant

If the engineer wishes to call for impregnation of components in their final form, a check should be made that a treatment plant is available and large enough to take the component. The major suppliers of preservatives will know the size and disposition of plants.

24.6.10 Moisture content readings

The readings of a moisture meter are affected by CCA preservatives and by fire retardants, therefore the moisture content of such treated timber can be measured only by destructive methods unless the manufacturer of the meter can give a correction factor. Organic-solvent treatments do not affect moisture meter readings to any significant extent (with the possible exception of those containing copper naphthanate).

24.6.11 Surface degradation

All species of preserved timber left exposed externally will be subject to surface degradation ('greying' or 'weathering'). Some preservative treatments will slow down the rate of degradation, in particular water-based vacuum pressure treatments with built-in water repellents.

24.6.12 Safety

Preservatives for use in the UK must be cleared for use under the Control of Pesticides Regulations 1986. Suppliers will produce safety information on their products and copies of this should be sought.

If moulding, cutting, etc., is carried out after preservation, great care should be taken to ensure safe disposal of the waste. It should not be used, for example, for animal litter or bedding, or used for fuel in barbecues, cooking stoves or grates.

24.7 PUBLICATIONS GIVING GUIDANCE OR RULES ON WHEN TO PRESERVE

It is important to note that timber preservatives introduced from the mid-1990s onwards will have been developed with European standards in mind. Therefore they may not be covered in certain British standards. Those British standards that are out of date are gradually being declared for obsolescence, or being updated to BS EN status.

24.7.1 European standards

BS EN specifications will, within a few years, entirely replace the existing BS specifications. These specifications use a system introduced in new European standards where a defined combination of penetration and retention of preservative must be achieved in the treated wood. The application process is not defined and any process may be used which achieves the desired penetration and retention combination.

The specification phrases commonly used when specifying preservative treated timber are: Preservative X conforms to the efficacy requirements of BS EN 599-1: 1997, and is treated in accordance with the penetration and retention guidance given in BS EN 351-1: 1996 to give a desired service life in the selected end use/hazard class as defined in BS EN 335-1: 1992.

BS EN 599-1: 1997 establishes the loadings in the analytical zone of treated timber for the end use/hazard class for which the preservative has been tested.

BS EN 351-1: 1996 provides a vocabulary for defining the preservative penetration and retention values, known as the analytical zone concept. Nine penetration classes are specified. A flow diagram for specifying preservative treatment is included.

Within the EN standards, no attempt is made to quantify the desired service life that could be expected from a particular preservative treatment as this will depend on the geographical location and the associated climate of the service environment. Individual countries therefore usually have their own national interpretative documents. In the UK, this document is Draft for Development DD239: 1998 and, unlike other countries, the UK attempts to addresses the subject of desired service life by defining recommendations for 15, 30 and 60 years desired service life.

24.7.2 BS 5268: Part 5

BS 5268: Part 5: 1989, 'Structural use of timber', is part of the Code of Practice for the structural use of timber. Clause 5.2 details four hazard categories A, B, C and D (not to be confused with the performance categories A and B of BS 5589: 1978 'Preservation of timber'). These are:

A Where preservative treatment of timbers, even those with low inherent resistance to biological degradation, is unnecessary. This is because the conditions of use involve negligible risk, render the consequences acceptable, or make the cost of preservative treatment generally unfavourable.

B Where there is a low risk of decay or insect attack or where remedial action or replacement is simple. In such situations preservation, if adopted, may be regarded as an insurance against the cost of subsequent repairs.

C Where experience has shown that there is an unacceptable risk of decay, whether due to the nature of the design or the standard of workmanship and maintenance, or where there is a substantial risk of decay or insect attack which, if it occurs, would be difficult and expensive to remedy.

D Where timber is exposed to a continually hazardous environment and cannot be protected by design or where there is a high risk of decay or insect attack in structures, the collapse of which would constitute a serious danger to persons or property.

This wording does permit the specifier an element of discretion, but leaves an element of doubt. However, Table 3 (and 4) of BS 5268: Part 5 gives more precise guidance. The way of stating the extent to which various species should be preserved in various situations is being changed. The method now favoured is to quote a treatment schedule for the plant operator to use. However, BS 5268: Part 5: 1977 is written in a different way. For organic-solvent methods it simply states 'yes' or 'no', leaving the specifier/applier to refer to schedules given by the preservative manufacturer or the BWPA (British Wood Preserving Association). For CCA methods retentions in kg/m^3 are quoted. It is the intention to amend BS 5268: Part 5 in the future to give preservation treatment schedules (perhaps with retention levels also being given for CCA).

24.7.3 BS 5589

BS 5589: 1989 'Preservation of timber' is complementary to BS 5268: Part 5 in that it aims to cover non-structural external timber in buildings and some special uses such as fencing. It gives tables of use conditions (e.g. external timber in buildings and out of contact with the ground), leaves it to the specifier or user to decide on a desired service life (e.g. 60, 50, 30, 20, 15 years), and then gives suitable treatment schedules (or even immersion periods) for various species. Thus, if a specifier quotes BS 5589: 1989 it is necessary also to quote the desired service life.

Species such as whitewood are popular for external cladding (to which Table 2 of BS 5589 only gives a 30-year treatment schedule for whitewood due to its low treatability). It is worth noting that schedules are given in BWPA Code C6 for European whitewood and Canadian hemlock (and European redwood) for a 60-year service life.

24.7.4 Codes of the British Wood Preserving and Damp-proofing Association

Several useful publications are available from the BWPDA. Among these there are ten 'codes' for various applications, as listed below:

Cl (1975)	Timber for use as packing in cooling towers
C2 (1975)	Timber for use permanently or intermittently in contact with sea or fresh water
C3 (1975)	Fencing
C4 (1975)	Agricultural and horticultural timbers
C5 (1975)	External joinery and external fittings not in ground contact
C6 (1975)	External cladding
C7 (1979)	Prefabricated timber buildings for use in termite-infested areas
C8 (1979)	Constructional timbers
C9 (1982)	Timber-framed housing (the constructional frame of external walls)
C10 (1984)	Treatment of hardwood exterior joinery with organic-solvent preservatives by double vacuum.

BWPDA codes are offered to BSI for consideration as British standards, therefore, if the date of the relevant BS is later than the date of the BWPA code, it is probably better to specify to the BS. NHBC refers to C9 in its Practice Note 5 on timber-framed housing. One of the advantages of code C9 is that it gives the 'agreed' treatment schedules for various components (e.g. for exterior stud walls), providing that equipment able to withstand 2 bar pressure is available.

Preservatives tested in accordance with EN standards can also be specified through the BWPDA. The EN system allows suppliers to place products on the European market based on their own claims of performance subject, where applicable, to national approval in respect of safe use. (In the UK the Control of Pesticides Regulations 1986 apply.) The BWPDA publishes a list of products, supplied by BWPDA members, for which the claims of compliance have been audited by BWPDA.

24.7.5 The Building Regulations

The Building Regulations for England and Wales 1976 require softwood timber in certain defined areas of the Home Counties of England in the construction of a roof or fixed within a roof to be treated with a suitable preservative to prevent infestation by the house longhorn beetle (Regulations B3 and B4).

Regulation A16(1) and Schedule 5 combine to require external softwood cladding of Douglas fir, hemlock, larch, European redwood, Scots pine, Sitka spruce and whitewood or European spruce to be preserved in accordance with Table 5 of Schedule 5. The Building Regulations (Northern Ireland) 1973 have similar requirements.

The Building Standards (Scotland) Regulations 1981 have a deemed-to-satisfy provision which is relevant to the preservation of weatherboarding fixed direct to studs. Most softwoods can be used if (with the exception of western red cedar,

Table 24.6

	CCA	Double-vacuum organic solvent
Lintels in brick or blockwork external walls	✓	
Battens as fixings for claddings	✓	✓
Any embedded timber	✓	
Joists in flat roofs	✓	✓
Joists with ends built into solid (non-cavity) walls	✓	✓
Door frames (for external doors)	✓	✓
Windows	✓	✓
Surrounds to metal windows	✓	✓
External doors other than flush doors	✓	✓
External timber features other than fencing	✓	✓

which does not need to be preserved) they are preserved in accordance with BS 5589. (See Schedule 11G9(6) and Schedule 14, Part I.S.) Where weatherboarding is fixed in a position where it is readily accessible for inspection and maintenance or renewal it can be claimed that Regulation B2(ii) makes preservation unnecessary, particularly if the weatherboarding is fixed to battens (although probably it makes sense to preserve).

24.7.6 The National House-Building Council Manual

In their handbook and in Practice Note 5 for timber-framed housing the NHBC call for preservation as required by the Building Regulations and, in addition for preservation of the items shown in Table 24.6. NHBC also refer to BWPA code C9 for the preservation of the constructional timber frame of external walls of timber-framed housing.

24.7.7 Further reading

BS 5268-5: 1989 'Code of practice for the preservative treatment of structural timber'.
BS 5589: 1989 'Code of practice for preservation of timber'.
BS 1282: 1975 'Guide to the choice, use and application of wood preservatives'.
BS 4072: 1974 'Specification for wood preservation by means of waterborne copper/chrome/arsenic compositions'.
BS 1282: 1998 'Guide to the choice, use and application of wood preservatives'.
BS 4072-1: 1987 'Wood preservation by means of copper/chromium/arsenic compositions – Part 1: Specification for preservatives'.
BS 5707: 1997 'Specification for preparations of wood preservatives in organic solvents'.

Table 24.7

Service class	Location	Average moisture content likely to be attained in service conditions (%)	Moisture content that should not be exceeded in individual pieces at time of erection (%)
3	External use, fully exposed	20 or more	–
2	Covered and generally unheated	18	24
2	Covered and generally heated	15	20
1	Internal uses, in continously heated building	12	20

BS EN 335-1: 1992 'Hazard classes of wood and wood-based products against biological attack – Part 1: Classification of hazard classes'.

BS EN 350-1: 1994 'Durability of wood and wood-based products – Natural durability of solid wood – Part 1: Guide to the principles of testing and classification of the natural durability of wood'.

BS EN 350-2: 1994 'Durability of wood and wood-based products – Natural durability of solid wood – Part 2: Guide to natural durability and treatability of selected wood species of importance in Europe'.

BS EN 351-1: 1996 'Durability of wood and wood-based products – Preservative treated solid wood – Part 1: Classification of preservative penetration and retention'.

BS EN 599-1: 1997 'Durability of wood and wood-based products – Performance of preventive wood preservatives as determined by biological tests – Part 1: Specification according to hazard class'.

24.8 MOISTURE CONTENT

24.8.1 Equilibrium moisture content in service

Table 24.7 gives guidance on moisture content for end-use categories as given in BS 5268-2 for structural components according to service class.

Ideally, all timber should have a moisture content at manufacture and installation close to the moisture content it will have in service. If timber has been correctly dried once, and is subjected to rain for relatively short periods during delivery or erection, it is most unlikely that anything other than the outer 2–3 mm will be affected. A normal progressive or chamber kiln takes several days to extract moisture from timber and this gives a measure of how little effect rain will have. The designer should not be misled by high surface readings into assuming that all the timber is equally wet throughout the thickness.

24.8.2 Moisture content at time of manufacture

The moisture content at the time of manufacture should be within a few percent of the service conditions. In addition, the type of connection used may set the upper limit. If the timber is being joined by simple nailing which is only lightly loaded, a moisture content of up to 22% at time of manufacture is probably satisfactory. With most mechanical connectors (Chapter 18) an upper limit of 20% is likely to be satisfactory. With a glued joint (Chapter 19) a slightly lower limit is more appropriate, but depends on the type of glue and the method of curing. Some WBP glues have been used successfully at a timber moisture content of 22% while radio-frequency curing requires a moisture content of around 12%.

In the manufacture of a component from timber which has been delivered 'green' to the factory and is still in the process of drying 'naturally', much more care is required in checking the moisture content than for timber which has been kiln dried at the sawmill (usually to 18% + 4% – 2%) or air dried at the sawmill (usually to 22% or less) some weeks or months before. Softwoods 'move' less during any subsequent changes in moisture content than they do during initial drying from the green condition.

24.8.3 Moisture content at time of erection

Generally speaking, the moisture content at time of erection can be as high as 19–24% providing that there is natural air circulation which will permit the timber to continue to dry.

If timber is erected at 22% moisture content or over, or the moisture content after erection becomes higher than around 18–20% due to site conditions, the sections should not be completely enclosed until the moisture content has dropped to below 20%.

24.8.4 Measuring moisture content

In the range of moisture content of interest to structural engineers (up to 22%) small portable moisture meters have sufficient accuracy. If the engineer has reason to believe that the initial moisture content is likely to be higher than the surface reading, deep probes (20–25 mm long) can be used. With these, the 'moisture gradient' can be traced through a piece.

Most codes and textbooks on timber technology detail a way of measuring the moisture content by weighing a sample piece as presented, drying it in an oven to constant weight, then expressing the weight of water removed as a percentage of the oven-dry weight. This method obviously destroys the piece and is only of use to a structural engineer if wishing to check the factory quality control before or during manufacture of components, or in extreme cases of site 'trouble-shooting'.

Often one hears of the oven-dry method referred to as being more accurate than the use of a moisture meter. This is not so; they have different uses. A properly calibrated meter gives an accurate method of measuring the moisture content at a point and can therefore be used to trace the source of moisture, or the pattern of

Table 24.8 Relative humidity

Dry-bulb temperature (°C)	Difference between dry-bulb temperature and wet-bulb depression (°C)											
	1	2	3	4	5	6	7	8	9	10	11	12
1	83	66	49	33	17							
2	84	68	52	37	22	7						
3	84	70	55	40	26	12						
4	85	71	57	43	30	16						
5	86	72	58	45	33	20	7					
6	86	73	60	48	35	24	11					
7	87	74	62	50	38	26	15					
8	87	75	63	51	40	29	19	8				
9	88	76	64	53	42	32	22	12				
10	88	77	66	55	44	34	24	15	6			
11	89	78	67	56	46	36	27	18	9			
12	89	78	68	58	48	39	29	21	12			
13	89	79	69	59	50	41	32	23	15	7		
14	90	79	70	60	51	42	34	26	18	10		
15	90	80	71	61	53	44	36	27	20	13	6	
16	90	81	71	63	54	46	38	30	23	15	8	
17	90	81	72	64	55	47	40	32	25	18	11	
18	91	82	73	65	57	49	41	34	27	20	14	7
19	91	82	74	65	58	50	43	36	29	22	16	10
20	91	83	74	66	59	51	44	37	31	24	18	13
21	91	83	75	67	60	53	46	39	32	26	20	14
22	92	83	76	68	61	54	47	40	34	28	22	17
23	92	84	76	68	62	55	48	42	36	30	24	19
24	92	84	77	69	62	56	49	43	37	31	26	20
25	92	84	77	70	63	57	50	44	39	33	28	22
26	92	85	78	71	64	58	51	46	40	34	29	24
27	92	85	78	71	65	58	52	47	41	36	31	26
28	93	85	78	72	65	59	53	48	42	37	32	27
29	93	86	79	72	66	60	54	49	43	38	33	28
30	93	86	79	73	67	61	55	50	44	39	35	30

The values given are for a 2 metres per second movement of air past the hygrometer.

moisture in a piece of timber. The oven-dry method is a destuctive method of giving an average moisture content. Hence the sample being dried must be kept small and the result can be influenced by the presence of resin in the sample.

24.8.5 Measuring relative humidity

Relative humidity is the amount of water in the air compared to the amount of moisture which the air would contain if fully saturated. The ratio is expressed as a percentage. The pressure which moisture exerts is closely related to the amount

Table 24.9 Equilibrium moisture content

Relative humidity (%)	Temperature (°C)							
	16	18	20	22	24	26	28	30
85	18	18	18	18	18	18	18	18
80	17	16	16	16	16	16	16	16
75	15	15	15	14	14	14	14	14
70	13	13	13	13	13	13	13	13
65	12	12	12	12	12	12	12	11
60	11	11	11	11	11	11	11	10
55	10	10	10	10	10	10	10	10
50	9	9	9	9	9	9	9	9
45	9	8	8	8	8	8	8	8

of moisture in the air, therefore the use of a dry- and wet-bulb thermometer gives a convenient method of establishing the relative humidity of a specific air condition.

Table 24.8 gives values of relative humidity for differences in degrees Celsius between the dry- and wet-bulb thermometer, based on a 2 metre per second movement of air past the hygrometer. The values are derived from Swedish sources.

Lightweight meters that measure relative humidity are available. In view of the importance of humidity rather than temperature in setting the moisture content, it is rather surprising that they are used so infrequently.

24.8.6 Relative humidity/temperature/moisture content

The moisture content of timber depends on conditions of relative humidity and air temperature. With constant humidity and temperature, the timber will assume an equilbrium moisture content. Changes in humidity and temperature do not lead to an instantaneous measurable change in the moisture content of the timber. Moisture content in buildings is more dependent on humidity than on temperature, therefore the service moisture content of timber can be established with reasonable accuracy for a building in which the relative humidity is fairly constant, even though the temperature varies. In textbooks it is usual to present the relationship in the form of curves of moisture content. For convenience, moisture content figures to the nearest degree are given in Table 24.9 for the limited temperature range likely to be of most interest to engineers.

Chapter 25

Considerations for the Structural Use of Hardwood

25.1 INTRODUCTION

In Chapter 1 it was made clear that the contents of this manual have concentrated on softwoods because these are the timbers mainly used in timber engineering. However, hardwoods are used and do fulfil a useful role. They can have the advantage of greater strength and durability than most softwoods, although the designer must not assume that all hardwoods are either more durable or stronger than softwoods. As well as possible or actual advantages, there can be disadvantages. By and large, a designer should not specify a hardwood without having investigated the properties of interest, and the availability of sizes and lengths.

25.2 SPECIES/GRADES/STRENGTH CLASSES

Hardwoods should be stress graded in accordance with BS 5757: 1997 'Specification for visual strength grades of hardwood'. The species listed in BS 5268-2 are then allocated to a particular strength class according to stress grade. Table 25.1, to which the following notes apply, summarizes the range of hardwood currently referenced in BS 5268-2.

(1) Stress graded in accordance with BS 5757: 1997.
(2) Allocated strength class for design of timber components.
(3) Allocated strength class for design of joints. Alternatively joints may be designed using the formulae given in Annex G using the characteristic density given in Table G1 of Annex G or Table 15 of BS 5268-2.
(4) Grades THA and THB for oak are only obtainable in cross-section sizes with no dimension less than 100 mm and with cross-section areas greater than 20 000 mm^2, i.e. 100 mm × 200 mm minimum section.
(5) Grades TH1 for sweet chestnut and TH2 for oak are not allocated to a strength class. Grade stresses and moduli of elasticity should be taken from Table 12a of BS 5268-2. This table is given for service classes 1 and 2. Timbers that are 100 mm or more in thickness are difficult to dry and it is therefore recommended that such sections should be considered as service class 3. This requires the values in Table 12a to be modified by the factors given in Table 16 of BS 5268-2.

Table 25.1 Hardwoods referenced in BS 5268: Part 2

Standard name	Stress grade(1)	Strength class(2)	Joints(3)
Iroko	HS	D40	D40
Jarrah	HS	D40	D40
Teak	HS	D40	D40
Merbau	HS	D50	D50
Opepe	HS	D50	D50
Karri	HS	D50	D50
Keruing	HS	D50	D50
Ekki	HS	D60	D60
Kapur	HS	D60	D60
Kempas	HS	D60	D60
Balau	HS	D70	D70
Greenheart	HS	D70	D70
Sweet chestnut	TH1	Not allocated(5)	C24
Oak	TH1	D30	D30
	TH2	Not allocated(5)	C24
	THA(4)	D40	D40
	THB(4)	D30	D30

25.3 PROPERTIES/CHARACTERISTICS

Most sawn sections of hardwood are relatively free from knots and wane, and are fairly straight grained; however, there is often a tendency to distort and split, particularly if sections are re-sawn and particularly as the *h/b* ratio increases.

Table 25.2 has been prepared as a starting point to assist a designer to choose a suitable species, or check if a suggested species is likely to be suitable for an intended use. The comments have been extracted from reference books on hardwoods such as the *Handbook of Hardwoods* by the Building Research Establishment. It is emphasized that Table 25.2 should be considered to be only indicative of properties. Descriptions such as 'satisfactory', 'pale', etc. are not precise. Before using a species for the first time, a designer should discuss properties with a hardwood specialist, preferably someone who has used the species and is also aware of availability and cost.

However, even with these reservations, Table 25.2 can be a useful starting point. For example, if a designer is looking for a species to use for glulam members (which requires to have stresses allocated by BS 5268, and needs to be dried, machined, relatively free of distortion, and to be glued) it looks as though iroko is a possibility, whereas ekki and greenheart are unlikely starters despite their advantages in heavy marine uses.

25.4 MOISTURE CONTENT

If the designer wishes to use 'dry' stresses, it is necessary to check that the timber to be supplied will be dried to around 20% or less before erection, as hardwoods can take rather long to dry out insitu. For thicknesses of 100 mm and over, it will normally be necessary to design using service class 3.

25.5 CONNECTIONS

Many hardwoods can be glued with normal structural glues, but rather more care is necessary than with softwoods. In addition to all the quality control requirements detailed in Chapter 19, the surface must be reasonably free from resin. Also, particularly if the fastenings which are used to hold the pieces in place during curing are to be removed after curing, the manufacturer must ensure that individual pieces can be pulled into place easily for gluing. If excessive force is necessary, it is quite possible that the piece will burst open the glue line when subsequently it tries to regain its previous shape. In glulam or similar, it is prudent to consider using thinner laminations to prevent this action. When considering nailing a hardwood it may be necessary to pre-drill (see Table 25.2 for guidance). For nailing, BS 5268-2 advises that all hardwoods in strength classes D30 to D70 will usually require pre-drilling.

BS 5268-2 advises that toothed plate connectors can be used in all hardwoods of strength classes D30 to D40 provided full embedment of the teeth can be achieved. Toothed plate connectors are not suitable for strength classes D50 to D70.

25.6 DESIGN DATA FOR OAK

The use of oak for purlins and traditional trusses has gained favour in recent years. Table 25.3 may be of assistance to the designer for initial design purposes.

Table 25.2

	Iroko	Jarrah	Teak	Merbau	Opepe
Weight (kg/m³) at 12% m.c.	Approx. 640	690–1040	610–690	740–900	Average 740
Durability of heartwood	Very durable	Very durable	Very durable	Durable	Very durable
Amenability to preservation: heartwood	Extremely resistant	Extremely resistant	Extremely resistant	Extremely resistant	Moderately resistant
sapwood	Permeable	Permeable			Permeable
Drying	Dries well with little splitting	Care must be taken to limit distortion	Dries well but slowly	Dries well	Variable. There can be serious splitting and distortion
Movement characteristics	Small	Medium	Small	Small	Small
Machinability	Satisfactory with experience	Satisfactory with experience	Relatively easy	Reported as variable	Satisfactory with experience
Nailability (also see § 25.7)	Satisfactory	Difficult	Pre-boring recommended	Pre-boring advisable	Pre-boring necessary
Gluability	Good	Good	Good		Good
Resinous (or gum) nature	No trouble reported	May contain gum streaks or pockets	No trouble reported	Unconfirmed reports of resinous nature	No trouble reported
Sapwood identification	Distinct from heartwood	Pale	Light/pale	Pale yellow	Whitish/pale
Other comment	Has been used as a substitute for teak. Not recommended for heavy duty flooring	High resistance to wear but inclined to splinter under heavy wear	A valuable timber, now expensive	Liable to stain in contact with iron in wet conditions	

Table 25.2 *(continued)*

Karri	Keruing	Balau	Ekki	Kapur	Kempas	Greenheart
Average 880	720–800	Usually less than 800	950–1100	720–800	770–1000	Approx. 1030
Durable	Moderately durable	Very durable to mod. dur.	Very durable	Very durable	Durable	Very durable
Extremely resistant	Mod./resist. to resistant	Extremely resistant	Extremely resistant	Extremely resistant	Resistant	
Permeable	Mod. resist.	Permeable		Permeable		
Pronounced tendency to check in thick pieces and to distort in thick pieces	Dries slowly and distortion may occur	Care necessary to prevent splitting and distortion	Dries slowly with splitting and some distortion likely	Dries slowly but well	Normally dries well	Dries slowly with degrade (splits)
Large	Medium to large	Small to medium	Medium	Medium	Stable when dry	Medium
Satisfactory with special care	Satisfactory with experience and care		Difficult	Rather difficult	Somewhat difficult	Can be difficult due to high density
Difficult	Satisfactory	Pre-boring necessary	Pre-boring necessary	Satisfactory	Pre-boring advisable	Unsuitable
Good	Variable		Variable			Variable to fairly good
No trouble reported	Known to exude resin	Resin canals present	No trouble reported	Non-resinous	No trouble reported	No trouble reported
	Grey	Paler than the heartwood	Paler than the heartwood	Pale	White or pale yellow	Pale yellow or green
		It is a Shorea and can be variable. Possible confusion with Red Balau		Acidic, can stain fabrics and corrode some metals	Slightly acidic	Noted for its strength

Table 25.3 Oak grade TH2 (service class 3): Principal members (BS 5268: Part 2: 1996)

Section		No. of units	Depth factor	Load-sharing factor		Section properties		Long term		Medium term		
b (mm)	h (mm)	N	K_7	K_8	K_9	Z ($\times 10^6$ mm^3)	I ($\times 10^6$ mm^4)	Shear (kN)	Moment (kN m)	Shear (kN)	Moment (kN m)	EI (kN m^2)
100	200	1	1.05	1.00	1.00	0.667	66.7	24.00	4.35	30.00	5.44	373
120	200	1	1.05	1.00	1.00	0.800	80.0	28.80	5.22	36.00	6.52	448
120	250	1	1.02	1.00	1.00	1.250	156.3	36.00	7.96	45.00	9.95	875
120	300	1	1.01	1.00	1.00	1.800	270.0	43.20	11.30	54.00	14.12	1512
150	150	1	1.08	1.00	1.00	0.563	42.2	27.00	3.79	33.75	4.74	236
150	200	1	1.05	1.00	1.00	1.000	100.0	36.00	6.52	45.00	8.16	560
150	220	1	1.03	1.00	1.00	1.210	133.1	39.60	7.81	49.50	9.77	745
150	240	1	1.02	1.00	1.00	1.440	172.8	43.20	9.21	54.00	11.51	968
150	260	1	1.02	1.00	1.00	1.690	219.7	46.80	10.71	58.50	13.39	1230
150	280	1	1.01	1.00	1.00	1.960	274.4	50.40	12.32	63.00	15.40	1537
150	300	1	1.01	1.00	1.00	2.250	337.5	54.00	14.12	67.50	17.65	1890
150	320	1	0.99	1.00	1.00	2.560	409.6	57.60	15.82	72.00	19.78	2294
150	340	1	0.98	1.00	1.00	2.890	491.3	61.20	17.62	76.50	22.02	2751
200	200	1	1.05	1.00	1.00	1.333	133.3	48.00	8.70	60.00	10.87	747
200	250	1	1.02	1.00	1.00	2.083	260.4	60.00	13.26	75.00	16.58	1458
200	300	1	1.01	1.00	1.00	3.000	450.0	72.00	18.83	90.00	23.54	2520
250	250	1	1.02	1.00	1.00	1.604	325.5	75.00	16.58	93.75	20.72	1823
250	300	1	1.01	1.00	1.00	3.750	562.5	90.00	23.54	112.50	29.42	3150
300	300	1	1.01	1.00	1.00	4.500	675.0	108.00	28.25	135.00	35.31	3780

Coefficients

Duration of load (clause 2.8, Table 14 of BS 5268-2)

Long term $K_3 = 1.00$

Medium term $K_3 = 1.25$

Depth factor (clause 2.10.6)

$K_7 = (300/h)^{0.11}$ for $72 < h < 300$

$K_7 = 0.81(h^2 + 92300)/(h^2 + 56800)$ for $h > 300$

Derivation of capacities

Shear ($K_2 = 0.9$)

Grade stress = 2 N/mm^2

Capacity = $(2bh)/3$ × grade stress × $K_3 \times K_8 \times K_2$

Moment ($K_2 = 0.8$)

Grade stress = 7.8 N/mm^2

Capacity = Z × grade stress × $K_3 \times K_7 \times K_8 \times K_2$

Deflection ($K_2 = 0.8$)

$E_{min} = 7000$ N/mm^2

$E = E_{min} \times K_2$

Chapter 26
Prototype Testing

26.1 GENERAL

Testing of prototypes or parts of construction can be useful to a designer in many ways, some of which are listed below. Where a test or tests is being carried out to satisfy BS 5268-2 then section 8 of that Code should be studied before testing is commenced.

Testing may be used to study the performance and arrive at the failure load and the load–deflection curve of a component. The test may be carried out to check a design which has already been prepared or as an alternative to design, for example when a component is a redundant framework. Usually prototype testing is carried out by or for an individual company or designer, but in the case of trussed rafters the information from several hundred tests carried out at a few testing stations for many companies has been pooled. Although satisfactory designs can be carried out mathematically for trussed rafters, the accumulated test information has justified increasing the span of standard trusses.

The permissible stresses in BS 5268-2 are largely set at the lower 5 in 100 exclusion limit, and if a designer wishes to justify a higher strength for a particular component, it is usually possible to do so by testing, particularly if two or more members occur in cross section. It is a matter of balancing the benefits with the cost of testing.

With a composite section constructed without glued joints, with the joints taking horizontal shear, or any framework constructed with mechanical joints, although it may well be possible to calculate the strength it is almost certain that the deflection will have to be found by test, either of the whole component, or of joints to arrive at the load slip characteristics.

A designer may use testing to check that a roof or floor construction is capable of lateral distribution of loading.

Codes of Practice are not always exactly applicable to all components or constructions, and the designer may wish to use a test to prove that a particular clause may be amended or disregarded in a particular application.

There are still cases in which the basic information required to produce a design is not available or is available only in a very conservative form (e.g. information to calculate sway of a wall panel braced with a different board material on each side). In these cases the designer may have to revert to testing.

In certain constructions (e.g. thin plywood decking), it is the 'feel' rather than the strength which is the limiting criterion. In such cases testing can be preferable to a mathematical design.

Prototype testing can be expensive, and great care should be taken in planning the test to ensure that it represents the actual conditions, restraint and loading as far as practical and to ensure that the correct readings are taken. Before the test is started, it is advisable for the supervising engineer to calculate how the component can be anticipated to deflect, perform and fail, and to check against a load–deflection curve as the test proceeds to see that nothing untoward is happening. A record of the test as it proceeds must be made.

Any testing of full-size components can be dangerous. A supervising engineer must be appointed and must ensure that the method of applying the load, the strength limit of the test rig, etc. will not place the testing staff at risk.

Part of the object of a prototype test is to find where and describe how failure occurs. Once failure has occurred or has started to occur at one point, this can lead to failure at other points, particularly if a triangulated framework is being tested. To prevent any secondary failures occurring which might make it impossible to see where the first failure started, the test rig should be designed so that the component is supported once it begins to fail.

26.2 TEST FACTOR OF ACCEPTANCE

From the notes in Chapter 2 on variability, it will be obvious that the timber from which the prototype component is made should contain defects as close to the grade (lower) limit as possible. Even then it is not certain that the chosen material will be the weakest that could be built into production components. Therefore in analysing a prototype test to failure, before accepting the component as satisfactory, a fairly high factor against failure is required. Where more than one timber/plywood component is being tested, BS 5268-2 requires a factor of 1.25 × K_{85} where K_{85} varies according to duration of load applicable to the design components and is shown in Table 26.1.

Where only one timber/plywood component is being tested BS 5268-2 requires that this same factor be maintained for at least 15 minutes.

The factors apply to service classes 1 and 2 only.

26.3 TEST PROCEDURE

The test should be divided into three separate parts.

1. ***Pre-load test.*** Firstly, loading should be applied for a short period of time and then released, load–deflection readings being recorded. The purpose of

Table 26.1 Strength test factors

Duration of loading	K_{85}	Test factor = 1.25 × K_{85}
Long	2	2.5
Medium	1.79	2.23
Short	1.52	1.9
Very short	1.3	1.62

this first test is to take up any slack at supports and any initial slip at connections. By measuring the height of the component at supports, any vertical movement at the supports can be eliminated from analysis of the load–deflection curve over the span. BS 5268-2 requires that for this 'preload' test the design dead loading should be applied for 30 minutes and then released.

2. ***Deflection test.*** Secondly, and straight away, the dead load should be applied again, maintained for a short period (15 minutes in BS 5268-2) and then the loading should be increased to the maximum design load. BS 5268-2 requires this further loading to be applied over a period of 30–45 minutes. This load is left in place for 24 hours and then released. The deflection readings and condition of the component should approximate to the required performance, but BS 5268-2 stipulates that the deflection after 24 hours should be only 80% of the calculated deflection and that the rate of deflection with time over the 24 hours should decrease and certainly not increase at any time during this period. Deflection readings are taken immediately before and after load release.
3. ***Strength test.*** The final stage of the test is to reload up to the maximum design load, checking deflection then, in the case of BS 5268-2, to load until the appropriate test factor (as given in Table 26.1) times the design load is applied (unless failure occurs before this). If the component is still intact, the testing engineer has the discretion to continue loading until failure occurs. If the component fails prematurely the engineer has the option of strengthening the component and retesting.

The testing method described above does not apply to simple joints or individual members, nor to quality control testing, for which a simpler procedure is necessary and which should comply with any appropriate standard.

Chapter 27
Design to Eurocode 5

Abdy Kermani BSc, MSc, PhD, CEng, MIStructE, FIWSc
Centre for Timber Engineering, Napier University

27.1 INTRODUCTION

Eurocodes are a set of unified codes of practice for designing building and civil engineering structures, which are developed by member states of the European Union and are published by the European Committee for Standardization (CEN) for use in all member countries. They are designed to provide a framework for harmonized specifications for use in the construction industry, with a prime objective of improving the competitiveness of the European construction and its associated industries.

The Eurocodes consist of ten codes covering the basis of design, actions on structures, the main structural materials and geotechnical and earthquake design. They establish a set of common technical rules for the design of building and civil engineering works that will ultimately replace the differing standards (codes) in European member states. The Eurocodes have been available as European prestandards (EuroNorm Vornorms or ENVs) for some years and now are being converted to the full European Standards (EuroNorms or ENs). In October 2001 the first two standards were converted to the full Eurocodes (ENs). These are: EN 1990 'Eurocode: Basis of structural design', which is the head Eurocode covering information and guidance on the principles and requirements for safety, serviceability and durability relating to all kinds of building and civil engineering structures; and EN 1991 'Eurocode 1: Actions on structures', which provides detailed information and guidance on all actions (loads) that should normally be considered in the design of building and civil engineering structures. Several other codes or parts of codes are scheduled to come into operation in 2002 and all should be converted by 2005. The ten Eurocodes are as follows:

EN 1990	(EC)	*Eurocode*: Basis of structural design
EN 1991	(EC1)	*Eurocode 1*: Actions on structures
EN 1992	(EC2)	*Eurocode 2*: Design of concrete structures
EN 1993	(EC3)	*Eurocode 3*: Design of steel structures
EN 1994	(EC4)	*Eurocode 4*: Design of composite steel and concrete structures
EN 1995	(EC5)	*Eurocode 5*: Design of timber structures
EN 1996	(EC6)	*Eurocode 6*: Design of masonry structures
EN 1997	(EC7)	*Eurocode 7*: Geotechnical design
EN 1998	(EC8)	*Eurocode 8*: Design of structures for earthquake regions
EN 1999	(EC9)	*Eurocode 9*: Design of aluminium structures.

Eurocode 5 (EN 1995 'Design of timber structures') will comprise two parts, the first part being subdivided into two sections. These are:

EN 1995-1-1 'Common rules and rules for buildings'
EN 1995-1-2 'Structural fire design'
EN 1995-2 'General rules: supplementary rules for bridges'.

The draft edition of Eurocode 5: Part 1.1 for timber structures was first published in the UK in late 1994 as DD ENV 1995-1-1: 1994, in conjunction with a National Application Document (NAD) for the United Kingdom, which contained supplementary information to facilitate its use nationally during the ENV period. The second draft was produced in October 2000 for comments. The most recent draft pre-standard was then published as final draft prEN 1995-1-1 in February 2002. The publication of the definitive EN 1995-1-1 is planned for late 2002 and is expected to be used in parallel with BS 5268 until the year 2008 (overlap period). It is also expected that by 2005 all nationally determined parameters will be calibrated and adopted so that the full implementation of EN 1995-1-1 can take place in around 2008. Meanwhile BS 5268 is to be termed 'obsolete' by 2005 and then withdrawn prior to full implementation of EN 1995.

In general, Eurocode 5 (EC5) contains all the rules necessary for the design of timber structures, but unlike BS 5268: Part 2 it does not provide material properties and other necessary design information. Such information is available in supporting European standards, for example BS EN 338: 1995 which contains the material properties such as bending and shear strength, etc., for the 15 strength classes of timber (i.e. C14 to D70). Therefore to design with EC5 it is necessary to refer to several additional documents.

This chapter is based on final draft prEN 1995-1-1: 2002 with the National Application Document (NAD) from DD ENV 1995-1-1: 1994 as at the time of writing, they are the most up-to-date editions of EC5 and its NAD.

27.2 SYMBOLS AND NOTATIONS

In Eurocode 5 the following symbols and subscripts are used to identify section properties of timber elements, applied loading (action) conditions, design forces, stresses and strengths.

Forces (actions)

F *force* due to either a *direct action* (applied load) or an *indirect action* such as imposed deformation due to settlement of supports or temperature effects
G *permanent actions* due to self-weight, fixtures, prestressing force and indirect action (e.g. settlement of supports)
Q *variable actions* due to imposed floor loads, snow loads, wind loads and indirect action (e.g. temperature effects)
A *accidental actions* due to, for example, explosions, fire or impact from vehicles
R resistance or load-carrying capacity
M bending moment

Geometrical and mechanical properties

a spacing distance
α angle of grain
A cross-sectional area
b breadth or width of a section or area
d diameter
E stiffness property, i.e. modulus of elasticity
G modulus of rigidity or shear modulus
h depth
I second moment of area
L length, span
m mass
n number
λ slenderness ratio
ρ density
X strength property
Z section modulus

Stresses

σ normal stress (bending, compression or tensile stress)
τ shear stress

Deflection

u deflection or joint slip

Coefficients

γ partial safety factor for actions and materials
k modification factor
Ψ coefficient for representative values of actions

Subscripts

ax axial
α grain direction with respect to applied load
c compression
cr critical
d design value
ef effective
ext external
Ed design value of a force
f flange
fin final
G permanent action
inf lower (inferior)
inst instantaneous
k characteristic
l lower
m bending
M material (property)

max	maximum
min	minimum
mod	modification
nom	nominal
Q	variable action
Rd	design resistance (capacity)
Rk	characteristic resistance (capacity)
ser	serviceability
sup	upper (superior)
sys	system strength (load sharing)
t	tension
tor	torsion
u	ultimate
v	shear
w	web
0	precamber (deflection) or timber grain parallel to direction of load/stress
0.05	fifth percentile value of a strength property, determined by standard tests

27.2.1 Decimal point

In EC5, in common with other Eurocodes based on standard ISO practice, decimal points are displayed as commas; for example, 1.2 is shown as 1,2. However, this text retains the current UK practice.

27.3 DESIGN PHILOSOPHY

EC5 along with other Eurocodes uses the concept of **limit states** design philosophy. Limit states are states beyond which a structure no longer satisfies the design performance requirements. The head Eurocode EN 1990 'Basis of structural design' specifies that the overall performance of structures should satisfy two basic requirements. The first is **ultimate limit states** (i.e. *safety*), usually expressed in terms of load-bearing capacity, and the second is **serviceability limit states** (i.e. *deformation and vibration* limits), which refers to the ability of a structural system and its elements to perform satisfactorily in normal use.

Ultimate limit states are those associated with total or partial collapse or other forms of structural failure. They concern the safety of the structure and its contents and the safety of people.

Serviceability limit states correspond to conditions beyond which specified service requirements for a structure or structural element are no longer met. They concern the functioning of the construction works or parts of them, the comfort of people and the appearance.

In general, it is well understood that the violation of ultimate limit states, i.e. safety criteria, may cause substantial damage and risk to human life, whereas exceeding the serviceability limits, i.e. excessive deformation and/or vibration, rarely leads to such severe consequences. However, while this may suggest that serviceability is relatively unimportant, in practice it is often the key to ensuring an economical and trouble-free structure.

Generally, it is necessary to check that design criteria for both ultimate limit states and serviceability limit states are satisfied. For checking ultimate limit states, the *characteristic values* of both the loads and the material properties are modified by specified partial safety factors that reflect the reliability of the values that they modify. These factors increase the values of the loads and decrease the values of the material properties. To check serviceability limit states EC5 requires that both instantaneous and time-dependent (creep) deflections are calculated and in addition, designers are required to demonstrate that vibration is not excessive (e.g. in floors).

The *characteristic values* are generally fifth percentile values derived directly from a statistical analysis of laboratory test results. The characteristic strength is the value below which the strength lies in only a small percentage of cases (not more than 5%), and the *characteristic load* is the value above which the load lies in only a small percentage of cases.

27.3.1 Principles and rules of application

Throughout EC5, depending on the character of the individual clauses, a distinction is made between **principles** and **rules of application**.

- *Principles* are statements, definitions, requirements or analytical models for which no alternative is permitted unless specifically stated. Principles are preceded by the letter P.
- *Rules of application* are generally recognized rules or design procedures which follow the principles and satisfy their requirements.

Designers can use alternative design methods to those recommended in the Eurocodes provided they can show that the alternatives comply with the relevant principles and are at least equivalent with regard to the structural safety, serviceability and durability that would be expected when using the Eurocode. It should be noted that if an alternative design rule is substituted for an application rule, the resulting design should not be claimed to be wholly in accordance with the Eurocode even though the design may satisfy its principles.

27.4 ACTIONS

Actions is the Eurocode terminology (in accordance with EN 1990) used to define: (1) direct loads (forces) applied to the structures and (2) indirect forces due to imposed deformations such as temperature effects or settlement. Actions are covered in EN 1991 'Eurocode 1: Actions on structures'.

Actions are classified as three types: (1) *permanent actions* denoted by the letter G, are all the dead loads acting on the structure, including the self-weight, finishes and fixtures and also indirect loads such as support settlements; (2) *variable actions* denoted by the letter Q, are the imposed floor loads, wind and snow loads and also

Table 27.1 Reduction factor Ψ for variable actions on buildings

Variable action	Ψ_0 indicative EN 1990	Ψ_0 UK*	Ψ_1 indicative EN 1990	Ψ_1 UK*	Ψ_2 indicative EN 1990	Ψ_2 UK*
Imposed floor loads in buildings						
dwellings	0.7	0.5	0.5	0.4	0.3	0.2
offices	0.7	0.7	0.5	0.6	0.3	0.3
parking	0.7	0.7	0.7	0.7	0.6	0.6
Imposed roof loads in buildings	0.6	0.7	0.2	0.2	0.0	0.0
Wind loads on buildings	0.6	0.7	0.2	0.2	0.0	0.0

*Based on UK NAD, DD ENV 1995-1-1: 1994.

indirect loads such as temperature effects; and (3) *accidental actions* denoted by the letter A, include effects of explosions, fire and impact from vehicles.

The characteristic value of an action is its main representative value. Guidance on obtaining the characteristic values of actions is given in the head Eurocode EN 1990 'Basis of structural design'. For example, the self-weight of a structure can be represented by a single characteristic value G_k if the variability of G is small. In situations where the statistical distribution indicates that the variability of G is not small, two values are used: a lower value $G_{k,inf}$ and an upper value $G_{k,sup}$.

According to the head Eurocode, a variable action has the following four representative values: (1) the characteristic value Q_k, (2) the combination value $\Psi_0 Q_k$, (3) the frequent value $\Psi_1 Q_k$ and (4) the quasi-permanent value $\Psi_2 Q_k$.

The combination value $\Psi_0 Q_k$ is used to verify ultimate limit states and irreversible serviceability limit states by considering the reduced probability of occurrence of several independent variable actions simultaneously. The frequent value $\Psi_1 Q_k$, which gives a greater reduction than Ψ_0, is used to verify ultimate limit states in accidental design situations, and also reversible serviceability limit states. The quasi-permanent value $\Psi_2 Q_k$ which gives the greatest reduction of all, is used only in accidental design situations for verification of ultimate limit states.

The indicative reduction factors Ψ for variable actions are given in the head Eurocode and are reproduced here in Table 27.1, together with those adopted in the UK NAD, based on DD ENV 1995-1-1: 1994.

27.4.1 Design values of actions

The term *design* is used for factored loading and member resistance and is generally obtained by using the characteristic or representative values in combination with partial safety factors, γ. The design value of a combination of actions is calculated from the appropriate expressions given in the head Eurocode EN 1990. When more than one variable action is involved, the design value is the most critical value of the expression, taking each variable action in turn as the main one and the others as secondary ones. This critical value is termed the *design load* for ultimate limit states, or the *service load* for serviceability limit states calculations.

Ultimate limit states

According to EN 1990, for the ultimate limit states verification, three types of combination of actions must be investigated: fundamental, accidental and seismic (if relevant).

In fundamental (persistent and transient) design situations, the design values of actions, F_d, are symbolically represented as:

$$F_d = \sum_{j\geq 1} \gamma_{G,j} G_{k,j} + \gamma_p P + \gamma_{Q,1} Q_{k,1} + \sum_{i>1} \gamma_{Q,i} \Psi_{0,i} Q_{k,i} \tag{27.1}$$

where $\Sigma\gamma_{G,j}\, G_{k,j}$ is the sum of factored permanent actions, $\gamma_p P$ is the factored pre-stressing action, $\gamma_{Q,1}\, Q_{k,1}$ is the factored dominant variable actions and $\Sigma\gamma_{Q,i}\, \Psi_{0,i}\, Q_{k,i}$ is the combination of other factored variable actions. Note that all characteristic values for actions are modified by partial safety factors (γ), see Table 27.2, and, the letters *j* and *i*, shown in the expressions, identify individual permanent and variable actions.

In accidental design situations, the design values of actions, F_d, are symbolically represented as:

$$F_d = \sum_{j\geq 1} G_{k,j} + P + A_d + (\Psi_{1,1} \text{ or } \Psi_{2,1}) Q_{k,1} + \sum_{i>1} \Psi_{2,i} Q_{k,i} \tag{27.2}$$

where $\Sigma G_{k,j}$ is the sum of permanent actions, P is the pre-stressing action, A_d is the design accidental action, $(\Psi_{1,1}$ or $\Psi_{2,1})\, Q_{k,1}$ is the frequent or quasi-permanent value of the dominant variable action and $\Sigma\Psi_{2,i}\, Q_{k,i}$ is all quasi-permanent values of other variable actions. Note that partial safety factors are not included.

EN 1990 recommends that for buildings, the partial factors for the ultimate limit states, for the permanent and variable situations, are considered as $\gamma_G = 1.35$ and $\gamma_Q = 1.5$, but these can be altered by the national annex.

Serviceability limit states

According to EN 1990, for the serviceability limit states verification, three types of combination of actions must be investigated: rare, frequent and quasi-permanent.

In the rare combination, where exceeding a limit state causes a permanent local damage or an unacceptable deformation, design loads at the serviceability limit state, $F_{d,ser}$, are symbolically represented as:

$$F_{d,ser} = \sum_{j\geq 1} G_{k,j} + P + Q_{k,1} + \sum_{i>1} \Psi_{0,i} Q_{k,i} \tag{27.3}$$

where $\Sigma G_{k,j}$ is the sum of permanent actions, P is the pre-stressing action, $Q_{k,1}$ is the dominant variable actions and $\Sigma\Psi_{0,i}\, Q_{k,i}$ is the combination of other variable actions. Note that all partial safety factors for serviceability limit states are equal to 1. The frequent combination refers to situations when exceeding a limit state causes large deformations, temporary vibrations or even local damage, and is symbolically represented as:

$$F_{d,ser} = \sum_{j\geq 1} G_{k,j} + P + \Psi_{1,1} Q_{k,1} + \sum_{i>1} \Psi_{2,i} Q_{k,i} \tag{27.4}$$

Table 27.2 Partial safety factors for actions in building structures for persistent and transient design situations

	Permanent (γ_G)		Variable (γ_Q) One with its characteristic value		Variable (γ_Q) Others with their combination value	
	EN 1990	EC5*	EN 1990	EC5*	EN1990	EC5*
Normal values						
Favourable	1.0	1.0		0		0
Unfavourable	1.35	1.35	1.5	1.5	1.5	1.5
Reduced values†						
Favourable		1.0		0		0
Unfavourable		1.2		1.35		1.35

*DD ENV 1995-1-1: 1994. †For specified small structures.

where $\Sigma G_{k,j}$ is the sum of permanent actions, P is the pre-stressing action, $\Psi_{1,1} Q_{k,1}$ is the dominant variable action and $\Sigma\Psi_{2,i} Q_{k,i}$ is all quasi-permanent values of other variable actions.

The quasi-permanent combination applies to situations where long-term effects are of importance and is represented by:

$$F_{d,ser} = \sum_{j\geq 1} G_{k,j} + P + \sum_{i>1} \Psi_{2,i} Q_{k,i} \qquad (27.5)$$

In Table 27.2 the values of the partial safety factors γ_G and γ_Q used to modify permanent and variable loads respectively, are shown. In this table the values of γ_G and γ_Q as given by the head Eurocode EN 1990 are compared with those of DD ENV 1995-1-1: 1994 and its UK NAD, which are, at the time of writing, the most up-to-date values available for use in timber structures.

27.5 MATERIAL PROPERTIES

Section 3 of prEN 1995-1-1 (final draft EC5: Part 1.1) deals with the material properties and defines the strength and stiffness parameters, stress–strain relations and gives values for modification factors for strength and deformation under various service classes and/or load duration classes. EC5, unlike BS 5268: Part 2: 1996 does not contain the material properties values and, as mentioned earlier, this information can be found in a supporting standard, i.e. in Table 1 of BS EN 338: 1995, reproduced here as Table 27.3. A comparison of this table with Table 7 of BS 5268: Part 2: 1996 highlights the difference between the characteristic values and the grade stress values familiar to users of BS 5268. The characteristic values, which are generally fifth percentile values, are considerably higher than the BS 5268 grade values which have been reduced for long-term duration and already include relevant safety factors. In addition BS EN 338 does not include grade TR26 timber. This is an extra grade which has been included in BS 5268 only.

Table 27.3 Characteristic values for structural timber strength classes (based on Table 1, BS EN 338: 1995)

	Softwoods									Hardwoods					
Strength class	C14	C16	C18	C22	C24	C27	C30	C35	C40	D30	D35	D40	D50	D60	D70
Strength properties (N/mm^2)															
Bending ($f_{m,k}$)	14	16	18	22	24	27	30	35	40	30	35	40	50	60	70
Tension ‖ ($f_{t,0,k}$)	8	10	11	13	14	16	18	21	24	18	21	24	30	36	42
Tension ⊥ ($f_{t,90,k}$)	0.3	0.3	0.3	0.3	0.4	0.4	0.4	0.4	0.4	0.6	0.6	0.6	0.6	0.7	0.9
Compression ‖ ($f_{c,0,k}$)	16	17	18	20	21	22	23	25	26	23	25	26	29	32	34
Compression ⊥ ($f_{c,90,k}$)	4.3	4.6	4.8	5.1	5.3	5.6	5.7	6.0	6.3	8.0	8.4	8.8	9.7	27.5	13.5
Shear ($f_{v,k}$)	1.7	1.8	2.0	2.4	2.5	2.8	3.0	3.4	3.8	3.0	3.4	3.8	4.6	5.3	6.0
Stiffness properties (kN/mm^2)															
Mean modulus of elasticity ‖ ($E_{0,mean}$)	7	8	9	10	11	12	12	13	14	10	10	11	14	17	20
Fifth percentile modulus of elasticity ‖ ($E_{0.05}$)	4.7	5.4	6.0	6.7	7.4	8.0	8.0	8.7	9.4	8.0	8.7	9.4	11.8	14.3	16.8
Mean modulus of elasticity ⊥ ($E_{90,mean}$)	0.23	0.27	0.30	0.33	0.37	0.40	0.40	0.43	0.47	0.64	0.69	0.75	0.93	1.13	1.33
Mean shear modulus (G_{mean})	0.44	0.50	0.56	0.63	0.69	0.75	0.75	0.81	0.88	0.60	0.65	0.7	0.88	1.06	1.25
Density (kg/m^3)															
Characteristic density (ρ_k)	290	310	320	340	350	370	380	400	420	530	560	590	650	700	900
Average density (ρ_{mean})	350	370	380	410	420	450	460	480	500	640	670	700	780	840	1080

27.5.1 Design values

The characteristic strength values given in Table 1 of BS EN 338, see Table 27.3, are measured in specimens of certain dimensions conditioned at 20 °C and 65% relative humidity, in tests lasting approximately five minutes; and therefore they are applicable to such loading and service conditions. For other loading and service conditions, the characteristic values should be modified using relevant modification factors, which are described later.

EC5 in Section 2.4 specifies that the design value X_d of a strength property is calculated from its characteristic value, X_k, using the following equation:

$$\text{strength properties,} \quad X_d = k_{mod} \frac{X_k}{\gamma_M} \tag{27.6}$$

Similarly, the design value E_d of a member stiffness property is calculated from the following equation:

$$\text{stiffness properties,} \quad E_d = \frac{E_{0.05}}{\gamma_M} \tag{27.7}$$

EC5 also states that the design value R_d of a resistance (i.e. load-carrying capacity) should be calculated as:

$$\text{design resistance,} \quad R_d = k_{mod} \frac{R_k}{\gamma_M} \tag{27.8}$$

where γ_M = the partial safety factor for material property and its values are given in Table 2.2 of EC5, reproduced here as Table 27.4.

k_{mod} = the strength modification factor for service class and duration of load. k_{mod} takes into account five different load–duration classes: permanent, long-term, medium-term, short-term and instantaneous, relating to service classes 1, 2 and 3 as defined in clauses 2.3.1.2, 2.3.1.3 and 3.1.3 of EC5. A summary of these clauses and Tables 2.1 and 3.1 of EC5 is reproduced here as Table 27.5. For metal components k_{mod} should be taken as 1, i.e. $k_{mod} = 1.0$.

Table 27.4 Partial coefficients for material properties (based on Table 2.2, EC5)

Limit states	γ_M
Ultimate limit states	
Fundamental combinations	
Solid timber	1.3
LVL, OSB, plywood, particle/fibre boards	1.2
Other wood-based materials	1.3
Glued laminated timber	1.25
Connections	1.3
Punched metal plates	1.25
Accidental combinations	1.0
Serviceability limit states	1.0

Table 27.5 Modification factor k_{mod} for service classes and duration of load. Values of k_{mod} for solid, glued-laminated timber, LVL and plywood (Table 3.1, EC5)

	Service class†		
Load duration class*	1	2	3
Permanent	0.60	0.60	0.50
Long term	0.70	0.70	0.55
Medium term	0.80	0.80	0.65
Short term	0.90	0.90	0.70
Instantaneous	1.10	1.10	0.90

* Load–duration classes (Table 2.1, EC5):
Permanent refers to order of duration of more than 10 years, e.g. self-weight,
Long term refers to order of duration of 6 months to 10 years, e.g. storage,
Medium term refers to order of duration of 1 week to 6 months, e.g. imposed floor load, snow,
Short term refers to order of duration of less than 1 week, e.g. snow, wind load,
Instantaneous refers to sudden loading, e.g. wind and accidental load.

† Service classes (Clause 2.3.1.3, EC5):
Service class 1 relates to a typical service condition of 20 °C, 65% RH with moisture content ≤12%,
Service class 2 relates to a typical service condition of 20 °C, 85% RH with moisture content ≤20%,
Service class 3 relates to a service condition leading to a higher moisture content than 20%.

For a connection which is constituted of two different timber based elements with strength modification factors of $k_{mod,1}$ and $k_{mod,2}$, the k_{mod} value should be calculated as:

$$k_{mod} = \sqrt{k_{mod,1} k_{mod,2}} \tag{27.9}$$

$E_{0.05}$ = the fifth percentile value of a stiffness property, determined by standard tests

R_k = the characteristic value of a load-carrying capacity.

The characteristic strength values given in Table 1 of BS EN 338, see Table 27.3, are related to a depth in bending and width in tension of 150 mm or more. For a depth in bending or width in tension of solid timber, h, less than 150 mm, EC5 in clause 3.2 recommends that the characteristic values for $f_{m,k}$ and $f_{t,0,k}$ may be increased by the factor k_h where:

$$k_h = \text{the lesser of} \begin{cases} k_h = \left(\dfrac{150}{h}\right)^{0.2} \\ k_h = 1.3 \end{cases} \tag{27.10}$$

27.6 ULTIMATE LIMIT STATES

Section 6 of EC5: Part 1.1 sets out the design procedures for members of solid timber, glued-laminated timber or wood-based structural products of constant cross section with regard to calculation for their strength properties (ultimate limit states) and treats them in the same way. Design procedures relevant to flexural and axially loaded members and dowel-type joints are described below.

27.6.1 Bending

Members should be designed so that the following conditions are satisfied:

$$\frac{\sigma_{m,y,d}}{f_{m,y,d}} + k_m \frac{\sigma_{m,z,d}}{f_{m,z,d}} \leq 1 \tag{27.11a}$$

$$k_m \frac{\sigma_{m,y,d}}{f_{m,y,d}} + \frac{\sigma_{m,z,d}}{f_{m,z,d}} \leq 1 \tag{27.11b}$$

In these equations:

- k_m is a modification factor for vatiation in material properties and for bending stress redistribution and is detailed in clause 6.1.5(2) of EC5 as:
  For solid, glued-laminated timber and LVL:
  for rectangular sections, $k_m = 0.7$
  for other cross sections, $k_m = 1.0$
  For other wood-based structural products: $k_m = 1.0$
- $\sigma_{m,y,d}$ and $\sigma_{m,z,d}$ are the design bending stresses about the principal axes y–y and z–z as shown in Fig. 27.1, where

$$\sigma_{m,y,d} = \frac{M_{y,d}}{W_y}$$

in which $M_{y,d}$ = design bending moment about the y–y axis and $W_y = bh^2/6$ is the appropriate section modulus,

$$\sigma_{m,z,d} = \frac{M_{z,d}}{W_z}$$

in which $M_{z,d}$ = design bending moment about the z–z axis and $W_z = hb^2/6$ is the appropriate section modulus.

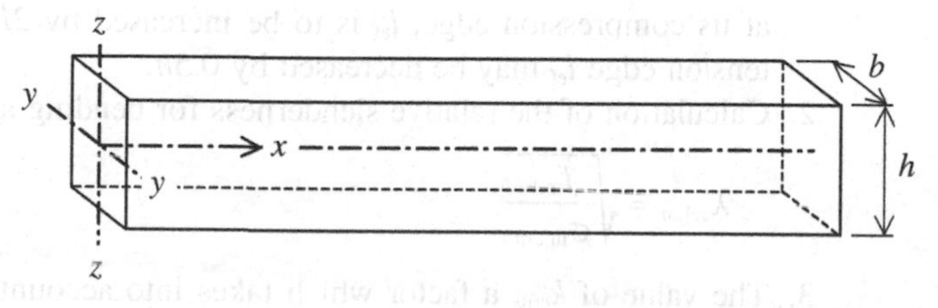

Fig. 27.1 Beam axes.

Also in eqs (27.11) the terms $f_{m,y,d}$ and $f_{m,z,d}$ are the design bending strengths about the y–y and z–z axes, where:

$$f_{m,y/z,d} = \frac{k_{mod}k_h k_{crit}k_{sys}f_{m,k}}{\gamma_M} \tag{27.12}$$

in which:

- k_{mod} is the modification factor for load duration and service classes as given in Table 3.1 of EC5 (see Table 27.5)
- k_h is the modification factor for bending depth (clause 3.2)
- k_{crit} is the modification factor for reducing bending strength of a beam where there is a possibility of lateral buckling. For a beam which is laterally restrained throughout the length of its compression edge and its ends are prevented from torsional rotation $k_{crit} = 1.0$. For other conditions, clause 6.3.3 of EC5 recommends the following:

1. Calculation of the critical bending stress $\sigma_{m,crit}$, as given by the following equation for softwood with solid rectangular section of breadth b and depth h:

$$\sigma_{m,crit} = \frac{0.78b^2}{l_{ef}h}E_{0.05} \tag{27.13}$$

where $E_{0.05}$ is the fifth percentile value of modulus of elasticity parallel to grain (Table 1, BS EN 338: 1995), see Table 27.3.

For other sections, the critical bending stress may be calculated according to the classical theory of stability, using fifth percentile stiffness values.

The effective length, l_{ef}, is governed by the support conditions and load configuration, and, based on Table 6.1 of EC5, is considered as:

for simply supported beams:

$l_{ef} = 1.0l$ for a beam subjected to a constant moment
$l_{ef} = 0.9l$ for a beam subjected to a constant distributed load
$l_{ef} = 0.8l$ for a beam subjected to a concentrated force at midspan.

for cantilevered beams:

$l_{ef} = 0.5l$ for a beam subjected to a distributed load
$l_{ef} = 0.8l$ for a beam subjected to a concentrated force at the free end.

Note that the relation between l_{ef} and l is valid for a beam with torsional restrained supports and loaded at the centre of gravity. If the beam is loaded at its compression edge, l_{ef} is to be increased by $2h$, and if loaded at its tension edge l_{ef} may be decreased by $0.5h$.

2. Calculation of the relative slenderness for bending as:

$$\lambda_{rel,m} = \sqrt{\frac{f_{m,k}}{\sigma_{m,crit}}} \tag{27.14}$$

3. The value of k_{crit}, a factor which takes into account the reduced strength due to lateral buckling, is then determined from:

$k_{crit} = 1$ for $\lambda_{rel,m} \leq 0.75$
$k_{crit} = 1.56 - 0.75\,\lambda_{rel,m}$ for $0.75 < \lambda_{rel,m} \leq 1.40$
$k_{crit} = 1 / \lambda^2_{rel,m}$ for $1.40 < \lambda_{rel,m}$

- k_{sys} is the modification factor for system strength (load sharing systems), see clause 6.7 of EC5 for details of the conditions outlined for system strength ($k_{sys} = 1.1$)
- $f_{m,k}$ is the characteristic bending strength (Table 1, BS EN 338: 1995), see Table 27.3, and
- γ_M is the partial coefficient for material properties (Table 2.2, EC5), see Table 27.4.

27.6.2 Shear

Members should be designed so that the following condition is satisfied (clause 6.1.6, EC5):

$$\tau_d \leq f_{v,d} \tag{27.15}$$

In eq. (27.15):

- τ_d is the design shear stress, and for beams with a rectangular cross section, is given by:

$$\tau_d = \frac{3\,V_d}{2\,A} \tag{27.16}$$

  where V_d is the design shear force (maximum reaction or concentrated load) and A is the cross-sectional area, where $A = bh$. For beams with a notched end $A = bh_{ef}$, see Fig. 27.2
- $f_{v,d}$ is the design shear strength, which is given by:

$$f_{v,d} = \frac{k_{mod} k_{sys} k_v f_{v,k}}{\gamma_M} \tag{27.17}$$

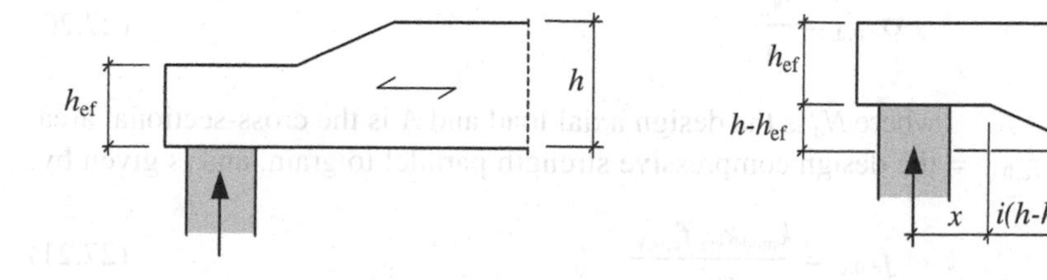

Fig. 27.2 End-notched beams.

In eq. (27.17):

- $f_{v,k}$ is the characteristic shear strength (Table 1, BS EN 338: 1995)
- γ_M is partial coefficient for material properties (Table 2.2, EC5)
- k_{mod} is the modification factor for load duration and service classes (Table 3.1, EC5)
- k_{sys} is the modification factor for system strength (load-sharing systems), ($k_{sys} = 1.1$), (clause 6.7, EC5)
- k_v is the modification factor for shear in members with notched ends, see Fig. 27.2, where
  - for beams with no notched ends, $k_v = 1.0$
  - for beams notched at the opposite side of the support, $k_v = 1.0$
  - for beams notched at the support side,

$$k_v = \text{the lesser of: } \begin{cases} k_v = \dfrac{k_n\left(1 + \dfrac{1.1 i^{1.5}}{\sqrt{h}}\right)}{\sqrt{h}\left(\sqrt{\alpha(1-\alpha)} + 0.8\dfrac{x}{h}\sqrt{\dfrac{1}{\alpha} - \alpha^2}\right)} \\ k_v = 1.0 \end{cases} \tag{27.18}$$

where k_n = 5.0 for solid timber
= 6.5 for glued-laminated timber
= 4.5 for LVL
h = beam depth (mm)
x = distance from line of action to the corner
$\alpha = h_{ef}/h$
i = notch inclination.

27.6.3 Compression or tension parallel to grain

For members subjected to axial compression only, the following condition should be satisfied, provided there is no tendency for buckling to occur (clause 6.1.3, EC5):

$$\sigma_{c,0,d} \leq f_{c,0,d} \tag{27.19}$$

where $\sigma_{c,0,d}$ = the design compressive stress parallel to grain, and is given by:

$$\sigma_{c,0,d} = \frac{N_d}{A} \tag{27.20}$$

where N_d is the design axial load and A is the cross-sectional area.
$f_{c,0,d}$ = the design compressive strength parallel to grain, and is given by:

$$f_{c,0,d} = \frac{k_{mod} k_{sys} f_{c,0,k}}{\gamma_M} \tag{27.21}$$

where $f_{c,0,k}$ = the characteristic compressive strength parallel to grain (Table 1, BS EN 338: 1995)

γ_M = partial coefficient for material properties (Table 2.2, EC5)
k_{mod} = the modification factor for load – duration and service classes (Table 3.1, EC5), and
k_{sys} = the modification factor for system strength (load-sharing systems), (k_{sys} = 1.1), (clause 6.7, EC5).

For members subjected to *axial tension* a similar condition and procedure apply (clause 6.1.1, EC5).

27.6.4 Compression perpendicular to grain (bearing)

For compression perpendicular to grain the following condition should be satisfied (clause 6.1.4, EC5):

$$\sigma_{c,90,d} \leq k_{c,90} f_{c,90,d} \tag{27.22}$$

where $\sigma_{c,90,d}$ = the design compressive stress in the contact area perpendicular to grain
$f_{c,90,d}$ = the design compressive strength perpendicular to grain, and is given by:

$$f_{c,90,d} = \frac{k_{mod} k_{sys} f_{c,90,k}}{\gamma_M} \tag{27.23}$$

in which:
$f_{c,90,k}$ = the characteristic compressive strength perpendicular to grain (Table 1, BS EN 338: 1995)
γ_M = partial coefficient for material properties (Table 2.2, EC5)
k_{mod} = the modification factor for load – duration and service classes (Table 3.1, EC5)
k_{sys} = the modification factor for system strength (load-sharing systems), (i.e. k_{sys} = 1.1), (clause 6.7, EC5)
$k_{c,90}$ = the modification factor for bearing length, which takes into account the load configuration, possibility of splitting and degree of compressive deformation. The value of $k_{c,90}$ should be taken as 1.0 unless the conditions and the load and support configurations specified in clauses 6.1.4(3) to 6.1.4(5) of EC5 apply, in which case a value up to 4.0 is permitted.

27.6.5 Members subjected to combined bending and axial tension

For members subjected to combined bending and axial tension, the following conditions should be satisfied (clause 6.2.2, EC5):

$$\frac{\sigma_{t,0,d}}{f_{t,0,d}} + \frac{\sigma_{m,y,d}}{f_{m,y,d}} + k_m \frac{\sigma_{m,z,d}}{f_{m,z,d}} \leq 1.0 \tag{27.24a}$$

$$\frac{\sigma_{t,0,d}}{f_{t,0,d}} + k_m \frac{\sigma_{m,y,d}}{f_{m,y,d}} + \frac{\sigma_{m,z,d}}{f_{m,z,d}} \leq 1.0 \tag{27.24b}$$

where $\sigma_{t,0,d}$ is the design tensile stress and $f_{t,0,d}$ is the design tensile strength parallel to grain; $\sigma_{m,y,d}$ and $\sigma_{m,z,d}$ are the respective design bending stresses due to any lateral or eccentric loads (see section 27.6.1) and $f_{m,y,d}$ and $f_{m,z,d}$ are the design bending strengths. For solid, glued-laminated timber and LVL, $k_m = 0.7$ for rectangular sections and $k_m = 1.0$ for other cross sections (clause 6.1.5, EC5).

27.6.6 Columns subjected to combined bending and axial compression

For the design of columns subjected to combined bending and axial compression (clause 6.2.3, EC5) and also for the design of slender columns in general, the following requirements should be considered:

1. Calculation of the relative slenderness ratios about the y–y and z–z axes of the column, as defined by:

$$\lambda_{rel,y} = \frac{\lambda_y}{\pi}\sqrt{\frac{f_{c,0,k}}{E_{0.05}}} \tag{27.25a}$$

$$\lambda_{rel,z} = \frac{\lambda_z}{\pi}\sqrt{\frac{f_{c,0,k}}{E_{0.05}}} \tag{27.25b}$$

where $\lambda_y = l_{ef,y}/i_y$, $\lambda_z = l_{ef,z}/i_z$ and $E_{0.05}$ is the fifth percentile modulus of elasticity parallel to the grain (Table 1, BS EN 338: 1995).

2. For both $\lambda_{rel,y} \leq 0.3$ and $\lambda_{rel,z} \leq 0.3$ the stresses should satisfy the following conditions:

$$\left(\frac{\sigma_{c,0,d}}{f_{c,0,d}}\right)^2 + \frac{\sigma_{m,y,d}}{f_{m,y,d}} + k_m \frac{\sigma_{m,z,d}}{f_{m,z,d}} \leq 1.0 \tag{27.26a}$$

$$\left(\frac{\sigma_{c,0,d}}{f_{c,0,d}}\right)^2 + k_m \frac{\sigma_{m,y,d}}{f_{m,y,d}} + \frac{\sigma_{m,z,d}}{f_{m,z,d}} \leq 1.0 \tag{27.26b}$$

where $\sigma_{c,0,d}$ is the design compressive stress and $f_{c,0,d}$ is the design compressive strength parallel to grain; $\sigma_{m,y,d}$ and $\sigma_{m,z,d}$ are the respective design bending stresses due to any lateral or eccentric loads (see section 27.6.1) and $f_{m,y,d}$ and $f_{m,z,d}$ are the design bending strengths. For solid, glued-laminated timber and LVL, $k_m = 0.7$ for rectangular sections and $k_m = 1.0$ for other cross sections (clause 6.1.5, EC5).

3. For all other cases (i.e. $\lambda_{rel,y} > 0.3$ and/or $\lambda_{rel,z} > 0.3$) the stresses, which will be increased due to deflection, should satisfy the following conditions:

$$\frac{\sigma_{c,0,d}}{k_{c,y}f_{c,0,d}}+k_m\frac{\sigma_{m,z,d}}{f_{m,z,d}}+\frac{\sigma_{m,y,d}}{f_{m,y,d}}\leq 1.0 \quad (27.27a)$$

$$\frac{\sigma_{c,0,d}}{k_{c,z}f_{c,0,d}}+\frac{\sigma_{m,z,d}}{f_{m,z,d}}+k_m\frac{\sigma_{m,y,d}}{f_{m,y,d}}\leq 1.0 \quad (27.27b)$$

where $k_{c,y}$ and $k_{c,z}$ are modification factors of compression members and are given by:

$$k_{c,y}=\frac{1}{k_y+\sqrt{k_y^2-\lambda_{rel,y}^2}} \quad \text{(similarly for } k_{c,z}) \quad (27.28a)$$

$$k_y=0.5(1+\beta_c(\lambda_{rel,y}-0.3)+\lambda_{rel,y}^2) \quad \text{(similarly for } k_z) \quad (27.28b)$$

where $\beta_c = 0.2$ for solid timber and $\beta_c = 0.1$ for glued-laminated timber and LVL.

27.6.7 Dowel-type fastener joints

EC5 section 8 deals with the ultimate limit states design criteria for joints made with dowel-type fasteners such as nails, screws, staples, bolts and dowels, and also connectored-type joints, i.e. toothed plates, split rings, etc.

EC5 provides formulae, based on Johansen's equations, to determine the load-carrying capacity of timber-to-timber, panel-to-timber and steel-to-timber joints, reflecting all possible failure modes. Load-carrying capacity formulae for timber-to-timber and panel-to-timber joints, with reference to their failure modes, are compiled in Table 27.6, and for steel-to-timber joints in Table 27.7.

As an example, for a two-member timber-to-timber or panel-to-timber joint there are six possible modes of failure. One or both of the members may fail, or the fastener may fail in one or both members, see failure modes (a) to (f) in Table 27.6. EC5 therefore provides six different formulae to calculate the failure load for each mode and recommends that the lowest value is taken as the failure load of the connection.

Therefore the characteristic load-carrying capacity, $F_{v,Rk}$, for a fastener per shear plane is given as the lesser of $F_{v,Rk,a}$ to $F_{v,Rk,f}$ (reproduced below from Table 27.6):

$$F_{v,Rk,a}=f_{h,1,k}t_1d \quad (27.29a)$$

$$F_{v,Rk,b}=f_{h,2,k}t_2d \quad (27.29b)$$

$$F_{v,Rk,c}=\frac{f_{h,1,k}t_1d}{1+\beta}\left[\sqrt{\beta+2\beta^2\left[1+\frac{t_2}{t_1}+\left(\frac{t_2}{t_1}\right)^2\right]+\beta^3\left(\frac{t_2}{t_1}\right)^2}-\beta\left(1+\frac{t_2}{t_1}\right)\right] \quad (27.29c)$$

$$F_{v,Rk,d}=1.05\frac{f_{h,1,k}t_1d}{2+\beta}\left[\sqrt{2\beta(1+\beta)+\frac{4\beta(2+\beta)M_{y,Rk}}{f_{h,1,k}t_1^2d}}-\beta\right]+\frac{F_{ax,Rk}}{4} \quad (27.29d)$$

$$F_{v,Rk,e}=1.05\frac{f_{h,1,k}t_2d}{1+2\beta}\left[\sqrt{2\beta^2(1+\beta)+\frac{4\beta(1+2\beta)M_{y,Rk}}{f_{h,1,k}t_2^2d}}-\beta\right]+\frac{F_{ax,Rk}}{4} \quad (27.29e)$$

$$F_{v,Rk,f}=1.15\sqrt{\frac{2\beta}{1+\beta}}\sqrt{2M_{y,Rk}f_{h,1,k}d}+\frac{F_{ax,Rk}}{4} \quad (27.29f)$$

Table 27.6 Load-carrying capacity of timber-to-timber and panel-to-timber joints (based on clause 8.2.1, EC5)

Joints in single shear

Failure modes

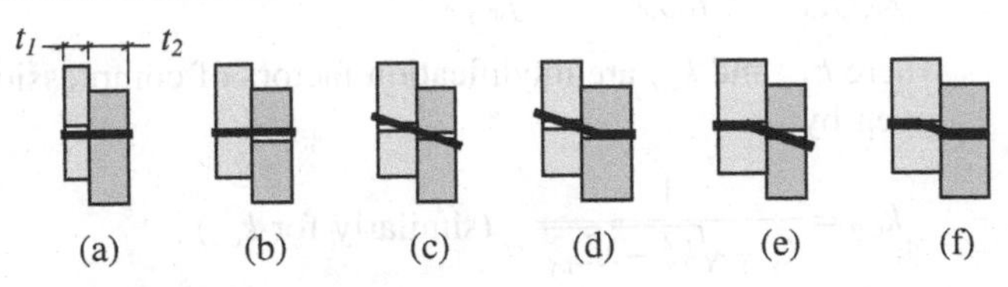

Characteristic load-carrying capacity per dowel, $F_{v,Rk}$, is the lesser of:

$$F_{v,Rk,a} = f_{h,1,k} t_1 d \quad (27.30a)$$

$$F_{v,Rk,b} = f_{h,2,k} t_2 d \quad (27.30b)$$

$$F_{v,Rk,c} = \frac{f_{h,1,k} t_1 d}{1+\beta}\left[\sqrt{\beta + 2\beta^2\left[1+\frac{t_2}{t_1}+\left(\frac{t_2}{t_1}\right)^2\right]+\beta^3\left(\frac{t_2}{t_1}\right)^2} - \beta\left(1+\frac{t_2}{t_1}\right)\right] \quad (27.30c)$$

$$F_{v,Rk,d} = 1.05\frac{f_{h,1,k} t_1 d}{2+\beta}\left[\sqrt{2\beta(1+\beta)+\frac{4\beta(2+\beta)M_{y,Rk}}{f_{h,1,k} t_1^2 d}} - \beta\right] + \frac{F_{ax,Rk}}{4} \quad (27.30d)$$

$$F_{v,Rk,e} = 1.05\frac{f_{h,1,k} t_2 d}{1+2\beta}\left[\sqrt{2\beta^2(1+\beta)+\frac{4\beta(1+2\beta)M_{y,Rk}}{f_{h,1,k} t_2^2 d}} - \beta\right] + \frac{F_{ax,Rk}}{4} \quad (27.30e)$$

$$F_{v,Rk,f} = 1.15\sqrt{\frac{2\beta}{1+\beta}}\sqrt{2M_{y,Rk} f_{h,1,k} d} + \frac{F_{ax,Rk}}{4} \quad (27.30f)$$

Joints in double shear

Failure modes

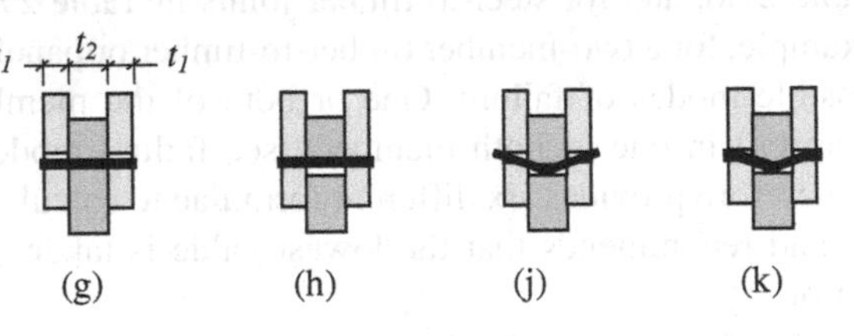

Characteristic load-carrying capacity per dowel per shear plane, $F_{v,Rk}$, is the lesser of:

$$F_{v,Rk,g} = f_{h,1,k} t_1 d \quad (27.30g)$$

$$F_{v,Rk,h} = 0.5 f_{h,2,k} t_2 d \quad (27.30h)$$

$$F_{v,Rk,j} = 1.05\frac{f_{h,1,k} t_1 d}{2+\beta}\left[\sqrt{2\beta(1+\beta)+\frac{4\beta(2+\beta)M_{y,Rk}}{f_{h,1,k} t_1^2 d}} - \beta\right] + \frac{F_{ax,Rk}}{4} \quad (27.30j)$$

$$F_{v,Rk,k} = 1.15\sqrt{\frac{2\beta}{1+\beta}}\sqrt{2M_{y,Rk} f_{h,1,k} d} + \frac{F_{ax,Rk}}{4} \quad (27.30k)$$

where t_1, t_2 = timber or board thickness or connector penetration

$f_{h,i,k}$ = the characteristic embedding strengths in timber member i
[The characteristic embedding strength values for different connection types are extracted from clauses 8.3 to 8.7 of EC5 and are compiled in Table 27.8. As an example, for a nailed

Table 27.7 Load-carrying capacity of steel-to-timber joints (based on clause 8.2.2, EC5)

Joints in single shear

Failure modes

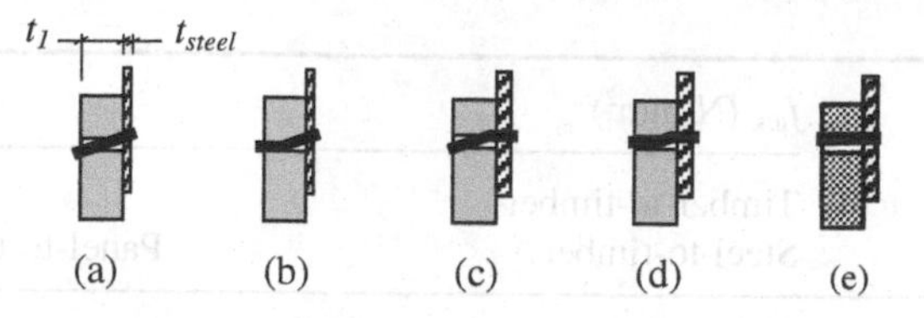

Characteristic load-carrying capacity per dowel, $F_{v,Rk}$, is the lesser of:

(1) Thin steel plate ($t_{steel} \leq 0.5d$)

$$F_{v,Rk,a} = 0.4 f_{h,k} t_1 d \tag{27.31a}$$

$$F_{v,Rk,b} = 1.15\sqrt{2M_{y,Rk} f_{h,k} d} \tag{27.31b}$$

(2) Thick steel plate ($t_{steel} \geq d$)

$$F_{v,Rk,c} = f_{h,k} t_1 d\left[\sqrt{2 + \frac{4M_{y,Rk}}{f_{h,k} t_1^2 d}} - 1\right] + \frac{F_{ax,Rk}}{4} \tag{27.31c}$$

$$F_{v,Rk,d} = 2.3\sqrt{M_{y,Rk} f_{h,k} d} + \frac{F_{ax,Rk}}{4} \tag{27.31d}$$

$$F_{v,Rk,e} = f_{h,k} t_1 d \tag{27.31e}$$

(3) Intermediate thick steel plate ($0.5d < t_{steel} < d$): use linearly interpolated values between (1) and (2).

Joints in double shear

Failure modes

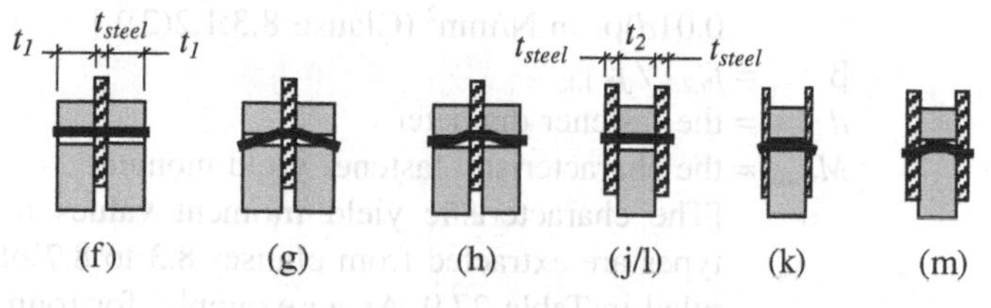

Characteristic load-carrying capacity per dowel per shear plane, $F_{v,Rk}$, is the lesser of:

(1) Steel plate of any thickness as the centre member

$$F_{v,Rk,f} = f_{h,1,k} t_1 d \tag{27.31f}$$

$$F_{v,Rk,g} = f_{h,1,k} t_1 d\left[\sqrt{2 + \frac{4M_{y,Rk}}{f_{h,1,k} t_1^2 d}} - 1\right] + \frac{F_{ax,Rk}}{4} \tag{27.31g}$$

$$F_{v,Rk,h} = 2.3\sqrt{M_{y,Rk} f_{h,1,k} d} + \frac{F_{ax,Rk}}{4} \tag{27.31h}$$

(2) Thin steel side plates ($t_{steel} \leq 0.5d$)

$$F_{v,Rk,j} = 0.5 f_{h,2,k} t_2 d \tag{27.31j}$$

$$F_{v,Rk,k} = 1.15\sqrt{2M_{y,Rk} f_{h,2,k} d} + \frac{F_{ax,Rk}}{4} \tag{27.31k}$$

(3) Thick steel side plates ($t_{steel} \geq d$)

$$F_{v,Rk,l} = 0.5 f_{h,2,k} t_2 d \tag{27.31l}$$

$$F_{v,Rk,m} = 2.3\sqrt{M_{y,Rk} f_{h,2,k} d} + \frac{F_{ax,Rk}}{4} \tag{27.31m}$$

(4) Intermediate thick steel side plates ($0.5d < t_{steel} < d$): use linearly interpolated values between (2) and (3).

Table 27.8 Values of the characteristic embedding strength, $f_{h,k}$ (based on Clauses 8.3 to 8.7, EC5)

Fastener type	$f_{h,k}$ (N/mm²) Timber-to-timber Steel-to-timber		Panel-to-timber	
Nails ($d \leq 6$ mm) and staples	$0.082\ \rho_k d^{-0.3}$	without pre-drilling	$0.11\ \rho_k d^{-0.3}$	for plywood
			$30t^{0.6}d^{-0.3}$	for hardboard
	$0.082(1 - 0.01d)\rho_k$	pre-drilled		
Bolts, dowels and nails ($d > 6$ mm)				
parallel to grain	$0.082(1 - 0.01d)\rho_k$		$0.11(1 - 0.01d)\rho_k$	
at an angle α to grain	$f_{h,0,k}/(k_{90}\sin^2\alpha + \cos^2\alpha)$			
Screws (see EC5, clause 8.7.1(3) and (4))	Rules for nails apply		Rules for nails apply	

ρ_k = the joint member characteristics density, given in Table 1, BS EN 338: 1995 (in kg/m³).
d = the fastener diameter (in mm). For screws d is the smooth shank diameter.
t = the panel thickness.
$k_{90} = 1.35 + 0.015d$ for softwoods
$= 0.90 + 0.015d$ for hardwoods.

timber-to-timber joint with pre-drilled holes $f_{h,k} = 0.082(1 - 0.01d)\rho_k$ in N/mm² (Clause 8.3.1.2(2)).]

β $= f_{h,2,k}/f_{h,1,k}$

d = the fastener diameter

$M_{y,Rk}$ = the characteristic fastener yield moment
[The characteristic yield moment values for different fastener types are extracted from clauses 8.3 to 8.7 of EC5 and are compiled in Table 27.9. As an example, for round wire nails $M_{y,Rk} = (f_u/600)180d^{2.6}$ in Nmm (clause 8.3.1.1, EC5); where f_u is the tensile strength of the wire in N/mm².]

$F_{ax,Rk}$ = the characteristic axially withdrawal capacity of the fastener. For single shear fasteners, this is the lower of the capacities of the two members, EC5 clause 8.3.2. For example, for smooth nails:

$$F_{ax,Rk} = \text{lesser of} \begin{cases} f_{ax,k}dt_{pen} & \text{and} \\ f_{ax,k}dt + f_{head,k}d_h^2 & \text{for } d_h \geq 2d \end{cases}$$

where $f_{ax,k}$ = the characteristic pointside withdrawal strength
$= 20 \times 10^6 \rho_k^2$

$f_{head,k}$ = the characteristic headside pull-through strength
$= 70 \times 10^6 \rho_k^2$

t_{per} = the pointisde penetration length

t = the thickness of the head side member

d_h = the fastener head diameter.

It should be noted that, as the load-carrying capacity of steel-to-timber connections near the ends of the timber member is affected by failure along the circumference

Table 27.9 Values of fastener yield moment, $M_{y,Rk}$ (based on clauses 8.3 to 8.7, EC5)

Fastener type	$M_{y,Rk}$ (N mm)
Nails – round wire	$(f_u/600)180d^{2.6}$
Nails – square	$(f_u/600)270d^{2.6}$
Bolts and dowels	$0.3f_{u,k}d^{2.6}$
Screws ($d \leq 6$ mm)	Rules for nails apply*
Screws ($d > 6$ mm)	Rules for bolts apply*
Staples – round wire	$240d^{2.6}$

f_u = the tensile strength of the wire
$f_{u,k}$ = the characteristic tensile strength of the bolt (for 4.6 grade steel $f_{u,k}$ = 400 N/mm²)
d = the fastener diameter (in mm).
*For smooth shank screws, effective diameter d_{ef} = smooth shank diameter d.

of the fastener group, the recommendations detailed in clause 8.2.2(4) of EC5 should be considered.

EC5 recommends that the effective characteristic load-carrying capacity of a multiple fastener $F_{v,ef,Rk}$, should be calculated from:

$$F_{v,ef,Rk} = mn_{ef}F_{v,Rk} \tag{27.32}$$

where $F_{v,ef,Rk}$ = the effective characteristic load-carrying capacity of the joint
m = the number of fastener rows in line with the force
$F_{v,Rk}$ = the characteristic load-carrying capacity per fastener
n_{ef} = a factor for the effective number of fasteners in a line parallel to the grain, and is given by:

1. *For nailed connections*:

$$n_{ef} = n^{k_{ef}} \tag{27.33}$$

where n_{ef} is the effective number of nails in a row, n is the number of nails in a row, and k_{ef} is a factor which depends on the ratio of nail spacing a_1 to nail diameter d, given as:

$k_{ef} = 1.0$ for $a_1 \geq 14d$
$k_{ef} = 0.85$ for $a_1 = 10d$
$k_{ef} = 0.7$ for $a_1 = 7d$
$k_{ef} = 0.5$ for $a_1 = 5d$ (in pre-drilled cases only).

For intermediate spacings, linear interpolation is permitted.

2. *For bolted connections*:

$$n_{ef} = \text{lesser of}\left(n \text{ and } n^{0.9}\sqrt[4]{\frac{a_1}{13d}}\right) \tag{27.34}$$

where n_{ef} is the effective number of bolts in a row and n is the number of bolts of diameter d in a row.

EC5 in clause 8.1.3(2) specifies that in order to prevent the possibility of splitting caused by the tension force component (of a force acting at an angle α to the grain, see Fig. 27.4), i.e. $F_{Ed} \sin\alpha$ perpendicular to the grain, the design-splitting capacity, $F_{90,Rd}$, of the (middle) timber member should satisfy the following requirement:

$$F_{90,Rd} \geq \max \text{ of } (F_{v,Ed,1} \text{ and } F_{v,Ed,2}) \tag{27.35}$$

where $F_{90,Rd}$ = the design-splitting capacity, calculated from the characteristic splitting capacity $F_{90,Rk}$ (i.e. $F_{90,Rd} = k_{mod}F_{90,Rk}/\gamma_M$) see section 27.5.1.
For softwoods, the characteristic splitting capacity $F_{90,Rk}$ in N, for the joint details shown in Fig. 27.3, is given in EC5, for dowel type joints, as:

$$F_{90,Rk} = 14b\sqrt{\frac{h_e}{1-h_e/h}} \tag{27.36}$$

The terms b, h and h_e (all in mm) are as illustrated in Fig. 27.3.

$F_{v,Ed,1}$, $F_{v,Ed,2}$ = the design shear forces on either side of the joint, see Fig. 27.3.

For nailed connections, EC5 clause 8.3.1 recommends that there should be at least two nails in a joint, that smooth nails should have a point-side penetration of at least $8d$ ($6d$ for improved nails), and nails should be driven into pre-drilled holes in timber with a characteristic density, ρ_k, of 500 kg/m^3 or more and when the diameter d of the nail exceeds 8 mm. The recommendations for minimum nail spacings and end distances are given in Table 8.2 of EC5, reproduced here as Table 27.10.

In a timber-to-timber connection, the minimum thickness, t, of timber members without pre-drilled holes should be:

$$t = \max \text{ of } \left[7d \text{ and } (13d - 30)\frac{\rho_k}{400}\right] \tag{27.37}$$

In a three-member connection, nails may overlap in the central member provided that the thickness of the central member, t, is greater than the nail point-side penetration, t_2, plus $4d$ (i.e. $t > t_2 + 4d$).

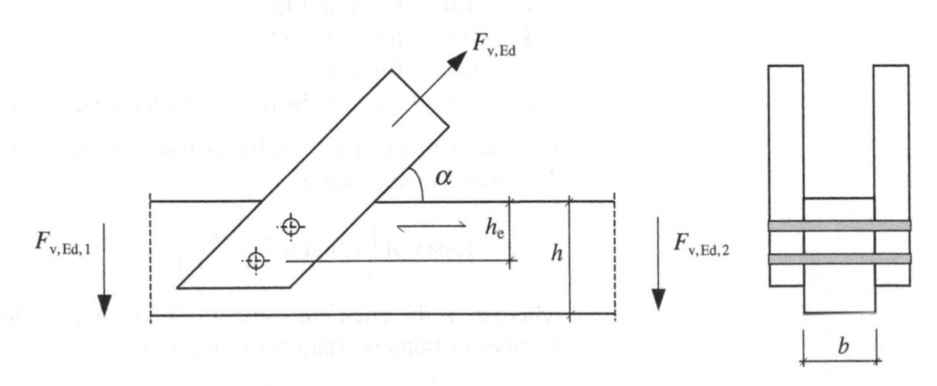

Fig. 27.3 Shear stress in the jointed timber.

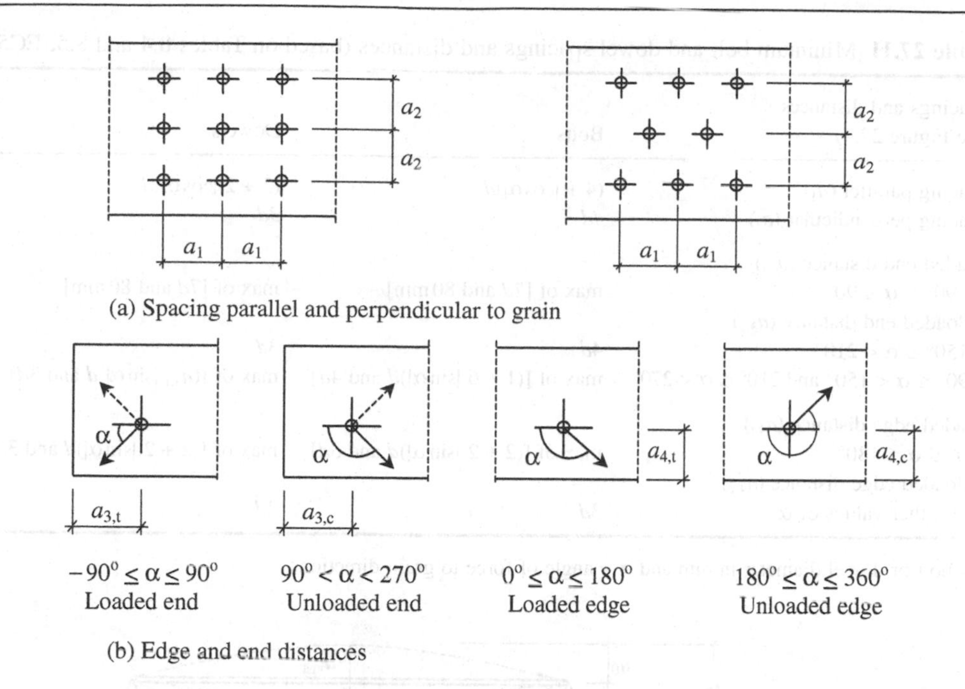

Fig. 27.4 Fastener spacings and distances.

Table 27.10 Minimum nail spacings and distances (based on Table 8.2, EC5)

Spacings and distances (see Fig. 27.4)	Without pre-drilled holes $\rho_k \le 420\,kg/m^3$	$420 < \rho_k < 500\,kg/m^3$	Pre-drilled holes
Spacing parallel (a_1)	$d < 5$ mm: $(5 + 5\,\lvert\cos\alpha\rvert)d$ $d \ge 5$ mm: $(5 + 7\,\lvert\cos\alpha\rvert)d$	$(7 + 8\,\lvert\cos\alpha\rvert)d$	$(4 + \lvert\cos\alpha\rvert)d$
Spacing perpendicular (a_2)	$5d$	$7d$	$(3 + 4\,\lvert\sin\alpha\rvert)d$
Loaded end distance ($a_{3,t}$)	$(10 + 5\,\lvert\cos\alpha\rvert)d$	$(15 + 5\,\lvert\cos\alpha\rvert)d$	$(7 + 5\,\lvert\cos\alpha\rvert)d$
Unloaded end distance ($a_{3,c}$)	$10d$	$15d$	$7d$
Loaded edge distance ($a_{4,t}$)	$d < 5$ mm: $(5 + 2\,\lvert\sin\alpha\rvert)d$ $d \ge 5$ mm: $(5 + 5\,\lvert\sin\alpha\rvert)d$	$d < 5$ mm: $(7 + 2\,\lvert\sin\alpha\rvert)d$ $d \ge 5$ mm: $(7 + 5\,\lvert\sin\alpha\rvert)d$	$d < 5$ mm: $(3 + 2\,\lvert\sin\alpha\rvert)d$ $d \ge 5$ mm: $(3 + 4\,\lvert\sin\alpha\rvert)d$
Unloaded edge distance ($a_{4,c}$)	$5d$	$7d$	$3d$

d = nail diameter in mm and α = angle of force to grain direction.

The recommendations for minimum bolt and dowel spacings and end distances are given in Tables 8.4 and 8.5 of EC5 respectively, reproduced here as Table 27.11.

Chapter 8 of EC5 also provides formulae for calculating resistance to axial loading in nailed, bolted and screwed joints, see clauses 8.3.2 and 3, 8.5.2 and 8.7.2 respectively. These cover withdrawal of the fastener from its head-side (pull-out), pulling through of the fastener head (pull-through) and also combined axial and lateral loading.

Table 27.11 Minimum bolt and dowel spacings and distances (based on Tables 8.4 and 8.5, EC5)

Spacings and distances (see Figure 27.4)	Bolts	Dowels
Spacing parallel (a_1)	$(4 + \lvert\cos\alpha\rvert)d$	$(3 + 2\lvert\cos\alpha\rvert)d$
Spacing perpendicular (a_2)	$4d$	$3d$
Loaded end distance ($a_{3,t}$)		
$-90° \le \alpha \le 90°$	max of [$7d$ and 80 mm]	max of [$7d$ and 80 mm]
Unloaded end distance ($a_{3,c}$)		
$150° \le \alpha < 210°$	$4d$	$3d$
$90° \le \alpha < 150°$ and $210° \le \alpha < 270°$	max of [$(1 + 6\lvert\sin\alpha\rvert)d$ and $4d$]	max of [$(a_{3,t}\lvert\sin\alpha\rvert)d$ and $3d$]
Loaded edge distance ($a_{4,t}$)		
$0° \le \alpha \le 180°$	max of [$(2 + 2\lvert\sin\alpha\rvert)d$ and $3d$]	max of [$(2 + 2\lvert\sin\alpha\rvert)d$ and $3d$]
Unloaded edge distance ($a_{4,c}$)		
all other values of α	$3d$	$3d$

d = bolt or dowel diameter in mm and α = angle of force to grain direction.

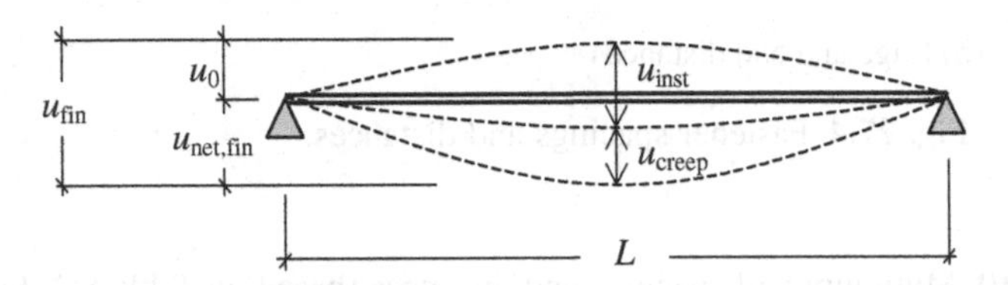

Fig. 27.5 Components of deflection.

27.7 SERVICEABILITY LIMIT STATES

Section 7 of EC5: Part 1.1 sets out serviceability requirements with regard to limiting values for deflections and vibrations which are often the governing factors in selection of suitable section sizes. EC5 in clause 7.1 also provides design procedure for calculation of slip for joints made with dowel-type fasteners such as nails, screws, bolts and dowels and also connectors such as shear plate, ring and toothed connectors.

27.7.1 Deflections

In order to prevent the possibility of damage to surfacing materials, ceilings, partitions and finishes, and to the functional needs as well as any requirements of appearance, clause 7.2 of EC5 recommends a number of limiting values of deflection for flexural and laterally loaded members. The components of deflection are shown in Fig. 27.5, where the symbols are defined as follows:

u_0 = the precamber (if applied)
u_{inst} = the instantaneous deflection
u_{creep} = the creep deflection
u_{fin} = the final deflection

$u_{net,fin}$ = the net final deflection below the straight line joining the supports, given by:

$$u_{net,fin} = u_{inst} + u_{creep} - u_0 = u_{fin} - u_0 \quad (27.38)$$

A summary of EC5 recommendations for limiting deflections is given in Table 27.12.

The following procedure for calculation of the instantaneous and final deflections may be used:

1. The instantaneous deflection of a solid timber member acting alone may be calculated using the appropriate fifth percentile modulus of elasticity, $E_{0.05}$ and/or fifth percentile shear modulus $G_{0.05}$, where $G_{0.05} = E_{0.05}/16$.
2. The instantaneous deflection of a member made of two or more sections fastened together to act as a single member, a member in a load-sharing system, glulam and a composite-section member may be calculated using the mean elastic or shear moduli, i.e. E_{mean} and G_{mean}.
3. The final mean value $E_{mean,fin}$ of a stiffness property can be calculated as:

$$E_{mean,fin} = \frac{E_{mean}}{(1 + \Psi_2 k_{def})} \quad (27.39)$$

4. The final deflection at the end of the design life of a component, u_{fin}, is calculated as:

$$u_{fin} = u_{inst}(1 + \psi_2 k_{def}) \quad (27.40)$$

where ψ_2 = a factor for the quasi-permanent value of a variable action, see section 27.4; for permanent actions $\psi_2 = 1.0$.

k_{def} = the modification factor which takes into account the effect of the stiffness parameters of the load and moisture content. The values of k_{def} are given in Table 3.2 of EC5, see Table 27.13.

For joints, the deformation factor k_{def} should be doubled.

For a joint constituted of two timber elements with deformation factors of $k_{def,1}$ and $k_{def,2}$:

$$k_{def} = \sqrt{k_{def,1} k_{def,2}} \quad (27.41)$$

Table 27.12 General deflection recommendations (based on Table 7.2, EC5)

Components of deflection	Maximum allowable deflection: Beams	Maximum allowable deflection: Cantilevers
Instantaneous deflection due to variable actions, u_{inst}	$L/300$ to $L/500$	$L/150$ to $L/250$
Final deflection due to variable actions, $u_{net,fin}$	$L/250$ to $L/350$	$L/125$ to $L/175$
Final deflection due to all applied actions including precamber, u_{fin}	$L/150$ to $L/300$	$L/75$ to $L/150$

L is the beam span or the length of a cantilever.

Table 27.13 Values of k_{def} for solid and glued-laminated timber, LVL, plywood, OSB and particle boards (based on Table 3.2, EC5)

Material	Service class* 1	2	3
Solid timber, glued-laminated timber and LVL	0.60	0.80	2.00
Plywood (EN 636 Part 3)	0.80	1.00	2.50
OSB (EN 300 for OSB/3 and 4)	1.50	2.25	–
Particleboard (EN 312 Part 7)	1.50	2.25	–

* Service classes are defined in Table 27.5.

27.7.2 Vibrations

EC5 clause 7.3 deals with vibrations in structures and states that actions which are anticipated to occur frequently should not produce vibrations which might impair the functioning of the structure or cause unacceptable discomfort to the users. Clause 7.3.3 of EC5 provides recommendations and detailed rules for limiting vibrations in residential floors.

The UK NAD (DD ENV 1995-1-1: 1994) recommends that for residential UK timber floors the requirements of EC5 can be met by ensuring that the total instantaneous deflection of the floor joists under load does not exceed 14 mm or $L/333$, whichever is the lesser, where L is the span in mm.

27.7.3 Joint slip

EC5 clause 7.1 provides a design procedure for the calculation of slip for joints made with metal fasteners such as nails, screws, bolts and dowels and shear plates and toothed connectors. In general, the serviceability requirements are to ensure that any deformation caused by slip does not impair the satisfactory functioning of the structure with regard to strength, attached materials and appearance.

The procedure for calculation of slip for joints made with dowel-type fasteners is summarized as follows:

1. The instantaneous slip, u_{inst} in mm, is calculated as:

for serviceability verification $$u_{inst,ser} = \frac{F_d}{K_{ser}} \quad (27.42a)$$

for strength verification $$u_{inst,u} = \frac{3F_d}{2K_{ser}} \quad (27.42b)$$

where $u_{inst,ser}$ = the instantaneous slip at the serviceability limit state
$u_{inst,u}$ = the instantaneous slip at the ultimate limit state
F_d = the design load of the joint, in N, per fastener per shear plane
K_{ser} = the instantaneous slip modulus, in N/mm, and its values are given in Table 7.1 of EC5, reproduced here as Table 27.14.

Table 27.14 Values of slip moduli K_{ser} per fastener per shear plane (based on Table 7.1, EC5)

Fastener type	K_{ser}(N/mm) Timber-to-timber Panel-to-timber	Steel-to-timber Concrete-to-timber
Bolts (without clearance)* and dowels Screws Nails (pre-drilled)	$\frac{1}{35}\rho_m{}^{1.5}d$	$\frac{2}{35}\rho_m{}^{1.5}d$
Nails (no pre-drilling)	$\frac{1}{45}\rho_m{}^{1.5}d^{0.8}$	$\frac{2}{45}\rho_m{}^{1.5}d^{0.8}$
Staples	$\frac{1}{120}\rho_m{}^{1.5}d^{0.8}$	$\frac{1}{60}\rho_m{}^{1.5}d^{0.8}$

ρ_m is the joint member mean density, given in Table 1, BS EN 338: 1995 (in kg/m^3) and d is the fastener diameter (in mm).
*Bolt clearance should be added separately to the joint slip.

The slip modulus for the ultimate limit state, K_u should be taken as:

$$K_u = \frac{2}{3}K_{ser} \tag{27.43}$$

2. If the mean densities of the two joined members ($\rho_{m,1}$ and $\rho_{m,2}$) are different then ρ_m in formulae given in Table 7.1 of EC5, see Table 27.14, should be calculated as:

$$\rho_m = \sqrt{\rho_{m,1}\rho_{m,2}} \tag{27.44}$$

3. The final joint slip, u_{fin}, should be calculated as:

$$u_{fin} = u_{inst}(1+\psi_2 k_{def}) \tag{27.45}$$

where k_{def} is the modification factor for deformation and is given in Table 3.2 of EC5.

For a joint made from members with different creep properties ($k_{def,1}$ and $k_{def,2}$), the final joint slip should be calculated as:

$$u_{fin} = u_{inst}(1+\psi_2\sqrt{k_{def,1}k_{def,2}}) \tag{27.46}$$

4. For joints made with bolts, EC5 recommends that the clearance, $u_{clearance}$ (in mm), should be added to joint slip values. Thus:

$$u_{inst,bolt} = u_{clearance} + \frac{F_d}{K_{ser}} \tag{27.47}$$

and

$$u_{fin,bolt} = u_{clearance} + u_{inst}(1+\psi_2 k_{def}) \tag{27.48}$$

27.8 BIBLIOGRAPHY

1. *Timber Engineering – STEP 1* (1995) Centrum Hout, The Netherlands.
2. *Timber Engineering – STEP 2* (1995) Centrum Hout, The Netherlands.
3. TRADA (1994) *Eurocode 5 Design Guidance.*

27.9 DESIGN EXAMPLES

In general design calculations to Eurocodes are best carried out with computer programming support; for example, to examine the various combinations of load duration and loading category to determine the worst-case situation or, to determine load-carrying capacity of dowel-type connections.

The following examples, which are produced using the Mathcad® computer software program, are included to illustrate some of the key aspects of design to Eurocode 5.

EXAMPLE 27.1 DESIGN OF FLOOR JOISTS

A timber floor spanning 3.8 m centre to centre is to be designed using timber joists at 600 mm centres (Fig. 27.6). The floor is subjected to an imposed load of 1.5 kN/m^2 and carries a dead loading, excluding self-weight, of 0.30 kN/m^2. Carry out design checks to show that a series of 44 mm × 225 mm deep sawn section timber in strength class C22 under service class 1 is suitable.

1. Geometrical properties

Breadth of beam section	$b = 44\,\text{mm}$
Depth of beam section	$h = 225\,\text{mm}$
Span: centre to centre of supports	$L = 3800\,\text{mm}$
Bearing length	$l_{\text{bearing}} = 75\,\text{mm}$
Joist spacing	$J_{\text{spacing}} = 600\,\text{mm}$

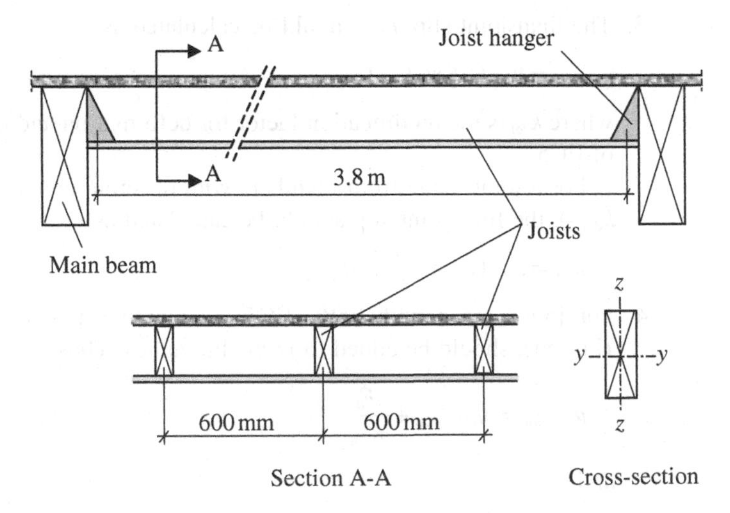

Fig. 27.6 Timber floor joists.

Cross-sectional area $A = bh = 9.9 \times 10^3\,\text{mm}^2$

Second moment of area about y–y axis $I_y = \dfrac{bh^3}{12} = 4.18 \times 10^7\,\text{mm}^4$

Section modulus about y–y axis $W_y = \dfrac{bh^2}{6} = 3.71 \times 10^5\,\text{mm}^3$

2. Timber strength properties

BS EN 338: 1995, Table 1: C22 timber

Characteristic bending strength $f_{m,k} = 22\,\text{N/mm}^2$

Characteristic shear strength $f_{v,k} = 2.4\,\text{N/mm}^2$

Characteristic compression perpendicular to grain $f_{c,90,k} = 5.1\,\text{N/mm}^2$

Mean modulus of elasticity parallel to grain $E_{0,mean} = 10\,\text{kN/mm}^2$

5th percentile modulus of elasticity parallel to grain $E_{0.05} = 6.7\,\text{kN/mm}^2$

Mean shear modulus $G_{mean} = 0.63\,\text{kN/mm}^2$

Average density $\rho_{mean} = 410\,\text{kg/m}^3$

3. Partial safety factors

EN 1990 and ENV 1995-1-1: 1994, UK NAD

Permanent actions $\gamma_G = 1.35$

Variable actions $\gamma_Q = 1.5$

EC5: Part 1.1, Table 2.2

Material factor for timber $\gamma_M = 1.3$

4. Actions

Applied dead load $F_{dead} = 0.3\,\text{kN/m}^2$

Self-weight $F_{self} = \rho_{mean} gbh = 0.04\,\text{kN/m}$

Total characteristic permanent load per metre length $G_k = F_{dead} J_{spacing} + F_{self} = 0.22\,\text{kN/m}$

Imposed load $F_{imposed} = 1.5\,\text{kN/m}^2$

Characteristic variable (imposed) load per metre length $Q_k = F_{imposed} J_{spacing} = 0.9\,\text{kN/m}$

For permanent duration, design load $F_{d,perm} = \gamma_G G_k = 0.3\,\text{kN/m}$

For medium-term duration, design load $F_{d,med} = \gamma_G G_k + \gamma_Q Q_k = 1.65\,\text{kN/m}$

Thus the medium-term duration will produce the most critical case. Therefore $F_d = F_{d,med}$

Design action $F_d = \gamma_G G_k + \gamma_Q Q_k = 1.65\,\text{kN/m}$

Note that the factor ψ_0 for representative values of variable actions in combination is not used in this example (see Section 27.4.1) as there is only one variable load.

ULTIMATE LIMIT STATES

5. Modification factors

Factor for medium-duration loading and service class 1 (k_{mod}, Table 3.1) — $k_{mod} = 0.8$

Size factor (k_h, clause 3.2) for $h > 150$ mm — $k_h = 1.0$

System strength (load sharing) applies (k_{sys}, clause 6.7) — $k_{sys} = 1.1$

Lateral stability (k_{crit}, clause 6.3.3):

Effective length, L_{ef}, for constant distributed load (Table 6.1, EC5) — $L_{ef} = 0.9L = 3.42$ m

Critical bending stress (EC5, clause 6.3.3) — $\sigma_{m,crit} = \dfrac{0.78E_{0.05}b^2}{L_{ef}h} = 13.15$ N/mm^2

Relative slenderness for bending — $\lambda_{rel,m} = \sqrt{\dfrac{f_{m,k}}{\sigma_{m,crit}}} = 1.29$

Lateral instability factor — $k_{crit} = 1.56 - 0.75\lambda_{rel,m} = 0.59$

Bearing factor ($k_{c,90}$, clause 6.1.4) — $k_{c,90} = 1$

Factor for shear in members with notched end (k_v, clause 6.5) — $k_v = 1$ for no notch

6. Bending strength

Design bending moment — $M_d = \dfrac{F_d L^2}{8} = 2.97$ kN m

Design bending stress — $\sigma_{m,y,d} = \dfrac{M_d}{W_y} = 8.01$ N/mm^2

Design bending strength

$$f_{m,y,d} = \frac{k_{mod}k_h k_{crit}k_{sys}f_{m,k}}{\gamma_M} = 8.78 \text{ N/mm}^2$$

Bending strength satisfactory

7. Shear strength

Design shear load — $V_d = \dfrac{F_d L}{2} = 3.13$ kN

Design shear stress — $\sigma_{v,d} = \dfrac{3V_d}{2A} = 0.47$ N/mm^2

Design shear strength — $f_{v,d} = \dfrac{k_{mod}k_v k_{sys}f_{v,k}}{\gamma_M} = 1.62$ N/mm^2

Shear strength satisfactory

8. Bearing strength

Design bearing load $V_d = \frac{F_d L}{2} = 3.13\,\text{kN}$

Design bearing stress $\sigma_{c,90,d} = \frac{V_d}{b l_{bearing}} = 0.95\,\text{N/mm}^2$

Design bearing strength

$$f_{c,90,d} = \frac{k_{mod} k_{c,90} k_{sys} f_{c,90,k}}{\gamma_M} = 3.45\,\text{N/mm}^2$$

Bearing strength satisfactory

SERVICEABILITY LIMIT STATES

9. Deflection

prNV 1995-1-1: 2001, clause 7.2

For serviceability limit states (ENV 1995-1-1: 1994, UK NAD) $\gamma_M = 1.0$

Bending moment due to permanent loads $M_G = \frac{G_k L^2}{8} = 0.4\,\text{kN m}$

Bending moment due to variable loads $M_Q = \frac{Q_k L^2}{8} = 1.62\,\text{kN m}$

1. Instantaneous deflection:
   for bending and shear due to permanent action, G_k:

$$u_{inst,perm} = \frac{5 G_k L^4}{384 E_{0,mean} I_y} + \frac{19.2 M_G}{(bh) E_{0,mean}} = 1.51\,\text{mm}$$

   for bending and shear due to variable action, Q_k:

$$u_{inst,var} = \frac{5 Q_k L^4}{384 E_{0,mean} I_y} + \frac{19.2 M_Q}{(bh) E_{0,mean}} = 6.17\,\text{mm}$$

   Maximum allowable deflection due to variable action $u_{inst,adm} = \frac{L}{350} = 10.86\,\text{mm}$

   $u_{inst,var} < u_{inst,adm}$ Satisfactory

2. Final deflection due to variable actions
   Modification factor for deformation, under service class 1 (k_{def}, Table 3.2, EC5) $k_{def} = 0.6$
   EN 1990: Reduction factor for variable action, Ψ_2 (see Table 27.1) $\Psi_{2,var} = 0.2$

$$u_{fin} = u_{inst,var}(1 + \Psi_{2,var} k_{def}) = 6.91\,\text{mm}$$

Maximum allowable deflection $u_{net,fin,adm} = \frac{L}{250} = 15.2\,mm$

$u_{fin} < u_{net,fin,adm}$ Satisfactory

3. Final deflection due to all actions (permanent and variable)
prEN 1995-1-1, clause 2.2.3(4)
Reduction factor for permanent action, Ψ_2 (see Table 27.1) $\Psi_{2,perm} = 1$

$$u_{fin} = u_{inst,perm}(1+\Psi_{2,perm}k_{def}) + u_{inst,var}(1+\Psi_{2,var}k_{def}) = 9.31\,mm$$

Assume maximum allowable deflection $u_{fin,adm} = \frac{L}{200} = 19\,mm$

$u_{fin} < u_{fin,adm}$ Satisfactory

10. Vibration
ENV 1995-1-1: 1994, UK NAD

Total instantaneous deformations should be less than the lesser of: $u_{inst,var} + u_{inst,perm} = 7.67\,mm$

$\frac{L}{333} = 11.41\,mm$ and $14\,mm$

Vibration satisfactory

Therefore 44 mm × 225 mm sawn timber sections in strength class C22 are satisfactory.

EXAMPLE 27.2 DESIGN OF AN ECCENTRICALLY LOADED COLUMN

For the design data given below, check that a 100 mm × 250 mm sawn section is adequate as a column if the load is applied 40 mm eccentric about its *y–y* axis (Fig. 27.7). The column is 3.75 m high and has its ends restrained in position but not in direction.

Design data
Timber: C22
Service class: 2
Permanent load: 15 kN
Variable load (medium term): 17 kN

1. Geometrical properties

Column length $L = 3.75\,m$

Effective length $L_{ef} = 1.0L = 3.75\,m$

Width of the section $b = 100\,mm$

Depth of the section $d = 250\,mm$

Cross-sectional area $A = bd = 2.5 \times 10^4\,mm^2$

Second moment of area about *y–y* axis $I_y = \frac{1}{12}bd^3 = 1.3 \times 10^8\,mm^4$

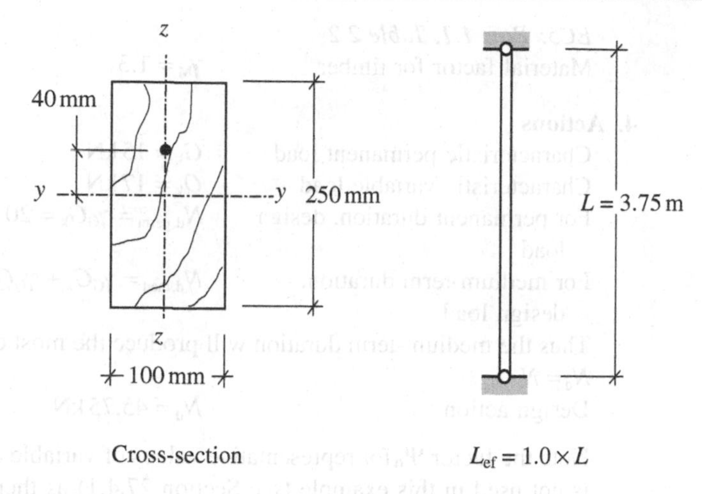

Fig. 27.7 Column details.

Section modulus about y–y axis $\quad W_y = \dfrac{bd^2}{6} = 1.04 \times 10^6\,\text{mm}^3$

Radius of gyration about y–y axis $\quad i_y = \sqrt{\dfrac{I_y}{A}} = 72.17\,\text{mm}$

Slenderness ratio about y–y axis $\quad \lambda_y = \dfrac{L_{ef}}{i_y} = 51.96$

Second moment of area about z–z axis $\quad I_z = \dfrac{1}{12}db^3 = 2.08 \times 10^7\,\text{mm}^4$

Radius of gyration about z–z axis $\quad i_z = \sqrt{\dfrac{I_z}{A}} = 28.87\,\text{mm}$

Slenderness ratio about z–z axis $\quad \lambda_z = \dfrac{L_{ef}}{i_z} = 129.9$

2. Timber strength properties

BS EN 338: 1995, Table 1: C22 timber

Characteristic bending strength $\quad f_{m,k} = 22\,\text{N/mm}^2$

Characteristic compression parallel to grain $\quad f_{c,0,k} = 20\,\text{N/mm}^2$

5th percentile modulus of elasticity parallel to grain $\quad E_{0.05} = 6.7\,\text{kN/mm}^2$

3. Partial safety factors

EN 1990 and ENV 1995-1-1: 1994, UK NAD

Permanent actions $\quad \gamma_G = 1.35$

Variable actions $\quad \gamma_Q = 1.5$

EC5: Part 1.1, Table 2.2
Material factor for timber $\gamma_M = 1.3$

4. Actions

Characteristic permanent load $G_k = 15\,kN$
Characteristic variable load $Q_k = 17\,kN$
For permanent duration, design load $N_{d,perm} = \gamma_G G_k = 20.25\,kN$
For medium-term duration, design load $N_{d,med} = \gamma_G G_k + \gamma_Q Q_k = 45.75\,kN$
Thus the medium-term duration will produce the most critical case. Therefore $N_d = N_{d,med}$
Design action $N_d = 45.75\,kN$

Note the factor Ψ_0 for representative values of variable actions in combination is not used in this example (see Section 27.4.1) as there is only one variable load.

Eccentricity $e_y = 40\,mm$
Design moment due to eccentricity about y–y axis $M_d = e_y N_d = 1.83\,kN\,m$

5. Modification factors

Factor for medium-duration loading and service class 2 (k_{mod}, Table 3.1) $k_{mod} = 0.8$
System strength (load sharing) does not apply (k_{sys}, clause 6.7) $k_{sys} = 1.0$

6. Bending strength

Design bending moment about y–y axis $M_d = 1.83\,kN\,m$

Design bending stress about y–y axis $\sigma_{m,y,d} = \dfrac{M_d}{W_y} = 1.76\,N/mm^2$

No bending moment about z–z axis, thus $\sigma_{m,z,d} = 0\,N/mm^2$

Design bending strength $f_{m,y,d} = \dfrac{k_{mod} k_{sys} f_{m,k}}{\gamma_M} = 13.54\,N/mm^2$

$f_{m,z,d} = f_{m,y,d}$ (not required)

7. Compression strength

Design compressive stress $\sigma_{c,0,d} = \dfrac{N_d}{A} = 1.83\,N/mm^2$

Design compression strength $f_{c,0,d} = \dfrac{k_{mod} k_{sys} f_{c,0,k}}{\gamma_M} = 12.31\,N/mm^2$

Buckling resistance (EC5, clause 6.3.2):

Relative slenderness ratios: $\lambda_{rel,y} = \dfrac{\lambda_y}{\pi}\sqrt{\dfrac{f_{c,0,k}}{E_{0.05}}} = 0.9$

$$\lambda_{rel,z} = \frac{\lambda_z}{\pi}\sqrt{\frac{f_{c,0,k}}{E_{0.05}}} = 2.26$$

Both relative slenderness ratios are >0.3, hence conditions in clause 6.3.2(3) apply:

For a solid timber section: $\beta_c = 0.2$

For a rectangular section (clause 5.1.6): $k_m = 0.7$

Thus

$$k_y = 0.5\left[1 + \beta_c(\lambda_{rel,y} - 0.3) + \lambda_{rel,y}^{\,2}\right] = 0.97$$

and

$$k_z = 0.5\left[1 + \beta_c(\lambda_{rel,z} - 0.3) + \lambda_{rel,z}^{\,2}\right] = 3.25$$

Hence, $k_{c,y} = \dfrac{1}{k_y + \sqrt{k_y^2 - \lambda_{rel,y}^{\,2}}} = 0.76$

and $k_{c,z} = \dfrac{1}{k_z + \sqrt{k_z^2 - \lambda_{rel,z}^{\,2}}} = 0.18$

Check the following conditions:

$$\frac{\sigma_{c,0,d}}{k_{c,z} f_{c,0,d}} + \frac{\sigma_{m,z,d}}{f_{m,z,d}} + k_m \frac{\sigma_{m,y,d}}{f_{m,y,d}} = 0.92 \quad <1.0 \quad \text{Satisfactory}$$

$$\frac{\sigma_{c,0,d}}{k_{c,y} f_{c,0,d}} + k_m \frac{\sigma_{m,z,d}}{f_{m,z,d}} + \frac{\sigma_{m,y,d}}{f_{m,y,d}} = 0.33 \quad <1.0 \quad \text{Satisfactory}$$

Therefore a 100 mm × 250 mm sawn section timber in strength class C22 is satisfactory.

EXAMPLE 27.3 DESIGN OF A TIMBER-TO-TIMBER NAILED TENSION SPLICE JOINT

A timber-to-timber tension splice joint comprises two 47 mm × 120 mm inner members and two 33 mm × 120 mm side members of strength class C22 timber in service class 2 (Fig. 27.8). It is proposed to use 3.35 mm diameter, 65 mm long round wire nails without pre-drilling. The joint is subjected to a permanent load of 2 kN and a medium-term variable load of 3 kN. Determine the required number of nails with a suitable nailing pattern.

1. **Geometrical properties**

Thickness of side members $t_1 = 33\,\text{mm}$

Thickness of inner members $t_2 = 47\,\text{mm}$

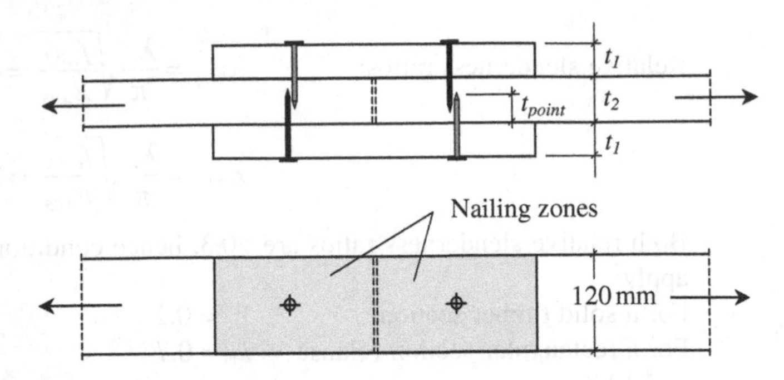

Fig. 27.8 Splice joint.

Width of timber members	$h = 120\,\text{mm}$
Cross-sectional area of side members	$A_{side} = ht_1 = 3.96 \times 10^3\,\text{mm}^2$
Cross-sectional area of inner members	$A_{inner} = ht_2 = 5.64 \times 10^3\,\text{mm}^2$
Nail diameter	$d = 3.35\,\text{mm}$
Nail length	$l_{nail} = 65\,\text{mm}$
Head diameter of nails	$d_h = 2d = 6.70\,\text{mm}$
Nail pointside penetration	$t_{point} = l_{nail} - t_1 = 32\,\text{mm}$
Clauses 8.3.1.2, EC5	
Minimum allowable pointside penetration	$t_{point,adm} = 8d = 26.8\,\text{mm}$
For overlap nailing without pre-drilling	$t_2 - t_{point} = 15\,\text{mm} > 4d = 13.4\,\text{mm}$

Both pointside penetration and overlap nailing are satisfactory

2. **Timber strength properties**

BS EN 338: 1995, Table 1

Characteristic tension parallel to grain	$f_{t,0,k} = 13\,\text{N/mm}^2$
Mean density	$\rho_m = 410\,\text{kg/m}^3$
Characteristic density	$\rho_k = 340\,\text{kg/m}^3$

3. **Partial safety factors**

EN 1990 and ENV 1995-1-1: 1994, UK NAD

Permanent actions	$\gamma_G = 1.35$
Variable actions	$\gamma_Q = 1.5$
EC5: Part 1.1, Table 2.2	
Material factor for timber	$\gamma_{M,timber} = 1.3$
Material factor for connections	$\gamma_{M,connection} = 1.3$

4. **Modification factors**

Factor for medium-duration loading and service class 2 (k_{mod}, Table 3.1)
$k_{mod} = 0.8$

Size factor (k_h, clause 3.2(3)), for $h < 150\,\text{mm}$, is the lesser of:

$$k_h = 1.3 \quad \text{and} \quad k_h = \left(\frac{150\ \text{mm}}{h}\right)^{0.2} = 1.05$$

ULTIMATE LIMIT STATES

5. Actions

Characteristic permanent load $G_k = 2\,kN$

Characteristic variable load $Q_k = 3\,kN$

Design action $F_d = \gamma_G G_k + \gamma_Q Q_k = 7.2\,kN$

6. Tension strength of timber

Design tension stress parallel to grain in side members $\sigma_{t,0,d} = \dfrac{F_d}{2A_{side}} = 0.91\,N/mm^2$

Design tension stress parallel to grain in inner members $\sigma_{t,0,d} = \dfrac{F_d}{A_{inner}} = 1.28\,N/mm^2$

Design tension strength parallel to grain $f_{t,0,d} = \dfrac{k_{mod} k_h f_{t,0,k}}{\gamma_{M,timber}} = 8.37\,N/mm^2$

Tension strength satisfactory

7. Embedding strength of timber

Characteristic embedding strength without pre-drilling

Since the following is an empirical equation (with hidden units in the coefficient part), for Mathcad to produce correct units, multiply it by the units shown inside the parentheses; or alternatively use the equation without units for ρ_k and d. Thus:

$$f_{h,k} = 0.082\rho_k d^{-0.3}(s^{-2}\,m^{2.3} \times 1.26 \times 10^5) = 19.42\ N/mm^2$$

Characteristic embedding strength

for headside timber: $f_{h,1,k} = f_{h,k} = 19.42\,N/mm^2$

for pointside timber: $f_{h,2,k} = f_{h,k} = 19.42\,N/mm^2$

8. Yield moment of nails

Yield moment of a nail (using a similar treatment for the units as above)

Assume nails with tensile strength of $f_u = 600\,N/mm^2$

$$M_{y,Rk} = \frac{f_u}{600} 180 d^{2.6}(kg/m^{0.6}s^2 \times 6.31 \times 10^4) = 4.17 \times 10^3\ N\,mm$$

9. Withdrawal resistance

EC5, clause 8.3.2

Characteristic pointside withdrawal strength (using similar treatment for the units as above)

$f_{ax,k} = 20 \times 10^{-6}\rho_k^2(m^6kg^{-2}N\,mm^{-2})$
$= 2.31\,N/mm^{-2}$

Characteristic headside pull-through strenth for $d_h \geq 2d$

$f_{head,k} = 70 \times 10^{-6}\rho_k^2(m^6kg^{-2}N\,mm^{-2})$
$= 8.09\,N/mm^{-2}$

Characteristic withdrawal capacity of round wire nails, is the lesser of:

$$F_{ax,Rk,1} = f_{ax,k}dt_{point} = 247.85\,N$$

and $$F_{ax,Rk,2} = f_{ax,k}dt_1 + f_{head,k}d_h^2 = 618.9\,N$$

Thus $$F_{ax,Rk} = \min\,(F_{ax,Rk,1}F_{ax,Rk,2}) = 247.85\,N$$

10. Load-carrying capacity

EC5, clause 8.2.1

For a timber-to-timber joint with nails in single shear, characteristic resistance per shear plane, $F_{v,Rk}$, is the lesser of $F_{v,Rk,a}$ to $F_{v,Rk,f}$ (see Table 27.6), where:

pointside penetration length $t_2 = t_{point}$

and the ratio, β $\beta = \frac{f_{h,2,k}}{f_{h,1,k}} = 1$

Failure mode (a)

$$F_{v,Rk,a} = f_{h,1,k}t_1d = 2.15 \times 10^3\,N$$

Failure mode (b)

$$F_{v,Rk,b} = f_{h,2,k}t_2d = 2.08 \times 10^3\,N$$

Failure mode (c)

$$F_{v,Rk,c} = \frac{f_{h,1,k}dt_1}{1+\beta}\left[\sqrt{\beta + 2\beta^2\left[1+\frac{t_2}{t_1}+\left(\frac{t_2}{t_1}\right)^2\right]+\beta^3\left(\frac{t_2}{t_1}\right)^2} - \beta\left(1+\frac{t_2}{t_1}\right)\right]$$

$$= 875.77\,N$$

Failure mode (d)

$$F_{v,Rk,d} = 1.05\frac{f_{h,1,k}dt_1}{2+\beta}\left[\sqrt{2\beta(1+\beta)+\frac{4\beta(2+\beta)M_{y,Rk}}{f_{h,1,k}dt_1^2}} - \beta\right] + \frac{F_{ax,Rk}}{4} = 940.57\,N$$

Failure mode (e)

$$F_{v,Rk,e} = 1.05\frac{f_{h,1,k}dt_2}{1+2\beta}\left[\sqrt{2\beta^2(1+\beta)+\frac{4\beta(1+2\beta)M_{y,Rk}}{f_{h,1,k}dt_2^2}} - \beta\right] + \frac{F_{ax,Rk}}{4} = 921.45\,N$$

Failure mode (f)

$$F_{v,Rk,f} = 1.15\sqrt{\frac{2\beta}{1+\beta}}\sqrt{2M_{y,Rk}f_{h,1,k}d} + \frac{F_{ax,Rk}}{4} = 909.23\,N$$

Therefore the characteristic resistance per nail is the minimum of:

$$F_{v,Rk} = (F_{v,Rk,a}F_{v,Rk,b}F_{v,Rk,c}F_{v,Rk,d}F_{v,Rk,e}F_{v,Rk,f})$$

i.e. $\min(F_{v,Rk}) = 875.77\,N$

The design resistance (ultimate load) per nail $$F_{v,Rd} = k_{mod}\frac{\min(F_{v,Rk})}{\gamma_{M,connection}} = 538.94\,N$$

Number of nails per side required $$N_{nails,reqd} = \frac{F_d}{F_{v,Rd}} = 13.36$$

For a symmetrical nailing pattern
adopt 16 nails per side. Thus $N_{nails} = 16$

11. Nail spacing (Fig. 27.9)

EC5, Table 8.2

Nail diameter $d = 3.35$

Angle to grain $\alpha = 0$

Minimum spacing parallel
for $d < 5\,mm$ $a_1 = (5 + 5\,|\cos\alpha|)d = 33.5\,mm$

EC5, clause 8.3.1.1(8)

To eliminate the effect of
number of nails in a row
(i.e. $k_{ef} = 1.0$), use: $a_1 = 14d = 46.9\,mm$

Minimum spacing perpendicular $a_2 = 5d = 16.75\,mm$

Minimum loaded end distance $a_{3,t} = (10 + 5\,|\cos\alpha|)d = 50.25\,mm$

Minimum unloaded end distance $a_{3,c} = 10d = 33.5\,mm$ (not required)

Minimum loaded edge distance
for $d < 5\,mm$ $a_{4,t} = (5 + 5\,|\sin\alpha|)d = 16.75\,mm$
(not required)

Minimum unloaded edge distance $a_{4,c} = 5d = 16.75\,mm$

SERVICEABILITY LIMIT STATES

12. Joint slip

Nail diameter and timber mean
density of: $d = 3.35\,mm$ and $\rho_m = 410\,kg/m^3$

Slip modulus, K_{ser} (Table 7.1, EC5) and using a similar treatment for the units of this empirical equation, for nails without pre-drilling

$$K_{ser} = \frac{1}{45}\rho_m^{1.5} d^{0.8} (kg^{-1.5}\,m^{2.7}\,N \times 10^{5.4}) = 485.29\,N/mm$$

Design load for serviceability
limit states $F_{d,ser} = G_k + Q_k = 5 \times 10^3\,N$

Load per nail $F_{nail} = \dfrac{F_{d,ser}}{N_{nails}} = 312.5\,N$

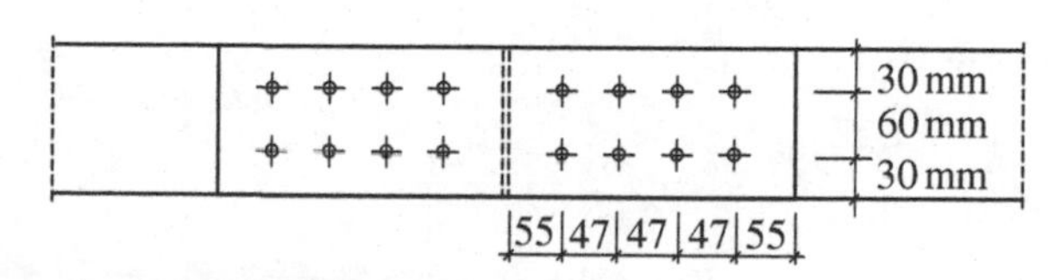

Fig. 27.9 Nail spacings and distances.

Instantaneous slip:

at serviceability limit state $$u_{inst,ser} = \frac{F_{nail}}{K_{ser}} = 0.64\,mm$$

at ultimate limit state $$u_{inst,u} = \frac{1.5F_{nail}}{K_{ser}} = 0.97\,mm$$

Assuming all nails to slip by the same amount, each inner member will move by 0.64 mm and 0.97 mm relative to side members at respective limit states. Therefore the total instantaneous slip is:

at serviceability limit state $u_{total,ser} = 2u_{inst,ser} = 1.28\,mm$
at ultimate limit state $u_{total,u} = 2u_{inst,u} = 1.94\,mm$

Modification factor for deformation, under service class 2 (k_{def}, Table 3.2, EC5)

$$k_{def} = 0.8$$

EN 1990 'Reduction factors for permanent action', $\Psi_{2,perm}$, and variable action, $\Psi_{2,var}$ (see also Table 27.1) are:

$$\Psi_{2,perm} = 1 \quad \text{and} \quad \Psi_{2,var} = 0.3$$

Final joint slip due to all actions (permanent and variable):

at serviceability limit state:

$$u_{fin,ser} = 2\left[u_{inst,ser}\frac{G_k}{F_{d,ser}}(1+\Psi_{2,perm}k_{def}) + u_{inst,ser}\frac{Q_k}{F_{d,ser}}(1+\Psi_{2,var}k_{def})\right]$$
$$= 1.89\,mm$$

at ultimate limit state:

$$u_{fin,u} = 2\left[u_{inst,u}\frac{G_k}{F_{d,ser}}(1+\Psi_{2,perm}k_{def}) + u_{inst,u}\frac{Q_k}{F_{d,ser}}(1+\Psi_{2,var}k_{def})\right]$$
$$= 2.83\,mm$$

Chapter 28

Miscellaneous Tables

28.1 WEIGHTS OF BUILDING MATERIALS

Selected weights of building materials are tabulated below. When the weight of a manufactured item is known, that value should be used in preference to the weight given in this section. 'Schedules of weights of building materials' are given in BS 648: 1964.

Although the unit of mass is the kilogram (kg), because an engineer is interested in loads on structures the weights tabulated below are given in kilonewtons (kN). To convert kilonewtons to kilograms (force) multiply by 101.97.

		Thickness (mm)	Loading (kN/m^2)
Asbestos cement sheeting	76 mm pitch	5.5	0.15
	146 mm pitch	6.4	0.16
Asphalt		20	0.46
Felt underlay for asphalt		2	0.02
Chipboard		25	0.19
Chippings			0.20*
3 layers of bitumen felt		6	0.11
Fibreboard insulation		12	0.04
		18	0.06
Glassfibre		80	0.016
Mineral wool		80	0.02
Plaster (gypsum lime)		12	0.30
Plasterboard (no skim coat)		9.5	0.10
		12.7	0.13
Plaster (skim coat)		–	0.05
Plywood		6.5	0.03
		9	0.05
		12	0.07
Screed. Sand/Cement		25	0.58
Vermiculite		25	0.12
Timber boarding		12	0.07
Water		25	0.25

	Thickness (mm)	Loading (kN/m²)
Wood wool slab. Standard	38	0.21
	50	0.25
	63	0.30
	76	0.33
Extra density	50	0.32
	63	0.37
	76	0.42
Wood wool slab. Channel reinforced	50	0.30
	76	0.38
Interlocking	50	0.36
	76	0.47

*Minimum. It is important to check the actual weight which will occur.

Tile Weights (as Laid Up Slope)

	Lap or gauge (mm)	Loading (kN/m²)
Broughton Moor		
Best		0.48
Seconds		0.56
Thirds		0.64
Special Peggies		0.52
Second Peggies		0.59
Hardrow		
457 mm × 305 mm	76 lap	0.79
457 mm × 457 mm	102 lap	0.79
711 mm × 457 mm	127 lap	0.96
Marley		
Anglia	75 lap	0.47
	100 lap	0.51
Bold Roll	75 lap	0.47
	100 lap	0.51
Ludlow Major	75 lap	0.45
	100 lap	0.49
Ludlow Plus	75 lap	0.47
	100 lap	0.51
Mendip	75 lap	0.47
	100 lap	0.51
Modern	75 lap	0.54
	100 lap	0.58
Plain	100 gauge	0.73
	90 gauge	0.80

Tile Weights (as Laid Up Slope) (*continued*)

	Lap or gauge (mm)	Loading (kN/m^2)
Wessex	75 lap	0.54
	100 lap	0.58
Yeoman	75 lap	0.50
	100 lap	0.54
Redland		
Delta	345 gauge	0.59
Double Pantile	345 gauge	0.50
Double Roman	345 gauge	0.46
Interlocking	280 gauge	0.50
Plain	100 gauge	0.80
Regent	345 gauge	0.46
Renown	345 gauge	0.46
Stonewold	355 gauge	0.54
Speakers		
Eternit		0.21
Duchess		0.21
Countess		0.21
Ladies		0.23
Others		
Asbestos slate		0.23
Bangor slate. Best		0.27
Seconds		0.35
Thirds		0.49
Westmorland. Best		0.44
Seconds		0.55
Thirds		0.70
Western Red Cedar Shingles		0.07–0.12

28.2 BENDING AND DEFLECTION FORMULAE

The groups of bending and deflection formulae presented in Figs 28.1–28.11, for simply supported beams and continuous beams, are extracted from *Steel Designers' Manual* by permission of Blackwell Science.

For simple span beams the additional deflection attributable to shear should be considered.

For continuous beams moment and reaction coefficients should be adjusted to allow for shear deflection. As quoted they apply to designs where only bending deflection is considered.

In this manual, total load is usually given the symbol F, however W or P is retained in Figs 28.1–28.11 from the *Steel Designers' Manual*. Likewise d is retained as the symbol for deflection.

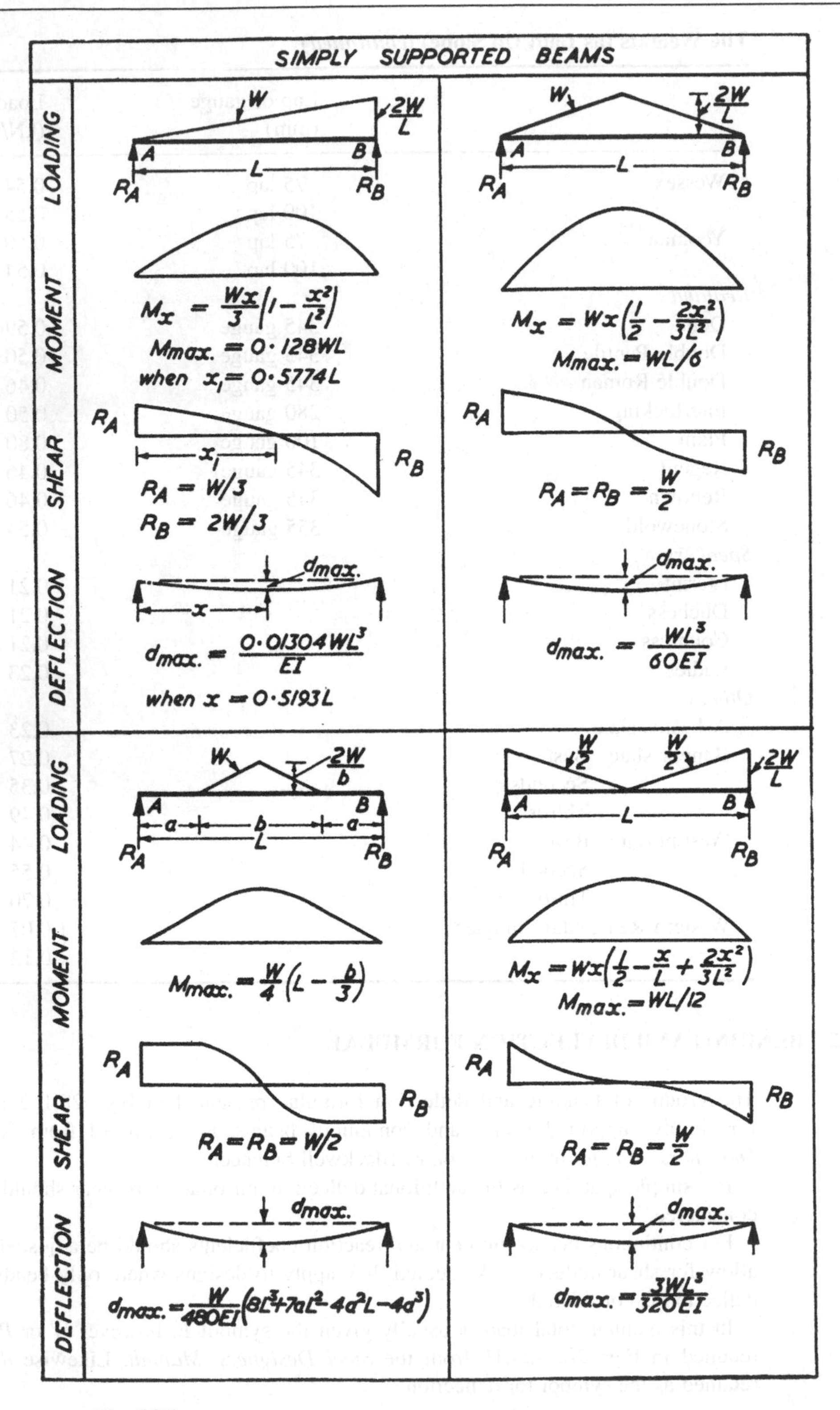

SIMPLY SUPPORTED BEAMS
LOADING
MOMENT
SHEAR
DEFLECTION
W
2W/L
A
B
L
R_A
R_B
$M_x = \frac{Wx}{3}\left(1 - \frac{x^2}{L^2}\right)$
$M_{max.} = 0{\cdot}128WL$
when $x_1 = 0{\cdot}5774L$
x_1
$R_A = W/3$
$R_B = 2W/3$
$d_{max.}$
x
$d_{max.} = \frac{0{\cdot}01304WL^3}{EI}$
when $x = 0{\cdot}5193L$
$M_x = Wx\left(\frac{1}{2} - \frac{2x^2}{3L^2}\right)$
$M_{max.} = WL/6$
$R_A = R_B = \frac{W}{2}$
$d_{max.} = \frac{WL^3}{60EI}$
2W/b
a
b
$M_{max.} = \frac{W}{4}\left(L - \frac{b}{3}\right)$
$R_A = R_B = W/2$
$d_{max.} = \frac{W}{480EI}(8L^3 + 7aL^2 - 4a^2L - 4a^3)$
W/2
$M_x = Wx\left(\frac{1}{2} - \frac{x}{L} + \frac{2x^2}{3L^2}\right)$
$M_{max.} = WL/12$
$d_{max.} = \frac{3WL^3}{320EI}$

Fig. 28.1

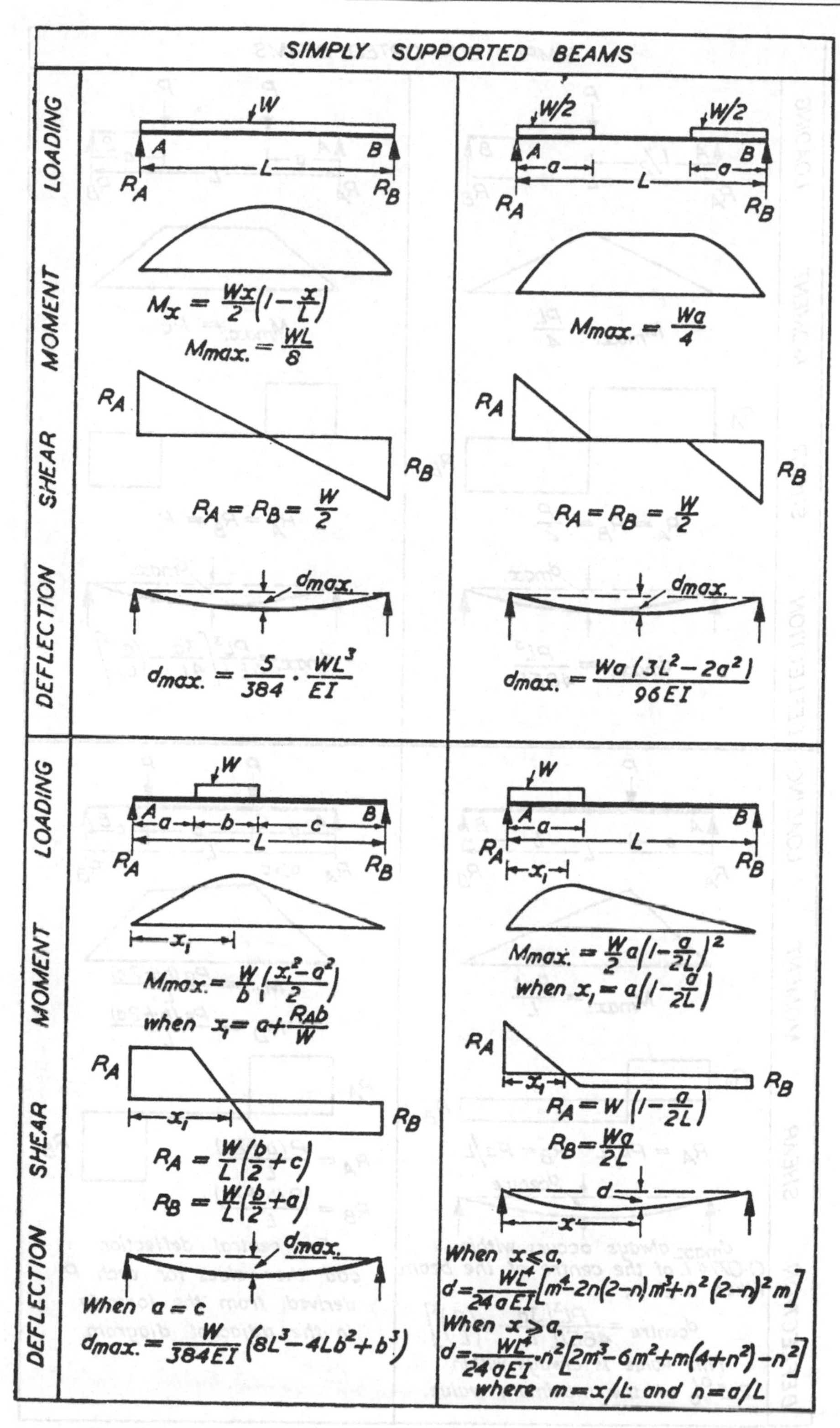
SIMPLY SUPPORTED BEAMS
LOADING
MOMENT
SHEAR
DEFLECTION
W
A
B
L
R_A
R_B
$M_x = \frac{Wx}{2}\left(1-\frac{x}{L}\right)$
$M_{max.} = \frac{WL}{8}$
$R_A = R_B = \frac{W}{2}$
$d_{max.}$
$d_{max.} = \frac{5}{384} \cdot \frac{WL^3}{EI}$
W/2
a
$M_{max.} = \frac{Wa}{4}$
$d_{max.} = \frac{Wa(3L^2-2a^2)}{96EI}$
b
c
x_1
$M_{max.} = \frac{W}{b}\left(\frac{x_1^2-a^2}{2}\right)$
when $x_1 = a + \frac{R_A b}{W}$
$R_A = \frac{W}{L}\left(\frac{b}{2}+c\right)$
$R_B = \frac{W}{L}\left(\frac{b}{2}+a\right)$
When $a = c$
$d_{max.} = \frac{W}{384EI}(8L^3-4Lb^2+b^3)$
$M_{max.} = \frac{W}{2}a\left(1-\frac{a}{2L}\right)^2$
when $x_1 = a\left(1-\frac{a}{2L}\right)$
$R_A = W\left(1-\frac{a}{2L}\right)$
$R_B = \frac{Wa}{2L}$
d
x
When $x \leq a$,
$d = \frac{WL^4}{24aEI}\left[m^4-2n(2-n)m^3+n^2(2-n)^2m\right]$
When $x \geq a$,
$d = \frac{WL^4}{24aEI} \cdot n^2\left[2m^3-6m^2+m(4+n^2)-n^2\right]$
where $m = x/L$ and $n = a/L$

Fig. 28.2

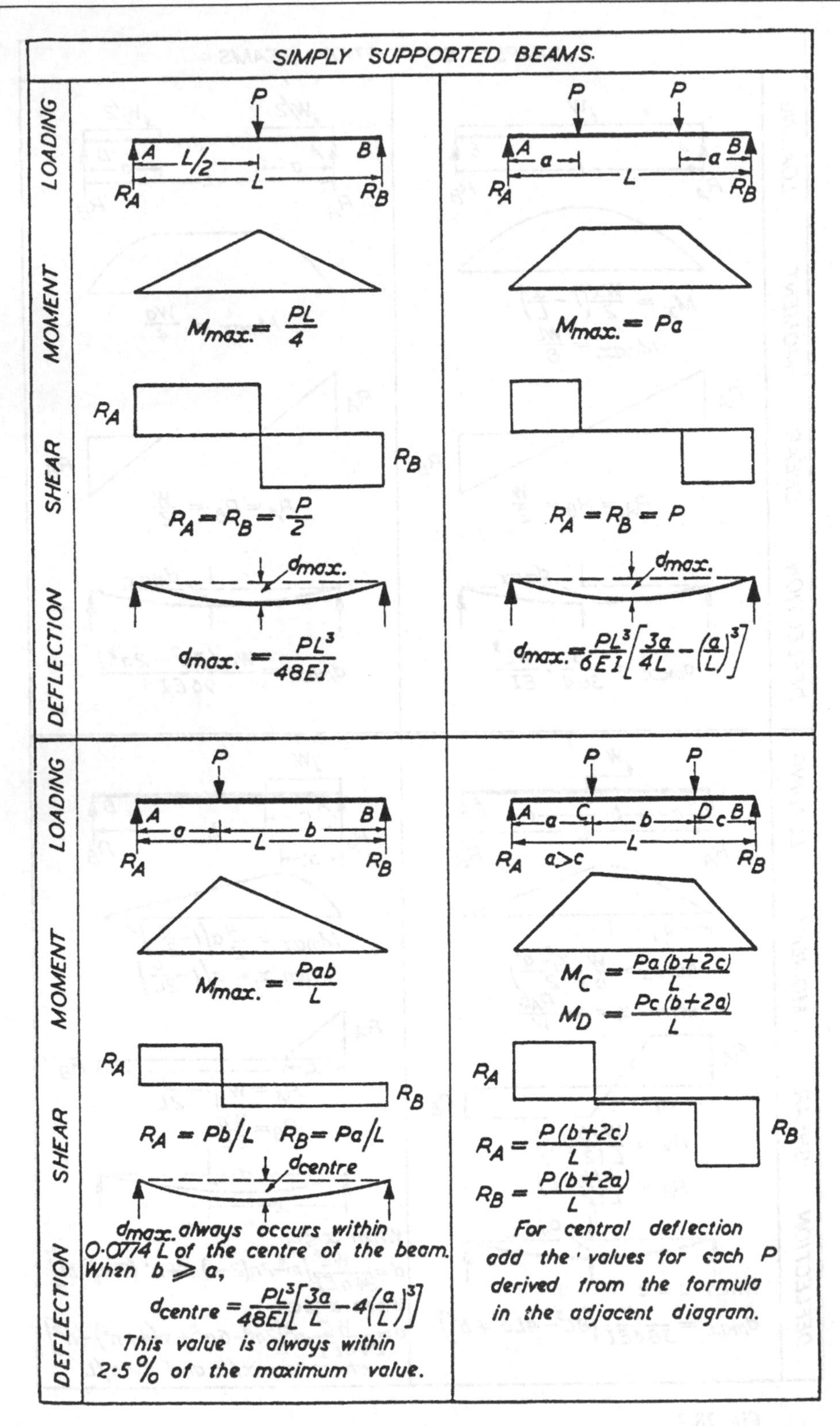
SIMPLY SUPPORTED BEAMS
LOADING
MOMENT
SHEAR
DEFLECTION
P
A
B
L/2
L
RA
RB
Mmax. = PL/4
RA = RB = P/2
dmax.
dmax. = PL³/48EI
P P
a
a
L
Mmax. = Pa
RA = RB = P
dmax. = PL³/6EI [3a/4L − (a/L)³]
a
b
Mmax. = Pab/L
RA = Pb/L RB = Pa/L
dcentre
dmax. always occurs within 0·0774 L of the centre of the beam.
When b ⩾ a,
dcentre = PL³/48EI [3a/L − 4(a/L)³]
This value is always within 2·5% of the maximum value.
C
D
c
a>c
MC = Pa(b+2c)/L
MD = Pc(b+2a)/L
RA = P(b+2c)/L
RB = P(b+2a)/L
For central deflection add the values for each P derived from the formula in the adjacent diagram.

Fig. 28.3

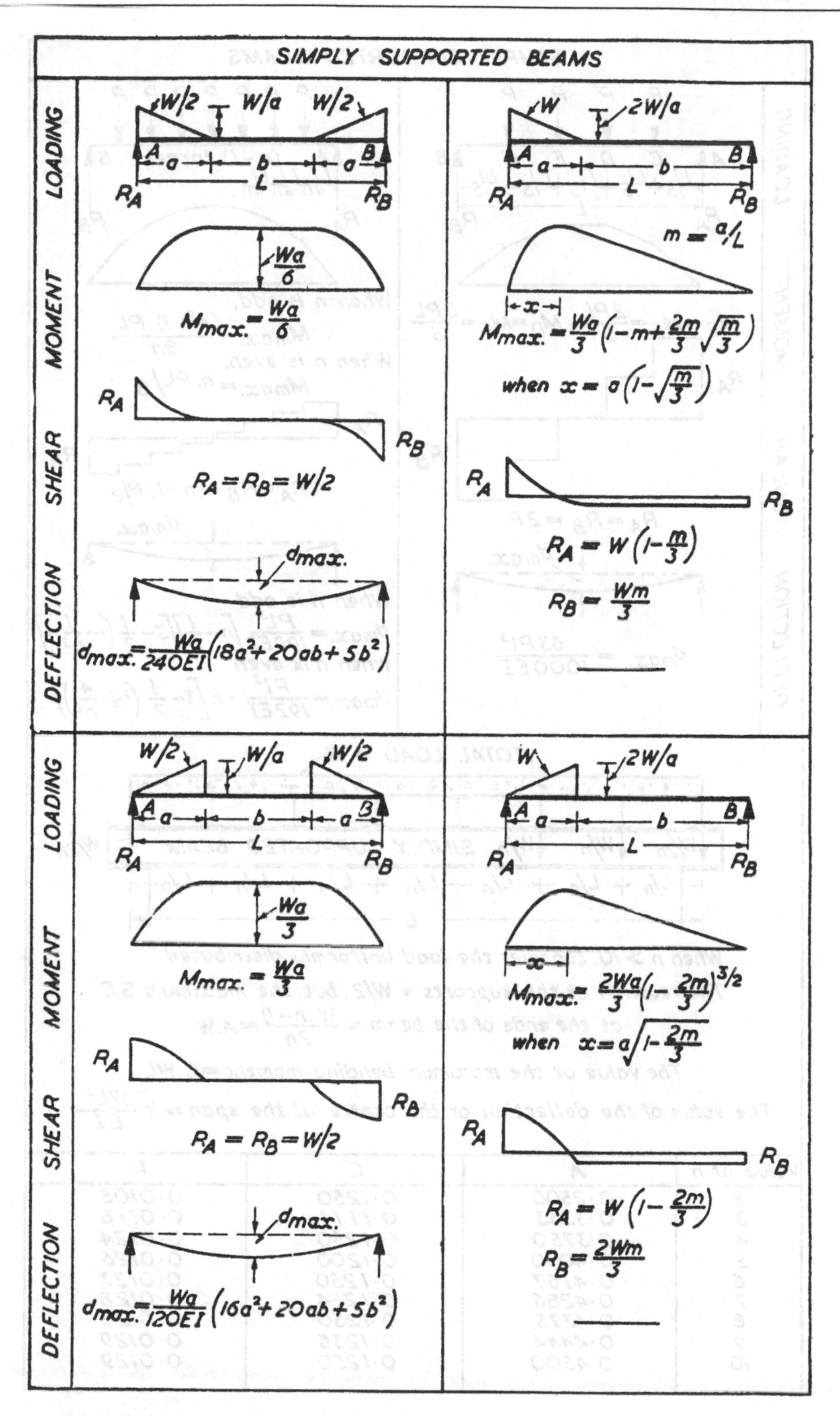
SIMPLY SUPPORTED BEAMS
LOADING
MOMENT
SHEAR
DEFLECTION
W/2
W/a
W/2
A
a
b
a
B
L
R_A
R_B
$\frac{Wa}{6}$
$M_{max.} = \frac{Wa}{6}$
$R_A = R_B = W/2$
$d_{max.}$
$d_{max.} = \frac{Wa}{240EI}(18a^2 + 20ab + 5b^2)$
W
2W/a
$m = a/L$
x
$M_{max.} = \frac{Wa}{3}\left(1 - m + \frac{2m}{3}\sqrt{\frac{m}{3}}\right)$
when $x = a\left(1 - \sqrt{\frac{m}{3}}\right)$
$R_A = W\left(1 - \frac{m}{3}\right)$
$R_B = \frac{Wm}{3}$
$\frac{Wa}{3}$
$M_{max.} = \frac{Wa}{3}$
$d_{max.} = \frac{Wa}{120EI}(16a^2 + 20ab + 5b^2)$
$M_{max.} = \frac{2Wa}{3}\left(1 - \frac{2m}{3}\right)^{3/2}$
when $x = a\sqrt{1 - \frac{2m}{3}}$
$R_A = W\left(1 - \frac{2m}{3}\right)$
$R_B = \frac{2Wm}{3}$

Fig. 28.4

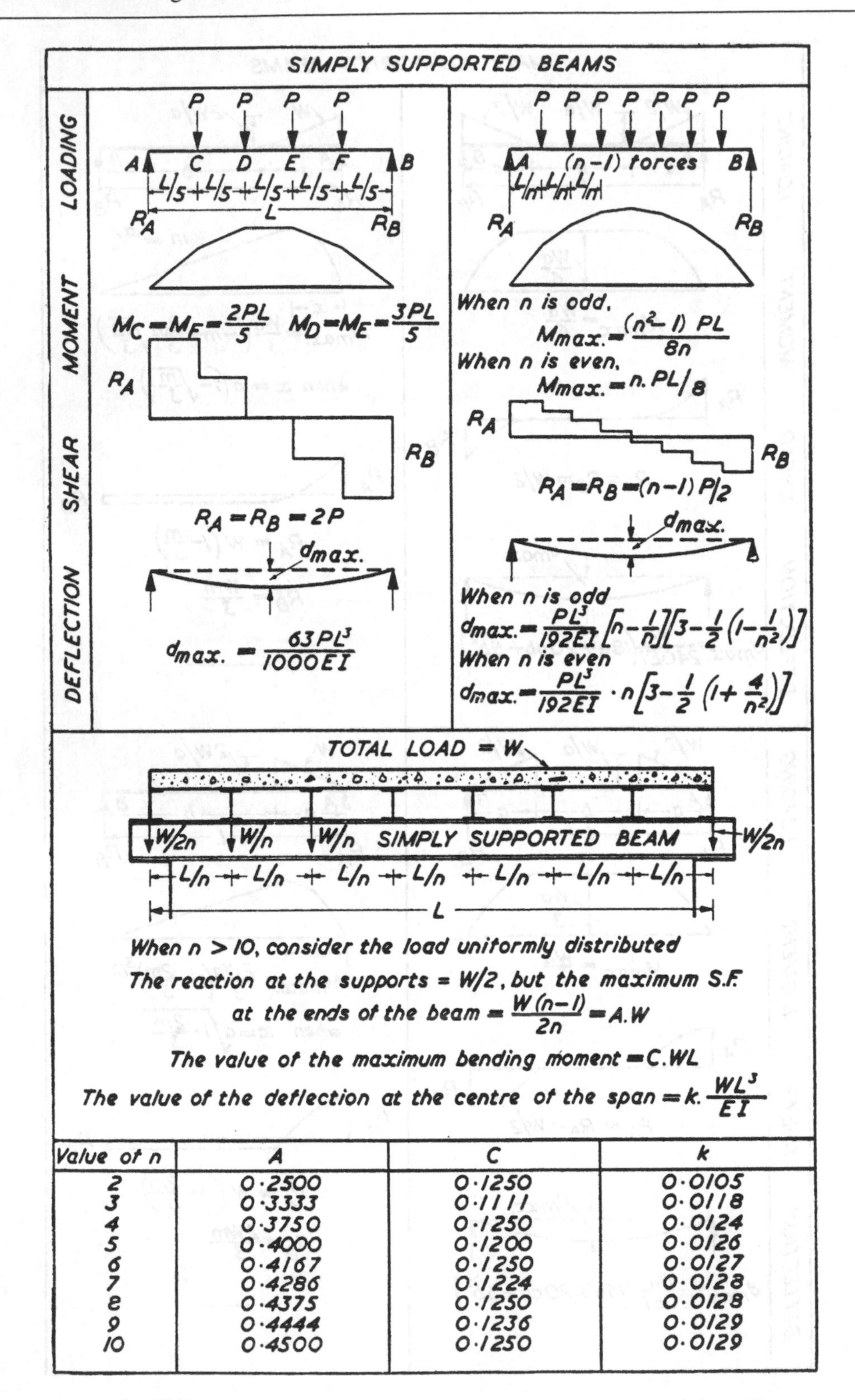

Value of n	A	C	k
2	0·2500	0·1250	0·0105
3	0·3333	0·1111	0·0118
4	0·3750	0·1250	0·0124
5	0·4000	0·1200	0·0126
6	0·4167	0·1250	0·0127
7	0·4286	0·1224	0·0128
8	0·4375	0·1250	0·0128
9	0·4444	0·1236	0·0129
10	0·4500	0·1250	0·0129

Fig. 28.5

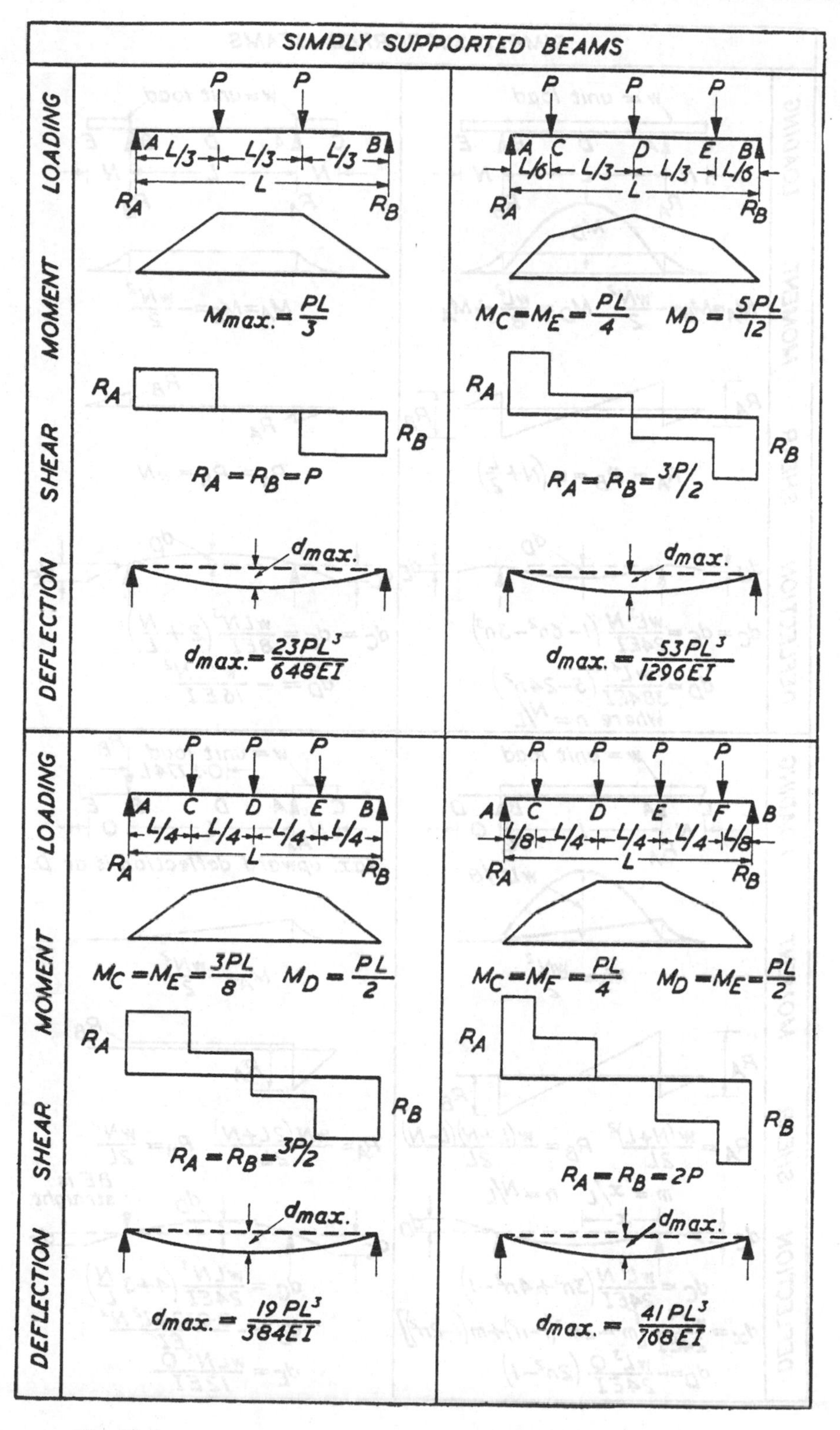
SIMPLY SUPPORTED BEAMS
LOADING
MOMENT
SHEAR
DEFLECTION
P P
A L/3 L/3 L/3 B
L
R_A R_B
M_max. = PL/3
R_A = R_B = P
d_max.
d_max. = 23PL³/648EI
P P P
A C D E B
L/6 L/3 L/3 L/6
L
M_C = M_E = PL/4 M_D = 5PL/12
R_A = R_B = 3P/2
d_max. = 53PL³/1296EI
P P P
A C D E B
L/4 L/4 L/4 L/4
L
M_C = M_E = 3PL/8 M_D = PL/2
R_A = R_B = 3P/2
d_max. = 19PL³/384EI
P P P P
A C D E F B
L/8 L/4 L/4 L/4 L/8
L
M_C = M_F = PL/4 M_D = M_E = PL/2
R_A = R_B = 2P
d_max. = 41PL³/768EI

Fig. 28.6

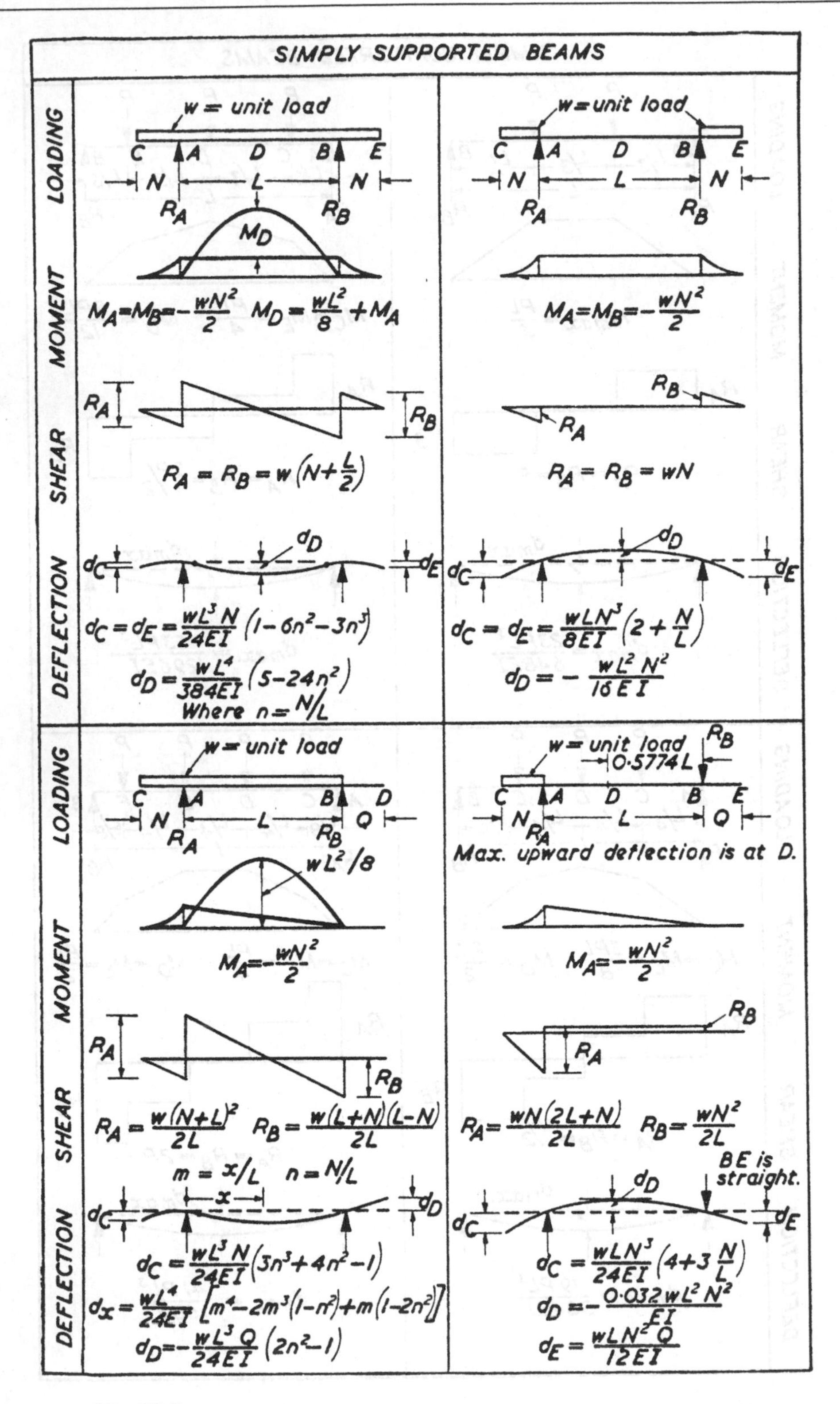

SIMPLY SUPPORTED BEAMS
LOADING
MOMENT
SHEAR
DEFLECTION
w = unit load
C A D B E
N L N
R_A R_B
M_D
$M_A = M_B = -\frac{wN^2}{2}$ $M_D = \frac{wL^2}{8} + M_A$
$R_A = R_B = w\left(N + \frac{L}{2}\right)$
d_C d_D d_E
$d_C = d_E = \frac{wL^3 N}{24EI}(1 - 6n^2 - 3n^3)$
$d_D = \frac{wL^4}{384EI}(5 - 24n^2)$
Where $n = N/L$
w = unit load
C A D B E
N L N
R_A R_B
$M_A = M_B = -\frac{wN^2}{2}$
$R_A = R_B = wN$
d_C d_D d_E
$d_C = d_E = \frac{wLN^3}{8EI}\left(2 + \frac{N}{L}\right)$
$d_D = -\frac{wL^2 N^2}{16EI}$
w = unit load
C A B D
N L Q
R_A R_B
$wL^2/8$
$M_A = -\frac{wN^2}{2}$
$R_A = \frac{w(N+L)^2}{2L}$ $R_B = \frac{w(L+N)(L-N)}{2L}$
$m = x/L$ $n = N/L$
x
d_C d_D
$d_C = \frac{wL^3 N}{24EI}(3n^3 + 4n^2 - 1)$
$d_x = \frac{wL^4}{24EI}\left[m^4 - 2m^3(1 - n^2) + m(1 - 2n^2)\right]$
$d_D = -\frac{wL^3 Q}{24EI}(2n^2 - 1)$
w = unit load
0·5774 L
C A D B E
N L Q
R_A R_B
Max. upward deflection is at D.
$M_A = -\frac{wN^2}{2}$
$R_A = \frac{wN(2L+N)}{2L}$ $R_B = \frac{wN^2}{2L}$
BE is straight.
d_C d_D d_E
$d_C = \frac{wLN^3}{24EI}\left(4 + 3\frac{N}{L}\right)$
$d_D = -\frac{0{\cdot}032\, wL^2 N^2}{EI}$
$d_E = \frac{wLN^2 Q}{12EI}$

Fig. 28.7

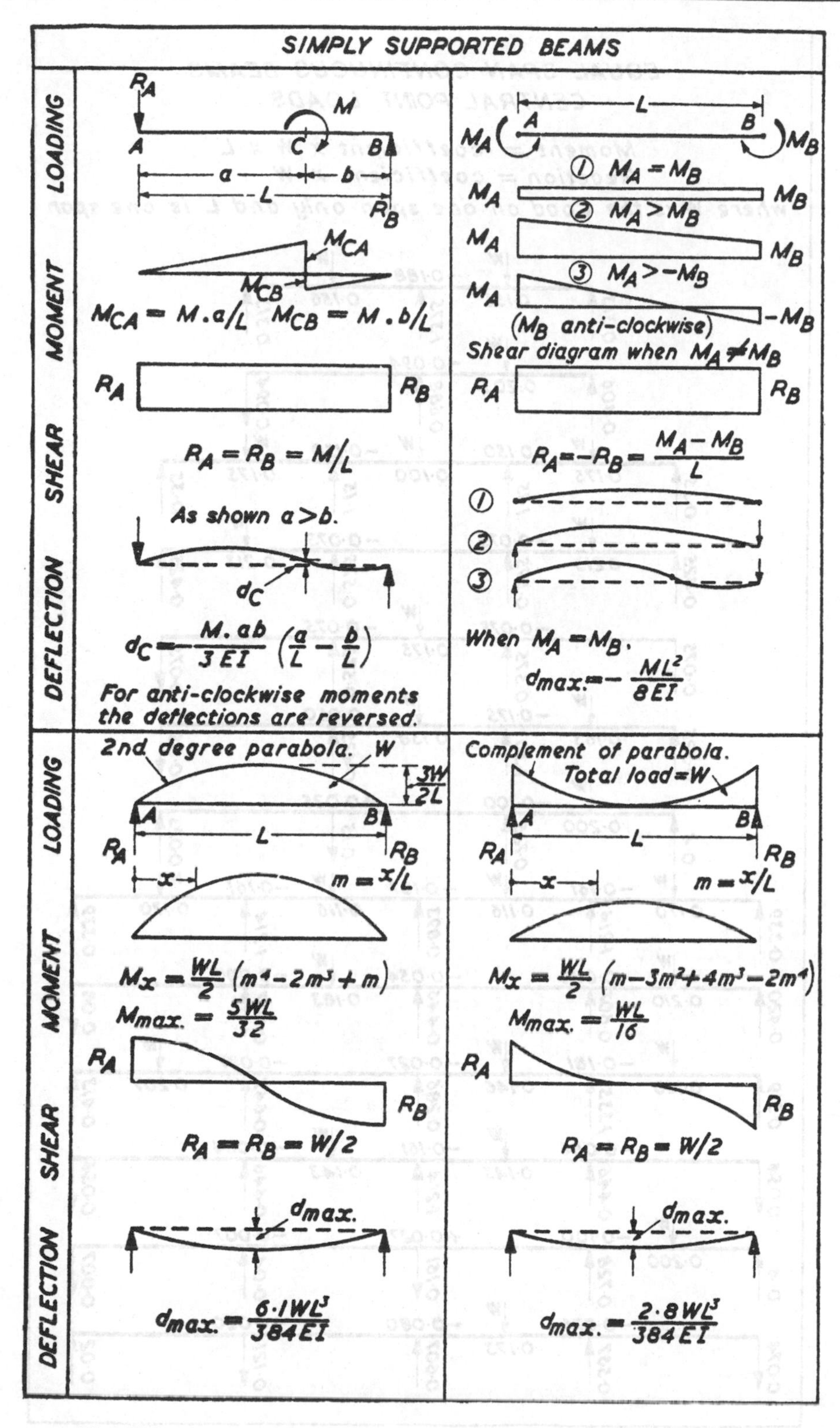

SIMPLY SUPPORTED BEAMS
LOADING
MOMENT
SHEAR
DEFLECTION
$M_{CA} = M.a/L$ $M_{CB} = M.b/L$
$R_A = R_B = M/L$
As shown a>b.
$d_C = -\frac{M.ab}{3EI}\left(\frac{a}{L} - \frac{b}{L}\right)$
For anti-clockwise moments the deflections are reversed.
① $M_A = M_B$
② $M_A > M_B$
③ $M_A > -M_B$
(M_B anti-clockwise)
Shear diagram when $M_A \neq M_B$
$R_A = -R_B = \frac{M_A - M_B}{L}$
When $M_A = M_B$,
$d_{max} = -\frac{ML^2}{8EI}$
2nd degree parabola.
$\frac{3W}{2L}$
$m = x/L$
$M_x = \frac{WL}{2}(m^4 - 2m^3 + m)$
$M_{max.} = \frac{5WL}{32}$
$R_A = R_B = W/2$
$d_{max.} = \frac{6.1WL^3}{384EI}$
Complement of parabola.
Total load=W
$M_x = \frac{WL}{2}(m - 3m^2 + 4m^3 - 2m^4)$
$M_{max.} = \frac{WL}{16}$
$R_A = R_B = W/2$
$d_{max.} = \frac{2.8WL^3}{384EI}$

Fig. 28.8

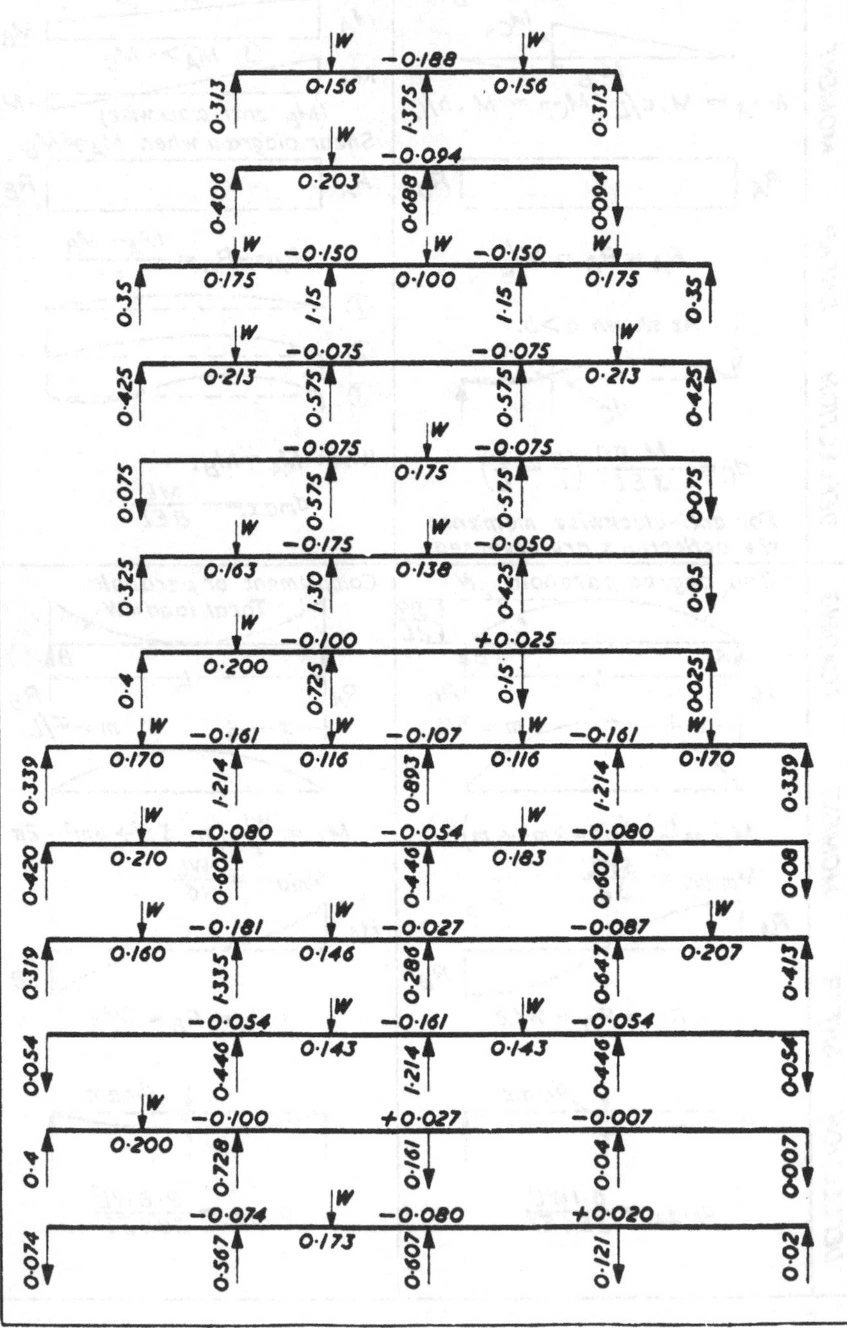
EQUAL SPAN CONTINUOUS BEAMS
CENTRAL POINT LOADS
Moment = coefficient x W x L
Reaction = coefficient x W
where W is the Load on one span only and L is one span
W W
−0·188
0·156 0·156
0·313 1·375 0·313
W
−0·094
0·203
0·406 0·688 0·094
W W W
−0·150 −0·150
0·175 0·100 0·175
0·35 1·15 1·15 0·35
W W
−0·075 −0·075
0·213 0·213
0·425 0·575 0·575 0·425
W
−0·075 −0·075
0·175
0·075 0·575 0·575 0·075
W W
−0·175 −0·050
0·163 0·138
0·325 1·30 0·425 0·05
W
−0·100 +0·025
0·200
0·4 0·725 0·15 0·025
W W W W
−0·161 −0·107 −0·161
0·170 0·116 0·116 0·170
0·339 1·214 0·893 1·214 0·339
W W
−0·080 −0·054 −0·080
0·210 0·183
0·420 0·607 0·446 0·607 0·08
W W W
−0·181 −0·027 −0·087
0·160 0·146 0·207
0·319 1·335 0·286 0·647 0·413
W W
−0·054 −0·161 −0·054
0·143 0·143
0·054 0·446 1·214 0·446 0·054
W
−0·100 +0·027 −0·007
0·200
0·4 0·728 0·161 0·04 0·007
W
−0·074 −0·080 +0·020
0·173
0·074 0·567 0·607 0·121 0·02

Fig. 28.9

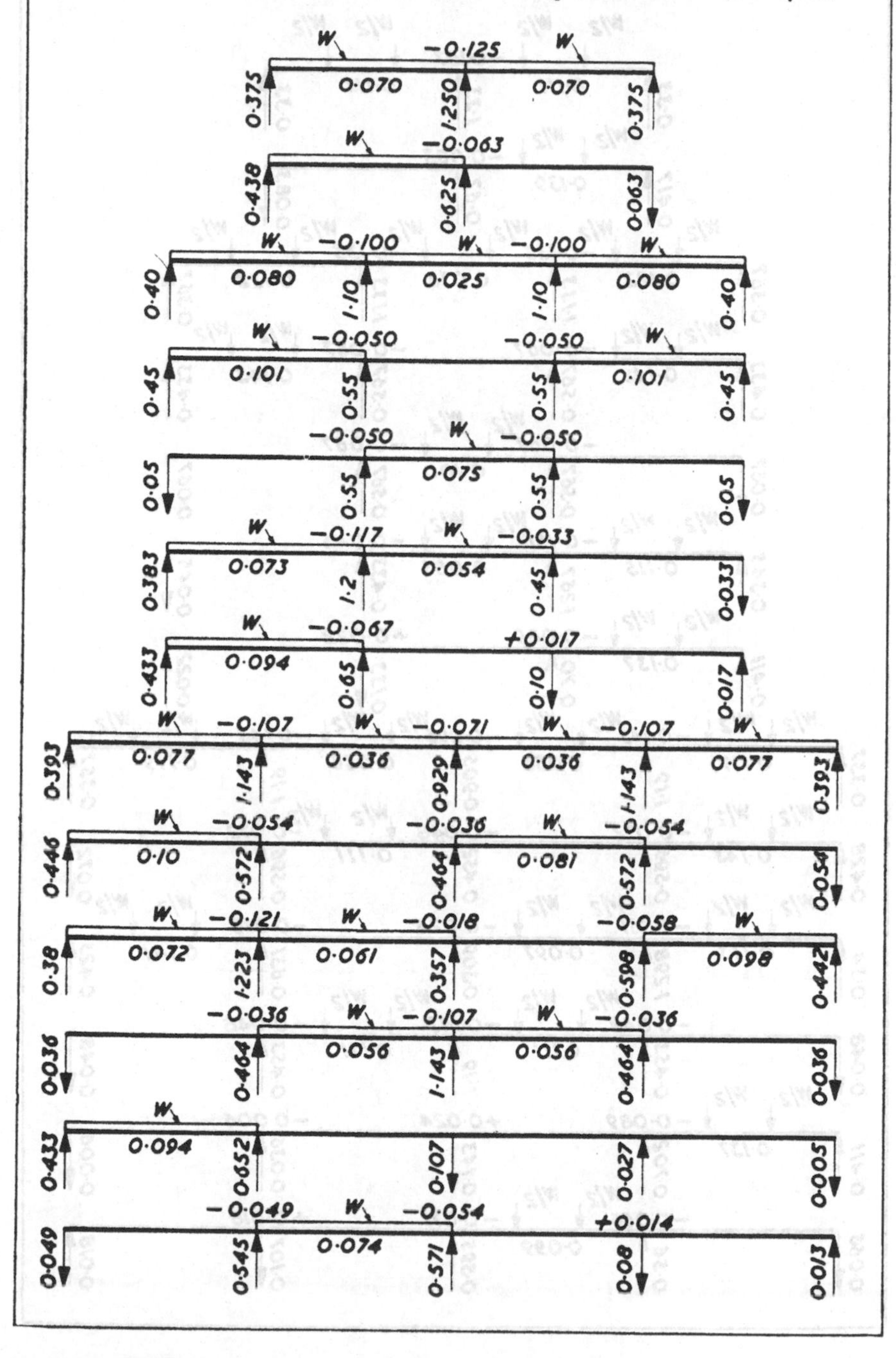
EQUAL SPAN CONTINUOUS BEAMS
UNIFORMLY DISTRIBUTED LOADS
Moment = coefficient x W x L
Reaction = coefficient x W
where W is the U.D.L. on one span only and L is one span
W
−0·125
W
0·375
0·070
1·250
0·070
0·375
W
−0·063
0·438
0·625
0·063
W
−0·100
W
−0·100
W
0·40
0·080
1·10
0·025
1·10
0·080
0·40
W
−0·050
−0·050
W
0·45
0·101
0·55
0·55
0·101
0·45
−0·050
W
−0·050
0·05
0·55
0·075
0·55
0·05
W
−0·117
W
−0·033
0·383
0·073
1·2
0·054
0·45
0·033
W
−0·067
+0·017
0·433
0·094
0·65
0·10
0·017
W
−0·107
W
−0·071
W
−0·107
W
0·393
0·077
1·143
0·036
0·929
0·036
1·143
0·077
0·393
W
−0·054
−0·036
W
−0·054
0·446
0·10
0·572
0·464
0·081
0·572
0·054
W
−0·121
W
−0·018
−0·058
W
0·38
0·072
1·223
0·061
0·357
0·598
0·098
0·442
−0·036
W
−0·107
W
−0·036
0·036
0·464
0·056
1·143
0·056
0·464
0·036
W
0·433
0·094
0·652
0·107
0·027
0·005
−0·049
W
−0·054
+0·014
0·049
0·545
0·074
0·571
0·08
0·013

Fig. 28.10

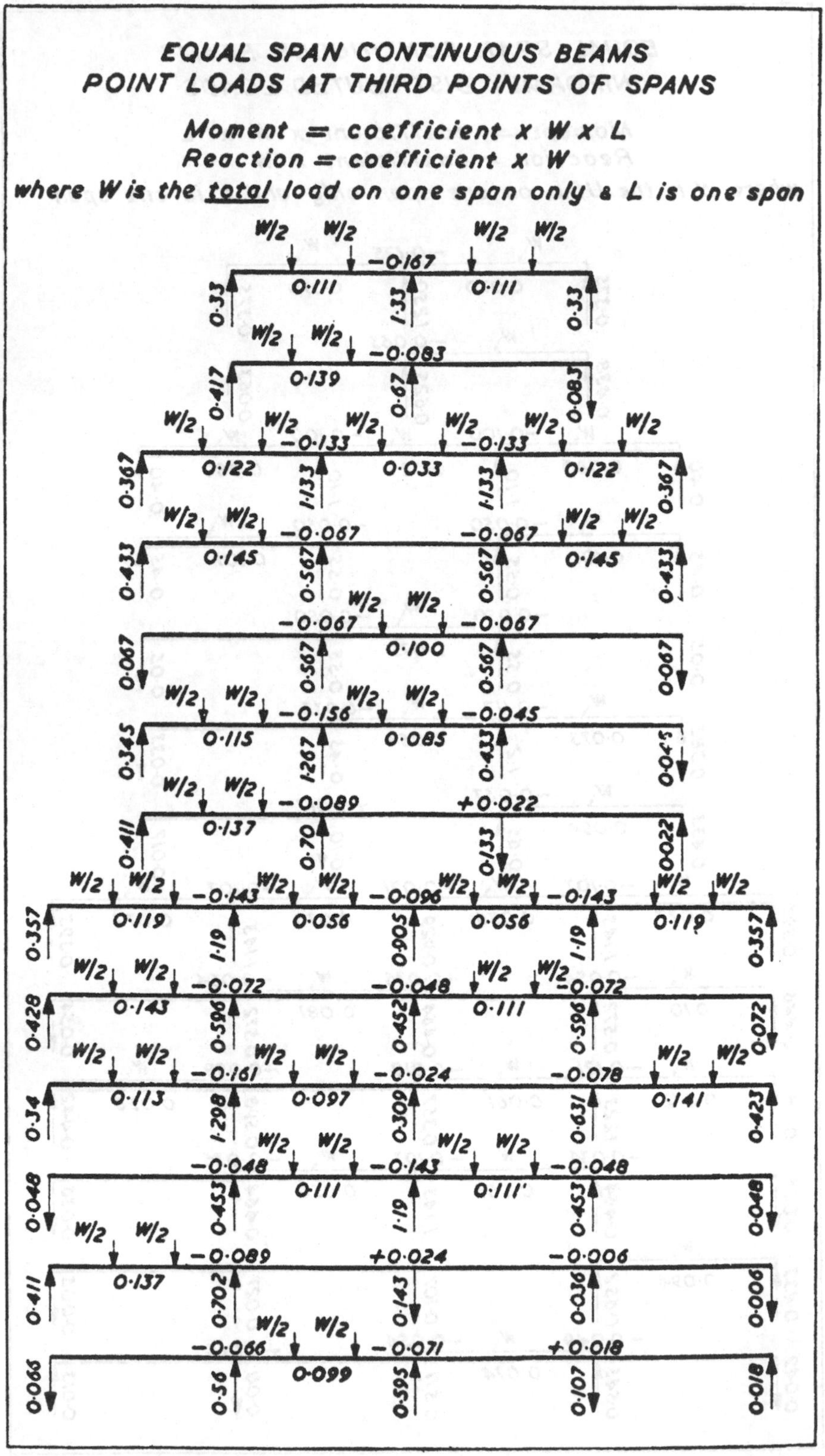

EQUAL SPAN CONTINUOUS BEAMS
POINT LOADS AT THIRD POINTS OF SPANS

Moment = coefficient x W x L
Reaction = coefficient x W
where W is the total load on one span only & L is one span

W/2 W/2 −0.167 W/2 W/2
0.33 0.111 1.33 0.111 0.33

W/2 W/2 −0.083
0.417 0.139 0.67 0.083

W/2 W/2 −0.133 W/2 W/2 −0.133 W/2 W/2
0.367 0.122 1.133 0.033 1.133 0.122 0.367

W/2 W/2 −0.067 −0.067 W/2 W/2
0.433 0.145 0.567 0.567 0.145 0.433

W/2 W/2
−0.067 −0.067
0.067 0.567 0.100 0.567 0.067

W/2 W/2 −0.156 W/2 W/2 −0.045
0.345 0.115 1.267 0.085 0.433 0.045

W/2 W/2 −0.089 +0.022
0.411 0.137 0.70 0.133 0.022

W/2 W/2 −0.143 W/2 W/2 −0.096 W/2 W/2 −0.143 W/2 W/2
0.357 0.119 1.19 0.056 0.905 0.056 1.19 0.119 0.357

W/2 W/2 −0.072 −0.048 W/2 W/2 −0.072
0.428 0.143 0.596 0.452 0.111 0.596 0.072

W/2 W/2 −0.161 W/2 W/2 −0.024 −0.078 W/2 W/2
0.34 0.113 1.298 0.097 0.309 0.631 0.141 0.423

W/2 W/2 W/2 W/2
−0.048 −0.143 −0.048
0.048 0.453 0.111 1.19 0.111 0.453 0.048

W/2 W/2 −0.089 +0.024 −0.006
0.411 0.137 0.702 0.143 0.036 0.006

W/2 W/2
−0.066 −0.071 +0.018
0.066 0.56 0.099 0.595 0.107 0.018

Fig. 28.11

In the two left-hand examples shown in Fig. 28.7 the designer must establish if the deflections at the end of the cantilevers and the deflections in the centre of the span are upwards or downwards. This is not clear from the diagrams/formulae. (See section 12.6.9.)

In calculating intermediate values of bending moment the designer should recall the proportions of a parabola as sketched below:

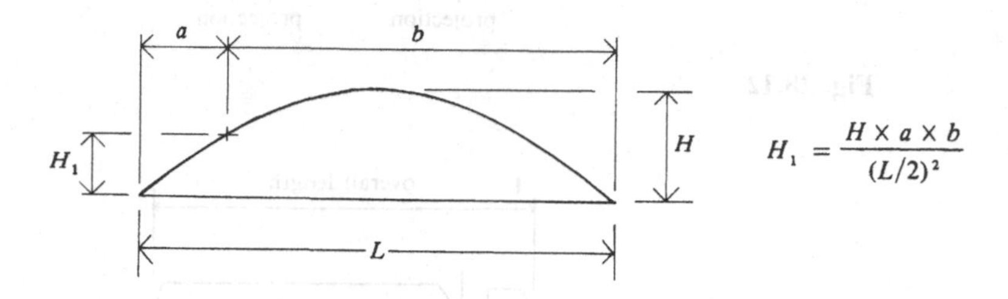

28.3 PERMISSIBLE LORRY OVERHANGS

Although the transporting of components is not usually the actual responsibility of a design engineer, it is reasonable for the transport manager or road haulage company to expect that the engineer has given some thought to delivery when setting the maximum size of components. Guidance is given in this section on the sizes which can be transported without special requirements. There are additional requirements, including requirements for central London.

If the site to which the components are to be delivered is some distance from main roads, the engineer would be well advised to check at the design stage with the highway authorities through whose areas the vehicle will pass, for details of weight or height restriction of any bridges, or details of narrow sections of road and bends which might restrict the height, width or length of the acceptable load.

Where it is necessary to give notification to the police in each police area through which a vehicle will pass, at least two clear days' notice is required (excluding Saturdays, Sundays and bank holidays). Police area headquarters are a source of information on route guidance. Occasionally a special event (e.g. a County Show) will mean that a vehicle is re-routed or delayed.

Weight

Compared to some loads (e.g. steelwork) timber is not particularly heavy, and the type of vehicle used to transport timber components will usually be able to have a total laden weight of up to 38 000 kg.

Width (Fig. 28.12)

Providing the overhang on either side does not exceed 305 mm and the width of the load does not exceed 2.9 m there are no special requirements.

If the projection on either side of an 'indivisible' load exceeds 305 mm or the width of the load exceeds 2.9 m police notification is required. If a load in this width category can be divided in width, it must be divided. If the width is in excess of 4.3 m, a written note must be given to the Secretary of State at the Department of Transport.

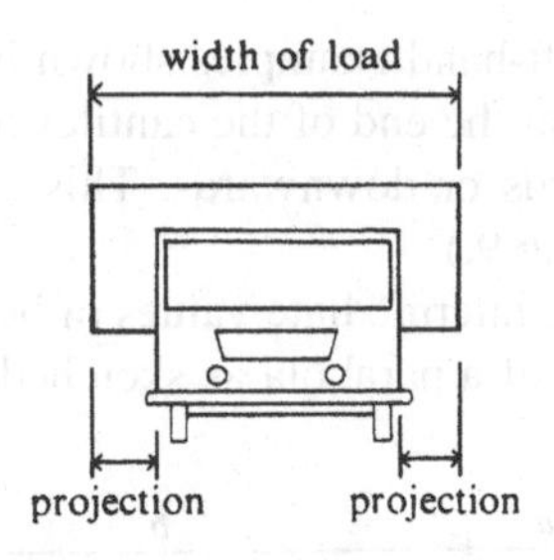

Fig. 28.12

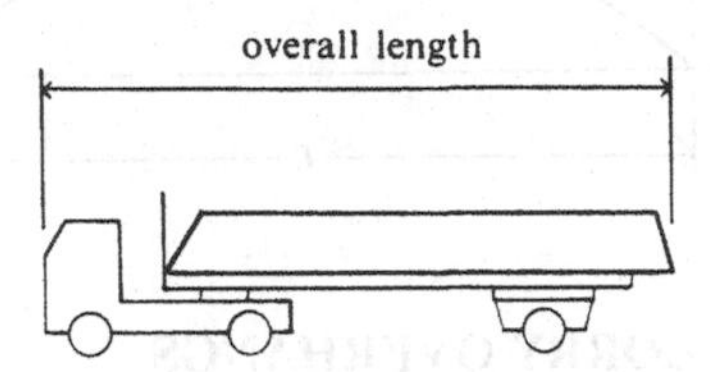

Fig. 28.13

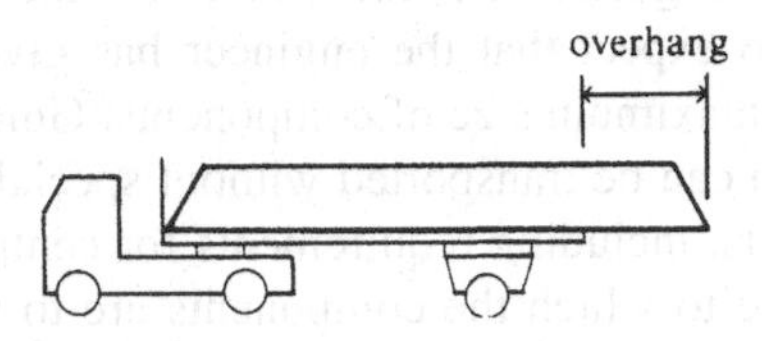

Fig. 28.14

Length (Fig. 28.13)
Providing the overall length of a standard articulated vehicle does not exceed 15.5 m, or 17.33 m over the vehicle and load, there are no special requirements. If this length is exceeded then police notification is required and the driver must be accompanied by an attendant.

In the case of an articulated vehicle specially constructed for abnormally long loads, providing the length overall of the load and trailer (but not the drawing vehicle) does not exceed 18.3 m there are no special requirements. If this length is exceeded, then police notification is required and the driver must be accompanied by an attendant.

In the case of a combination of vehicles and load, providing the overall length including projections does not exceed 25.9 m there are no special requirements. If this length is exceeded, then police notification is required and the driver must be accompanied by an attendant.

Rear overhang (Fig. 28.14)
Providing the overhang does not exceed 1.07 m there are no special requirements.

If the overhang exceeds 1.07 m the overhang must be made 'clearly visible'.

If the overhang exceeds 1.83 m standard end marker boards or a reflective marker are required.

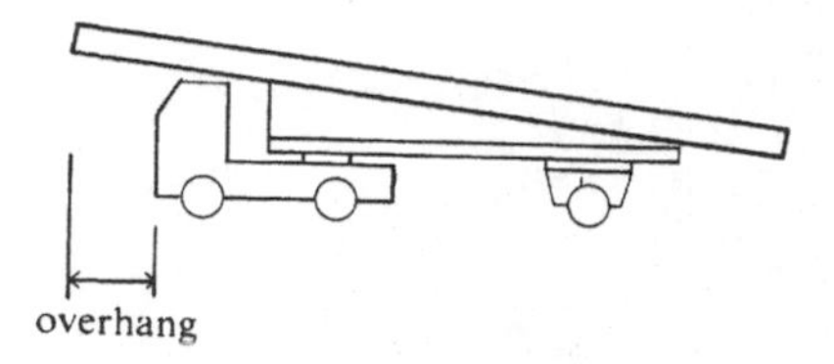

Fig. 28.15

If the overhang is over 3.05 m standard end marker boards are required, the police must be notified and the driver must be accompanied by an attendant.

If the overhang is over 5.1 m additional side marker boards are required within 3.6 m of the normal marker boards.

Marker boards must be illuminated after lighting-up time.

Front overhang (Fig. 28.15)

Providing the overhang does not exceed 1.83 m there are no special requirements.

If the overhang exceeds 1.83 m standard end and side marker boards are required and the driver must be accompanied by an attendant.

If the overhang exceeds 3.05 m standard end and side marker boards are required, the police must be notified and the driver must be accompanied by an attendant.

If the overhang is over 4.5 m additional side marker boards are required within 2.4 m of the normal marker boards.

Height

The minimum clearance height under motorway bridges is 5.03 m. On minor roads the clearance is likely to be much less and, if a loaded lorry requires a clearance approaching 3.6 m and will be travelling on minor roads, it is advisable to contact the highway authorities or police areas through which the load will pass for details of clearance under all over-bridges. Very occasionally a bridge with much less clearance is encountered.

Index